Young Measures on Topological Spaces

Mathematics and Its Applications

Volume 571

Young Measures on Topological Spaces

With Applications in Control Theory and Probability Theory

by

Charles Castaing

Université Montpellier II,
Montpellier, France

Paul Raynaud de Fitte

Université de Rouen,
Mont-Saint-Aignan, France

and

Michel Valadier

Université Montpellier II,
Montpellier, France

KLUWER ACADEMIC PUBLISHERS
DORDRECHT / BOSTON / LONDON

A C.I.P. Catalogue record for this book is available from the Library of Congress.

ISBN 978-90-481-6552-0 (PB)
ISBN 978-1-4020-1964-7(e-book)

Published by Kluwer Academic Publishers,
P.O. Box 17, 3300 AA Dordrecht, The Netherlands.

Sold and distributed in North, Central and South America
by Kluwer Academic Publishers,
101 Philip Drive, Norwell, MA 02061, U.S.A.

In all other countries, sold and distributed
by Kluwer Academic Publishers,
P.O. Box 322, 3300 AH Dordrecht, The Netherlands.

Printed on acid-free paper

à Juliane, Marie-Blanche et Marie-Hélène

Contents

Preface

Classical examples of more and more oscillating real–valued functions on a domain Ω of $\mathbb{R}^N$ are the functions $u_n(x) = \sin(nx_1)$ with $x = (x_1, \ldots, x_n)$ or the so-called Rademacher functions on $]0,1[$, $u_n(x) = \mathfrak{r}_n(x) = \mathrm{sgn}(\sin(2^{n+1}\pi x))$ (see later 3.1.4). They may appear as the gradients ∇v_n of minimizing sequences $(v_n)_{n\in\mathbb{N}}$ in some variational problems. In these examples, the function u_n converges in some sense to a measure μ on $\Omega \times \mathbb{R}$, called *Young measure.* In Functional Analysis formulation, this is the narrow convergence to μ of the image of the Lebesgue measure on Ω by $\omega \mapsto (\omega, u_n(\omega))$. In the disintegrated form $(\mu_\omega)_{\omega\in\Omega}$, the *parametrized measure* μ_ω captures the possible scattering of the u_n around ω.

Curiously if $(X_n)_{n\in\mathbb{N}}$ is a sequence of random variables deriving from independent ones, the n-th one may appear more and more far from the k first ones as if it was oscillating (think of orthonormal vectors in L^2 which converge weakly to 0). More precisely when the laws $\mathrm{L}(X_n)$ narrowly converge to some probability measure ϖ, it often happens that for any k and any A in the algebra generated by $X_1, \ldots, X_k$, the conditional law $\mathrm{L}(X_n|A)$ still converges to ϖ (see Chapter 9) which means

$$\forall \varphi \in \mathrm{C}_b(\mathbb{R}) \quad \frac{1}{\mathrm{P}(A)} \int_A \varphi(X_n(\omega))\, d\mathrm{P}(\omega) \longrightarrow \int_{\mathbb{R}} \varphi \, d\varpi$$

or equivalently, $\underline{\delta}_{X_n}$ denoting the image of P by $\omega \mapsto (\omega, X_n(\omega))$,

$$\int_{\Omega\times\mathbb{R}} (\mathbb{1}_A \otimes \varphi)\, d\underline{\delta}_{X_n} \longrightarrow \int_{\Omega\times\mathbb{R}} (\mathbb{1}_A \otimes \varphi)\, d[\mathrm{P} \otimes \varpi]\,.$$

This is exactly the same convergence as the one raised in the first paragraph (excepted that the limit measure is not always a product).

Many authors wrote on Young measures in Control and Calculus of Variations: Young [You37], McShane [McS40], Gamkrelidze [Gam62, Gam78], Warga [War72], Ghouila-Houri [GH67], Tartar [Tar78], Ekeland [Eke72], Berliocchi and Lasry [BL71, BL73], Balder in [Bal95, Bal00a, Bal00b] and many other papers cited in this book. On the probabilistic side, we refer to Rényi [Rén58, Rén66, Rén63] for applications in limit theorems, Baxter and Chacon [BC77] and Meyer [Mey78] for relaxed stopping times, Pellaumail for weak solutions of stochastic differential equations [Pel80, Pel81] (see also Jacod and Mémin [JM81b, JM81a]).

The topology on the space of Young measures is usually called "narrow topology" or "weak topology" in the study of Young measures in Functional Analysis. But this topology is called "stable" in Probability Theory, and the word "stable" has the advantage that it avoids confusions with the usual narrow topology on the space of measures on a topological space (see more details on this discussion page 22). So, we choose here "stable topology".

For a long time Young measures were considered only when the functions u_n take their values in a compact subset of an Euclidean space. Berliocchi and Lasry introduced locally compact spaces and Balder extended the Prohorov theorem to these parametrized measures that are Young measures. Then during a long period authors considered that the good space was a Polish or a metrizable Suslin space. Between 1985 and 1990 several works (mainly due to Balder) treated the case of a separable reflexive Banach with the weak topology which is not metrizable. For an example of Young measures on a function space, see [MV02]. In this book the extension of the compactness Topsøe criterion to Young measures allows a significant progress. We will consider basically (but always adding some technical topological hypotheses) a general Hausdorff topological space.

Contents Our aim is not to give a short introduction to Young measures. Instead, we present the results in a general setting, in the hope it will be useful for further developments of the theory. Due to the general framework four stable topologies are introduced in Chapter 2. Dudley's results are used to study in Chapter 3 convergence in probability of the functions $\omega \mapsto \mu_\omega^\alpha$ where α is the index of a net and $(\mu_\omega^\alpha)_{\omega\in\Omega}$ denotes the disintegration of μ^α (which is assumed to exist). A general fiber product theorem and a parametrized Kantorovich-Rubinštein theorem are provided. The heart is Chapter 4 where the Topsøe criterion is extended to Young measures. Chapter 6 is devoted to vector valued functions, the biting Lemma, weak compactness results in $L^1_{\mathbb{E}}$ and Visintin's theorem in several infinite dimensional frameworks. Chapter 7 develops several relaxation results in Control and evolution problems. Chapter 8 gives some results of Calculus of Variations: the lower semicontinuity theorem and Reshetnyak's theorem, and deals with the fiber product of Young measures and its applications to control problems: essentially we establish the link between the value function which occurs in these problems and the viscosity solution of the associated Hamilton–Jacobi–Bellman equation. Finally Chapter 9 gives some results from Probability Theory which involve stable convergence.

There are many directions that we did not investigate... Specially the result of Kinderlehrer and Pedregal about Young measures generated by gradients of vector valued functions (as this necessarily happens in some physical problems where the functions describe the deformation of a 3-dimensional material): see the books of Roubíček [Rou97] and Pedregal [Ped97], see also [Syc98, Syc99] and the forthcoming book [ABM]. We did not either investigate some generalizations of Young measures which should be very useful in an infinite dimensional setting, especially in the case of nonseparable spaces; for example, we did not consider Young measures with cylindrical values in a Banach space, nor Young measures which have only finitely additive values (such measures are considered by Fattorini [Fat99]).

Definitions are spread all along the text, without any numbering. The reader should consult the Subject Index and the Index of Notations to find their precise location.

Aknowledgements We warmly thank Lionel Thibault for his incredibly careful and efficient reading of our manuscript (all shortcomings of this book, if any, must have been added later). We are also greatly indebted to Ahmed Bouziad for his precious "S.O.S. topology" service, free and available 24h a day.

Chapter 1

Generalities, preliminary results

This chapter contains some results, but mainly definitions that can be skipped at first reading: The reader can access these definitions by using the Subject Index and the Index of Notations.

We denote by $\mathbb{R}$ the set of real numbers and by $\mathbb{N}$ the set $\{0, 1, 2, \ldots\}$ of natural integers. The set $\mathbb{N} \setminus \{0\} = \{1, 2, \ldots\}$ is denoted by $\mathbb{N}^*$.

1.1 General topology

Throughout, $\mathbb{T}$ denotes a topological space, the topology of which is denoted by $\tau_{\mathbb{T}}$. Unless explicitely stated, $\mathbb{T}$ is assumed to be Hausdorff. One reason to deal with spaces which satisfy at least the separation axiom T_2 is that we need the compact subsets of $\mathbb{T}$ to be closed. The set of closed subsets of $\mathbb{T}$ is denoted by $\mathcal{F}(\mathbb{T})$, for short $\mathcal{F}$ if no confusion can arise. Similarly, the set of open subsets of $\mathbb{T}$ is denoted by $\mathcal{G}(\mathbb{T})$ or $\mathcal{G}$ and the set of compact subsets of $\mathbb{T}$ is denoted by $\mathcal{K}(\mathbb{T})$ or $\mathcal{K}$. Recall that a *semidistance* on $\mathbb{T}$ is a function $d : \mathbb{T} \times \mathbb{T} \to [0, +\infty[$ satisfying, for all $t, s, r \in \mathbb{T}$, (i) $d(t, t) = 0$, (ii) $d(t, s) = d(s, t)$ and (iii) $d(t, r) \leq d(t, s) + d(s, r)$. (if furthermore $d(t, s) = 0$ implies $t = s$, we say that d is a *distance*).

The Borel σ–algebra of a topological space $\mathbb{T}$ is denoted by $\mathcal{B}_{\mathbb{T}}$.

If $\mathfrak{T}$ is an ordered set with order relation $\leq$, we say that $\mathfrak{T}$ is *upwards filtering* if, for any $\mathfrak{s}, \mathfrak{t} \in \mathfrak{T}$, there exists $\mathfrak{u} \in \mathfrak{T}$ such that $\mathfrak{s} \leq \mathfrak{u}$ and $\mathfrak{t} \leq \mathfrak{u}$. We say that $\mathfrak{T}$ is *σ–upwards filtering* if, for any sequence $(\mathfrak{t}_n)_n$ of elements of $\mathfrak{T}$, there exists $\mathfrak{u} \in \mathfrak{T}$ such that $\mathfrak{t}_n \leq \mathfrak{u}$ for every n. Similar definitions hold for *downwards filtering* or *σ–downwards filtering*.

If $\mathcal{D}$ is a set of continuous semidistances on $\mathbb{T}$, in expressions such as "$\mathcal{D}$ is upwards filtering", we refer to the topological order between the induced topologies. For example, if we say that $\mathcal{D}$ is σ–upwards filtering, this means that, for any sequence $(d_n)_n$ of elements of $\mathcal{D}$, there exists an element d of $\mathcal{D}$ which is topologically finer than the continuous semidistance $\sum_{n\geq 1} d_n \wedge 2^{-n}$ (where $a \wedge b$ denotes the infimum of a and b). However, a formula such as $\delta \geq d$, where δ and d are semidistances on $\mathbb{T}$, means that we have $\delta(t,s) \geq d(t,s)$ for all $(t,s) \in \mathbb{T} \times \mathbb{T}$.

The Hausdorff topological space $\mathbb{T}$ is said to be

- *regular* if, for any $t \in \mathbb{T}$ and any closed subset F of $\mathbb{T}$ which does not contain t, there exist two disjoint open subsets U and V such that $t \in U$ and $F \subset V$,

- *completely regular* if, for any $t \in \mathbb{T}$ and any closed subset F of $\mathbb{T}$ which does not contain t, there exists a continuous function $f : \mathbb{T} \to [0,1]$ such that $f(t) = 0$ and $f = 1$ on F (equivalently, $\mathbb{T}$ is *uniformizable*, that is, the topology $\tau_{\mathbb{T}}$ can be defined by a set of semidistances),

- *normal* if, for any two disjoint closed subsets F_1 and F_2 of $\mathbb{T}$, there exist two disjoint open subsets U_1 and U_2 such that $F_1 \subset U_1$ and $F_2 \subset U_2$, (equivalently, from Urysohn Lemma, for any two disjoint closed subsets F_1 and F_2 of $\mathbb{T}$, there exists a continuous function $f : \mathbb{T} \to [0,1]$ such that $f = 0$ on F_1 and $f = 1$ on F_2),

- *Lindelöf* (resp. *hereditarily Lindelöf*) if every open cover of $\mathbb{T}$ (resp. of any open set of $\mathbb{T}$) has a countable subcover,

- *(separably) submetrizable* if there exists a (separable) metrizable topology τ_0 on $\mathbb{T}$ which is coarser than $\tau_{\mathbb{T}}$,

- *perfectly normal* if $\mathbb{T}$ is a normal space and if each closed subset of $\mathbb{T}$ is a G_δ *set*, that is, a countable intersection of open sets. Equivalently, $\mathbb{T}$ is perfectly normal if and only if $\mathbb{T}$ is a topological space such that, for every closed subset F of $\mathbb{T}$, there exists a continuous function $f : \mathbb{T} \to \mathbb{R}$ such that $F = f^{-1}(0)$ (see [Eng89, Theorem 1.5.19]). Thus, if $\mathbb{T}$ is perfectly normal, its Borel σ–algebra is generated by the set $C_b(\mathbb{T})$ of bounded continuous functions on $\mathbb{T}$. Obviously, every metric space is perfectly normal.

Remark 1.1.1 If $\mathbb{T}$ is completely regular and submetrizable, the topology of $\mathbb{T}$ can be defined by a set of continuous *distances* (instead of semidistances). Indeed, let $\mathcal{D}$ be a set of semidistances which induces the topology of $\mathbb{T}$. Let d_0 be a continuous distance on $\mathbb{T}$. Then the set $\mathcal{D}' = \{d \vee d_0; d \in \mathcal{D}\}$ is a set of continuous distances which induces the topology of $\mathbb{T}$ (where $a \vee b$ denotes the supremum of a and b). Furthermore, if $\mathcal{D}$ is upwards filtering (resp. σ–upwards filtering), then $\mathcal{D}'$ is upwards filtering (resp. σ–upwards filtering).

Hereditarily Lindelöf spaces We list some properties of hereditarily Lindelöf spaces that we shall often use in this book. Proofs can be found in [Bou74] or [Sch73].

1. Any continuous image of a Lindelöf (resp. hereditarily Lindelöf) space is Lindelöf (resp. hereditarily Lindelöf).
2. Any regular Lindelöf space is paracompact, hence normal.
3. Any regular hereditarily Lindelöf space is perfectly normal.
4. If $\mathbb{T} \times \mathbb{T}$ is hereditarily Lindelöf and if the continuous functions separate the points of $\mathbb{T}$, there exists a sequence of continuous functions which separate the points of $\mathbb{T}$. In particular, $\mathbb{T}$ is submetrizable.

Here is another useful property.

Lemma 1.1.2 *Assume that $\mathbb{T}$ is hereditarily Lindelöf and regular (in particular, $\mathbb{T}$ is completely regular). Let $\mathcal{D}$ be a set of continuous semidistances on $\mathbb{T}$ which induces the topology of $\mathbb{T}$ and is σ–upwards filtering.*

1. *Let $G \in \mathcal{G}$. There exists $d \in \mathcal{D}$ such that G is d–open.*
2. *Let $f : \mathbb{T} \to \mathbb{R}$ be continuous. Then there exists $d \in \mathcal{D}$ such that f is d–continuous.*

Proof.

1 For each $t \in G$, as $\mathcal{D}$ induces the topology of $\mathbb{T}$, there exists $d \in \mathcal{D}$ and an open d–ball B_d with center t which is contained in G. From the hereditary Lindelöf property, we can extract a sequence $B_{d_1}, \dots, B_{d_n}, \dots$ such that $G = \cup_n B_{d_n}$. Then take for d any element of $\mathcal{D}$ which is topologically finer than each d_n $(n \geq 1)$: Each B_{d_n} is d–open, thus G is d–open.

2 Let $(]a_n, b_n[)_{n \geq 1}$ be an enumeration of all open intervals of $\mathbb{R}$ with rational endpoints. From Part 1, there exists for each n an element d_n of $\mathcal{D}$ such that $f^{-1}]a_n, b_n[$ is d_n–open. Then take for d any element of $\mathcal{D}$ which is topologically finer than each d_n $(n \geq 1)$. □

Remark 1.1.3 (hereditary Lindelöf property and separability) It is not known whether, under the usual axioms of Logic (ZFC), there exist hereditary Lindelöf spaces which are not separable. Examples are known under the continuum hypothesis (see [Eng89, Remark in Problem 3.12.7(c)] and the survey [Roi84]).

Suslin spaces The (Hausdorff) space $\mathbb{T}$ is *Suslin* if there exists a Polish space $\mathbb{S}$ and a continuous surjective mapping from $\mathbb{S}$ to $\mathbb{T}$ (recall that a mapping φ from a set $\mathbb{A}$ to a set $\mathbb{B}$ is said to be *surjective* if we have $\varphi(\mathbb{A}) = \mathbb{B}$, and that a topological space $\mathbb{S}$ is said to be *Polish* if it is separable and if there exists a distance d which is compatible with the topology of $\mathbb{S}$ and such that $(\mathbb{S}, d)$ is complete). Let us list below the main properties of Suslin spaces that we use in this book. Assume that $\mathbb{T}$ is Suslin. The following hold (see [Sch73, Bou74]):

1. $\mathbb{T}$ is separable.

2. $\mathbb{T} \times \mathbb{T}$ is hereditarily Lindelöf (in particular, $\mathbb{T}$ is hereditarily Lindelöf).

3. Thus, if $\mathbb{T}$ is regular, it is perfectly normal and separably submetrizable.

4. Consequently, a compact space is metrizable if and only if it is Suslin. Thus, all compact subsets of a Suslin space are metrizable.

5. If $\mathbb{T}$ is regular, its topology is defined by the family of all $\tau_{\mathbb{T}}$–continuous distances on $\mathbb{T}$ (see Remark 1.1.1).

6. The Borel σ–algebra $\mathcal{B}_{\mathbb{T}}$ of $\mathbb{T}$ is countably generated.

7. $\mathbb{T}$ is a Radon space, that is, every finite Borel measure on $\mathbb{T}$ is Radon (see Section 1.3 about Radon measures and spaces).

8. Any Hausdorff topology on $\mathbb{T}$ which is coarser than $\tau_{\mathbb{T}}$ has the same Borel sets as $\tau_{\mathbb{T}}$ (consequently, any Suslin topology which is comparable with $\tau_{\mathbb{T}}$ has the same Borel sets as $\tau_{\mathbb{T}}$).

9. Any Suslin topology which is comparable with $\tau_{\mathbb{T}}$ has the same Radon measures as $\tau_{\mathbb{T}}$.

10. We shall also use sometimes the following property: If $\tau_{\mathbb{T}}$ is Suslin regular, there exists on $\mathbb{T}$ a Suslin metrizable topology $\tau'_{\mathbb{T}}$ on $\mathbb{T}$ which is finer than $\tau_{\mathbb{T}}$ [SP76].

In Remark 1.2.1, we list some measurability properties of random subsets of a Suslin space.

A useful class of Suslin spaces is that of Lusin spaces (recall that a Hausdorff topological space $\mathbb{T}$ is *Lusin* if there exists a Polish topology on $\mathbb{T}$ which is finer than the original one, see [Bou74] or [Sch73]).

The class of Suslin spaces and the class of Lusin spaces are stable under countable products, countable topological sums, and countable unions.

A particularly useful class of Lusin spaces is that of submetrizable k_ω–spaces (see Section 4.4 for definitions and basic properties).

Cosmic spaces The space $\mathbb{T}$ is said to be *cosmic* ("Continuous-image Of Separable Metric", [Mic66]) if there exists a separable metric space $\mathbb{S}$ and a continuous surjective mapping from $\mathbb{S}$ to $\mathbb{T}$. Thus the class of cosmic spaces contains the separable metrizable spaces and the Suslin spaces.

A *network* of the topological space $\mathbb{T}$ is a collection $\mathcal{N}$ of subsets of $\mathbb{T}$ such that each open subset U of $\mathbb{T}$ is the union of the elements of $\mathcal{N}$ which are contained in U. The following propositions are equivalent for the Hausdorff topological space $\mathbb{T}$ (see [Cal82, Cal84, Mic66]):

(i) $\mathbb{T}$ is cosmic.

(ii) There exists a (non–necessarily Hausdorff) topological space $\mathbb{S}$ with countable base and a continuous surjective mapping from $\mathbb{S}$ to $\mathbb{T}$.

(iii) $\mathbb{T}$ has a countable network.

Cosmic spaces are hereditarily Lindelöf. Recall that regular Lindelöf spaces are normal [Eng89, Theorem 3.8.2]. Furthermore, any regular cosmic space is submetrizable (see [Cal82]).

An immediate consequence of (iii) is that, if $\mathbb{T}$ is a regular cosmic space, then it has a countable network the elements of which are closed, thus Borel (such a network will be called a *countable Borel network*). Thus the Borel σ–algebra of any regular cosmic space is countably generated.

An important result about regular cosmic spaces was obtained independently by Bešlagić and Calbrix [Cal82, Beš83, Cal84]: Every regular cosmic space can be embedded in a Lusin space. We use this result in Proposition 1.3.2 and in Chapter 3. Unfortunately, it is not known whether this Lusin space can be chosen among regular spaces. A positive answer to that question would have interesting consequences on the general theory of Young measures: see e.g. Part F of Theorem 2.1.3 in the light of Corollary 2.1.8, and see other results of Section 2.1 where it is assumed that $\mathbb{T}$ is a dense subset of a *regular* Suslin space.

Vietoris topology We endow the set $\mathcal{F}$ with the *Vietoris topology*, that is, the topology generated by the sets

$$U^- = \{F \in \mathcal{F};\ F \cap U \neq \emptyset\}$$

and

$$U^+ = \{F \in \mathcal{F};\ F \subset U\},$$

where U runs over the open subsets of $\mathbb{T}$. A base of the Vietoris topology is given by the sets

$$[U_1, \dots, U_n] := \{F \in \mathcal{F};\ \forall\, i = 1, \dots, n,\ \ F \cap U_i \neq \emptyset, \quad F \subset \cup_{1 \le i \le n} U_i\},$$

where $\{U_1, \dots, U_n\}$ runs over the set of finite subsets of $\mathcal{G}$. This is an exercise left to the reader. It rests on the formula

$$U_1^- \cap \dots \cap U_n^- \cap V^+ = [V, U_1 \cap V, \dots, U_n \cap V].$$

The restriction to $\mathcal{K}$ of the Vietoris topology is called the *Vietoris topology on $\mathcal{K}$*. The reader can find further information on the Vietoris topology in e.g. [Bee93, Mic51, Chr74].

1.2 Random elements, random sets, integrands

Throughout, $(\Omega, \mathcal{S}, \mathrm{P})$ is a probability space (but in some cases, it will be explicitely stated that P is an arbitrary measure on $(\Omega, \mathcal{S})$). A *random element* of a topological space $\mathbb{T}$ is an equivalence class (for the equality P-almost everywhere) of Borel–measurable mappings $\Omega \to \mathbb{T}$. The set of random elements of $\mathbb{T}$ is denoted by $\mathfrak{X}(\Omega, \mathbb{T})$ or $\mathfrak{X}(\mathbb{T})$, or $\mathfrak{X}$ if no confusion can arise. The *law* of an element X of $\mathfrak{X}(\mathbb{T})$, (that is, the image by $\mathfrak{X}$ of the measure P), is denoted by $\mathcal{L}(X)$.

Random sets We shall need the notions of measurable random set and integrand to define the stable topologies on spaces of Young measures. A *random subset* of a topological space $\mathbb{T}$ is an element of $\mathcal{S} \otimes \mathcal{B}_{\mathbb{T}}$. If A is a subset of $\Omega \times \mathbb{T}$ and if $\omega \in \Omega$, we denote by $A(\omega)$ the set $\{t \in \mathbb{T};\ (\omega, t) \in A\}$, and we call it the *value* of A at ω. A *selection* of A is a mapping $\sigma : \Omega \to \mathbb{T}$ such that $\sigma(\omega) \in A(\omega)$ for all $\omega \in \Omega$.

A random subset A is said to be *open (closed, compact)* if its values are open (closed, compact).

Clearly, a subset A of $\Omega \times \mathbb{T}$ may be seen as a *multifunction* $\omega \mapsto A(\omega)$. This multifunction is said to be *graph–measurable* if $A \in \mathcal{S} \otimes \mathcal{B}_{\mathbb{T}}$, that is, if A is a random set. The *graph* of a multifunction $\omega \mapsto A(\omega)$ is the set $\operatorname{gph}(A) = \{(\omega, t) \in \Omega \times \mathbb{T};\ t \in A(\omega)\}$. Usually, we identify a multifunction with its graph, and thus we use the same notation to denote a random set A and the associated multifunction. Other notions of measurability for closed valued multifunctions are presented in e.g. [CV77].

In particular, it is interesting for us to know when a closed valued multifunction F is Borel measurable for the Vietoris topology. We say that a closed valued multifunction F is *V–measurable*, or simply *measurable*, if, for any $U \in \mathcal{G}$, the sets

$$F^-(U) := \{\omega \in \Omega;\ F(\omega) \cap U \neq \emptyset\}$$

and

$$F^+(U) := \{\omega \in \Omega;\ F(\omega) \subset U\}$$

belong to $\mathcal{S}$, and we say that F is *LV–measurable* if, for any $U \in \mathcal{G}$, $F^-(U)$ belongs to $\mathcal{S}$. Thus V–measurability means measurability w.r.t. the Vietoris topology, whereas LV–measurability means measurability w.r.t. the *lower Vietoris topology*, generated by the sets U^-, $U \in \mathcal{G}$. The Borel σ–algebra of the lower Vietoris topology is called the *Effros σ–algebra*.

The sets of random subsets, open random subsets, closed random subsets and compact random subsets of $\mathbb{T}$ are respectively denoted by $\underline{\mathcal{B}}(\mathbb{T})$, $\underline{\mathcal{G}}(\mathbb{T})$, $\underline{\mathcal{F}}(\mathbb{T})$, $\underline{\mathcal{K}}(\mathbb{T})$, or $\underline{\mathcal{B}}$, $\underline{\mathcal{G}}$, $\underline{\mathcal{F}}$, $\underline{\mathcal{K}}$ if no confusion may arise, and their elements are simply called *random sets, random open sets, random closed sets and random compact sets.* Note that $\underline{\mathcal{B}}$ is the σ–algebra $\mathcal{S} \otimes \mathcal{B}_{\mathbb{T}}$. Furthermore, it is generated by the random open sets of the form $G = A \times U$ with $A \in \mathcal{S}$ and U an open subset of $\mathbb{T}$, thus $\underline{\mathcal{B}}$ is generated by $\underline{\mathcal{G}}$.

Remark 1.2.1 In the case when $\mathbb{T}$ is Suslin, the elements of $\underline{\mathcal{B}}$ have some nice measurability properties, that we list below. For each measure $\mu \in \mathcal{M}^+(\Omega)$, we denote by $\mathcal{S}^*_\mu$ the μ–completion of $\mathcal{S}$. We denote by $\mathcal{S}^*$ the *universal completion* of $\mathcal{S}$, that is, $\mathcal{S}^* = \cap_{\mu \in \mathcal{M}^+(\mathbb{T})} \mathcal{S}_\mu$. Assume that $\mathbb{T}$ is Suslin, and let $A \in \underline{\mathcal{B}}$.

(*i*) From a result of Freedmann and Neveu [Deb66, Theorem 3.4], the set $\{\omega \in \Omega;\ A(\omega) \neq \emptyset\} = \pi_\Omega(A)$ is an element of $\mathcal{S}^*$, where π_Ω denotes the canonical projection $\Omega \times \mathbb{T} \to \Omega$. We shall refer to this theorem as the *Projection Theorem.* Another proof of this result can be found in [SB74] and in the proofs of Theorems III.23 and III.22 of [CV77].

(*ii*) If A has nonempty values, then, from a result of M. F. Sainte Beuve ([SB74], [CV77, Theorem III.22]), generalizing results of Aumann and von Neumann, the set A admits an $\mathcal{S}^*$–measurable selection. Consequently (see [CV77, Theorem III.22]), A admits an $\mathcal{S}^*$–measurable *Castaing representation*, that is, there exists a sequence $(\sigma_n)_{n \in \mathbb{N}}$ of $\mathcal{S}^*$–measurable selections of A such that, for every $\omega \in \Omega$, the closure $\overline{A(\omega)}$ of $A(\omega)$ satisfies

$$\overline{A(\omega)} = \overline{\{\sigma_n(\omega);\ n \in \mathbb{N}\}}.$$

We shall refer to this result as the *Sainte Beuve–von Neumann–Aumann Selection Theorem.*

(*iii*) From a theorem of Debreu ([Deb66], Theorem 4.4), the random set A is $\mathcal{S}^*$–V–measurable. Indeed, Debreu proved the result for A with nonempty values, but, using the Projection Theorem, the results extends to A with possibly empty values. We can also deduce this result directly from the Projection Theorem, because, for any open subset U of $\mathbb{T}$, we have

$$A^-(U) = \pi_\Omega\left(A \cap (\Omega \times U)\right),$$
$$A^+(U) = \left(\pi_\Omega\left(A \cap (\Omega \times U^c)\right)\right)^c$$

(where A^c denotes the complement of A).

(*iv*) From the preceding property, each element F of $\underline{\mathcal{F}}$ may be viewed as an $\mathcal{S}^*$–measurable mapping $\omega \mapsto F(\omega)$, $\Omega \to \mathcal{F}$ (recall that $\mathcal{F}$ is endowed with the Vietoris topology).

(*v*) Consequently, the mapping $\overline{A} : \omega \mapsto \overline{A(\omega)}$ is LV–measurable w.r.t. $\mathcal{S}^*$. Indeed, we have, for any open subset U of $\mathbb{T}$,

$$\left(\overline{A}\right)^- U = \cup_{n \in \mathbb{N}} \sigma_n^{-1}(U).$$

More generally, if $\mathbb{T}$ is separably submetrizable, we also have the following property:

(*vi*) Let τ_0 be a separable metrizable topology on $\mathbb{T}$ which is coarser than $\tau_{\mathbb{T}}$. Let $K \subset \Omega \times \mathbb{T}$ such that $K(\omega) := \{t \in \mathbb{T};\ (\omega, t) \in K\}$ is a compact subset of $\mathbb{T}$ for every $\omega \in \Omega$ and such that $\omega \mapsto K(\omega)$, $\Omega \to \mathcal{F}$ is measurable (for the Vietoris topology). Then $K \in \underline{\mathcal{K}}$. Indeed, for every $\omega \in \Omega$, $K(\omega)$ is compact (thus closed) for τ_0, and $\omega \mapsto K(\omega)$ is still measurable for the Vietoris topology associated with τ_0. From [CV77, Proposition III.13], we have $K \in \mathcal{S} \otimes \mathcal{B}_{\tau_0} \subset \mathcal{S} \otimes \mathcal{B}_{\tau_{\mathbb{T}}} = \underline{\mathcal{B}}$.

Lemma 1.2.2 *Assume that $\mathbb{T}$ is a Suslin space and that $(\Omega, \mathcal{S}, \mathrm{P})$ is complete. Let $F \in \underline{\mathcal{F}}$. Let d be a continuous semidistance on $\mathbb{T}$ and let $\varepsilon > 0$. The set*

$$G^{d,\varepsilon} = \{(\omega, t) \in \Omega \times \mathbb{T};\ F(\omega) \neq \emptyset \text{ and } d(t, F(\omega)) < \varepsilon\}$$

*is a random open set. Consequently, the d–*closure $\overline{F}^d = \cap_{k \geq 1} G^{d,1/k}$ *of F is a random closed set.*

Proof. From the Projection Theorem, the set $\Omega' = \{\omega \in \Omega;\ F(\omega) \neq \emptyset\}$ is $\mathcal{S}^*$–measurable (where $\mathcal{S}^*$ is the universal completion of $\mathcal{S}$). As we have assumed $(\Omega, \mathcal{S}, \mathrm{P})$ to be complete, we have here $\mathcal{S}^* = \mathcal{S}$. We can thus assume without loss of generality that $\Omega' = \Omega$.

Again from the Projection Theorem, [CV77, Theorem III.22], as $\mathbb{T}$ is Suslin and as we have $\mathcal{S}^* = \mathcal{S}$, F admits an $\mathcal{S}$–measurable Castaing representation, i.e. there exists a sequence $(\sigma_n)_{n \in \mathbb{N}}$ of $\mathcal{S}$–measurable mappings $\Omega \to \mathbb{T}$ such that $F(\omega)$ is the closure of $\{\sigma_n(\omega);\ n \in \mathbb{N}\}$ for every $\omega \in \Omega$. For each $n \in \mathbb{N}$, the mapping $(\omega, t) \mapsto d(\sigma_n(\omega), t)$ is measurable, and we have

$$G^{d,\varepsilon} = \cup_{n \in \mathbb{N}} \{(\omega, t) \in \Omega \times \mathbb{T};\ d(t, \sigma_n(\omega)) < \varepsilon\},$$

thus $G^{d,\varepsilon}$ is in $\underline{\mathcal{G}}$. □

Integrands An *integrand* on $\Omega \times \mathbb{T}$ is an $\mathcal{S} \otimes \mathcal{B}_{\mathbb{T}}$–measurable function $f : \Omega \times \mathbb{T} \to [-\infty, +\infty]$. Note that integrands generalize random sets, because a subset A of $\Omega \times \mathbb{T}$ is an element of $\mathcal{S} \otimes \mathcal{B}_{\mathbb{T}}$ if and only if its indicator function $\mathbb{1}_A$ is an integrand (recall that the *indicator function* $\mathbb{1}_A$ of a A is defined by $\mathbb{1}_A(x) = 1$ if $x \in A$, $\mathbb{1}_A(x) = 0$ if $x \in A^c$). Conversely, any function $f : \Omega \times \mathbb{T} \to$ can be identified with its *epigraph*

$$\mathrm{Epi}\,[f] := \{(\omega, (t, r)) \in \Omega \times (\mathbb{T} \times [-\infty, +\infty]);\ f(\omega, t) \leq r\}.$$

It is easy to check that f is an integrand if and only if $\mathrm{Epi}\,[f]$ is a random subset of $\mathbb{T} \times [-\infty, +\infty]$. Thus integrands can also be seen as special random sets.

An integrand f is said to be L^1*-bounded* if there exists a P–integrable function ϕ such that $|f(\omega, .)| \leq \phi(\omega)$ for each $\omega \in \Omega$. We say that f is *lower semicontinuous (l.s.c.)* if $f(\omega, .)$ is l.s.c. for each $\omega \in \Omega$. Thus l.s.c. integrands generalize open random sets. L.s.c. integrands are also called *normal integrands.*

We define similarly uppersemicontinuous (u.s.c.) integrands, continuous integrands, *etc.* Note that a *bounded integrand* f satisfies $|f(\omega, t)| \leq M$ for some

$M \geq 0$, for all $t \in \mathbb{T}$ and *for all* $\omega \in \Omega$: It is not sufficient that $f(\omega, .)$ be bounded for each ω.

Following a well established tradition, we call *Carathéodory integrand* any function f on $\Omega \times \mathbb{T}$ such that, for each $\omega \in \Omega$, $f(\omega, .)$ is continuous and, for each $t \in \mathbb{T}$, $f(., t)$ is $\mathcal{S}$–measurable. Thus a Carathéodory integrand is not necessarily $\mathcal{S} \otimes \mathcal{B}_{\mathbb{T}}$–measurable, that is, it is not always an integrand in our terminology. However, the following result shows that, in most interesting cases, a Carathéodory integrand is just a continuous integrand. In the sequel, we shall speak of Carathéodory integrands only when they are integrands.

Lemma 1.2.3 (Carathéodory integrands) *Assume that $\mathbb{T}$ is Lusin or regular Suslin. Let f be a mapping from $\Omega \times \mathbb{T}$ to a metric space U such that, for each $\omega \in \Omega$, $f(\omega, .)$ is continuous and, for each $t \in \mathbb{T}$, $f(., t)$ is $\mathcal{S}$–measurable. Then f is $\mathcal{S} \otimes \mathcal{B}_{\mathbb{T}}$–measurable.*

Proof. Let $\tau'_{\mathbb{T}}$ be a Suslin metrizable topology which is finer than the topology $\tau_{\mathbb{T}}$ of $\mathbb{T}$ [SP76]. For each $\omega \in \Omega$, the function $f(\omega, .)$ is $\tau'_{\mathbb{T}}$–continuous. Let d be a distance that is compatible with $\tau'_{\mathbb{T}}$. We then proceed as in [CV77, Lemma III.14]. Let $(t_n)_{n \in \mathbb{N}}$ be a dense sequence in $\mathbb{T}$. For each integer $p \geq 1$, let $(B_n^p)_n$ be a measurable partition of $\mathbb{T}$ into parts of d–diameter $\leq 1/p$ such that, for each n, B_n^p contains t_n. Set $f_p(\omega, t) = f(\omega, t_n)$ if $t \in B_n^p$. From continuity of $f(\omega, .)$, the sequence $(f_p)_p$ of measurable functions converges pointwise to f. As U is metric, this entails that f is $\mathcal{S} \otimes \mathcal{B}_{\tau'_{\mathbb{T}}}$–measurable. But, as $\tau'_{\mathbb{T}}$ is Suslin, we have $\mathcal{B}_{\tau'_{\mathbb{T}}} = \mathcal{B}_{\tau_{\mathbb{T}}}$. □

We will sometimes need to approximate arbitrary functions on $\Omega \times \mathbb{T}$ by u.s.c. or l.s.c. integrands. The following lemma is an extension to the Suslin case of a result which is essentially due to Balder ([Bal84a], Lemmas A5 and A6). We give the first part of it in the same form as it is given in [But89, Lemma 2.1.6]. This first part is also an immediate consequence of an old result of the third author ([Val70, 1.14] [Val71, Proposition 13]), applied to epigraphs. Recall that the space $\mathbb{T}$ is said to be *second countable* if it admits a countable basis. Recall also that the *outer measure* P^* associated with P is defined by $\mathrm{P}^*(E) = \inf\{\mathrm{P}(B);\, B \supset E, B \in \mathcal{S}\}$ for any subset E of Ω. Its restriction to $\mathcal{S}_{\mathrm{P}}^*$, which we shall also denote by P^*, is the unique measure on $\mathcal{S}_{\mathrm{P}}^*$ which extends P.

Lemma 1.2.4 (Measurable regularization ([Bal84a])) *Assume that $\mathbb{T}$ is a second countable Suslin space and let $f : \Omega \times \mathbb{T} \to [0, +\infty]$ be a function such that $f(\omega, .)$ is l.s.c. for P–a.e. $\omega \in \Omega$. There exists an l.s.c. integrand $\widetilde{f}$ such that*

1. *$\widetilde{f}(\omega, .) \geq f(\omega, .)$ for P–a.e. $\omega \in \Omega$,*

2. *for every integrand g such that $g(\omega, .) \geq f(\omega, .)$ P–a.e., we have $g(\omega, .) \geq \widetilde{f}(\omega, .)$ P–a.e.*

Furthermore, if f is $\mathcal{S}_{\mathrm{P}}^* \otimes \mathcal{B}_{\mathbb{T}}$–measurable, then $\widetilde{f}(\omega, .) = f(\omega, .)$ for P^*–a.e. $\omega \in \Omega$.

Following Balder [Bal95], if f is as above, we call *measurable regularization* of f any l.s.c. integrand $\widetilde{f}$ satisfying Conditions 1 and 2 of Lemma 1.2.4.

In applications of Lemma 1.2.4, $\mathbb{T}$ is a Suslin metrizable space, but there exist second countable Suslin spaces which are not metrizable. Indeed, any countable Hausdorff space $\mathbb{T}$ is Suslin, because the discrete topology on $\mathbb{T}$ is Polish. Thus any countable second countable Hausdorff space which is not regular provides an example of a nonmetrizable second countable Suslin space. Counterexamples 60, 61, 75, 100, 126 and 127 in [SS78] are examples of such spaces.

Proof of Lemma 1.2.4. We assume without loss of generality that $f(\omega,.)$ is l.s.c. for every $\omega \in \Omega$.

Let U be a countable basis of $\mathbb{T}$ and let J be the set of pairs (r,G) such that r is a rational number and $G \in U$. For each $j=(r,G)\in J$, we set $f_j = r\,\mathbb{1}_G$. For each $\omega \in \Omega$, we have

$$f(\omega,.) = \sup\{f_j;\ j\in J, f_j \le f(\omega,.)\}$$

(e.g. adapt the proof of Proposition 3 page IV.31 in [Bou71]). For each $j \in J$, let

$$E_j = \{\omega \in \Omega;\ f_j \le f(\omega,.)\}.$$

We have, for all $(\omega,t)\in \Omega\times\mathbb{T}$,

$$f(\omega,t) = \sup\{\,\mathbb{1}_{E_j}(\omega) f_j(t);\ j\in J\}.$$

For each $j \in J$, let $B_j \in \mathcal{S}$ be such that $E_j \subset B_j$ and $\mathrm{P}^*(E_j) = \mathrm{P}(B_j)$. We set

$$\widetilde{f} = \sup\{\,\mathbb{1}_{B_j}(\omega) f_j(t);\ j\in J\}.$$

The function $\widetilde{f}$ is an integrand and satisfies Condition 1. Let g be an integrand such that P–a.e. $g(\omega,.) \ge f(\omega,.)$. We assume w.l.g. that $g \ge f$. For each $j \in J$, let

$$C_j = \{\omega\in\Omega;\ f_j \le g(\omega,.)\}.$$

Each C_j is the complement of the projection on Ω of the $\mathcal{S}\otimes\mathcal{B}_{\mathbb{T}}$–measurable set $\{(\omega,t)\in\Omega\times\mathbb{T};\ f_j(t) > g(\omega,t)\}$. Therefore, from the Projection Theorem, since $\mathbb{T}$ is Suslin, each C_j is $\mathcal{S}^*_{\mathrm{P}}$–measurable. Furthermore, we have, for all $(\omega,t)\in\Omega\times\mathbb{T}$,

$$g(\omega,t) \ge \sup\{\,\mathbb{1}_{C_j}(\omega) f_j(t);\ j\in J\}.$$

As $g \ge f$, we have $E_j \subset C_j$ for each $j \in J$. From the definition of B_j, we thus have

$$\mathrm{P}^*(B_j \setminus C_j) = 0.$$

As J is countable, there exists a measurable subset Ω' of Ω such that $\mathrm{P}(\Omega')=1$ and $B_j\cap\Omega' \subset C_j\cap\Omega'$ for each $j\in J$. For each $\omega\in\Omega'$, we thus have

$$\widetilde{f}(\omega,.) \le g(\omega,.).$$

Finally, assume that f is $\mathcal{S}_{\mathrm{P}}^* \otimes \mathcal{B}_{\mathbb{T}}$–measurable. Then, for every $j \in J$, E_j is $\mathcal{S}_{\mathrm{P}}^*$–measurable and we have $\mathrm{P}^*(B_j \setminus E_j) = 0$, that is, there exists a P–negligible set $\mathcal{N}_j$ such that $E_j \cup \mathcal{N}_j = B_j$. Let $\mathcal{N} = \cup_{j \in J} \mathcal{N}_j$. The set $\mathcal{N}$ is negligible and we have $\widetilde{f}(\omega, .) = f(\omega, .)$ for every $\omega \in \Omega \setminus \mathcal{N}$. □

Here is a variant, for random sets, of Lemma 1.2.4.

Lemma 1.2.5 *Assume that $\mathbb{T}$ is a second countable Suslin space and let $A \subset \Omega \times \mathbb{T}$ be such that, for a.e. $\omega \in \Omega$, $A(\omega) \in \mathcal{G}$. Then there exists $G \in \underline{\mathcal{G}}$ such that*

- *$A(\omega) \subset G(\omega)$ for* P*–almost every $\omega \in \Omega$,*
- *if $B \in \underline{\mathcal{G}}$ satisfies $A(\omega, .) \subset B(\omega, .)$ for* P*–a.e. $\omega \in \Omega$, then $G(\omega, .) \subset B(\omega, .)$ for* P*–a.e. $\omega \in \Omega$.*

Furthermore, if $A \in \mathcal{S}_{\mathrm{P}}^* \otimes \mathcal{B}_{\mathbb{T}}$, then $A(\omega) = G(\omega)$ for P*–almost every $\omega \in \Omega$.*

Proof. We assume w.l.g. that $A(\omega) \in \mathcal{G}$ for every $\omega \in \Omega$. Let $f = \mathbb{1}_A$ and construct $\widetilde{f}$ as in the proof of Lemma 1.2.4. The range of f is contained in $\{0, 1\}$, thus

$$f(\omega, t) = \sup \left\{ \mathbb{1}_{E_j}(\omega) f_j(t);\ j = (1, A), A \in U \right\},$$

which yields

$$\widetilde{f}(\omega, t) = \sup \left\{ \mathbb{1}_{B_j}(\omega) f_j(t);\ j = (1, A), A \in U \right\}.$$

Thus the range of $\widetilde{f}$ is also contained in $\{0, 1\}$, which proves that $\widetilde{f}$ is the indicator function of an element G of $\underline{\mathcal{G}}$. □

Let us now give an easy but useful extension of the second part of Lemma 1.2.4. If $\mathbb{E}$ is a separable Banach space, with topological dual $\mathbb{E}^*$, the following result can be applied in the case when $\mathbb{T} = (\mathbb{E}, \sigma(\mathbb{E}, \mathbb{E}^*))$ or when $\mathbb{T} = (\mathbb{E}^*, \sigma(\mathbb{E}^*, \mathbb{E}))$ because the unit balls of $\mathbb{E}$ and $\mathbb{E}^*$ are Suslin metrizable for the topologies induced by $\sigma(\mathbb{E}, \mathbb{E}^*)$ and $\sigma(\mathbb{E}^*, \mathbb{E})$ (see Examples 4.3.15 and 4.4.1 for more general situations). Note that, in these cases, $\mathbb{T}$ is a Lusin space because the identity mapping from $(\mathbb{E}, \|.\|)$ to $(\mathbb{E}, \sigma(\mathbb{E}, \mathbb{E}^*))$ is continuous, but $\mathbb{T}$ is not second countable.

Lemma 1.2.6 *Assume that $\mathbb{T}$ is the union of a sequence $(\mathbb{T}_n)_{n \in \mathbb{N}}$ of second countable Suslin spaces (for the topology induced by $\mathbb{T}$), which are Borel subsets of $\mathbb{T}$. Let $f : \Omega \times \mathbb{T} \to [0, +\infty]$ be an $\mathcal{S}_{\mathrm{P}}^* \otimes \mathcal{B}_{\mathbb{T}}$–measurable function such that $f(\omega, .)$ is l.s.c. for* P*–a.e. $\omega \in \Omega$. Then there exists an $\mathcal{S} \otimes \mathcal{B}_{\mathbb{T}}$–measurable regularization $\widetilde{f}$ of f such that $\widetilde{f}(\omega, .) = f(\omega, .)$ for* P*–a.e. $\omega \in \Omega$.*

Furthermore, if $f(\omega, .)$ is continuous for P*–a.e. $\omega \in \Omega$, we can choose $\widetilde{f}$ such that $\widetilde{f}(\omega, .)$ is continuous for* P*–a.e. $\omega \in \Omega$.*

If $A \in \mathcal{S}_{\mathrm{P}}^* \otimes \mathcal{B}_{\mathbb{T}}$ satisfies $A(\omega) \in \mathcal{G}$ for each $\omega \in \Omega$, there exists $G \in \underline{\mathcal{G}}$ such that $A(\omega) = G(\omega)$ for P*–almost every $\omega \in \Omega$.*

Proof. For each $n \in \mathbb{N}$, let f_n be the restriction of f on $\Omega \times \mathbb{T}_n$ and let $\widetilde{f}_n : \Omega \times \mathbb{T}_n$ be its measurable regularization. There exists a P–negligible set $\mathcal{N}_n \in \mathcal{S}$ such that $\widetilde{f}_n(\omega, .) = f_n(\omega, .)$ for every $\omega \notin \mathcal{N}_n$. Let $\mathcal{N} = \cup_n \mathcal{N}_n$ and let us set

$$\widetilde{f}(\omega, t) = \begin{cases} f(\omega, t) & \text{if} \quad \omega \in \Omega \setminus \mathcal{N}, \\ 0 & \text{if} \quad \omega \in \mathcal{N}. \end{cases}$$

Then $\widetilde{f}$ is an l.s.c. integrand and $\widetilde{f}(\omega, .) = f(\omega, .)$ for every $\omega \notin \mathcal{N}$, and thus $\widetilde{f}$ is a measurable regularization of f. Furthermore, if the set $\{\omega \in \Omega;\ f(\omega, .)$ is not continuous$\}$ is contained in a measurable negligible set $\mathcal{N}'$, then we can replace $\widetilde{f}$ by $\widetilde{f}\,\mathbb{1}_{(\Omega \setminus \mathcal{N}') \times \mathbb{T}}$.

We then deduce the statements about $A \in \mathcal{S}_{\mathrm{P}}^* \otimes \mathcal{B}_{\mathbb{T}}$ as in the proof of Lemma 1.2.5. □

1.3 Narrow and weak convergence of measures on a topological space

As in the rest of this chapter, we give here definitions and results that will be needed in the sequel. Complements on the topic of this section can be found in Bogachev's very clear and useful survey [Bog98b].

By "measure", we always mean "bounded nonnegative σ–additive measure". The set of measures on a measure space $(\Omega, \mathcal{S})$ is denoted by $\mathcal{M}^+(\Omega, \mathcal{S})$. For any $\alpha \geq 0$, we denote by $\mathcal{M}^{+,\alpha}(\Omega, \mathcal{S})$ the set of elements μ of $\mathcal{M}^+(\Omega, \mathcal{S})$ such that $\mu(\Omega) = \alpha$. An element of $\mathcal{M}^{+,1}(\Omega, \mathcal{S})$ is also called a *(probability) law*. The set of Borel bounded measures on $\mathbb{T}$ is denoted by $\mathcal{M}^+(\mathbb{T})$, or $\mathcal{M}^+(\mathbb{T}, \tau_{\mathbb{T}})$ if we need to precise which Borel σ–algebra is taken into account. Similarly, we use the notations $\mathcal{M}^{+,\alpha}(\mathbb{T})$ and $\mathcal{M}^{+,\alpha}(\mathbb{T}, \tau_{\mathbb{T}})$.

An element μ of $\mathcal{M}^+(\mathbb{T})$ is *tight* if, for any $\varepsilon > 0$, there exists a compact subset K of $\mathbb{T}$ such that $\mu(K^c) < \varepsilon$, and μ is *Radon* if, for any $B \in \mathcal{B}_{\mathbb{T}}$, $\mu(B) = \sup\{\mu(K);\ K$ is compact, $K \subset B\}$. We denote respectively by $\mathcal{M}_t^+(\mathbb{T})$ and $\mathcal{M}_R^+(\mathbb{T})$ the sets of tight and Radon measures on $\mathbb{T}$. We have $\mathcal{M}_R^+(\mathbb{T}) \subset \mathcal{M}_t^+(\mathbb{T})$. The topological space $\mathbb{T}$ is a *Radon space* if every measure on $\mathbb{T}$ is Radon (see [Sch73] about Radon spaces and Radon measures). It is easy to check that $\mathcal{M}_R^+(\mathbb{T}) = \mathcal{M}_t^+(\mathbb{T})$ if and only if every compact subset K of $\mathbb{T}$ is a Radon space. This is for example the case when $\mathbb{T}$ is submetrizable, because, in this case, each compact subset of $\mathbb{T}$ is metrizable, and because each metrizable compact space is Radon (see [Sch73], Proposition 6 page 117).

An element μ of $\mathcal{M}^+(\mathbb{T})$ is *τ–regular* if, for every downwards filtering net $(F_\alpha)_{\alpha \in \mathbb{A}}$ in $\mathcal{F}(\mathbb{T})$, we have $\mu(\cap_\alpha F_\alpha) = \inf_\alpha \mu(F_\alpha)$. We then have $\mu(\inf_\alpha f_\alpha) = \inf_\alpha \mu(f_\alpha)$ for any downwards filtering uniformly bounded net $(f_\alpha)_{\alpha \in \mathbb{A}}$ of u.s.c. functions on $\mathbb{T}$ [Top70b, P15 page XIII]. Every Radon measure is τ–regular. If $\mathbb{T}$ is hereditarily Lindelöf, it is very easy to show that every measure on $\mathbb{T}$ is τ–regular. We denote by $\mathcal{M}_\tau^+(\mathbb{T})$ the space of τ–regular measures on $\mathbb{T}$.

Of course, we also use notations such as $\mathcal{M}_t^{+,\alpha}(\mathbb{T})$, $\mathcal{M}_R^{+,\alpha}(\mathbb{T})$ or $\mathcal{M}_\tau^{+,\alpha}(\mathbb{T})$.

For a recent unified treatment of measure theory in topological spaces, with an extensive study of the notion of regularity of a measure, see also [Kön97, Kön].

If t is an element of $\mathbb{T}$, we denote by δ_t the Dirac mass at t. It is sometimes convenient to identify δ_t with t and to write formulae like $\mathbb{T} \subset \mathcal{M}^+(\mathbb{T})$.

We endow $\mathcal{M}^+(\mathbb{T})$ with the *narrow topology*, which is the coarsest topology for which the function

$$\begin{cases} \mathcal{M}^+(\mathbb{T}) & \to & \mathbb{R} \\ \mu & \mapsto & \mu(f) \end{cases}$$

is upper semi–continuous for every bounded upper semi–continuous function $f : \mathbb{T} \to \mathbb{R}$ (see [Top70b]). A net $(\mu_\alpha)_{\alpha \in \mathbb{A}}$ of measures on $\mathbb{T}$ narrowly converges to a measure μ_∞ if and only if, for each open subset U of $\mathbb{T}$, $\liminf_{\alpha\to\infty} \mu_\alpha(U) \geq \mu_\infty(U)$, and if $\lim_{\alpha\to\infty} \mu_\alpha(\mathbb{T}) = \mu_\infty(\mathbb{T})$. If this is the case, then $\lim_{\alpha\to\infty} \mu_\alpha(f) = \mu_\infty(f)$ for every bounded continuous function f. The topology induced on $\mathbb{T}$ by the narrow topology (via the identification $t \mapsto \delta_t$) is $\tau_\mathbb{T}$.

We shall sometimes consider a coarser topology on $\mathcal{M}^+(\mathbb{T})$, namely the *weak topology* [1], which is the coarsest topology such that, for every bounded continuous function $f : \mathbb{T} \to \mathbb{R}$, the function

$$\begin{cases} \mathcal{M}^+(\mathbb{T}) & \to & \mathbb{R} \\ \mu & \mapsto & \mu(f) \end{cases}$$

is continuous. From the Portmanteau Theorem [Top70b, Theorem 8.1], if $\mathbb{T}$ is completely regular and if $\mu_\infty \in \mathcal{M}_\tau^+(\mathbb{T})$, a net $(\mu_\alpha)_{\alpha\in\mathbb{A}}$ of elements of $\mathcal{M}^+(\mathbb{T})$ converges to μ_∞ for the narrow topology if and only if it converges to μ_∞ for the weak topology. Consequently, the weak topology and the narrow topology coincide on $\mathcal{M}_\tau^+(\mathbb{T})$ if and only if $\mathbb{T}$ is completely regular. Indeed, if $\mathbb{T}$ is completely regular, the equivalence of both topologies on $\mathcal{M}_\tau^+(\mathbb{T})$ is given by the Portmanteau Theorem [Top70b, Theorem 8.1]. Conversely, note that the topology induced on $\mathbb{T}$ by the weak topology is the coarsest topology on $\mathbb{T}$ such that the bounded $\tau_\mathbb{T}$–continuous functions are continuous. But this topology coincides with $\tau_\mathbb{T}$ if and only if $\mathbb{T}$ is completely regular (see [Eng89, Example 8.1.19]).

Note that the set $C_b(\mathbb{T})$ of bounded continuous functions on an arbitrary Hausdorff topological space $\mathbb{T}$ may be very poor, even if $\mathbb{T}$ is regular, e.g. it may be equal to the set of constant functions as in [SS78, Example 92] (Hewitt's condensed corkscrew). In such a case, the weak topology is only the indiscrete topology $\{\emptyset, \mathcal{M}^{+,1}(\mathbb{T})\}$. Another interesting example where $C_b(\mathbb{T})$ is the set of constants is provided by [SS78, Example 75] (irrational slope topology): The space $\mathbb{T}$ is Hausdorff nonregular, Lusin (because $\mathbb{T}$ is countable and every countable union of Lusin spaces is Lusin), and its compact subsets are metrizable (because $\mathbb{T}$ is first countable, see [Eng89, Theorem 4.2.8]). However, some results for an arbitrary

[1] Topsøe [Top70b], who seems to be the inventor of the narrow topology, calls it the "weak topology". Our terminology is that of Schwartz [Sch73]: this allows us to name "weak topology" the topology induced by $\sigma(C_b(\mathbb{T})', C_b(\mathbb{T}))$.

topological space $\mathbb{T}$ involve only $C_b(\mathbb{T})$, not the whole topology $\tau_{\mathbb{T}}$. Considering only the narrow topology would oblige us to restrict ourselves unnecessarily to the frame of completely regular spaces. In particular, in Chapter 3 (which is devoted to convergence in probability of Young measures), the weak topology will be much more convenient to use. But, *by default, $\mathcal{M}^+(\mathbb{T})$ is endowed with the narrow topology.*

Let A be a subset of $\mathbb{T}$. The space $\mathcal{M}^+(A)$ can be considered as a subspace of $\mathcal{M}^+(\mathbb{T})$. Indeed, let $\mathcal{M}^+_A(\mathbb{T})$ be the set of elements μ of $\mathcal{M}^+(\mathbb{T})$ such that $\mu_*(A^c) = 0$ (where μ_* denotes the inner measure associated with μ). The Borel σ-algebra of A is $\mathcal{B}_A = \{B \cap A;\, B \in \mathcal{B}_{\mathbb{T}}\}$. With each Borel measure $\mu \in \mathcal{M}^+(A)$, we can associate the Borel measure $\widehat{\mu} \in \mathcal{M}^+_A(\mathbb{T})$ defined by

$$\forall B \in \mathcal{B}_{\mathbb{T}} \quad \widehat{\mu}(B) = \mu(B \cap A)$$

(thus, if ψ is the canonical injection from A to $\mathbb{T}$, we have $\widehat{\mu} = \psi_\sharp(\mu)$, where $\psi_\sharp\,\mu$ denotes the image by ψ of the measure μ). Conversely, if $\mu \in \mathcal{M}^+_A(\mathbb{T})$, the restriction $\breve{\mu}$ of the outer measure μ^* to $\mathcal{B}_A$ is a Borel measure on A. This is a mere exercise if one notices that

$$\forall B \in \mathcal{B}_{\mathbb{T}} \quad \breve{\mu}(B \cap A) = \mu(B). \tag{1.3.1}$$

This shows that $\breve{\widehat{\mu}} = \mu$ for each $\mu \in \mathcal{M}^+(A)$ and $\widehat{\breve{\nu}} = \nu$ for each $\nu \in \mathcal{M}^+_A(\mathbb{T})$, thus $\mu \mapsto \widehat{\mu}$ is a bijection from $\mathcal{M}^+(A)$ to $\mathcal{M}^+_A(\mathbb{T})$. Furthermore, let $(\mu_\alpha)_{\alpha\in\mathbb{A}}$ be a net in $\mathcal{M}^+(A)$ and let $\mu_\infty \in \mathcal{M}^+(A)$. Using again (1.3.1), we see that $(\mu_\alpha)_{\alpha\in\mathbb{A}}$ narrowly converges to μ_∞ in $\mathcal{M}^+(A)$ if and only if $(\widehat{\mu}_\alpha)_{\alpha\in\mathbb{A}}$ narrowly converges to $\widehat{\mu}_\infty$ in $\mathcal{M}^+_A(\mathbb{T})$. We have thus proved the first half of the following lemma. The second half is left to the reader.

Lemma 1.3.1 *For any subset A of $\mathbb{T}$, the space $\mathcal{M}^+(A)$ (endowed with the narrow topology) is homeomorphic to the subspace $\mathcal{M}^+_A(\mathbb{T})$ of $\mathcal{M}^+(\mathbb{T})$ through the mapping $\mu \mapsto \widehat{\mu}$.*

Furthermore, for any $\mu \in \mathcal{M}^+_A(\mathbb{T})$ and for any bounded measurable function $f : \mathbb{T} \to \mathbb{R}$, we have

$$\mu(f) = \breve{\mu}(f|_A).$$

In the sequel, we shall identify $\mathcal{M}^+(A)$ and $\mathcal{M}^+_A(\mathbb{T})$.

When $\mathbb{T}$ is completely regular, many topological properties of the space $\mathbb{T}$ also hold for $\mathcal{M}^{+,1}_\tau(\mathbb{T})$, endowed with the weak topology. We mention below the topological properties of $\mathcal{M}^{+,1}(\mathbb{T})$ that will be needed in the sequel. Other properties can be found in e.g. [Kou81, HK99] (see also the bibliography in [Bog98b, Section 8.5]).

The most obvious property is that, if $\mathbb{T}$ is completely regular, then the weak topology on $\mathcal{M}^{+,1}_\tau(\mathbb{T})$ is also completely regular. Indeed, it is induced by the semidistances $(\mu, \nu) \mapsto |\mu(f) - \nu(f)|$, where f runs over the set $C_b(\mathbb{T})$ of bounded continuous functions on $\mathbb{T}$.

If $\mathbb{T}$ is a metrizable space, then the weak topology on $\mathcal{M}^+(\mathbb{T})$ is metrizable (see Section 1.4 about the equality $\mathcal{M}^+_\tau(\mathbb{T}) = \mathcal{M}^+(\mathbb{T})$ when $\mathbb{T}$ is not separable). In particular, if d is a distance which is compatible with the topology of $\mathbb{T}$, then the weak topology on $\mathcal{M}^+(\mathbb{T})$ is generated by the *Dudley distance* $\Delta_{\mathrm{BL}}^{(d)}$ defined as follows (see [Dud66]): Let $\mathrm{BL}_1(\mathbb{T}, d)$ be the set of bounded Lipschitz functions $f : \mathbb{T} \to \mathbb{R}$ such that $\max(\|f\|_\infty, \|f\|_{\mathrm{L}}(d)) \le 1$, where $\|f\|_{\mathrm{L}}(d)$ is the infimum of all $r \ge 0$ such that f is r–Lipschitz for d. We then set, for $\mu, \nu \in \mathcal{M}^{+,1}(\mathbb{T})$,

$$\Delta_{\mathrm{BL}}^{(d)}(\mu, \nu) = \sup_{f \in \mathrm{BL}_1(\mathbb{T},d)} |\mu(f) - \nu(f)| .$$

If $\mathbb{T}$ is a separable metric space, there exists a totally bounded distance d which is compatible with $\mathbb{T}$ (see e.g. [Eng89, Par67], actually the condition of separability is also necessary). In this case, the space $\mathrm{C}_u(\mathbb{T}, d)$ of d–uniformly continuous functions on $\mathbb{T}$, endowed with the topology of uniform convergence, is separable. Similarly, $\mathrm{BL}_1(\mathbb{T}, d)$ is separable. There exists thus a countable set $\{f_k;\ k \in \mathbb{N}\}$ of bounded continuous functions on $\mathbb{T}$ (which may be chosen in $\mathrm{C}_u(\mathbb{T}, d)$ or in $\mathrm{BL}_1(\mathbb{T}, d)$) such that the topology of $\mathcal{M}^{+,1}(\mathbb{T}) = \mathcal{M}^{+,1}_\tau(\mathbb{T})$ is the coarsest topology such that the mappings $\nu \mapsto \nu(f_k)$ $(k \in \mathbb{N})$ are continuous (see [Par67, Dud76, Dud02]).

Let us also recall that, if $\mathbb{T}$ is Polish, then $\mathcal{M}^{+,1}(\mathbb{T}) = \mathcal{M}^{+,1}_\tau(\mathbb{T})$ is also Polish (see e.g. [Par67, Pro56]). Furthermore, if $\mathbb{S}$ is a Suslin space and $\pi : \mathbb{S} \to \mathbb{T}$ a surjective continuous mapping, then the mapping $\mu \mapsto \pi_\sharp\,\mu$ from $\mathcal{M}^{+,1}(\mathbb{S})$ to $\mathcal{M}^{+,1}(\mathbb{T})$ is surjective [Sch73, Theorem 12 page 126], where $\pi_\sharp\,\mu$ denotes the image by π of the measure μ. Thus, if $\mathbb{T}$ is Suslin (resp. Lusin), then $\mathcal{M}^{+,1}(\mathbb{T})$ is also Suslin (resp. Lusin) [Sch73, Theorem 7 page 385].

Thanks to Calbrix–Bešlagić Embedding Theorem (see page 5), we can deduce a similar conclusion for cosmic regular spaces.

Proposition 1.3.2 *If $\mathbb{T}$ is cosmic regular, so is $\mathcal{M}^{+,1}(\mathbb{T})$.*

Proof. From Calbrix–Bešlagić Embedding Theorem, $\mathbb{T}$ is a subspace of a Lusin space $\mathbb{S}$. We have $\mathcal{M}^{+,1}(\mathbb{T}) \subset \mathcal{M}^{+,1}(\mathbb{S})$ (see Lemma 1.3.1), and $\mathcal{M}^{+,1}(\mathbb{S})$ is Lusin, thus $\mathcal{M}^{+,1}(\mathbb{T})$ is cosmic. Furthermore, the space $\mathbb{T}$ is completely regular, thus $\mathcal{M}^{+,1}(\mathbb{T})$ is also completely regular. □

Let us return to the case when $\mathbb{T}$ is a completely regular space. Let $\mathcal{D}$ be a set of continuous semidistances on $\mathbb{T}$ which induces the topology $\tau_\mathbb{T}$ of $\mathbb{T}$. For each $d \in \mathcal{D}$, let $\mathbb{S}_d$ be the quotient metric space $(\mathbb{T}/d, \widehat{d})$ and let π_d be the canonical projection $\mathbb{T} \to \mathbb{S}_d$. The topology $\tau_\mathbb{T}$ is the *initial topology* [Bou71] for the system $(\mathbb{S}_d, \pi_d)_{d \in \mathcal{D}}$, that is, $\tau_\mathbb{T}$ is the coarsest topology such that the mappings π_d are continuous. Equivalently, a net $(t_\alpha)_{\alpha \in \mathbb{A}}$ converges to an element t_∞ if and only if, for each $d \in \mathcal{D}$, the net $(\pi_d(t_\alpha))_{\alpha \in \mathbb{A}}$ converges to $\pi_d(t_\infty)$.

Lemma 1.3.3 *Assume that $\mathbb{T}$ is completely regular. Let $\mathcal{D}$ be an upwards filtering set of continuous semidistances on $\mathbb{T}$ which induces the topology of $\mathbb{T}$. With the* *above notations, the narrow topology on $\mathcal{M}_\tau^+(\mathbb{T})$ is the initial topology for the system $(\mathcal{M}_\tau^+(\mathbb{S}_d), \mu \mapsto (\pi_d)_\sharp(\mu))_{d\in\mathcal{D}}$.*

Proof. Let $(\mu_\alpha)_{\alpha\in\mathbb{A}}$ be a convergent net in $\mathcal{M}_\tau^+(\mathbb{T})$ and let μ_∞ be its limit. Let $d \in \mathcal{D}$. Let f be a bounded continuous function on $\mathbb{S}_d$. Then $f \circ \pi_d$ is a bounded continuous function on $\mathbb{T}$, thus

$$\lim_\alpha (\pi_d)_\sharp \mu_\alpha(f) = \lim_\alpha \mu_\alpha(f \circ \pi_d) = \mu_\infty(f \circ \pi_d) = (\pi_d)_\sharp \mu_\infty(f).$$

Thus $((\pi_d)_\sharp \mu_\alpha)_{\alpha\in\mathbb{A}}$ converges to $(\pi_d)_\sharp \mu_\infty$ in $\mathcal{M}_\tau^+(\mathbb{S}_d)$.

Conversely, assume that $((\pi_d)_\sharp \mu_\alpha)_{\alpha\in\mathbb{A}}$ converges to $(\pi_d)_\sharp \mu_\infty$ in $\mathcal{M}_\tau^+(\mathbb{S}_d)$ for each $d \in \mathcal{D}$. We make a slight adaptation of Topsøe's proof of Portmanteau Theorem for narrow convergence [Top70b]. Let $F \in \mathcal{F}(\mathbb{T})$. For each $d \in \mathcal{D}$, let us denote by $\mathrm{BL}(\mathbb{T}, d)$ the set of bounded d–Lipschitz functions on $\mathbb{T}$. Then we have $F = \bigcap_{d\in\mathcal{D}} \overline{F}^d$, where $\overline{F}^d$ denotes the closure of F with respect to d. But, for each $d \in \mathcal{D}$, we have

$$\mathbb{1}_{\overline{F}^d} = \inf_{n\geq 1} f_n^d,$$

where

$$f_n^d(t) = 1 - n\left(d(t,F) \wedge \frac{1}{n}\right).$$

As $f_n^d \in \mathrm{BL}(\mathbb{T}, d)$, we thus have $\mathbb{1}_F = \inf \mathcal{H}$, with

$$\mathcal{H} = \{f \in \mathrm{C}_b(\mathbb{T})\,;\, \exists d \in \mathcal{D} \quad f \in \mathrm{BL}(\mathbb{T}, d),\ \mathbb{1}_F \leq f \leq 1\}.$$

The set $\mathcal{H}$ is downwards filtering, thus

$$\begin{aligned}\mu^\infty(F) &= \inf_{f\in\mathcal{H}} \mu^\infty(f) = \inf_{f\in\mathcal{H}} \lim_\alpha \mu_\alpha(f) \\ &\geq \limsup_\alpha \mu_\alpha(F).\end{aligned}$$

□

Remark 1.3.4 (Initial topology and projective limit) With the hypothesis and notations of Lemma 1.3.3, if furthermore $\mathbb{T}$ is Suslin and $\mathcal{D}$ is an upwards filtering set of continuous *distances* which induces the topology $\tau_\mathbb{T}$, then the topological space $\mathcal{M}^+(\mathbb{T})$ is the projective limit (or inverse limit) of the system $(\mathcal{M}^+(\mathbb{S}_d), \mu \mapsto (\pi_d)_\sharp(\mu))_{d\in\mathcal{D}}$ (see e.g. [Eng89] about inverse limit of topological spaces, and [Par67, HJ91] about inverse limits of measures). Indeed, for each $d \in \mathcal{D}$, the set $\mathbb{S}_d$ coincides with $\mathbb{T}$, the mapping π_d is the identity mapping, and the Borel σ–algebra $\mathcal{B}_{\mathbb{S}_d}$ is the same as $\mathcal{B}_\mathbb{T}$, thus the sets $\mathcal{M}^{+,1}(\mathbb{S}_d)$ and $\mathcal{M}^{+,1}(\mathbb{T})$ are equal. Furthermore, as $\mathbb{T}$ and $\mathbb{S}_d$ are hereditarily Lindelöf, we have $\mathcal{M}_\tau^{+,1}(\mathbb{S}_d) = \mathcal{M}^{+,1}(\mathbb{S}_d) = \mathcal{M}^{+,1}(\mathbb{T}) = \mathcal{M}_\tau^{+,1}(\mathbb{T})$.

1.4 Measurable cardinals and separable Borel measures

We say that a cardinal number $\mathfrak{m}$ is *(real–valued) measurable* if, for any set $\mathbb{S}$ of cardinal $\mathfrak{m}$, there exists a probability measure on the algebra of all subsets of $\mathbb{S}$ such that every countable subset of $\mathbb{S}$ has measure 0.

Clearly $\aleph_0$ is not measurable and, if $\mathfrak{n}$ is not measurable and $\mathfrak{m} < \mathfrak{n}$, then $\mathfrak{m}$ is not measurable. It is known that it is consistent with the usual axioms of mathematics (ZFC) to assume that there exists no measurable cardinal. Furthermore, under the continuum hypothesis, the cardinal c is not measurable (see a short proof in [Dud02, Appendix C] or in [HJ99]). On the other hand, it seems unknown whether it is consistent with ZFC to assume that there exists a measurable cardinal. More details on measurable cardinals can be found in [Jec78].

The assumption that no measurable cardinal exists has nice consequences in probability theory. We say that a measure μ on the σ–algebra of a metric space $\mathbb{T}$ is *separable* if there exists a separable Borel subspace $\mathbb{T}_0$ of $\mathbb{T}$ such that $\mu(\mathbb{T}_0) = 1$. Marczewski and Sikorski [MS48] proved that, for a metric space $\mathbb{T}$, the following conditions are equivalent:

(*a*) There exists a dense subset of $\mathbb{T}$ with non–measurable cardinal.

(*b*) Every Borel probability measure on $\mathbb{T}$ is separable.

Proofs of this result, along with discussions about its consequences, can also be found in [Bil68, Appendix III] (where it is formulated in a slightly different way) or in [BK83].

So, we can safely assume that every law μ on any metric space $\mathbb{T}$ is separable, in particular μ is inner regular w.r.t. the totally bounded subsets of $\mathbb{T}$, that is, for every $B \in \mathcal{B}_{\mathbb{T}}$ and every $\varepsilon > 0$, there exists a totally bounded set $K \subset A$ such that $\mu(A \setminus K) < \varepsilon$ [Dud02, Theorem 7.1.4]. In particular, we have $\mathcal{M}_\tau^+(\mathbb{T}) = \mathcal{M}^+(\mathbb{T})$ and the weak and narrow topologies coincide on $\mathcal{M}^+(\mathbb{T})$.

The price to pay for this nice situation is that the Borel σ–algebra of a nonseparable metric space is often "too big", and there are rather "few" Borel probability measures on a nonseparable metric space $\mathbb{T}$, that is, one is likely to deal with measures on smaller σ–algebras (see e.g. [Dud76, vdVW96] for examples involving empirical processes).

Chapter 2

Young measures, the four stable topologies: S, M, N, W

2.1 Definitions, Portmanteau Theorem

We call *Young measures* the elements of the set

$$\mathcal{Y}^1(\Omega, \mathcal{S}, \mathrm{P}; \mathbb{T}) = \left\{\mu \in \mathcal{M}^{+,1}(\Omega \times \mathbb{T}, \mathcal{S} \otimes \mathcal{B}_{\mathbb{T}});\ \forall A \in \mathcal{S},\ \mu(A \times \mathbb{T}) = \mathrm{P}(A)\right\}.$$

When no confusion can arise, we omit some parts of the information and use notations such as $\mathcal{Y}^1(\mathbb{T})$ or $\mathcal{Y}^1(\mathrm{P})$ or simply $\mathcal{Y}^1$.

Young measures generalize random probability measures, random elements, and they also generalize probability laws on $\mathbb{T}$. Let us have a first look at these special subsets.

- $\mathcal{Y}^1_{\mathrm{dis}}(\mathbb{T})$: We denote by $\mathcal{Y}^1_{\mathrm{dis}}(\Omega, \mathcal{S}, \mathrm{P}; \mathbb{T})$ (for short, $\mathcal{Y}^1_{\mathrm{dis}}(\mathbb{T})$, $\mathcal{Y}^1_{\mathrm{dis}}(\mathrm{P})$ or $\mathcal{Y}^1_{\mathrm{dis}}$) the set $\mathfrak{X}\left(\mathcal{M}^{+,1}(\mathbb{T})\right)$ of random elements $\mu_. : \omega \mapsto \mu_\omega$ of $\mathcal{M}^{+,1}(\mathbb{T})$. It is very easy to check that a mapping $\mu_. : \omega \mapsto \mu_\omega$, $\Omega \to \mathcal{M}^{+,1}(\mathbb{T})$, is measurable if and only if, for each $B \in \mathcal{B}_{\mathbb{T}}$, the mapping $\omega \mapsto \mu_\omega(B)$ is measurable.

  With each $\mu_. \in \mathcal{Y}^1_{\mathrm{dis}}$, we can associate a unique probability measure

  $$\boldsymbol{\mu} = \int_\Omega \delta_\omega \otimes \mu_\omega \, d\,\mathrm{P}(\omega) \in \mathcal{Y}^1$$

  (where δ_ω denotes the Dirac measure concentrated on ω), that is, $\boldsymbol{\mu}(A \times B) = \int_A \mu_\omega(B)\, d\,\mathrm{P}(\omega)$ for all $(A, B) \in \mathcal{S} \times \mathcal{B}_{\mathbb{T}}$. The mapping $\mu_. \mapsto \boldsymbol{\mu}$ is injective, because we have defined random elements as equivalence classes. We shall thus identify $\mathcal{Y}^1_{\mathrm{dis}}$ with a subset of $\mathcal{Y}^1$ and call its elements *disintegrable Young measures.* If $\mathbb{T}$ is a Radon space (that is, if every law on $\mathbb{T}$ is Radon), then $\mathcal{Y}^1_{\text{dis}} = \mathcal{Y}^1$, thanks to the Disintegration Theorem (see [HJ71, Val72, Val73]).

- $\mathfrak{X}(\mathbb{T})$: With each random element X of $\mathbb{T}$, we can associate the disintegrable Young measure $\underline{\delta}_X : \omega \mapsto \delta_{X(\omega)}$. Thus $\mathfrak{X}(\mathbb{T})$ can be seen as a subset of $\mathcal{Y}^1_{\text{dis}}(\mathbb{T})$. We have
$$\underline{\delta}_X = \int \delta_\omega \otimes \delta_{X(\omega)}\, d\mathrm{P}(\omega),$$
the latter notation being more convenient when we need to emphasize the underlying measure P. Young measures of the form $\underline{\delta}_X$ will be called *degenerate Young measures.* As degenerate Young measures are parametrized Dirac measures, they are also sometimes called *Dirac Young measures.*

- $\mathcal{M}^{+,1}(\mathbb{T})$: The set $\mathcal{M}^{+,1}(\mathbb{T})$ can be embedded in $\mathcal{Y}^1_{\text{dis}}$ if we associate with each $\nu \in \mathcal{M}^{+,1}(\mathbb{T})$ the constant random measure $\omega \mapsto \nu$. The corresponding Young measure is $\mu = P \otimes \nu$. Such a measure μ is called a *homogeneous Young measure.*

- $\mathbb{T}$: If we identify $\mathfrak{X}(\mathbb{T})$ and $\mathcal{M}^{+,1}(\mathbb{T})$ with their corresponding subsets in $\mathcal{Y}^1_{\text{dis}}(\mathbb{T})$, then $\mathbb{T}$ will be naturally identified with $\mathfrak{X}(\mathbb{T}) \cap \mathcal{M}^{+,1}(\mathbb{T})$.

We shall consider these embeddings from a topological point of view in Section 2.2. The embedding of $\mathfrak{X}(\mathbb{T})$ in $\mathcal{Y}^1_{\text{dis}}(\mathbb{T})$ will be studied more in depth in Chapter 3, which is devoted to convergence in probability of random laws.

We endow the set $\mathcal{Y}^1$ with four different topologies, called *stable topologies,* which coincide in most applications: a strong one $\tau^{\mathrm{S}}_{\mathcal{Y}^1}$, two intermediate topologies $\tau^{\mathrm{M}}_{\mathcal{Y}^1}$ and $\tau^{\mathrm{N}}_{\mathcal{Y}^1}$, and a weak one $\tau^{\mathrm{W}}_{\mathcal{Y}^1}$. In the Portmanteau Theorem 2.1.3, we shall compare them precisely and give some simple necessary and sufficient conditions for a net to converge in one or another of these stable topologies.

- The topology $\tau^{\mathrm{S}}_{\mathcal{Y}^1}$ is the coarsest topology for which the functions
$$\begin{cases} \mathcal{Y}^1 & \to & \mathbb{R} \\ \mu & \mapsto & \mu(f) \end{cases}$$
are l.s.c. for every bounded l.s.c. integrand f on $\Omega \times \mathbb{T}$. This means that a net $(\mu^\alpha)_{\alpha\in\mathbb{A}}$ of elements of $\mathcal{Y}^1$ converges to a young measure μ^∞ if and only if, for each normal L^1-bounded integrand f, we have $\liminf_\alpha \mu^\alpha(f) \geq \mu^\infty(f)$. We call this topology *topology of S–stable convergence* (for short *S–stable topology*) on $\mathcal{Y}^1$. As usual, if we need to precise what is the underlying topological space, or the underlying probability, we use notations such as $\tau^{\mathrm{S}}_{\mathcal{Y}^1(\mathbb{T})}$ or $\tau^{\mathrm{S}}_{\mathcal{Y}^1(\mathrm{P})}$. A subbase for $\tau^{\mathrm{S}}_{\mathcal{Y}^1}$ is given by the sets
$$\mathcal{U}(f,r) = \{\mu \in \mathcal{Y}^1;\ \mu(f) > r\}$$
where f is a normal L^1-bounded integrand and $r \in \mathbb{R}$ (that is, each open subset of $\mathcal{Y}^1$ is a union of finite intersections of sets $\mathcal{U}(f,r)$).

- The topology $\tau^{\mathrm{M}}_{\mathcal{Y}^1}$ is the coarsest topology for which the functions

$$\begin{cases} \mathcal{Y}^1 & \to & \mathbb{R} \\ \mu & \mapsto & \mu(f) \end{cases}$$

are continuous for every bounded continuous integrand f on $\Omega \times \mathbb{T}$. We call it *topology of M–stable convergence* (or *M–stable topology*) on $\mathcal{Y}^1$. Thus $(\mu^\alpha)_{\alpha\in\mathbb{A}}$ M–stably converges to μ^∞ if and only if, for any bounded continuous integrand f, we have $\lim_\alpha \mu^\alpha(f) = \mu^\infty(f)$.

- The topology $\tau^{\mathrm{N}}_{\mathcal{Y}^1}$ is the coarsest topology for which the functions

$$\begin{cases} \mathcal{Y}^1 & \to & \mathbb{R} \\ \mu & \mapsto & \mu(\mathbb{1}_A \otimes f) \end{cases}$$

are l.s.c. for every $A \in \mathcal{S}$ and every bounded l.s.c. f on $\mathbb{T}$. We call it *topology of N–stable convergence* (or *N–stable topology*) on $\mathcal{Y}^1$. A net $(\mu^\alpha)_{\alpha\in\mathbb{A}}$ N–stably converges to μ^∞ if and only if, for each $A \in \mathcal{S}$, the net $(\mu^\alpha(A\times .))_{\alpha\in\mathbb{A}}$ of elements of $\mathcal{M}^+(\mathbb{T})$ narrowly converges to $\mu^\infty(A \times .)$.

- The topology $\tau^{\mathrm{W}}_{\mathcal{Y}^1}$ is the coarsest topology for which the functions

$$\begin{cases} \mathcal{Y}^1 & \to & \mathbb{R} \\ \mu & \mapsto & \mu(\mathbb{1}_A \otimes f) \end{cases}$$

are continuous for every $A \in \mathcal{S}$ and every bounded continuous function f on $\mathbb{T}$. We call it *topology of W–stable convergence* (or *W–stable topology*) on $\mathcal{Y}^1$. A net $(\mu^\alpha)_{\alpha\in\mathbb{A}}$ W–stably converges to μ^∞ if and only if, for each $A \in \mathcal{S}$, the net $(\mu^\alpha(A \times .))_{\alpha\in\mathbb{A}}$ of elements of $\mathcal{M}^+(\mathbb{T})$ weakly converges to $\mu^\infty(A \times .)$.

We thus have the following obvious arrows, which represent implications:

$$\begin{array}{ccc} \tau^{\mathrm{S}}_{\mathcal{Y}^1}\text{–convergence} & \longrightarrow & \tau^{\mathrm{M}}_{\mathcal{Y}^1}\text{–convergence} \\ \downarrow & & \downarrow \\ \tau^{\mathrm{N}}_{\mathcal{Y}^1}\text{–convergence} & \longrightarrow & \tau^{\mathrm{W}}_{\mathcal{Y}^1}\text{–convergence} \end{array}$$

(to remember the notations for the two intermediate topologies $\tau^{\mathrm{M}}_{\mathcal{Y}^1}$ and $\tau^{\mathrm{N}}_{\mathcal{Y}^1}$, the reader may think that "M" stands for "medium" and "N" for "narrow"). The topologies $\tau^{\mathrm{N}}_{\mathcal{Y}^1}$ and $\tau^{\mathrm{W}}_{\mathcal{Y}^1}$ coincide when $\mathbb{T}$ is completely regular and the elements of $\mathcal{M}^{+,1}(\mathbb{T})$ are τ–regular, e.g. $\mathbb{T}$ is Suslin regular or $\mathbb{T}$ is metrisable and contains a dense subspace with non–measurable cardinal, see Section 1.4 (recall that the weak and narrow topologies coincide on $\mathcal{M}^+_\tau(\mathbb{T})$ if and only if $\mathbb{T}$ is completely regular, see Section 1.3). Note also that, when $\mathbb{T}$ is not completely regular, $\tau^{\mathrm{W}}_{\mathcal{Y}^1}$ is not necessarily Hausdorff.

If $(\mu^\alpha)_{\alpha\in\mathbb{A}}$ is a net of elements of $\mathcal{Y}^1$ which $*$–stably converges to an element μ^∞ of $\mathcal{Y}^1$ $(* = \mathrm{S}, \mathrm{M}, \mathrm{N}, \mathrm{W})$, we write

$$\mu^\alpha \xrightarrow{*\text{–stably}} \mu^\infty.$$

For simplicity, if (X_α) is a net of random elements of $\mathbb{T}$, the expression "(X_α) $*$–stably converges to μ^∞" will be used for "$(\underline{\delta}_{X_\alpha})$ $*$–stably converges to μ^∞".

When all four stable topologies coincide, we skip the symbols S, M, N, W or $*$, and we write, e.g., that $(\mu^\alpha)_\alpha$ *stably converges* to μ^∞, or $\mu^\alpha \xrightarrow{\text{stably}} \mu^\infty$.

Remark 2.1.1 We shall give later conditions for these four topologies to coincide on $\mathcal{Y}^1$ or on some particular subsets of $\mathcal{Y}^1$. This will be done especially in Portmanteau Theorem for Stable Topology 2.1.3, where it is shown that they coincide if $\mathbb{T}$ is Suslin metrizable, and in Chapter 4 where it is shown e.g. (see Corollary 4.3.7) that, if $\mathbb{T}$ is a Suslin regular space, or a completely regular Prohorov space whose compact subsets are metrizable, these topologies share the same compact subsets, thus the same convergent sequences. If $\mathbb{T}$ has too few continuous functions, e.g. if all continuous functions are constant and $\mathbb{T}$ has more than one element (see Section 1.3), then obviously $\tau^{\mathrm{S}}_{\mathcal{Y}^1}$ is strictly finer than $\tau^{\mathrm{M}}_{\mathcal{Y}^1}$, and $\tau^{\mathrm{N}}_{\mathcal{Y}^1}$ is strictly finer than $\tau^{\mathrm{W}}_{\mathcal{Y}^1}$. *However, in the completely regular case, even if $\mathbb{T}$ is Suslin, we do not know whether the topologies $\tau^{\mathrm{S}}_{\mathcal{Y}^1}$, $\tau^{\mathrm{M}}_{\mathcal{Y}^1}$, $\tau^{\mathrm{N}}_{\mathcal{Y}^1}$ and $\tau^{\mathrm{W}}_{\mathcal{Y}^1}$ can differ.*

If in the above definitions we restrict the test functions to be $\mathcal{U}\otimes\mathcal{B}_{\mathbb{T}}$–measurable, for some sub–σ–algebra $\mathcal{U}$ of $\mathcal{S}$, we then speak of $\mathcal{U}$–$*$–stable convergence, $* = \mathrm{S}, \mathrm{M}, \mathrm{N}, \mathrm{W}$. If $\mathcal{U} \neq \mathcal{S}$, the topology of $\mathcal{U}$–$*$–stable convergence is obviously not Hausdorff. For example, if $A \in \mathcal{S}\setminus\mathcal{U}$ and if t_1 and t_2 are distinct elements of $\mathbb{T}$, the Young measures $\mathbb{1}_A\underline{\delta}_{t_1} + \mathbb{1}_{A^c}\underline{\delta}_{t_2}$ and $E^{\mathcal{U}}\left(\mathbb{1}_A\right)\underline{\delta}_{t_1} + E^{\mathcal{U}}\left(\mathbb{1}_{A^c}\right)\underline{\delta}_{t_2}$ are different, but they cannot be separated by any set of $\mathcal{U}\otimes\mathcal{B}_{\mathbb{T}}$–measurable functions. Actually, $\mathcal{U}$–$*$–stable convergence provides a bridge between $*$–stable convergence of Young measures and weak or narrow convergence of their margins on Ω: If $\mathcal{U}=\{\emptyset,\Omega\}$ and $* = \mathrm{W}, \mathrm{M}$ (resp. $* = \mathrm{S}, \mathrm{N}$), a net $(\mu^\alpha)_{\alpha\in\mathbb{A}}$ of elements of $\mathcal{Y}^1$ $\mathcal{U}$–$*$–stably converges to a Young measure μ^∞ if and only if the net of margins $(\mu^\alpha(\Omega\times .))_{\alpha\in\mathbb{A}}$ in $\mathcal{M}^{+,1}(\mathbb{T})$ weakly (resp. narrowly) converges to $\mu^\infty(\Omega\times .)$.

The terminology of "stable convergence" stems from Rényi [Rén63, RR58], who observed that this convergence behaves like narrow convergence plus some nice stability properties (see Proposition 9.2.2). We prefer to use this terminology rather than that of "narrow convergence of Young measures", to stress the difference with the narrow topology on the space of measures on a topological space. Although, in the definitions, $*$–stable convergence looks like a parametrized narrow or weak convergence, it will appear in Chapter 3 that $*$–stable convergence generalizes both narrow or weak convergence and convergence in probability (see the title of [JM81b]). Furthermore, it is very convenient to say that a sequence $(X_n)_n$ of random variables $*$–stably converges if the sequence $(\underline{\delta}_{X_n})_n$ $*$–stably converges. This is much more convenient than to say that $(X_n)_n$ "narrowly converges" (one would think that $(X_n)_n$ converges in law) or that $(X_n)_n$ "weakly converges" (one would think that $(X_n)_n$ converges for $\sigma(\mathrm{L}^1, \mathrm{L}^\infty)$).

The following lemma will be extended in Corollary 2.1.9, in the case $* = \mathrm{W}$.

Lemma 2.1.2 (Stable topology relative to the completed σ–algebra) *Assume that $\mathbb{T}$ is the union of a sequence $(\mathbb{T}_n)_{n\in\mathbb{N}}$ of second countable Suslin spaces (for the topology induced by $\mathbb{T}$) which are Borel subsets of $\mathbb{T}$, this is the case if e.g. $\mathbb{T} = (\mathbb{E}, \sigma(\mathbb{E}, \mathbb{E}^*))$ or $\mathbb{T} = (\mathbb{E}^*, \sigma(\mathbb{E}^*, \mathbb{E}))$ for some separable Banach space $\mathbb{E}$. Let us identify each $\mu \in \mathcal{Y}^1$ with its unique extension in $\mathcal{Y}^1(\Omega, \mathcal{S}^*_\mathrm{P}, \mathrm{P}^*)$. For $* = S, M, N, W$, the topology $\tau^*_{\mathcal{Y}^1}$ remains unchanged if we replace $(\Omega, \mathcal{S}, \mathrm{P})$ by $(\Omega, \mathcal{S}^*_\mathrm{P}, \mathrm{P}^*)$.*

Proof. Obvious in view of Lemma 1.2.6. □

The following key theorem is an adaptation of the classical Portmanteau Theorem for narrow convergence of laws on a topological space as in [Top70b] (see also [Bil68] for the metric case). Part G, where W–stable convergence implies S–stable convergence, is also known as the *Semicontinuity Theorem* (see also Proposition 2.1.12).

Recall that a portmanteau is a very large suitcase which can contain a lot of clothes. Using simple compactness arguments, we shall add extensions to this portmanteau in Theorems 2.1.13, 4.3.8 and 4.5.1 (see also Corollary 2.1.4). A probabilistic point of view will be presented in Proposition 9.2.2. A synthesis on the Semicontinuity Theorem is given in Table 1 page 112.

Theorem 2.1.3 (Portmanteau theorem for stable convergence of Young measures) *Recall that $\mathbb{T}$ is a Hausdorff topological space. Let $(\mu^\alpha)_{\alpha\in\mathbb{A}}$ be a net of elements of $\mathcal{Y}^1$ and let $\mu^\infty \in \mathcal{Y}^1$. Let $\mathcal{C}$ be a set of nonnegative $\mathcal{S}$–measurable bounded functions which is stable under multiplication of two elements, which contains the constant function 1, and which generates $\mathcal{S}$ (e.g. $\mathcal{C}$ may consist in the indicator functions of the elements of a subset of $\mathcal{S}$ which is stable under finite intersection, which contains Ω, and which generates $\mathcal{S}$). If $f \in \mathcal{C}$ and g is a function on $\mathbb{T}$, we denote by $f \otimes g$ the function on $\Omega \times \mathbb{T}$ defined by $(f \otimes g)(\omega, t) = f(\omega)g(t)$. For any $\mu \in \mathcal{Y}^1$ and any $f \in \mathcal{C}$, we denote by $\mu(f \otimes .)$ the measure $\nu \in \mathcal{M}^+(\mathbb{T})$ defined by $\nu(B) = \mu(f \otimes \mathbb{1}_B)$ for any $B \in \mathcal{B}_\mathbb{T}$.*

A) The following 6 conditions are equivalent:

1. $\mu^\alpha \xrightarrow{\mathcal{S}\text{-stably}} \mu^\infty$.
2. $\liminf_\alpha \mu^\alpha(f) \geq \mu^\infty(f)$ *for any normal* L^1*-bounded integrand* f.
3. $\limsup_\alpha \mu^\alpha(f) \leq \mu^\infty(f)$ *for any* L^1*-bounded u.s.c. integrand* f.
4. $\liminf_\alpha \mu^\alpha(f) \geq \mu^\infty(f)$ *for any normal* L^1*-bounded integrand* f *such that* $0 < f < 1$.
5. $\liminf_\alpha \mu^\alpha(G) \geq \mu^\infty(G)$ *for any* $G \in \underline{\mathcal{G}}$.

6. $\limsup_\alpha \mu^\alpha(F) \le \mu^\infty(F)$ *for any* $F \in \underline{\mathcal{F}}$.

B) *Furthermore, each of Conditions 1 to 6 implies Conditions 7 and 8 below, which are equivalent.*

7. $\mu^\alpha \xrightarrow{M\text{-stably}} \mu^\infty$.

8. $\lim_\alpha \mu^\alpha(f) = \mu^\infty(f)$ *for any* L^1*-bounded continuous integrand* f.

C) *Each of Conditions 1 to 6 implies Conditions 9, 10 and 11 below, which are equivalent.*

9. $\mu^\alpha \xrightarrow{N\text{-stably}} \mu^\infty$.

10. $\liminf_\alpha \mu^\alpha(f \otimes g) \ge \mu^\infty(f \otimes g)$ *for any* $f \in \mathcal{C}$ *and for any bounded l.s.c. function* g *on* $\mathbb{T}$.

11. *For every* $f \in \mathcal{C}$, *the net* $(\mu^\alpha(f \otimes .))_{\alpha \in \mathbb{A}}$ *in* $\mathcal{M}^+(\mathbb{T})$ *narrowly converges to* $\mu^\infty(f \otimes .)$.

D) *Each of Conditions 1 to 11 implies each of Conditions 12 to 14 below, which are equivalent. Furthermore, if* $\mathbb{T}$ *is completely regular, if* $\mathcal{U}$ *is any uniformity on* $\mathbb{T}$ *which is compatible with the topology of* $\mathbb{T}$, *and if* $\mu^\infty(\Omega \times .) \in \mathcal{M}^+_\tau(\mathbb{T})$, *then Conditions 9 to 15 are equivalent.*

12. $\mu^\alpha \xrightarrow{W\text{-stably}} \mu^\infty$.

13. $\lim_\alpha \mu^\alpha(f \otimes g) = \mu^\infty(f \otimes g)$ *for any* $f \in \mathcal{C}$ *and for any bounded continuous function* g *on* $\mathbb{T}$.

14. *For every* $f \in \mathcal{C}$, *the net* $(\mu^\alpha(f \otimes .))_{\alpha \in \mathbb{A}}$ *in* $\mathcal{M}^+(\mathbb{T})$ *weakly converges to* $\mu^\infty(f \otimes .)$.

15. $\lim_\alpha \mu^\alpha(f \otimes g) = \mu^\infty(f \otimes g)$ *for any* $f \in \mathcal{C}$ *and for any bounded* $\mathcal{U}$*–uniformly continuous function* g *on* $\mathbb{T}$.

E) *Assume that* $\mathbb{T}$ *is hereditarily Lindelöf and regular (thus completely regular). Assume furthermore that* $\mathbb{T}$ *is separable (or that* $\mathbb{T}$ *contains a dense subspace with non–measurable cardinal, see Section 1.4 and also Remark 1.1.3). Let* $\mathcal{D}$ *be a set of semidistances which induces the topology of* $\mathbb{T}$ *and which is* σ*–upwards filtering (if we assume furthermore that* $\mathbb{T}$ *is submetrizable, e.g.* $\mathbb{T}$ *is cosmic regular, then we may take for* $\mathcal{D}$ *a set of distances on* $\mathbb{T}$*). Then Conditions 9 to 15 are equivalent to Conditions 16 to 18 below.*

16. $\mu^\alpha \xrightarrow{W\text{-stably}} \mu^\infty$ *in* $\mathcal{Y}^1(\mathbb{T}, d)$ *for any* $d \in \mathcal{D}$.

17. $\lim_\alpha \mu^\alpha(f \otimes g) = \mu^\infty(f \otimes g)$ *for any* $f \in \mathcal{C}$, *for any* $d \in \mathcal{D}$ *and for any bounded* d*–uniformly continuous function* g *on* $\mathbb{T}$.

18. $\lim_\alpha \mu^\alpha(f \otimes g) = \mu^\infty(f \otimes g)$ *for any* $f \in \mathcal{C}$, *for any* $d \in \mathcal{D}$ *and for any bounded* d*–Lipschitz function* g *on* $\mathbb{T}$.

F) Assume that $\mathbb{T}$ *is Suslin regular. Let* $\mathcal{D}$ *be a set of distances which induces the topology of* $\mathbb{T}$ *and which is* σ*–upwards filtering. Then Conditions 9 to 18 are equivalent to Condition 19 below.*

19. $\mu^\alpha \xrightarrow{S\text{-stably}} \mu^\infty$ *in* $\mathcal{Y}^1(\mathbb{T}, d)$ *for any* $d \in \mathcal{D}$.

G) Assume that $\mathbb{T}$ *is metrizable and Suslin. Let* d *be a distance which is compatible with the topology of* $\mathbb{T}$. *Then Conditions 1 to 19 (with* $\mathcal{D} = \{d\}$*) are equivalent.*

We shall give several corollaries later, but let us immediately give the following one, to compare with Parts D and E. A similar result, without separability condition, will be given in Proposition 2.1.10.

Corollary 2.1.4 *Assume that* $\mathbb{T}$ *is completely regular and separable. Let* $(\mu^\alpha)_{\alpha\in\mathbb{A}}$ *be a net in* $\mathcal{Y}^1$ *which W–stably converges to a Young measure* $\mu^\infty \in \mathcal{Y}^1$. *Let* d *be a continuous semidistance on* $\mathbb{T}$ *and let* f *be a bounded* d*–uniformly continuous integrand. We have*

$$\lim_\alpha \mu^\alpha(f) = \mu^\infty(f).$$

Proof. Let $\mathbb{S}$ be the completion of the quotient space $\mathbb{T}/d$. We denote by π the canonical mapping $\mathbb{T} \to \mathbb{S}$. Define the corresponding mapping $\underline{\pi}$ on $\Omega \times \mathbb{T}$:

$$\underline{\pi} : \begin{cases} \Omega \times \mathbb{T} & \to \quad \Omega \times \mathbb{S} \\ (\omega, t) & \mapsto \quad (\omega, \pi(t)). \end{cases}$$

There exists a bounded uniformly continuous integrand $\hat{f}$ defined on $\Omega \times \mathbb{S}$ such that $f = \hat{f} \circ \underline{\pi}$. We have

$$\underline{\pi}_\sharp(\mu^\alpha) \xrightarrow{\text{W-stably}} \underline{\pi}_\sharp(\mu^\infty).$$

But $\mathbb{S}$ is Polish thus, from Part G of Theorem 2.1.3,

$$\underline{\pi}_\sharp(\mu^\alpha) \xrightarrow{\text{S-stably}} \underline{\pi}_\sharp(\mu^\infty).$$

In particular, we have

$$\lim_\alpha \mu^\alpha(f) = \lim_\alpha \underline{\pi}_\sharp(\mu^\alpha)(\hat{f}) = \underline{\pi}_\sharp(\mu^\infty)(\hat{f}) = \mu^\infty(f).$$

□

Proof of Theorem 2.1.3.

A) $1 \Leftrightarrow 2$ follows directly from the definition of S–stable convergence.

$2 \Leftrightarrow 3$, $2 \Rightarrow 4$, $5 \Leftrightarrow 6$ and $2 \Rightarrow 5$ are obvious.

$5 \Rightarrow 4$. Note that, if $g : \Omega \times \mathbb{T} \to \mathbb{R}$ is any measurable function such that $0 < g < 1$, we have, for any $\mu \in \mathcal{Y}^1$,

$$\frac{1}{n}\left(1 + \sum_{k=1}^{n-1} \mu\{g > k/n\}\right) \geq \mu(g) \geq \frac{1}{n}\sum_{k=1}^{n-1} \mu\{g > k/n\}.$$

Let f be a normal L^1-bounded integrand with $0 < f < 1$. We have, for any integer $n \geq 1$,

$$\begin{aligned}
\liminf_\alpha \mu^\alpha(f) &\geq \liminf_\alpha \left\{\frac{1}{n}\sum_{k=1}^{n-1} \mu^\alpha\{f > k/n\}\right\} \\
&\geq \frac{1}{n}\sum_{k=1}^{n-1} \liminf_\alpha \mu^\alpha\{f > k/n\} \\
&\geq \frac{1}{n}\sum_{k=1}^{n-1} \mu^\infty\{f > k/n\} \text{ (from Condition 5)} \\
&\geq \mu^\infty(f) - 1/n,
\end{aligned}$$

thus $\liminf_\alpha \mu^\alpha(f) \geq \mu^\infty(f)$.

$4 \Rightarrow 1$. If f is a normal L^1-bounded integrand such that $a < f < b$ for some fixed real numbers a and b, Condition 4 applied to $(f-a)/(b-a)$ yields the result. Assume now that $|f(\omega, .)| \leq \phi(\omega)$ for some integrable function $\phi : \Omega \to \mathbb{R}$ and for every $\omega \in \Omega$. Let $\varepsilon > 0$. There exists $c > 0$ such that $\int_{\phi > c} \phi \, d\mathrm{P} < \varepsilon$. We have, for any index $\alpha \in \mathbb{A} \cup \{\infty\}$,

$$|\mu^\alpha(f\,\mathbb{1}_{\phi > c})| \leq \mu^\alpha(\phi\,\mathbb{1}_{\phi > c}) = \mathrm{P}(\phi\,\mathbb{1}_{\phi > c}) < \varepsilon,$$

thus

$$\begin{aligned}
\liminf_\alpha \mu^\alpha(f) &\geq \liminf_\alpha \mu^\alpha(f\,\mathbb{1}_{\phi \leq c}) + \liminf_\alpha \mu^\alpha(f\,\mathbb{1}_{\phi > c}) \\
&\geq \mu^\infty(f\,\mathbb{1}_{\phi \leq c}) - \varepsilon \\
&= \mu^\infty(f) - \mu^\infty(f\,\mathbb{1}_{\phi > c}) - \varepsilon \\
&\geq \mu^\infty(f) - 2\varepsilon.
\end{aligned}$$

B) The implications (2 and 3) $\Rightarrow 8 \Rightarrow 7$ are obvious. The implication $7 \Rightarrow 8$ can be easily obtained by the same trucation method as for $4 \Rightarrow 1$.

C) The implication 2 $\Rightarrow$ 9 is obvious. Furthermore, in the case when $\mathcal{C} = \{ \mathbb{1}_A;\ A \in \mathcal{S}\}$, the equivalences 9 $\Leftrightarrow$ 10 $\Leftrightarrow$ 11 are immediate consequences of the definition of $\tau^{\mathrm{N}}_{\mathcal{Y}^1}$ and of the definition of narrow convergence on $\mathcal{M}^+(\mathbb{T})$. There remains to prove that Conditions 10 and 11 are independent of the choice of $\mathcal{C}$.

Let us denote by $\mathbb{M}\mathrm{eas}$ the set of bounded measurable functions on Ω. It is clear that Condition 11 is implied by

11′. For every $f \in \mathbb{M}\mathrm{eas}$, the net $(\mu^\alpha(f\otimes .))_{\alpha\in\mathbb{A}}$ in $\mathcal{M}^+(\mathbb{T})$ narrowly converges to $\mu^\infty(f \otimes .)$.

So, we only need to prove the implication (11 $\Rightarrow$ 11′).

We shall prove this implication by using a Functional Monotone Class Theorem. A set $\mathcal{E}$ of bounded measurable functions on Ω is called a *monotone vector space* [Sha88, Appendix A] if (i) it is a vector space over $\mathbb{R}$, (ii) it contains the constant functions and (iii) it is stable under monotone limits of uniformly bounded sequences. The Monotone Class Theorem asserts that if $\mathcal{E}$ is a monotone vector space and if $\mathcal{H}$ is a subset of $\mathcal{E}$ which is stable under multiplication of two elements, then $\mathcal{E}$ contains all bounded functions which are measurable for the σ–algebra generated by $\mathcal{H}$ (this theorem is proved in [DM75, Théorème 21, page 20], with the extra hypothesis that $\mathcal{E}$ is stable under uniform convergence, but it appears that any monotone vector space satisfies this condition, see [DM83, page 231] or [Sha88]). Let $\mathcal{E}$ be the set of elements f of $\mathbb{M}\mathrm{eas}$ such that $\lim_\alpha \mu^\alpha(f\otimes g) = \mu^\infty(f\otimes g)$ for any bounded continuous function g on $\mathbb{T}$. We have $\mathcal{C} \subset \mathcal{E}$, and $\mathcal{E}$ is a vector space which contains the constant functions. To prove that $\mathcal{E} = \mathbb{M}\mathrm{eas}$, we only need to prove that $\mathcal{E}$ is stable under monotone limits of uniformly bounded sequences. Let $(f_n)_{n\in\mathrm{N}}$ be an increasing sequence of elements of $\mathcal{E}$ such that $f := \sup_{n\in\mathrm{N}} f_n \in \mathbb{M}\mathrm{eas}$. By adding a constant, we assume w.l.g. that $f_0 \geq 0$. Let g be a continuous function on $\mathbb{T}$, with $0 \leq g \leq 1$. Let $\varepsilon > 0$. There exists an integer $n_0 \in \mathrm{N}$ such that $\mathrm{P}(f_{n_0}) + \varepsilon \geq \mathrm{P}(f)$. We thus have, for any $\mu \in \mathcal{Y}^1$,

$$0 \leq \mu(f\otimes g) - \mu(f_{n_0}\otimes g) = \mu((f - f_{n_0})\otimes g) \leq \mathrm{P}\,(f - f_{n_0}) \leq \varepsilon,$$

and therefore

$$\begin{aligned}
\limsup_\alpha \mu^\alpha(f\otimes g) &\le \lim_\alpha \mu^\alpha(f_{n_0}\otimes g)+\varepsilon\\
&\le \mu^\infty(f\otimes g)+\varepsilon\\
&= \sup_{n\in\mathbb{N}} \mu^\infty(f_n\otimes g)+\varepsilon\\
&= \sup_{n\in\mathbb{N}} \lim_\alpha \mu^\alpha(f_n\otimes g)+\varepsilon\\
&\le \liminf_\alpha \sup_{n\in\mathbb{N}} \mu^\alpha(f_n\otimes g)+\varepsilon\\
&= \liminf_\alpha \mu^\alpha(f\otimes g)+\varepsilon.
\end{aligned}$$

As ε is arbitrary, this proves that $f\in\mathcal{E}$. Thus $\mathcal{E}=\mathbb{M}\mathrm{eas}$.

D) The implications $7\Rightarrow 12$ and $9\Rightarrow 12$ are obvious.

The equivalence $13\Leftrightarrow 14$ is an immediate consequence of the definitions, as well as, if $\mathcal{C}=\{\mathbb{1}_A;\ A\in\mathcal{S}\}$, the equivalence $12\Leftrightarrow 13$.

The proof that Condition 14 does not depend on the choice of $\mathcal{C}$ is the same as in Part C.

If $\mathbb{T}$ is completely regular and if $\mu_\infty(\Omega\times .)\in\mathcal{M}^+_\tau(\mathbb{T})$, Condition 15 is equivalent to Conditions 11 and 14 by the classical Portmanteau Theorem for narrow convergence of Borel measures on a topological space (see page 13 and [Top70b, Theorem 8.1]).

E) As $\mathbb{T}$ is hereditarily Lindelöf, we have $\mu_\infty(\Omega\times .)\in\mathcal{M}^+_\tau(\mathbb{T})$, thus the conclusions of Part D hold true.

The equivalence of Conditions 17 and 18 with Condition 16 comes from the Portmanteau Theorem for narrow convergence of measures on a metric space: see e.g. [Dud76, Dud02] for the case when $\mathbb{T}$ is separable. In the nonseparable case, the equivalence 17⇔16 is provided by [Top70b, Theorem 8.1], and the equivalence 18⇔16 by [HJ91, Corollary 7.12].

The implication $12\Rightarrow 16$ is obvious. Assume 16, let $f\in\mathcal{C}$ and let $g:\mathbb{T}\to\mathbb{R}$ be bounded continuous. From Lemma 1.1.2, there exists a continuous semidistance $d\in\mathcal{D}$ such that g is d–continuous (from Remark 1.1.1, if $\mathbb{T}$ is submetrizable, we can take for d a distance). Thus, by Condition 16, we have $\lim_\alpha \mu^\alpha(f\otimes g)=\mu^\infty(f\otimes g)$, which proves 13.

We now prove G, which we will use in the proof of F. Note that, when $\mathbb{T}$ is Suslin, we have $\mathcal{M}^+(\mathbb{T})=\mathcal{M}^+_\tau(\mathbb{T})$, which ensures that $\mu^\infty(.\times\mathbb{T})\in\mathcal{M}^+_\tau(\mathbb{T})$.

G) (First step) In this step, $\mathbb{T}$ need not be Suslin (but then, we assume that $\mu^\infty(.\times\mathbb{T})\in\mathcal{M}^+_\tau(\mathbb{T})$). As $\mathbb{T}$ is separable metrizable, the results of Parts D and E apply. Thus Conditions 12 to 18 are equivalent. In particular, these conditions are independent of the choice of $\mathcal{C}$. We can thus assume without loss of generality that $\mathcal{C} = \{ \mathbb{1}_A;\ A \in \mathcal{S}\}$.

Let us prove that Condition 17 implies

7′. $\lim_\alpha \mu^\alpha(f) = \mu^\infty(f)$ for any bounded uniformly continuous integrand f.

As Condition 17 is equivalent to Condition 12, which is independent of the choice of d, we shall assume w.l.g. that $(\mathbb{T}, d)$ is totally bounded (see [Eng89, Par67] for the existence of such a distance). Then the space $\mathrm{C}_u(d)$, endowed with the topology of uniform convergence, is separable (this is the main argument in this part of the proof). Following the notations of [JM81b, JM83], we denote by $\mathbb{B}_{\mathrm{m}}$ the set of bounded measurable functions on $\Omega \times \mathbb{T}$ and we set

$$\mathbb{B}^1_{\mathrm{mc}} = \{ \mathbb{1}_A \otimes f;\ A \in \mathcal{S}, f \in \mathrm{C}_u(d)\},$$

$$\mathbb{B}^2_{\mathrm{mc}} = \{f = \sum_{n\in\mathbb{N}} \mathbb{1}_{A_n} \otimes f_n;$$

$$f \in \mathbb{B}_{\mathrm{m}},\ \forall n, m \in \mathbb{N} \quad \mathbb{1}_{A_n} \otimes f_n \in \mathbb{B}^1_{\mathrm{mc}},\ (n \neq m \Rightarrow A_n \cap A_m = \emptyset)\},$$

$$\mathbb{B}^3_{\mathrm{mc}} = \{f \in \mathbb{B}_{\mathrm{m}};\ \forall\omega \in \Omega \quad f(\omega, .) \in \mathrm{C}_u(d)\}.$$

So, Condition 17 reads

$\lim_\alpha \mu^\alpha(f) = \mu^\infty(f)$ for any $f \in \mathbb{B}^1_{\mathrm{mc}}$.

Let $f = \sum_{n\in\mathbb{N}} \mathbb{1}_{A_n} \otimes f_n \in \mathbb{B}^2_{\mathrm{mc}}$, with $(A_n)_{n\in\mathbb{N}}$ a measurable partition of Ω, and let us prove that $\lim_\alpha \mu^\alpha(f) = \mu^\infty(f)$.

Let $a = \sup_{(\omega,t)\in\Omega\times\mathbb{T}} |f(\omega, t)|$. For each $\varepsilon > 0$, choose an integer $n(\varepsilon)$ such that $\mathrm{P}(\cup_{i>n(\varepsilon)} A_i) < \varepsilon/a$. Let $g_\varepsilon = \sum_{i\le n(\varepsilon)} \mathbb{1}_{A_i} \otimes f_i$. For any $\mu \in \mathcal{Y}^1$, we have

$$|\mu(f) - \mu(g_\varepsilon)| = \sum_{i>n(\varepsilon)} |\mu(\mathbb{1}_{A_i} \otimes f_i)| \le \varepsilon.$$

But, for each $\varepsilon > 0$, we also have $\lim_\alpha \mu^\alpha(g_\varepsilon) = \mu^\infty(g_\varepsilon)$. Thus $\lim_\alpha \mu^\alpha(f) = \mu^\infty(f)$.

Now, let $f \in \mathbb{B}^3_{\mathrm{mc}}$. Let $(f_n)_{n\in\mathbb{N}}$ be a dense sequence in $\mathrm{C}_u(d)$ (recall that $(\mathbb{T}, d)$ is a totally bounded metric space). For each $\varepsilon > 0$ and each $n \in \mathbb{N}$, set

$$B^\varepsilon_n = \{\omega \in \Omega;\ \sup_{t\in\mathbb{T}} |f(\omega, t) - f_n(t)| \le \varepsilon\}.$$

Define $(A^\varepsilon_n)_{n\in\mathbb{N}}$ inductively by $A^\varepsilon_0 = B^\varepsilon_0$ and $A^\varepsilon_{n+1} = B^\varepsilon_{n+1} \setminus \cup_{0\le i\le n} B^\varepsilon_i$. Let $f^\varepsilon = \sum_{n\in\mathbb{N}} \mathbb{1}_{A^\varepsilon_n} \otimes f_n$. Then $f^\varepsilon \in \mathbb{B}^2_{\mathrm{mc}}$, thus $\lim_\alpha \mu^\alpha(f^\varepsilon) = \mu^\infty(f^\varepsilon)$. But we also have $\sup_{\Omega\times\mathbb{T}} |f - f^\varepsilon| \le \varepsilon$, thus, for any $\mu \in \mathcal{Y}^1$, $|\mu(f) - \mu(f^\varepsilon)| \le \varepsilon$. Thus $\lim_\alpha \mu^\alpha(f) = \mu^\infty(f)$, which proves Condition 7′.

G) (SECOND STEP) Note that, in our current setting, Condition 19 is equivalent to Condition 1. So, to complete the proof of Part G, we only need to prove the implication $7' \Rightarrow 6$.

Now, from Lemma 2.1.2, as $\mathbb{T}$ is Suslin and second countable, we can assume w.l.g. that $(\Omega, \mathcal{S}, \mathrm{P})$ is complete.

Assume $7'$ and let $F \in \underline{\mathcal{F}}$. We shall approximate the integrand $\mathbf{1}_F$ by elements f of $\mathbb{B}^3_{\mathrm{mc}}$ such that $f \geq \mathbf{1}_F$ (recall that $\mathbb{B}^3_{\mathrm{mc}}$ is the set of bounded d–uniformly continuous integrands on $\Omega \times \mathbb{T}$).

From the Projection Theorem, the set $\Omega' = \{\omega \in \Omega;\ F(\omega) \neq \emptyset\}$ is $\mathcal{S}$–measurable. We shall assume w.l.g. that $\Omega' = \Omega$.

Let $(\sigma_n)_{n\in\mathbb{N}}$ be a Castaing representation of F, that is, each σ_n is a measurable mapping from Ω to $\mathbb{T}$ and $F(\omega)$ is the closure of $\{\sigma_n(\omega);\ n \in \mathbb{N}\}$ for every $\omega \in \Omega$. Let $d \in \mathcal{D}$. The function

$$g_d : (\omega, t) \mapsto d(t, F(\omega)) = \inf_{n\in\mathbb{N}} d(t, \sigma_n(\omega))$$

is an integrand. From Lemma 1.2.2, for each $\varepsilon > 0$, the set

$$\begin{aligned} G^{d,\varepsilon} &= \{(\omega, t) \in \Omega \times \mathbb{T};\ d(t, F(\omega)) < \varepsilon\} \\ &= \cup_{n\in\mathbb{N}} \{(\omega, t) \in \Omega \times \mathbb{T};\ d(t, \sigma_n(\omega)) < \varepsilon\} \end{aligned}$$

is in $\underline{\mathcal{G}}$. Let $\varepsilon > 0$ and define an uniformly continuous integrand $f_{d,\varepsilon}$ by

$$f_{d,\varepsilon}(\omega, t) = \begin{cases} 1 & \text{if} \quad (\omega, t) \in F \\ (1/\varepsilon)(\varepsilon - g_d(\omega, t)) & \text{if} \quad 0 \leq g_d(\omega, t) \leq \varepsilon \\ 0 & \text{if} \quad g_d(\omega, t) > \varepsilon. \end{cases}$$

We thus have

$$\mathbf{1}_F \leq f_{d,\varepsilon} \leq \mathbf{1}_{G^{d,\varepsilon}},$$

and

$$\begin{aligned} \mu^\infty(F) &= \inf_{k\geq 1} \mu^\infty(G^{d,1/k}) \\ &\geq \inf_{k\geq 1} \mu^\infty(f_{d,1/k}) \\ &\geq \inf \{\mu^\infty(f);\ f \in \mathbb{B}^3_{\mathrm{mc}}, f \geq \mathbf{1}_F\} \\ &= \inf \left\{\lim_\alpha \mu^\alpha(f);\ f \in \mathbb{B}^3_{\mathrm{mc}}, f \geq \mathbf{1}_F\right\} \\ &\geq \limsup_\alpha \left(\inf \{\mu^\alpha(f);\ f \in \mathbb{B}^3_{\mathrm{mc}}, f \geq \mathbf{1}_F\}\right) \\ &\geq \limsup_\alpha \mu^\alpha(F), \end{aligned}$$

which proves Condition 6.

We can now prove F.

F) As $\mathbb{T}$ is cosmic regular, the results of E apply, thus Conditions 9 to 18 are equivalent. Furthermore, by Part G, for any $d \in \mathcal{D}$, 19 is equivalent to 16.

□

Remark and Definition 2.1.5 (Extension of $\tau^{\mathrm{S}}_{\mathcal{Y}^1}$ to the space of measures on $(\Omega \times \mathbb{T}, \mathcal{S} \otimes \mathcal{B}_{\mathbb{T}})$). Let us call *extended S–stable topology* and denote by $\tau^{\mathrm{S}}_{\mathcal{M}}$ the coarsest topology on $\mathcal{M}^+(\Omega \times \mathbb{T}, \mathcal{S} \otimes \mathcal{B}_{\mathbb{T}})$ such that the mapping $\mu \mapsto \mu(f)$ is l.s.c. for any l.s.c. bounded integrand f.

Similarly, for $* = \mathrm{M}, \mathrm{N}, \mathrm{W}$, we can define the *extended ∗–stable topology* $\tau^{*}_{\mathcal{M}}$.

Obviously, $\mathcal{Y}^1$ is closed in $\mathcal{M}^+(\Omega \times \mathbb{T}, \mathcal{S} \otimes \mathcal{B}_{\mathbb{T}})$ for $\tau^{*}_{\mathcal{M}}$, $* = \mathrm{S}, \mathrm{M}, \mathrm{N}, \mathrm{W}$.

Note that, in Parts A and B of the proof of Theorem 2.1.3, we did not use the fact that all elements of $\mathcal{Y}^1$ have the same marginal P on Ω; we used this hypothesis in C and D, when we applied the monotone class theorem, but we can dodge this part of the proof if we assume a stronger hypothesis on $\mathcal{C}$, e.g. that $\mathcal{C}$ contains all indicator functions $\mathbb{1}_A$ $(A \in \mathcal{S})$. If we modify the hypothesis on $\mathcal{C}$ in this way, Theorem 2.1.3 extends to $\mathcal{M}^{+,1}(\Omega \times \mathbb{T}, \mathcal{S} \otimes \mathcal{B}_{\mathbb{T}})$, with the topologies induced by $\tau^{*}_{\mathcal{M}}$, $* = \mathrm{S}, \mathrm{M}, \mathrm{N}, \mathrm{W}$.

Furthermore, it is easy to adapt the proof of Theorem 2.1.3 to $(\mathcal{M}^+(\Omega \times \mathbb{T}, \mathcal{S} \otimes \mathcal{B}_{\mathbb{T}}), \tau^{*}_{\mathcal{M}})$ (see [RdF03]). Besides the above mentionned modification, we only need to add in Conditions 5 and 6 that $\lim_\alpha \mu^\alpha(\Omega \times \mathbb{T}) = \mu^\infty(\Omega \times \mathbb{T})$.

The topologies $\tau^{\mathrm{S}}_{\mathcal{M}}$ and $\tau^{\mathrm{N}}_{\mathcal{M}}$ are particular cases of the *w–topology* defined by Topsøe [Top70a] on an abstract set endowed with some pavings $\underline{\mathcal{K}}$, $\underline{\mathcal{G}}$ and $\underline{\mathcal{F}}$. We shall apply Topsøe's results on compactness in chapter 4. In the case when $\mathbb{T}$ is separable metrizable, the topology $\tau^{\mathrm{W}}_{\mathcal{M}}$ was also investigated by Schäl [Sch75] (with sometimes topological assumptions on $(\Omega, \mathcal{S})$), under the name of *ws–topology*. In the case when $\mathbb{T}$ is Polish, the topology $\tau^{\mathrm{S}}_{\mathcal{M}} = \tau^{\mathrm{W}}_{\mathcal{M}}$ was also studied by Jacod and Mémin [JM81b] and by Galdéano [Gal97]. More recently, Balder [Bal01] studied $\tau^{\mathrm{W}}_{\mathcal{M}}$ in the case when $\mathbb{T}$ is Suslin regular.

Remark 2.1.6 (W–Stable topology as an initial topology) Actually, in Part E, we identified each measure μ^α $(\alpha \in \mathbb{A} \cup \{\infty\})$ with its restriction to the σ–algebra $\mathcal{S} \otimes \mathcal{B}_{(\mathbb{T},d)}$, which may be different from $\mathcal{S} \otimes \mathcal{B}_{\mathbb{T}}$ if $\mathbb{T}$ is not Suslin. Assume that $\mathbb{T}$ is separable, hereditarily Lindelöf and regular, and let $\mathcal{D}$ be a set of semidistances which induces the topology of $\mathbb{T}$ and which is σ–upwards filtering. For each $d \in \mathcal{D}$, let π_d be the canonical projection of $\mathbb{T}$ onto $(\mathbb{T}, d)$. Let $\underline{\pi}_d$ be the mapping $(\omega, t) \mapsto (\omega, \pi_d(t))$ defined on $\Omega \times \mathbb{T}$. Then a more precise writing of Condition 16 is

16′. $(\underline{\pi}_d)_\sharp \, \mu^\alpha \xrightarrow{\text{W-stably}} (\underline{\pi}_d)_\sharp \, \mu^\infty$ in $\mathcal{Y}^1(\mathbb{T}, d)$ for any $d \in \mathcal{D}$,

where $(\underline{\pi}_d)_\sharp \, \mu^\alpha$ $(\alpha \in \mathbb{A} \cup \{\infty\})$ denotes the measure image of μ^∞ by $\underline{\pi}_d$.

The equivalence 16′ ⇔ 12 means that $\tau^{\mathrm{W}}_{\mathcal{Y}^1}$ is the initial topology for the system $(\tau^{\mathrm{W}}_{\mathcal{Y}^1}(d), (\underline{\pi}_d)_\sharp)$, where d runs over $\mathcal{D}$.

Remark 2.1.7 (W–Stable convergence vs. S–stable convergence in the Suslin regular case) Parts F and G give a criterion for $\tau^{\mathrm{S}}_{\mathcal{Y}^1}$ and $\tau^{\mathrm{W}}_{\mathcal{Y}^1}$ to coincide when $\mathbb{T}$ is Suslin regular. Notice first that, when $\mathbb{T}$ is Suslin, for each continuous distance d on $\mathbb{T}$, the σ–algebra $\mathcal{B}_d$ coincides with $\mathcal{B}_{\tau_{\mathbb{T}}}$, thus the *sets* $\mathcal{Y}^1(d)$ and $\mathcal{Y}^1(\tau_{\mathbb{T}})$ coincide. Let Id be the identity mapping on $\mathcal{Y}^1$. From Remark 2.1.6, $\tau^{\mathrm{W}}_{\mathcal{Y}^1}$ is the initial topology (and even the projective limit) for the system $(\tau^{\mathrm{W}}_{\mathcal{Y}^1}(d), \mathrm{Id})$, where d runs over $\mathcal{D}$. Clearly, $\tau^{\mathrm{S}}_{\mathcal{Y}^1}$ is finer than the initial topology (and even the projective limit) for the system $(\tau^{\mathrm{S}}_{\mathcal{Y}^1}(d), \mathrm{Id})$, where d runs over $\mathcal{D}$. But, for each $d \in \mathcal{D}$, we have $\tau^{\mathrm{W}}_{\mathcal{Y}^1}(d) = \tau^{\mathrm{S}}_{\mathcal{Y}^1}(d)$ by Part G. Thus, *if $\mathbb{T}$ is Suslin regular, the topologies $\tau^{\mathrm{W}}_{\mathcal{Y}^1}$ and $\tau^{\mathrm{S}}_{\mathcal{Y}^1}$ coincide if and only if $\tau^{\mathrm{S}}_{\mathcal{Y}^1}$ is the initial topology for the system $(\tau^{\mathrm{S}}_{\mathcal{Y}^1}(d), \mathrm{Id})$, where d runs over $\mathcal{D}$.*

Corollary 2.1.8 (W–stable convergence for dense subspaces) *Assume that $\mathbb{T}$ is a dense subspace of a completely regular space $\mathbb{S}$. Assume furthermore that $\mathcal{M}^+(\mathbb{T}) = \mathcal{M}^+_\tau(\mathbb{T})$ (e.g. $\mathbb{T}$ is hereditarily Lindelöf). For each $\mu \in \mathcal{Y}^1(\mathbb{T})$, let $\widetilde{\mu} \in \mathcal{Y}^1(\mathbb{S})$ be the measure on $\Omega \times \mathbb{S}$ defined by $\widetilde{\mu}(A) = \mu(A \cap (\Omega \times \mathbb{T}))$ for any $A \in \mathcal{S} \otimes \mathcal{B}_{\mathbb{S}}$. The injection $\mu \mapsto \widetilde{\mu}$ from $\left(\mathcal{Y}^1(\mathbb{T}), \tau^{\mathrm{W}}_{\mathcal{Y}^1}(\mathbb{T})\right)$ to $\left(\mathcal{Y}^1(\mathbb{S}), \tau^{\mathrm{W}}_{\mathcal{Y}^1}(\mathbb{S})\right)$ is a topological embedding, that is, a net $(\mu^\alpha)_\alpha$ in $\mathcal{Y}^1(\mathbb{T})$ W–stably converges to an element μ^∞ of $\mathcal{Y}^1(\mathbb{T})$ if and only if the net $(\widetilde{\mu}^\alpha)_\alpha$ W–stably converges to $\widetilde{\mu}^\infty$ in $\mathcal{Y}^1(\mathbb{S})$.*

Proof. Observe that, for any $\mu \in \mathcal{Y}^1(\mathbb{T})$, the definition of $\widetilde{\mu}$ implies that the measure $\widetilde{\mu}(\Omega \times .)$ is in $\mathcal{M}^+_\tau(\mathbb{S})$. Let $\mathcal{U}$ be a uniformity on $\mathbb{S}$ which is compatible with the topology $\tau_{\mathbb{S}}$ of $\mathbb{S}$. For simplicity, the uniformity on $\mathbb{T}$ induced by $\mathcal{U}$ will also be denoted by $\mathcal{U}$. Let $\mathrm{C}_{bu}(\mathbb{S},\mathcal{U})$ (resp. $\mathrm{C}_{bu}(\mathbb{T},\mathcal{U})$) be the space of bounded $\mathcal{U}$–uniformly continuous functions on $\mathbb{S}$ (resp. on $\mathbb{T}$). Each element f of $\mathrm{C}_{bu}(\mathbb{T},\mathcal{U})$ has a unique extension $\widetilde{f} \in \mathrm{C}_{bu}(\mathbb{S},\mathcal{U})$ (e.g. [Bou71, Theorem 2, Page II.20]). Let $\underline{\psi}$ be the canonical injection from $\mathbb{T}$ to $\mathbb{S}$. For each $\mu \in \mathcal{Y}^1$ and each $\widetilde{\mu}$–integrable integrand f on $\Omega \times \mathbb{S}$, we have $\widetilde{\mu}(f) = \mu(f \circ \underline{\psi}) = \mu(f|_{\Omega \times \mathbb{T}})$. Let $(\mu^\alpha)_\alpha$ be a net in $\mathcal{Y}^1(\mathbb{T})$ and let $\mu^\infty \in \mathcal{Y}^1(\mathbb{T})$. Using the equivalence 12 $\Leftrightarrow$ 15 from Part D of Theorem 2.1.3 , we get

$$\begin{aligned}
\mu^\alpha \xrightarrow{\text{W-stably}} \mu^\infty &\Leftrightarrow \forall A \in \mathcal{S} \;\; \forall f \in \mathrm{C}_{bu}(\mathbb{T},\mathcal{U}) \;\; \mu^\alpha(\mathbb{1}_A \otimes f) \to \mu^\infty(\mathbb{1}_A \otimes f) \\
&\Leftrightarrow \forall A \in \mathcal{S} \;\; \forall f \in \mathrm{C}_{bu}(\mathbb{T},\mathcal{U}) \;\; \widetilde{\mu}^\alpha(\mathbb{1}_A \otimes \widetilde{f}) \to \widetilde{\mu}^\infty(\mathbb{1}_A \otimes \widetilde{f}) \\
&\Leftrightarrow \forall A \in \mathcal{S} \;\; \forall f \in \mathrm{C}_{bu}(\mathbb{S},\mathcal{U}) \;\; \widetilde{\mu}^\alpha(\mathbb{1}_A \otimes f) \to \widetilde{\mu}^\infty(\mathbb{1}_A \otimes f) \\
&\Leftrightarrow \widetilde{\mu}^\alpha \xrightarrow{\text{W-stably}} \widetilde{\mu}^\infty.
\end{aligned}$$

□

Theorem 2.1.3 and its Corollary 2.1.8 also allow us to extend Lemma 2.1.2 to dense subsets of regular Suslin spaces, for W–stable convergence.

Corollary 2.1.9 (W–Stable topology relative to the completed σ–algebra) *If $\mathbb{T}$ is a dense subspace of a Suslin regular space, then $\tau^{\mathrm{W}}_{\mathcal{Y}^1}(\Omega, \mathcal{S}, \mathrm{P})$ coincides with $\tau^{\mathrm{W}}_{\mathcal{Y}^1}(\Omega, \mathcal{S}^*_{\mathrm{P}}, \mathrm{P}^*)$.*

Proof. From Corollary 2.1.8, we can assume w.l.g. that $\mathbb{T}$ is Suslin regular. Let $\mathcal{D}$ be any σ–upwards filtering set of distances which induces the topology of $\mathbb{T}$. From the beginning of Remark 2.1.7, and with the same notations, $\tau^{\mathrm{W}}_{\mathcal{Y}^1}(\Omega, \mathcal{S}, \mathrm{P})$ is the initial topology for the system $(\tau^{\mathrm{W}}_{\mathcal{Y}^1}(\Omega, \mathcal{S}, \mathrm{P}, d), \mathrm{Id})$, where d runs over $\mathcal{D}$. But, from Lemma 2.1.2, for each $d \in \mathcal{D}$, the topological spaces $\tau^{\mathrm{W}}_{\mathcal{Y}^1}(\Omega, \mathcal{S}, \mathrm{P}; \mathbb{T}, d)$ and $\tau^{\mathrm{W}}_{\mathcal{Y}^1}(\Omega, \mathcal{S}^*_{\mathrm{P}}, \mathrm{P}^*; \mathbb{T}, d)$ coincide. □

The following result gives convergence of W–stably converging nets on some integrands which are not tensor products. Contrarily to Corollary 2.1.4, no separability condition is assumed (provided $\mathbb{T}$ contains a dense subspace with non–measurable cardinal, see Section 1.4).

Let d be a continuous semidistance on $\mathbb{T}$. Recall that $\mathrm{BL}_1(\mathbb{T}, d)$ is the set of bounded Lipschitz functions $f : \mathbb{T} \to [-1, 1]$ with Lipschitz modulus bounded by 1 (see page 15). We denote by $\underline{\mathrm{BL}}_1(\Omega, \mathbb{T}, d)$ the space of integrands f such that $f(\omega, .) \in \mathrm{BL}_1(\mathbb{T}, d)$ for all $\omega \in \Omega$ and by $\underline{\mathrm{BL}}'_1(\Omega, \mathbb{T}, d)$ the set of elements f of $\underline{\mathrm{BL}}_1(\Omega, \mathbb{T}, d)$ which have the form $f = \sum_{i=1}^n \mathbb{1}_{A_i} \otimes g_i$, where $(A_i)_{1 \le i \le n}$ is a measurable partition of Ω (which depends on f).

Proposition 2.1.10 (W–Stable convergence and Lipschitz integrands) *Let d be a continuous semidistance on $\mathbb{T}$. We assume that $\mathbb{T}$ contains a dense subspace with non–measurable cardinal (Section 1.4) or that there exists a separable subset $\mathbb{T}_0$ of $\mathbb{T}$ such that $\mu^\infty(\Omega \times \mathbb{T}_0) = 1$. Let $(\mu^\alpha)_{\alpha \in \mathbb{A}}$ be a net in $\mathcal{Y}^1$ which W–stably converges to some $\mu^\infty \in \mathcal{Y}^1$. Then we have*

$$\lim_\alpha \mu^\alpha(\varphi \circ f) = \mu^\infty(\varphi \circ f)$$

for each integrand $f \in \underline{\mathrm{BL}}_1(\Omega, \mathbb{T}, d)$ and for each continuous function $\varphi : [-1, 1] \to \mathbb{R}$.

Proof. Assuming that $\mathbb{T}$ contains a dense subspace with non–measurable cardinal or that $\mu^\infty(\Omega \times \mathbb{T}_0) = 1$ for some subspace $\mathbb{T}_0$ of $\mathbb{T}$, the measure $\mu^\infty(\Omega \times .)$ is inner regular w.r.t. the totally bounded subsets of $\mathbb{T}$. Let $\varepsilon > 0$. There exists a totally bounded subset K of $\mathbb{T}$ such that $\mu^\infty(\Omega \times K) > 1 - \varepsilon$. Recall that every Lipschitz function $f \in \mathrm{BL}_1(K)$ can be extended to a Lipschitz function $f \in \mathrm{BL}_1(\mathbb{T}, d)$ (see e.g. [Dud02, Theorem 6.1.1]).

For any continuous function f on $\mathbb{T}$ and any $B \subset \mathbb{T}$, let us denote $\|f\|_B := \sup_{x \in B} |f(x)|$. The set of restrictions to K of elements of $\mathrm{BL}_1(d)$ is totally bounded for $\|.\|_K$ (it is a subset of the compact space $\mathrm{BL}_1(\widehat{K}, d)$, where $\widehat{K}$ is the d–completion of K). For each $\eta > 0$, there exist thus $g_1, \dots, g_n \in \mathrm{BL}_1(d)$ such that, for each $f \in \mathrm{BL}_1(d)$, we have $\inf_{i=1,\dots,n} \|f - g_i\|_K \le \eta$.

Now, let $f \in \underline{\mathrm{BL}}_1(\Omega, \mathbb{T}, d)$ and let $\varphi : [-1,1] \to \mathbb{R}$ be a continuous function. The mapping φ is uniformly continuous, thus there exists $\eta > 0$ such that, for all $x, y \in [-1,1]$, if $|x-y| \leq \eta$ then $|\varphi(x) - \varphi(y)| \leq \varepsilon$. For each $\omega \in \Omega$, we can find $N(\omega) \in \{1, \ldots, n\}$ such that $\left\|f(\omega,.) - g_{N(\omega)}\right\|_K \leq \eta/3$. Furthermore, we can assume that N is measurable, because Lipschitz functions on K are determined by their values on a countable dense subset of K. For $i = 1, \ldots, n$, let $A_i = \{N(\omega) = i\}$, and let $g = \sum_{i=1}^n \mathbf{1}_{A_i} \otimes g_i$. We have $g \in \underline{\mathrm{BL}}_1'(\Omega, \mathbb{T}, d)$ and $\|f(\omega,.) - g(\omega,.)\|_K \leq \eta/3$ for every $\omega \in \Omega$.

Let $K^{\eta/3} = \{x \in \mathbb{T};\ d(x,K) < \eta/3\}$. For each $\omega \in \Omega$, as $f(\omega,.)$ and $g(\omega,.)$ are 1–Lipschitz, we have

$$\|f(\omega,.) - g(\omega,.)\|_{K^{\eta/3}} \leq 3\frac{\eta}{3} = \eta,$$

thus

$$\|\varphi \circ f(\omega,.) - \varphi \circ g(\omega,.)\|_{K^{\eta/3}} \leq \varepsilon. \tag{2.1.1}$$

Let $h : \mathbb{T} \to [0,1]$ be a Lipschitz mapping such that $h(x) = 1$ if $x \in K$ and $h(x) = 0$ if $x \notin K^{\eta/3}$ (we can take e.g. $h(x) = (1 - (3/\eta)d(x,K)) \vee 0)$. We have

$$\lim_\alpha \mu^\alpha(\mathbf{1}_\Omega \otimes (1-h)) = \mu^\infty(\mathbf{1}_\Omega \otimes (1-h)) \leq \mu^\infty(\Omega \times K^c) \leq \varepsilon,$$

thus there exists $\alpha_0 \in \mathbb{A}$ such that

$$\alpha \geq \alpha_0 \Rightarrow \mu^\alpha(\mathbf{1}_\Omega \otimes (1-h)) \leq 2\varepsilon. \tag{2.1.2}$$

On the other hand, let $M = \max_{x\in[-1,1]} |\varphi(x)|$. For every $\omega \in \Omega$, we have $\|\varphi \circ f(\omega,.) - \varphi \circ g(\omega,.)\|_{\mathbb{T}} \leq 2M$. From (2.1.1) and (2.1.2) we thus have, for $\alpha \geq \alpha_0$,

$$\begin{aligned}|(\mu^\alpha - \mu^\infty)(\varphi \circ f - \varphi \circ g)| &\leq (\mu^\alpha + \mu^\infty)(|(\varphi \circ f - \varphi \circ g)(\mathbf{1}_\Omega \otimes h)|) \\ &\quad + (\mu^\alpha + \mu^\infty)(|2M(\mathbf{1}_\Omega \otimes (1-h))|) \\ &\leq (\varepsilon + \varepsilon) + (4\varepsilon + 2\varepsilon)M = (2+6M)\varepsilon.\end{aligned}$$

But we also have $\lim_\alpha (\mu^\alpha - \mu^\infty)(\varphi \circ g) = 0$, thus, for α large enough,

$$|(\mu^\alpha - \mu^\infty)(\varphi \circ f)| \leq (3+6M)\varepsilon,$$

which yields the result because ε is arbitrary. □

We are going to deduce from Proposition 2.1.10 (with $\varphi(x) = x$) a new criterion of W–stable convergence. Let d be a continuous distance on $\mathbb{T}$. Let $\mathrm{BL}(\mathbb{T}, d)$ denote the vector space of real-valued bounded Lipschitz functions defined on $\mathbb{T}$. It is a Banach space for the norm

$$\|f\|_{\mathrm{BL}(\mathbb{T},d)} := \|f\|_\infty + \sup\{\frac{|f(s) - f(t)|}{d(s,t)};\ s \neq t\}.$$

We denote by $\underline{\mathrm{BL}}(\Omega, \mathbb{T}, d)$ the set of all integrands f on $\Omega \times \mathbb{T} \to \mathbb{R}$ such that there exists a measurable mapping $\phi : \Omega \to \mathbb{R}^+$ with $\|f(\omega, .)\|_{\mathrm{BL}(d)} \le \phi(\omega)$ for every $\omega \in \Omega$ (if $\mathbb{T}$ is separable, the mapping $\omega \to \|f(\omega, .)\|_{\mathrm{BL}(d)}$ is measurable, thus $\underline{\mathrm{BL}}(\Omega, \mathbb{T}, d)$ is simply the set of integrands f such that $f(\omega, .) \in \mathrm{BL}(\mathbb{T}, d)$ for each $\omega \in \Omega$).

We denote by $\underline{\mathrm{BL}}'(\Omega, \mathbb{T}, d)$ the set of elements f of $\underline{\mathrm{BL}}(\Omega, \mathbb{T}, d)$ which have the form

$$f(\omega, t) = \sum_{i=1}^{n} \mathbb{1}_{A_i}(\omega) f_i(t),$$

where $(A_1, \dots, A_n)$ is a measurable partition of Ω and each f_i is in $\mathrm{BL}(\mathbb{T}, d)$. Let $\mathrm{L}^1_{\mathrm{BL}(\mathbb{T},d)}$ be the space of Bochner integrable functions defined on $(\Omega, \mathcal{S}, \mathrm{P})$ with values in $\mathrm{BL}(\mathbb{T}, d)$. We have

$$\text{(2.1.3)} \qquad \underline{\mathrm{BL}}'(\Omega, \mathbb{T}, d) \subset \mathrm{L}^1_{\mathrm{BL}(\mathbb{T},d)} \subset \left\{ f \in \underline{\mathrm{BL}}(\Omega, \mathbb{T}, d);\ f \text{ is } \mathrm{L}^1\text{-bounded} \right\}.$$

Corollary 2.1.11 (W–Stable convergence and Lipschitz integrands) *Assume that $\mathbb{T}$ is metrizable and let d be a distance which is compatible with the topology of $\mathbb{T}$. Let $(\mu^\alpha)_{\alpha \in \mathbb{A}}$ be a net in $\mathcal{Y}^1$ and let $\mu^\infty \in \mathcal{Y}^1$. The following conditions are equivalent:*

(a) $\mu^\alpha \xrightarrow{W\text{-stably}} \mu^\infty$.

(b) *For each L^1-bounded integrand $f \in \underline{\mathrm{BL}}(\Omega, \mathbb{T}, d)$, we have $\lim_\alpha \mu^\alpha(f) = \mu^\infty(f)$.*

(c) *For each $f \in \mathrm{L}^1_{\mathrm{BL}(\mathbb{T},d)}$, we have $\lim_\alpha \mu^\alpha(f) = \mu^\infty(f)$.*

(d) *For each integrand $f \in \underline{\mathrm{BL}}'(\Omega, \mathbb{T}, d)$, we have $\lim_\alpha \mu^\alpha(f) = \mu^\infty(f)$.*

Proof.

The implications $(b) \Rightarrow (c) \Rightarrow (d)$ are clear from (2.1.3).

The implication $(d) \Rightarrow (a)$ comes from Part D of Portmanteau Theorem 2.1.3.

Assume (a). Let $\varepsilon > 0$. Let f be an L^1-bounded element of $\underline{\mathrm{BL}}(\Omega, \mathbb{T}, d)$. In order to show that $\lim_\alpha \mu^\alpha(f) = \mu^\infty(f)$, we have to control both $|f(\omega, t)|$ and the Lipschitz modulus of $f(\omega, .)$. From the definition of $\underline{\mathrm{BL}}(\Omega, \mathbb{T}, d)$, there exists a measurable function $\phi : \Omega \to \mathbb{R}^+$ such that $\|f(\omega, .)\|_{\mathrm{BL}(d)} \le \phi(\omega)$ for every $\omega \in \Omega$. Furthermore, as f is L^1-bounded, there exists a P–integrable function $\varphi : \Omega \to \mathbb{R}^+$ such that $|f(\omega, t)| \le \varphi(\omega)$ for each $(\omega, t) \in \Omega \times \mathbb{T}$. We can thus find $\Omega_\varepsilon \in \mathcal{S}$ and $M > 0$ such that

$$\mathrm{P}(\Omega \setminus \Omega_\varepsilon) < \varepsilon, \quad \phi \mathbb{1}_{\Omega_\varepsilon} \le M, \quad \text{and} \quad \int_{\Omega \setminus \Omega_\varepsilon} \varphi \, d\mathrm{P} < \varepsilon.$$

Define $g \in \underline{\mathrm{BL}}(\Omega, \mathbb{T}, d)$ by

$$g(\omega, t) = \begin{cases} \dfrac{1}{M} f(\omega, t) & \text{if } \omega \in \Omega_\varepsilon \\ 0 & \text{if } \omega \in \Omega \setminus \Omega_\varepsilon \end{cases}$$

We have $g \in \underline{\mathrm{BL}}_1(\Omega, \mathbb{T}, d)$, thus, from Proposition 2.1.10, $\lim_\alpha \mu^\alpha(g) = \mu^\infty(g)$. Moreover, for any $\mu \in \mathcal{Y}^1$, we have

$$|\mu(f - Mg)| \leq \varepsilon.$$

As ε is arbitrary, this proves the result. □

The following easy consequence of Theorem 2.1.3 and Proposition 2.1.10 is often useful.

Proposition 2.1.12 (Semicontinuity Theorem) *Let $(\mu^\alpha)_{\alpha \in \mathbb{A}}$ be a net of elements of $\mathcal{Y}^1$ and let $\mu^\infty \in \mathcal{Y}^1$. Assume that one of the following conditions is satisfied:*

(a) $\mu^\alpha \xrightarrow{W\text{-stably}} \mu^\infty$ and $f : \Omega \times \mathbb{T} \to \mathbb{R}^+$ is an integrand such that there exists a continuous distance d on $\mathbb{T}$ with $f(\omega, .) \in \mathrm{BL}_1(\mathbb{T}, d)$ for each $\omega \in \Omega$.

(b) $\mu^\alpha \xrightarrow{M\text{-stably}} \mu^\infty$ and $f : \Omega \times \mathbb{T} \to [0, +\infty]$ is a continuous integrand.

(c) $\mu^\alpha \xrightarrow{N\text{-stably}} \mu^\infty$, $g : \Omega \to [0, +\infty]$ is a measurable mapping, $h : \mathbb{T} \to \mathbb{R}^+$ is an l.s.c. mapping and f is the integrand $g \otimes h$.

(d) $\mu^\alpha \xrightarrow{S\text{-stably}} \mu^\infty$ and $f : \Omega \times \mathbb{T} \to [0, +\infty]$ is an l.s.c. integrand.

Then we have

$$\mu^\infty(f) \leq \liminf_\alpha \mu^\alpha(f).$$

Proof. From Proposition 2.1.10 (in the case (a)) or from Theorem 2.1.3, we have, for every $M \in [0, +\infty[$, $\mu^\infty(f \wedge M) \leq \liminf_\alpha \mu^\alpha(f \wedge M)$.

Assume first that $\mu^\infty(f) < +\infty$. Choose $M > 0$ such that $\mu^\infty(f \wedge M) \geq \mu^\infty(f) - \varepsilon$. We then have

$$\mu^\infty(f) \leq \mu^\infty(f \wedge M) + \varepsilon \leq \liminf_\alpha \mu^\alpha(f \wedge M) + \varepsilon \leq \liminf_\alpha \mu^\alpha(f) + \varepsilon$$

which yields the result, because ε is arbitrary.

Assume now that $\mu^\infty(f) = +\infty$. For each $A > 0$, there exists $M > 0$ such that $\mu^\infty(f \wedge M) > A$. We then have

$$A < \liminf_\alpha \mu^\alpha(f \wedge M) \leq \liminf_\alpha \mu^\alpha(f).$$

As A is arbitrary, this proves that $\liminf_\alpha \mu^\alpha(f) = +\infty$. □

In the preceding results, in particular in the Portmanteau Theorem 2.1.3, we considered mainly conditions on the topological space $\mathbb{T}$, not on $(\Omega, \mathcal{S})$. Now, from Part D, it is easy to deduce the following continuation of Theorem 2.1.3, in the case when Ω can be endowed with a topology. We show that, in some cases, W–stable convergence coincides with narrow convergence w.r.t. an appropriate topology on $\Omega \times \mathbb{T}$. This result generalizes [Val94, Theorem 3, 2)].

Before we state this result, we need to fix some more vocabulary. Let $\mathcal{C}$ be a set of subsets of Ω. We say that $\mathcal{C}$ *essentially generates* $\mathcal{S}$ if $\mathcal{S}$ is contained in the universal completion of the σ-algebra generated by $\mathcal{C}$. We say that $\mathcal{S}$ is *essentially countably generated* if there exists a countable set of subsets of Ω which essentially generates $\mathcal{S}$.

Theorem 2.1.13 (Portmanteau Theorem continued: W–stable convergence vs. narrow convergence on the product space) *Let $(\mu^\alpha)_{\alpha\in\mathbb{A}}$ be a net in $\mathcal{Y}^1$ and let $\mu^\infty \in \mathcal{Y}^1$. Assume that there exists a topology τ_Ω on Ω such that $\mathcal{B}_{\tau_\Omega} = \mathcal{S}$.*

A) If τ_Ω is perfectly normal, the following Conditions 1 and 2 are equivalent:

1. $\mu^\alpha \xrightarrow{W\text{–stably}} \mu^\infty$.
2. $\lim_\alpha \mu^\alpha(f \otimes g) = \mu^\infty(f \otimes g)$ *for any bounded τ_Ω–continuous function f on Ω and for any bounded $\tau_\mathbb{T}$–continuous function g on $\mathbb{T}$.*

B) If furthermore $\mathbb{T}$ is the union of a sequence of second countable Suslin spaces which are Borel subsets of $\mathbb{T}$, or if $\mathbb{T}$ is Suslin regular, we can replace in A the hypothesis $\mathcal{B}_{\tau_\Omega} = \mathcal{S}$ by

$$\mathcal{B}_{\tau_\Omega}{}^* = \mathcal{S}^*,$$

where, as usual, for any σ–algebra $\mathcal{A}$, we denote by $\mathcal{A}^*$ its universal completion.

C) Assume that $(\mathbb{T}, \tau_\mathbb{T})$ and (Ω, τ_Ω) are completely regular topological spaces such that $\mathcal{B}_{\tau_\Omega \otimes \tau_\mathbb{T}} = \mathcal{B}_{\tau_\Omega} \otimes \mathcal{B}_{\tau_\mathbb{T}}$ (this is the case if $\tau_\Omega \otimes \tau_\mathbb{T}$ is hereditarily Lindelöf, or if one of the spaces $(\mathbb{T}, \tau_\mathbb{T})$ and (Ω, τ_Ω) is cosmic, see the proof of this theorem). Assume furthermore that the elements of $\mathcal{M}^{+,1}(\Omega \times \mathbb{T}, \tau_\Omega \otimes \tau_\mathbb{T})$ are τ–regular (this is the case if $\tau_\Omega \otimes \tau_\mathbb{T}$ is hereditarily Lindelöf). Then Condition 2 is equivalent to each of Conditions 3 and 4 below.

3. $\lim_\alpha \mu^\alpha(f) = \mu^\infty(f)$ *for any bounded $\tau_\Omega \otimes \tau_\mathbb{T}$–continuous function f on $\Omega \times \mathbb{T}$.*
4. $\mu^\alpha \xrightarrow{narrow} \mu^\infty$ *in $\mathcal{M}^{+,1}(\Omega \times \mathbb{T}, \tau_\Omega \otimes \tau_\mathbb{T})$.*

D) In particular, $\tau^{\mathrm{W}}_{\mathcal{Y}^1}$ coincides with the topology induced on $\mathcal{Y}^1$ by the topology of narrow convergence on $\mathcal{M}^{+,1}(\Omega \times \mathbb{T}, \tau_\Omega \otimes \tau_\mathbb{T})$ in each of the following cases:

(i) $(\mathbb{T}, \tau_\mathbb{T})$ and (Ω, τ_Ω) are regular cosmic spaces and $\mathcal{S}$ is the Borel σ–algebra of τ_Ω,

(ii) $(\mathbb{T}, \tau_\mathbb{T})$ is a regular Suslin space, (Ω, τ_Ω) is a regular cosmic space, and $\mathcal{S}^*$ is essentially generated by τ_Ω.

Proof.

A) As τ_Ω is perfectly normal, the σ–algebra $\mathcal{S} = \mathcal{B}_{\tau_\Omega}$ is generated by $C_b(\Omega, \tau_\Omega)$. The result is thus a direct consequence of Part D of Theorem 2.1.3.

B) Obvious from Lemma 2.1.2 and Corollary 2.1.9.

C) This is a result on narrow and weak convergence. The equivalence between 3 and 4 comes from the fact that, by the classical Portmanteau Theorem [Top70b], as $\tau_\Omega \otimes \tau_\mathbb{T}$ is completely regular, the weak topology and the narrow topology on $\mathcal{M}_\tau^+(\Omega \times \mathbb{T}, \tau_\Omega \otimes \tau_\mathbb{T})$ coincide. The hypothesis $\mathcal{B}_{\tau_\Omega \otimes \tau_\mathbb{T}} = \mathcal{B}_{\tau_\Omega} \otimes \mathcal{B}_{\tau_\mathbb{T}}$ ensures that each μ^α $(\alpha \in \mathbb{A} \cup \{\infty\})$ is in $\mathcal{M}^{+,1}(\Omega \times \mathbb{T}, \tau_\Omega \otimes \tau_\mathbb{T})$. The proof that $\mathcal{B}_{\tau_\Omega \otimes \tau_\mathbb{T}} = \mathcal{B}_{\tau_\Omega} \otimes \mathcal{B}_{\tau_\mathbb{T}}$ holds when $\tau_\Omega \otimes \tau_\mathbb{T}$ is hereditarily Lindelöf, or when one of the spaces $(\mathbb{T}, \tau_\mathbb{T})$ and (Ω, τ_Ω) has a countable Borel network, can be found in [BC93, Exercice 3.18, with solution] (see also [FJW96] where it is proved that, if (Ω, τ_Ω) is a regular space, then $\mathcal{B}_{\tau_\Omega \otimes \tau_\mathbb{T}} = \mathcal{B}_{\tau_\Omega} \otimes \mathcal{B}_{\tau_\mathbb{T}}$ holds for any regular space $(\mathbb{T}, \tau_\mathbb{T})$ if and only if (Ω, τ_Ω) has a countable Borel network).

   If $(\mathbb{T}, \tau_\mathbb{T})$ and (Ω, τ_Ω) are metrizable, the equivalence between 3 and 2 is proved in [vdVW96, Corollary 1.4.5]. Assume now that $(\mathbb{T}, \tau_\mathbb{T})$ and (Ω, τ_Ω) are completely regular. Clearly, 3 implies 2. Assume 2 and let $\mathcal{D}_\Omega$ (resp. $\mathcal{D}_\mathbb{T}$) be an upward filtering set of continuous distances which induces the topology of Ω (resp. $\mathbb{T}$). For any $d \in \mathcal{D}_\Omega$ and any $d' \in \mathcal{D}_\mathbb{T}$, define the distance $d \otimes d'$ by $d \otimes d'((\omega_1, t_1), (\omega_2, t_2)) = d(\omega_1, \omega_2) \vee d'(t_1, t_2)$. Then $\mathcal{D} = \{d \otimes d';\, d \in \mathcal{D}_\Omega,\, d \in \mathcal{D}_\mathbb{T}\}$ is an upwards filtering set of distances that induces $\tau_\Omega \otimes \tau_\mathbb{T}$. From the metrizable case, we have, for any $d \otimes d' \in \mathcal{D}$,

$$\mu^\alpha \xrightarrow{\text{narrow}} \mu^\infty \text{ in } \mathcal{M}^{+,1}(\Omega \times \mathbb{T}, d \otimes d' \in \mathcal{D}).$$

   The conclusion follows from Lemma 1.3.3.

D) Obvious.

□

Remark 2.1.14 In Part D, $(\mathcal{Y}^1(\mathbb{T}), \tau^{\mathrm{W}}_{\mathcal{Y}^1})$ is a closed subspace of $\mathcal{M}^+(\Omega \times \mathbb{T})$ endowed with its narrow topology.

2.2 Special subspaces of Young measures, denseness of the space $\mathfrak{X}$ of random variables

In this section, we consider the topological subspaces $\mathcal{M}^{+,1}(\mathbb{T})$, $\mathbb{T}$ and $\mathfrak{X}(\mathbb{T})$ (see page 20 how they are imbedded in $\mathcal{Y}^1_{\text{dis}}$). The subspaces of p–integrable Young measures will be considered in another section (Section 2.4), because we shall endow them with new stable topologies wich are not induced by the topologies $\tau^*_{\mathcal{Y}^1}$.

For the spaces $\mathcal{M}^{+,1}(\mathbb{T})$ and $\mathbb{T}$, we compare their usual topologies with the topology induced by $\tau^*_{\mathcal{Y}^1}$, $* = \mathrm{S}, \mathrm{M}, \mathrm{N}, \mathrm{W}$: We shall see that $\tau_{\mathbb{T}}$ is the topology induced on $\mathbb{T}$ by $\tau^*_{\mathcal{Y}^1}$ and that, when $\mathbb{T}$ is completely regular, the narrow topology on $\mathcal{M}^{+,1}(\mathbb{T})$ is also induced by $\tau^*_{\mathcal{Y}^1}$.

For $\mathfrak{X}(\mathbb{T})$, it is only in Chapter 3 that we will compare its natural topology (that is, the topology of convergence in P–probability) with the topology induced by $\tau^*_{\mathcal{Y}^1}$, $* = \mathrm{S}, \mathrm{M}, \mathrm{N}, \mathrm{W}$. We prove in this section that, in a quite general case, $\mathfrak{X}(\mathbb{T})$ is dense in $\mathcal{Y}^1$ for $\tau^*_{\mathcal{Y}^1}$, $* = \mathrm{S}, \mathrm{M}, \mathrm{N}, \mathrm{W}$.

Proposition 2.2.1 (∗–Stable topology on $\mathcal{M}^{+,1}(\mathbb{T})$) *The narrow topology on $\mathcal{M}^{+,1}(\mathbb{T})$ is induced by $\tau^{\mathrm{N}}_{\mathcal{Y}^1}$ and the weak topology on $\mathcal{M}^{+,1}(\mathbb{T})$ is induced by $\tau^{\mathrm{W}}_{\mathcal{Y}^1}$.*

Proof. Let g be a bounded l.s.c. function on $\mathbb{T}$ and let $A \in \mathcal{S}$. Let $f = \mathbb{1}_A \otimes g$. Then the function $\widetilde{f} : \mathbb{T} \to \mathbb{R}$ defined by

$$\widetilde{f}(t) = \int_\Omega f(\omega, t)\, d\,\mathrm{P}(\omega)$$

is bounded l.s.c. Conversely, any bounded l.s.c. function g can be written $g = \widetilde{f}$, with $f(\omega, t) = g(t) = (\mathbb{1}_\Omega \otimes g)(\omega, t)$. Let $(\mu^\alpha)_{\alpha \in \mathbb{A}}$ be a net in $\mathcal{M}^{+,1}(\mathbb{T})$ and let $\mu^\infty \in \mathcal{M}^{+,1}(\mathbb{T})$. For the time of the proof let us distinguish between each μ^α $(\alpha \in \mathbb{A} \cup \{\infty\})$ and the constant disintegrable Young measure $\boldsymbol{\mu}^\alpha : \omega \mapsto \mu^\alpha$. We have, for each $\alpha \in \mathbb{A} \cup \{\infty\}$,

$$\boldsymbol{\mu}^\alpha(f) = \int_\Omega f(\omega, t)\, d\mu^\alpha(t)\, d\,\mathrm{P}(\omega) = \mu^\alpha(\widetilde{f}).$$

Thus $\liminf_{\alpha \in \mathbb{A}} \boldsymbol{\mu}^\alpha(f) \geq \boldsymbol{\mu}^\infty(f)$ if and only if $\liminf_{\alpha \in \mathbb{A}} \mu^\alpha(\widetilde{f}) \geq \mu^\infty(\widetilde{f})$, and one deduces easily the first part of the proposition.

The same reasoning shows that the topology induced on $\mathcal{M}^{+,1}(\mathbb{T})$ by $\tau^{\mathrm{W}}_{\mathcal{Y}^1}$ coincides with the weak topology. □

Corollary 2.2.2 (∗–Stable topology on $\mathbb{T}$) *The topology $\tau_{\mathbb{T}}$ is induced by $\tau^{\mathrm{N}}_{\mathcal{Y}^1}$. If $\mathbb{T}$ is completely regular, it is also induced by $\tau^{\mathrm{W}}_{\mathcal{Y}^1}$.*

Proof. Indeed, the restriction to $\mathbb{T}$ of the narrow topology on $\mathcal{M}^{+,1}(\mathbb{T})$ is $\tau_{\mathbb{T}}$. □

The subspace $\mathfrak{X}(\mathbb{T})$: Denseness Theorem The following result is essential in applications. Since the origins of the theory of Young measures, it is well-known that, if P is nonatomic and $\mathbb{T}$ is metrizable and compact, then $\mathfrak{X}$ is dense in $\mathcal{Y}^1$ [You37, pages 226–228] (actually, Young measures were constructed as the completion of $\mathfrak{X}$ in an appropriate uniformity). This result has been extended by Balder [Bal84b] to the case when $\mathbb{T}$ is a completely regular Suslin space and P is nonatomic. We give below a slight generalization, with $\mathbb{T}$ non necessarily regular.

Theorem 2.2.3 (Denseness Theorem) *Assume that $\mathbb{T}$ is Radon and that the compact subsets of $\mathbb{T}$ are metrizable (e.g. $\mathbb{T}$ is Suslin). Assume furthermore that* P *is nonatomic. Then $\mathfrak{X}$ is dense in $\mathcal{Y}^1$ for $\tau^*_{\mathcal{Y}^1}$, $* =$ S, M, N, W.*

Our proof of Theorem 2.2.3 relies on the denseness result in the case when $\mathbb{T}$ is Suslin metrizable. For the convenience of the reader, we also give the proof of Theorem 2.2.3 in this well-known case.

Proof.

Step 1: Suslin metrizable case. In this case, know by Part G of Theorem 2.1.3 that $\tau^{\mathrm{S}}_{\mathcal{Y}^1} = \tau^{\mathrm{M}}_{\mathcal{Y}^1} = \tau^{\mathrm{N}}_{\mathcal{Y}^1} = \tau^{\mathrm{W}}_{\mathcal{Y}^1}$. Let $\mu \in \mathcal{Y}^1$. Let d be a distance which is compatible with the topology of $\mathbb{T}$. We only need to show that, for any $\varepsilon > 0$, any finite subset $\{A_1, \dots, A_n\}$ of $\mathcal{S}$ and any subset $\{f_1, \dots, f_n\}$ of $\mathrm{BL}_1(\mathbb{T}, d)$, there exists a random element X of $\mathbb{T}$ such that, for each $i = 1, \dots, n$, we have $\left|\mu(\mathbb{1}_{A_i} \otimes f_i) - \int_{A_i} f_i \circ X \, d\mathrm{P}\right| < \varepsilon$. Considering the measurable partition of Ω generated by the set A_i, and taking a smaller $\varepsilon > 0$, we can assume that, for $i \neq j$, we have $A_i = A_j$ or $A_i \cap A_j = \emptyset$. Furthermore, we only need to define X on each A_i. So, the problem reduces to the search, for any $\varepsilon > 0$, any $A \in \mathcal{S}$ and any finite subset $\{f_1, \dots, f_n\}$ of $\mathrm{BL}_1(\mathbb{T}, d)$, of a random element X such that $\left|\mu(\mathbb{1}_A \otimes f_i) - \int_A f_i \circ X \, d\mathrm{P}\right| < \varepsilon$. Let $\nu := \mu(A \times .) \in \mathcal{M}^+(\mathbb{T})$. The set of convex combinations of Dirac measures is dense in $\mathcal{M}^{+,1}(\mathbb{T})$, thus there exist a finite subset $\{t_1, \dots, t_m\}$ of $\mathbb{T}$ and positive coefficients $\alpha_1, \dots \alpha_m$ such that $\sum_{k=1}^m \alpha_k = \nu(\mathbb{T}) = \mathrm{P}(A)$ and such that we have $\Delta^{(d)}_{\mathrm{BL}}(\nu, \sum_{k=1}^m \alpha_k \delta_{t_k}) < \varepsilon$ (where $\Delta^{(d)}_{\mathrm{BL}}$ is the Dudley distance). Now, as P has no atom, we can find a measurable partition $C_1, \dots, C_m$ of A such that $\mathrm{P}(C_k) = \alpha_k$ for $k = 1, \dots, m$. Then we only need to set $X(\omega) = t_k$ on C_k for $k = 1, \dots, m$.

Step 2: general case. We only need to prove the theorem for $* =$ S.

Let $\mu \in \mathcal{Y}^1$. Let V be a neighbourhood of μ in $\mathcal{Y}^1$ for $\tau^{\mathrm{S}}_{\mathcal{Y}^1}$. We can assume w.l.g. that

$$V = \{\nu \in \mathcal{Y}^1;\ \forall i = 1, \dots, n \quad \nu(f_i) > \mu(f_i) - \varepsilon\},$$

where n is a positive integer, $f_1, \dots, f_n$ are l.s.c. integrands such that $0 \le f_i \le 1$ for all $i = 1, \dots, n$ and $\varepsilon > 0$.

As $\mathbb{T}$ is Radon, the measure $\mu(\Omega \times .)$ is tight. Let $K \in \mathcal{K}$ be such that $\mu(\Omega \times K) > 1 - \varepsilon/2$. Let $\widetilde{\mu} = \mu|_{\Omega \times K}$ be the restriction of μ on $\Omega \times K$. Let $\eta = \mu(\Omega \times K)$. Let $\widetilde{\mathrm{P}} = \widetilde{\mu}(. \times \mathbb{T}) = \mu(. \times K)$. We have $\widetilde{\mathrm{P}}(\Omega) = \eta \in]1 - \varepsilon/2, 1]$, $\widetilde{\mathrm{P}} \le \mathrm{P}$ and $\widetilde{\mu} \le \mu$ (that is, $\widetilde{\mathrm{P}}(A) \le \mathrm{P}(A)$ for any $A \in \mathcal{S}$ and $\widetilde{\mu}(A) \le \mu(A)$ for any $A \in \underline{\mathcal{B}}$). In particular, $\widetilde{\mathrm{P}}$ is nonatomic.

Now, the measure $(1/\eta)\widetilde{\mu}$ is an element of $\mathcal{Y}^1(\Omega, \mathcal{S}, (1/\eta)\widetilde{\mathrm{P}}; K)$. Consider the following neighbourhood of $(1/\eta)\widetilde{\mu}$:

$$\widetilde{V} = \left\{\nu \in \mathcal{Y}^1(\Omega, \mathcal{S}, (1/\eta)\widetilde{\mathrm{P}}; K);\ \forall i = 1, \dots, n \quad \nu(f_i) > \widetilde{\mu}(f_i) - \varepsilon/2\right\}.$$

From the denseness result for metrizable compact spaces, there exists a random element X of K such that the measure $\int \delta_\omega \otimes \delta_{X(\omega)}\, d(1/\eta)\widetilde{\mathrm{P}}(\omega)$ is an element of $\widetilde{V}$, that is, for $i = 1, \dots, n$,

$$\frac{1}{\eta}\int f_i(\omega, X(\omega))\, d\widetilde{\mathrm{P}}(\omega) > \frac{1}{\eta}\mu(f_i \mathbb{1}_{\Omega\times K}) - \varepsilon/2.$$

As $0 \le f_i \le 1$, we thus have

$$\begin{aligned}\int f_i(\omega, X(\omega))\, d\mathrm{P}(\omega) &\ge \int f_i(\omega, X(\omega))\, d\widetilde{\mathrm{P}}(\omega)\\ &> \mu(f_i \mathbb{1}_{\Omega\times K}) - \eta\varepsilon/2\\ &> \mu(f_i) - \eta\varepsilon/2 - \varepsilon/2 \ge \mu(f_i) - \varepsilon.\end{aligned}$$

Thus $\int \delta_\omega \otimes \delta_{X(\omega)}\, d\mathrm{P}(\omega)$ is in V. □

Remark 2.2.4 If $\mathbb{T}$ is metrizable, another construction, using the martingale convergence theorem, can be found in [RdFZ02].

When random elements of $\mathbb{T}$ converge in $\mathcal{Y}^1$, it might be interesting to know whether the limit μ is also a random element of $\mathbb{T}$. The following criterion was given by Zięba.

Theorem 2.2.5 (Zięba Criterion [Zię85]) *Assume that $\mathcal{B}_{\mathbb{T}}$ is countably generated. Let $\mu \in \mathcal{Y}^1_{\mathrm{dis}}$. The following conditions are equivalent:*

(a) There exists $X \in \mathfrak{X}$ such that $\mu = \underline{\delta}_X$ (from the identifications we made, this can be written: $\mu \in \mathfrak{X}$).

(b) $\forall A \times B \in \mathcal{S} \otimes \mathcal{B}_{\mathbb{T}}$
$$\mu(A\times B) > 0 \Rightarrow \big(\exists A' \subset A \quad \mathrm{P}(A') > 0, \quad \forall \omega \in A' \quad \mu_\omega(B) = 1\big).$$

We shall need the following easy lemma.

Lemma 2.2.6 *Let $\nu \in \mathcal{M}^{+,1}(\mathbb{T})$. Assume that $\nu \notin \mathbb{T}$, that is, ν is not a Dirac measure. Assume furthermore that $\mathcal{B}_{\mathbb{T}}$ is generated by a sequence $(B_1, B_2, \dots)$. Then there exists an integer n such that $\nu(B_n) > 0$ and $\nu(B_n^c) > 0$.*

Proof. Indeed, assume that the conclusion of Lemma 2.2.6 is false. For each $n \ge 1$, let $\mathcal{C}_n$ be the finite partition generated by $B_1, \dots, B_n$. Then, for each n, one can find $C_n \in \mathcal{C}_n$ such that $\nu(C_n) = 1$. Necessarily $(C_n)_n$ is decreasing. Let $C = \cap_n C_n$. We have $\nu(C) = 1$. But C is an atom of $\mathcal{B}_{\mathbb{T}}$, thus, as $\mathbb{T}$ is Hausdorff, C is a singleton $\{t\}$, and we have $\nu = \delta_t$. □

Proof of Theorem 2.2.5. The necessary part is obvious. Assume now that $\mu \notin \mathfrak{X}$. Then there exists $A_0 \in \mathcal{S}$, $\mathrm{P}(A_0) > 0$, such that, for every $\omega \in A_0$, μ_ω is nondegenerate, that is, $\mu_\omega \notin \mathbb{T}$. Let $(B_1, B_2, \dots)$ be a sequence which generates $\mathcal{B}_\mathbb{T}$. From Lemma 2.2.6, the random variable

$$N(\omega) = \inf \{n \in \mathbb{N}^*; \ \mu_\omega(B_n) > 0 \text{ and } \mu_\omega(B_n^c) > 0\}$$

has finite values on A_0. There exists thus an integer n such that $\mathrm{P}\{N = n\} > 0$. Let $A = \{N = n\}$. Now, for each $A' \subset A$ such that $\mathrm{P}(A') > 0$, we have $\mu(A' \times B_n) > 0$ but $\mu_\omega(B_n) < 1$ for all $\omega \in A'$. Thus (b) is not satisfied. $\square$

Corollary 2.2.7 ([Zię85]) *Assume that $\mathcal{B}_\mathbb{T}$ is countably generated. Let $\mu \in \mathcal{Y}^1_{\mathrm{dis}}$. For any measure Q on $(\Omega, \mathcal{S})$ which is absolutely continuous w.r.t. P, let $dQ/d\mathrm{P}$ denote the Radon–Nikodým derivative of Q w.r.t. P. The following conditions are equivalent:*

(a) *There exists $X \in \mathfrak{X}$ such that $\mu = \underline{\delta}_X$.*

(b) *For each $B \in \mathcal{B}_\mathbb{T}$ such that $\mu(\Omega \times B) > 0$, let Q_B denote the measure $\mu(. \times B)$. Then, almost everywhere, $dQ_B/d\mathrm{P}$ takes its values in $\{0, 1\}$.*

(c) *For each $B \in \underline{\mathcal{B}}$ such that $\mu(B) > 0$, let $\underline{Q}_{\underline{B}}$ denote the measure $\mu((. \times T) \cap B)$. Then, almost everywhere, $d\underline{Q}_{\underline{B}}/d\mathrm{P}$ takes its values in $\{0, 1\}$.*

Proof. The implication $(a) \Rightarrow (c)$ is straightforward, with

$$\frac{d\underline{Q}_{\underline{B}}}{d\mathrm{P}}(\omega) = \mathbf{1}_B(\omega, X(\omega))$$

for $B \in \underline{\mathcal{B}}$, and (b) is a particular case of (c). Now, assume (b). Let $B \in \mathcal{B}_\mathbb{T}$ such that $\mu(\Omega \times B) > 0$. Let $h = dQ_B/d\mathrm{P}$, with values in $\{0, 1\}$. Let $A \in \mathcal{S}$ such that $\mu(A \times B) > 0$. Let $A' = \{\omega \in A; \ h(\omega) > 0\} = \{\omega \in A; \ h(\omega) = 1\}$. We have $\mathrm{P}(A') > 0$ thus $\mu_\omega(B) = 1$ for P–almost every $\omega \in A'$. From Theorem 2.2.5, (a) is thus satisfied. $\square$

2.3   Properties of $(\mathcal{Y}^1, \tau^{\mathrm{w}}_{\mathcal{Y}^1})$ related to the topology of $\mathbb{T}$

Metrizability   From Corollary 2.2.2, the space $\mathbb{T}$ is a subspace of $(\mathcal{Y}^1, \tau^{\mathrm{N}}_{\mathcal{Y}^1})$. Thus, for $(\mathcal{Y}^1, \tau^{\mathrm{N}}_{\mathcal{Y}^1})$ to be metrizable, it is necessary that $\mathbb{T}$ be metrizable.

Proposition 2.3.1 *Assume that $\mathbb{T}$ is separable and metrizable, and that $\mathcal{S}$ is essentially countably generated. Then $\tau^{\mathrm{w}}_{\mathcal{Y}^1} = \tau^{\mathrm{N}}_{\mathcal{Y}^1}$ is metrizable (thus, if furthermore $\mathbb{T}$ is Suslin, then $\tau^{\mathrm{s}}_{\mathcal{Y}^1} = \tau^{\mathrm{w}}_{\mathcal{Y}^1}$ is metrizable).*

Proof. By Corollary 2.1.9, we can assume w.l.g. that $\mathcal{S}$ is countably generated. Let d be a distance which is compatible with the topology of $\mathbb{T}$. Let $\Delta_{\mathrm{BL}}^{(d)}$ be the Dudley distance on $\mathcal{M}^+(\mathbb{T})$ associated with d. Let (A_n) be a sequence in $\mathcal{S}$ which generates $\mathcal{S}$. For each $n \in \mathbb{N}$, let $\mathcal{C}^n = \{C_0^n, \dots, C_{m(n)}^n\}$ be the partition of Ω generated by $A_0, \dots, A_n$ (we skip of it all subsets with P–probability 0). We rename the elements of $\mathcal{C} = \cup_n \mathcal{C}^n$ so as to have $\mathcal{C} = \{C_1, \dots, C_n, \dots\}$. We define a distance $\delta : \mathcal{Y}^1 \times \mathcal{Y}^1 \to [0, 1]$ by

$$\begin{aligned}\delta(\mu, \nu) &= \sum_{n \geq 1} 2^{-n} \sup_{f \in \mathrm{BL}_1(\mathbb{T},d)} |\mu(\mathbb{1}_{C_n} \otimes f) - \nu(\mathbb{1}_{C_n} \otimes f)| \\ &= \sum_{n \geq 1} 2^{-n} \Delta_{\mathrm{BL}}^{(d)}(\mu(C_n \times .), \nu(C_n \times .)).\end{aligned}$$

Let $(\mu^\alpha)_{\alpha \in \mathbb{A}}$ be a net in $\mathcal{Y}^1$ which W–stably converges to $\mu^\infty \in \mathcal{Y}^1$. Then, for each $n \in \mathbb{N}$, the net $(\mu^\alpha(C_n \times .))_{\alpha \in \mathbb{A}}$ converges in $\mathcal{M}^{+,1}(\mathbb{T})$ to $\mu^\infty(C_n \times .)$ (for the Dudley distance), thus the net $(\sup_{f \in \mathrm{BL}_1(\mathbb{T},d)} |\mu^\alpha(\mathbb{1}_{C_n} \otimes f) - \mu^\infty(\mathbb{1}_{C_n} \otimes f)|)_{\alpha \in \mathbb{A}}$, with values in $[0, 1]$, converges to 0. Therefore the net $(\delta(\mu^\alpha, \mu^\infty))_{\alpha \in \mathbb{A}}$ converges to 0.

Conversely, assume that $(\delta(\mu^\alpha, \mu^\infty))_{\alpha \in \mathbb{A}}$ converges to 0. Then Condition 18 in Theorem 2.1.3 is satisfied, thus $(\mu^\alpha)_{\alpha \in \mathbb{A}}$ W–stably converges to $\mu^\infty \in \mathcal{Y}^1$.

□

Submetrizability

Corollary 2.3.2 *Assume that there exists a separable metrizable topology τ_0 on $\mathbb{T}$ which is coarser than the original topology τ of $\mathbb{T}$, and such that τ and τ_0 have the same Borel sets. Assume furthermore that $\mathcal{S}$ is essentially countably generated. Then $\tau_{\mathcal{Y}^1}^{\mathrm{W}}$ is submetrizable.*

Proof. If τ and τ_0 have the same Borel sets, they have the same Young measures, and we have $\tau_{\mathcal{Y}^1}^{\mathrm{W}}(\tau_0) \subset \tau_{\mathcal{Y}^1}^{\mathrm{W}}(\tau)$. □

Suslin property and Polishness

Proposition 2.3.3 *Assume that $\mathbb{T}$ is Suslin regular and that $\mathcal{S}$ is* standard, *that is, there exists a Polish topology τ_Ω on Ω such that $\mathcal{S} = \mathcal{B}_{\tau_\Omega}$. Then $(\mathcal{Y}^1, \tau_{\mathcal{Y}^1}^{\mathrm{W}}) = (\mathcal{Y}^1, \tau_{\mathcal{Y}^1}^{\mathrm{N}})$ is Suslin regular. If furthermore $\mathbb{T}$ is Polish, then $(\mathcal{Y}^1, \tau_{\mathcal{Y}^1}^{\mathrm{W}}) = (\mathcal{Y}^1, \tau_{\mathcal{Y}^1}^{\mathrm{S}})$ is Polish.*

Proof. From Theorem 2.1.13 and Remark 2.1.14, the space $(\mathcal{Y}^1, \tau_{\mathcal{Y}^1}^{\mathrm{W}})$ is a closed subspace of $\mathcal{M}^{+,1}(\Omega \times \mathbb{T}, \tau_\Omega \otimes \tau_\mathbb{T})$ endowed with the narrow topology. But the space $\mathcal{M}^{+,1}(\Omega \times \mathbb{T}, \tau_\Omega \otimes \tau_\mathbb{T})$ is Suslin regular, thus $(\mathcal{Y}^1, \tau_{\mathcal{Y}^1}^{\mathrm{W}})$ is also Suslin regular. If furthermore $\mathbb{T}$ is Polish, then $\mathcal{M}^{+,1}(\Omega \times \mathbb{T}, \tau_\Omega \otimes \tau_\mathbb{T})$ is Polish, thus $(\mathcal{Y}^1, \tau_{\mathcal{Y}^1}^{\mathrm{W}})$ is Polish too. □

2.4 Integrable Young measures and $L^p_{\mathbb{E}}$ spaces

In this section, we are given a real number $p \geq 1$ and we assume that $\mathbb{T}$ is completely regular and that its topology is defined by a set $\mathcal{D}$ of semidistances. We are going to define the subspace $\mathcal{Y}^{1,p}_{\mathcal{D}}$ of p–integrable Young measures. On this space, new "stable" topologies are defined, using integrands satisfying some growth condition related to p instead of bounded integrands. Another topology on this space will be investigated (for $p = 1$) in Section 3.4.

The restriction to $\mathfrak{X}(\mathbb{T})$ yields the spaces $\mathfrak{X}^p_{(\mathbb{T},\mathcal{D})}$. The space $L^p_{\mathbb{E}}$ of Bochner p–integrable functions with values in a Banach space $\mathbb{E}$ is a particular case of $\mathfrak{X}^p_{(\mathbb{T},\mathcal{D})}$ space.

We say that a Young measure $\mu \in \mathcal{Y}^1$ is *p–integrable* (relatively to $\mathcal{D}$) if we have

$$\forall d \in \mathcal{D} \quad \int_{\Omega\times\mathbb{T}} d(a,t)^p \, d\mu(\omega,t) < +\infty$$

for some (equivalently for all) $a \in \mathbb{T}$. A p–integrable Young measure is also said to be *of order p*. If $p = 1$, we simply say that μ is *integrable*. We denote by $\mathcal{Y}^{1,p}_{\mathcal{D}}$ the set of p-integrable Young measures relatively to $\mathcal{D}$. If $\mathcal{D}$ has a single element d, we simplify notations by writing $\mathcal{Y}^{1,p}_{d}$ instead of $\mathcal{Y}^{1,p}_{\{d\}}$. Also, if d stems from a seminorm $\|.\|$, we shall use notations such as $\mathcal{Y}^{1,p}_{\|.\|}$.

The space $\mathcal{Y}^{1,p}_{\mathcal{D}} \cap \mathfrak{X}$ is naturally denoted by $\mathfrak{X}^p_{(\mathbb{T},\mathcal{D})}$ or $\mathfrak{X}^p_{\mathcal{D}}$ and its elements are called *p–integrable random elements* (relatively to $\mathcal{D}$). Obviously, if $\mathbb{E}$ is a Banach space and if d is the distance associated with the norm of $\mathbb{E}$, the space $\mathfrak{X}^p_{(\mathbb{E},d)}$ is nothing but the space $L^p_{\mathbb{E}}$ of Bochner $\mathbb{E}$–valued p–integrable functions. Similarly, Young measures extend Bochner integrable random elements of locally convex topological spaces.

We can define new stable topologies $\tau^{*}_{p,\mathcal{D}}$, on $\mathcal{Y}^{1,p}_{\mathcal{D}}$, with $* = \mathrm{S}, \mathrm{M}, \mathrm{N}, \mathrm{W}$, by replacing the boundedness condition on the test integrands by a condition of growth of order p. For example, let $\tau^{\mathrm{W}}_{p,\mathcal{D}}$ be the coarsest topology on $\mathcal{Y}^{1,p}_{\mathcal{D}}$ such that, for each $d \in \mathcal{D}$, the functions

$$\begin{cases} \mathcal{Y}^1 & \to \quad \mathbb{R} \\ \mu & \mapsto \quad \mu(\mathbf{1}_A \otimes f) \end{cases}$$

are continuous for every $A \in \mathcal{S}$ and every continuous function f on $\mathbb{T}$ satisfying

$$|f(t)| \leq 1 + d(a,t)^p$$

for some fixed $a \in \mathbb{T}$.

The topology $\tau^{\mathrm{W}}_{p,\|.\|}$, when $\mathbb{T} \subset \mathbb{R}^n$, has been considered (with different definitions) by Schonbek [Sch82], Ball [Bal89c], Kinderlehrer and Pedregal [KP94] (see also [Rou97] about these three papers), by Piccinini and Valadier [PV95], by Artstein [Art01a, Art01b], Artstein and Popa [AP03], and by Dedecker and Merlevède [DM02].

We shall see in Theorem 2.4.3 that, if $\mathbb{T}$ is a Banach space $(\mathbb{E}, \|.\|)$, the restriction of $\tau^{\mathrm{W}}_{p,\|.\|}$ to $\mathrm{L}^p_{\mathbb{E}}$ is the strong topology of the vector space $\mathrm{L}^p_{\mathbb{E}}$. Unfortunately, the sum of random elements and the multiplication of a random vector by a scalar do not admit continuous extensions on the space $\mathcal{Y}^{1,p}_{\|.\|}(\Omega, \mathcal{S}, \mathrm{P}; \mathbb{E})$ of integrable Young measures on $\mathbb{T}$. Multivalued extensions which keep some of the properties of these operations are constructed in [Art01a, Art01b, AP03].

If the elements of $\mathcal{D}$ are bounded, $\mathcal{Y}^{1,p}_{\mathcal{D}}$ is nothing but the familiar space $\mathcal{Y}^1$, and we have $\tau^{\mathrm{W}}_{p,\mathcal{D}} = \tau^{\mathrm{W}}_{\mathcal{Y}^1}$.

Let us now introduce a notion of uniform integrability which will help us to characterize $\tau^{\mathrm{W}}_{p,\mathcal{D}}$–convergence. We say that a set $\mathfrak{Y} \subset \mathcal{Y}^{1,p}_{\mathcal{D}}$ is *uniformly p–integrable* (relatively to $\mathcal{D}$) if, for every $d \in \mathcal{D}$,

$$\lim_{R \to +\infty} \sup_{\mu \in \mathfrak{Y}} \int_{\Omega \times \{d(a,.)>R\}} d(a,t)^p \, d\mu(\omega, t) = 0$$

for some (equivalently, for any) $a \in \mathbb{T}$. We say that a net $(\mu^\alpha)_{\alpha \in \mathbb{A}}$ of elements of $\mathcal{Y}^{1,p}_{\mathcal{D}}$ is *asymptotically uniformly p–integrable* (relatively to $\mathcal{D}$) if, for every $d \in \mathcal{D}$,

$$\lim_{R \to +\infty} \limsup_{\alpha} \int_{\Omega \times \{d(a,.)>R\}} d(a,t)^p \, d\mu^\alpha(\omega, t) = 0$$

for some (or any) $a \in \mathbb{T}$. If $\mathbb{A} = \mathbb{N}$, $(\mu^\alpha)_{\alpha \in \mathbb{A}}$ is asymptotically uniformly p–integrable if and only if it is uniformly p–integrable.

For every $d \in \mathcal{D}$, we denote by $\mathrm{Lip}_1(d)$ the set of d–Lipschitz functions on $\mathbb{T}$, with Lipschitz modulus not greater than 1.

Proposition 2.4.1 (Various characterizations of $\tau^{\mathrm{W}}_{p,\mathcal{D}}$–convergence) *Let $(\mu^\alpha)_{\alpha \in \mathbb{A}}$ be a net in $\mathcal{Y}^{1,p}_{\mathcal{D}}$, and let $\mu^\infty \in \mathcal{Y}^{1,p}_{\mathcal{D}}$. Let a be some fixed element of $\mathbb{T}$. We assume that $\mathbb{T}$ contains a dense subspace with non–measurable cardinal (Section 1.4) or that there exists a separable subset $\mathbb{T}_0$ of $\mathbb{T}$ such that $\mu^\infty(\Omega \times \mathbb{T}_0) = 1$. The following conditions are equivalent:*

1. *$(\mu^\alpha)_{\alpha \in \mathbb{A}}$ converges to μ^∞ for $\tau^{\mathrm{W}}_{p,\mathcal{D}}$.*

2. *$\mu^\alpha \xrightarrow{W\text{-stably}} \mu^\infty$ and $\lim_\alpha \int_{\Omega \times \mathbb{T}} d(a,t)^p \, d\mu^\alpha((\omega,t)) = \int_{\Omega \times \mathbb{T}} d(a,t)^p \, d\mu^\infty((\omega,t))$ for every $d \in \mathcal{D}$.*

3. *$\mu^\alpha \xrightarrow{W\text{-stably}} \mu^\infty$ and $(\mu^\alpha)_{\alpha \in \mathbb{A}}$ is asymptotically uniformly p–integrable.*

4. *$\lim_\alpha \mu^\alpha(f) = \mu^\infty(f)$ for every integrand $f : \Omega \times \mathbb{T} \to \mathbb{R}^+$ such that $f^{1/p}(\omega, .) \in \mathrm{Lip}_1(d)$ for each $\omega \in \Omega$ and $f(.,a) \in \mathrm{L}^1(\mathrm{P})$.*

5. *$\lim_\alpha \mu^\alpha(\mathbf{1}_A \otimes f) = \mu^\infty(\mathbf{1}_A \otimes f)$ for every $A \in \mathcal{S}$ and for every function $f : \mathbb{T} \to \mathbb{R}^+$ such that $f^{1/p} \in \mathrm{Lip}_1(d)$.*

Remark 2.4.2 Analogous characterizations for the other stable topologies $\tau^{*}_{p,\mathcal{D}}$ can be easily obtained, except for Conditions 4 and 5, which are specific to $\tau^{\mathrm{W}}_{p,\mathcal{D}}$. A proof of $3 \Rightarrow 1$ for $\tau^{\mathrm{S}}_{p,\mathcal{D}}$ in a particular case will be given in Lemma 6.2.2.

Note also that Condition 4 applies in particular to Hölder continuous integrands f (relatively to d) of order p with bounded modulus, that is, integrands f satisfying $\sup_{d(t,s)\neq 0} \dfrac{|f(\omega,t)-f(\omega,s)|}{d(t,s)^p} \leq M$ for some $M>0$ and for all $\omega \in \Omega$.

Proof of Proposition 2.4.1. We shall first prove $1 \Rightarrow 2 \Rightarrow 3 \Rightarrow 1$ and then $3 \Rightarrow 4 \Rightarrow 5 \Rightarrow 2$.

The implication $1 \Rightarrow 2$ is obvious.

Let us assume 2. Let $d \in \mathcal{D}$. Let $\varepsilon > 0$. There exists R such that

$$\int_{\Omega\times\{t\in\mathbb{T};\, d(a,t)>R\}} d(a,s)^p \, d\mu^{\infty}(\omega,s) < \varepsilon/2.$$

For every $\rho > 0$, let us denote $B_d(a,\rho) = \{t \in \mathbb{T};\, d(a,t) < \rho\}$. Let $h : \mathbb{T} \to [0,1]$ be a continuous function such that $\mathbb{1}_{B_d(a,R)} \leq h \leq \mathbb{1}_{B_d(a,R+1)}$. We have

$$\int_{\Omega\times\mathbb{T}} (1-h(t))\, d(a,t)^p \, d\mu^{\infty}(\omega,t) < \varepsilon/2.$$

From W–stable convergence of $(\mu^\alpha)_\alpha$ to μ^∞, we also have

$$\lim_\alpha \int_{\Omega\times\mathbb{T}} h(t)\, d(a,t)^p \, d\mu^{\alpha}(\omega,t) = \int_{\Omega\times\mathbb{T}} h(t)\, d(a,t)^p \, d\mu^{\infty}(\omega,t).$$

From Condition 2 and the above convergence,

$$\lim_\alpha \int_{\Omega\times\mathbb{T}} (1-h(t))\, d(a,t)^p \, d\mu^{\alpha}(\omega,t) = \int_{\Omega\times\mathbb{T}} (1-h(t))\, d(a,t)^p \, d\mu^{\infty}(\omega,t),$$

hence, for α large enough,

$$\int_{\Omega\times\{t\in\mathbb{T};\, d(a,t)>R+1\}} d(a,t)^p \, d\mu^{\alpha}(\omega,t) < \varepsilon,$$

which proves 3.

Assume now 3 and let us prove 1. Let $d \in \mathcal{D}$, let $A \in \mathcal{S}$ and let $f : \mathbb{T} \to \mathbb{R}$ be a continuous function such that $|f(t)| \leq 1 + d(a,t)^p$ for each $t \in \mathbb{T}$. Let $\varepsilon > 0$. There exists $R > 0$ and $\alpha_0 \in \mathbb{A}$ such that

$$\forall \alpha \in \mathbb{A}\cup\{\infty\} \quad \alpha \geq \alpha_0 \Rightarrow \int_{\Omega\times\{t\in\mathbb{T};\, d(a,t)>R\}} 1 + d(a,t)^p \, d\mu^{\alpha}(\omega,t) < \varepsilon. \tag{2.4.1}$$

Let $h : \mathbb{T} \to \mathbb{R}$ be a continuous function such that $\mathbb{1}_{\{d(a,.)\leq R\}} \leq h \leq \mathbb{1}_{\{d(a,.)\leq R+1\}}$. Let $\widetilde{h} = \mathbb{1}_\Omega \otimes h$. By W–stable convergence, we have

$$\lim_\alpha \mu^\alpha((\mathbb{1}_A \otimes f)\,\widetilde{h}) = \mu^\infty((\mathbb{1}_A \otimes f)\,\widetilde{h}). \tag{2.4.2}$$

But we also have

$$|\mathbb{1}_A \otimes f|\,(1-\widetilde{h}) \leq (1 + d(a,.)^p)\,\mathbb{1}_{\{d(a,.)>R\}}$$

thus, from (2.4.1),

$$\forall \alpha \in \mathbb{A} \cup \{\infty\} \quad \alpha \geq \alpha_0 \Rightarrow \mu^\alpha(|(\mathbb{1}_A \otimes f)(1-\widetilde{h})|) \leq \varepsilon. \tag{2.4.3}$$

Condition 1 follows immediately from (2.4.2) and (2.4.3).

We now prove 3 $\Rightarrow$ 4. Assume again 3. Let $d \in \mathcal{D}$ and let $f : \Omega \times \mathbb{T} \to \mathbb{R}^+$ be an integrand such that $f^{1/p}(\omega, .) \in \mathrm{Lip}_1(d)$ for each $\omega \in \Omega$ and $f(., a) \in \mathrm{L}^1(\mathrm{P})$. Let $\varepsilon > 0$. Let $R > 0$ and $\alpha \in \mathbb{A}$ such that, for each $\alpha \in \mathbb{A}$ satisfying $\alpha \geq \alpha_0$ or for $\alpha = \infty$,

$$\int_{\Omega\times\{d(a,.)>R\}} d(a,t)^p \, d\mu^\alpha(\omega, t) < \frac{\varepsilon}{2^{p-1}}. \tag{2.4.4}$$

We can assume also that R is large enough such that

$$\int_{\Omega\times\{d(a,.)>R\}} f(\omega, a)\, d\mu^\infty(\omega, t) < \frac{\varepsilon}{2^p}. \tag{2.4.5}$$

Let $h : \mathbb{T} \to \mathbb{R}$ be a continuous function such that $\mathbb{1}_{B(a,R]} \leq h \leq \mathbb{1}_{B(a,R+1]}$, where $B\,(a,\rho] = \{t \in \mathbb{T};\ d\,(t,a) \leq \rho\}$. We can assume furthermore that $h^{1/p} \in \mathrm{BL}_1(\mathbb{T}, d)$. For all $(\omega, t) \in \Omega \times \mathbb{T}$, we have $f^{1/p}(\omega,t) \leq f^{1/p}(\omega, a) + d(a,t)$, thus, from convexity of $x \mapsto |x|^p$,

$$f(\omega, t) \leq 2^{p-1} f(\omega, a) + 2^{p-1} d(a,t)^p. \tag{2.4.6}$$

Let $\Omega_\varepsilon \in \mathcal{S}$ such that $f(., a)$ is bounded on Ω_ε and such that

$$\int_{\Omega_\varepsilon^c} 2^{p-1} f(\omega, a) + 2^{p-1}(R+1)^p \, d\,\mathrm{P}(\omega) < \varepsilon. \tag{2.4.7}$$

We have, for all $\alpha \in \mathbb{A}$,

$$\begin{aligned} |(\mu^\infty - \mu^\alpha)\,(f)| \leq \quad & |(\mu^\infty - \mu^\alpha)\,(f(\mathbb{1}_{\Omega_\varepsilon} \otimes h))| \\ & + \left[\mu^\infty(f(\mathbb{1}_{\Omega_\varepsilon^c} \otimes h)) + \mu^\alpha(f(\mathbb{1}_{\Omega_\varepsilon^c} \otimes h))\right] \\ & + \left[\mu^\infty(f(\mathbb{1}_\Omega \otimes (1-h)) + \mu^\alpha(f(\mathbb{1}_\Omega \otimes (1-h))\right]. \end{aligned} \tag{2.4.8}$$

Let us estimate the first part of the right hand side of (2.4.8). Let

$$M = \sup_{\omega\in\Omega_\varepsilon} 2^{p-1} f(\omega, a) + 2^{p-1}(R+1)^p.$$

From (2.4.6), we have

$$M \geq \sup_{(\omega,t)\in\Omega_\varepsilon\times B(a,R+1]} f(\omega, t).$$

For each $\omega \in \Omega_\varepsilon$, $f^{1/p}(\omega, .)(\mathbb{1}_{\Omega_\varepsilon} \otimes h)^{1/p}(\omega, .)$ is bounded by $M^{1/p}$ and d–Lispchitz with Lipschitz modulus $1 + M^{1/p}$. From Proposition 2.1.10, with $\varphi(x) = x^p$, we thus have

(2.4.9)
$$\lim_\alpha |(\mu^\infty - \mu^\alpha)(f(\mathbb{1}_{\Omega_\varepsilon} \otimes h))| = \lim_\alpha \left|(\mu^\infty - \mu^\alpha)\left(\varphi\circ (f(\mathbb{1}_{\Omega_\varepsilon} \otimes h))^{1/p}\right)\right| = 0.$$

For the second part of the right hand side of (2.4.8), we have, from (2.4.6) and (2.4.7),

(2.4.10)
$$\mu^\infty(f(\mathbb{1}_{\Omega_\varepsilon^c} \otimes h)) + \mu^\alpha(f(\mathbb{1}_{\Omega_\varepsilon^c} \otimes h)) \leq 2\int_{\Omega_\varepsilon^c} 2^{p-1} f(\omega, a) + 2^{p-1}(R+1)^p \, d\mathrm{P}(\omega) < 2\varepsilon.$$

For the third part in (2.4.8), we have, by W–stable convergence and (2.4.5),

$$\lim_\alpha \int f(., a) \otimes (1-h)\, d\mu^\alpha = \int f(., a) \otimes (1-h)\, d\mu^\infty < \frac{\varepsilon}{2^p},$$

thus we can take α_0 large enough such that, for all $\alpha \geq \alpha_0$,

$$\int_{\Omega\times\mathbb{T}} f(\omega, a)(1-h(t))\, d\mu^\alpha(\omega, t) < \frac{\varepsilon}{2^{p-1}}. \tag{2.4.11}$$

Using (2.4.4), (2.4.6), and (2.4.11), we have, for every $\alpha \in \mathbb{A}$ such that $\alpha \geq \alpha_0$ and for $\alpha = \infty$,

$$\begin{aligned}
&\mu^\alpha(f(\mathbb{1}_\Omega \otimes (1-h))) \\
&\quad\leq 2^{p-1}\int_{\Omega\times\mathbb{T}} f(\omega, a)(1-h(t))\, d\mu^\alpha(\omega, t) \\
&\quad\quad + 2^{p-1}\int_{\Omega\times\{d(a,.)>R\}} d(t, a)^p \, d\mu^\alpha(\omega, t) \\
&\quad\leq 2\varepsilon,
\end{aligned}$$

thus, for $\alpha \geq \alpha_0$,

$$\mu^\infty(f(\mathbb{1}_\Omega \otimes (1-h))) + \mu^\alpha(f(\mathbb{1}_\Omega \otimes (1-h))) \leq 4\varepsilon. \tag{2.4.12}$$

Substituting (2.4.9), (2.4.10) and (2.4.12) in (2.4.8) yields

$$\lim_{\alpha} \mu^{\alpha}(f) = \mu^{\infty}(f),$$

which proves 4.

The implication 4 $\Rightarrow$ 5 is obvious. Finally, assume 5 and let us prove 2. Let $d \in \mathcal{D}$. Applying 5 to $\mathbb{1}_{\Omega} \otimes d(a,.)^p$ shows that $\lim_{\alpha} \int_{\Omega\times\mathbb{T}} d(a,t)^p \, d\mu^{\alpha}((\omega,t)) = \int_{\Omega\times\mathbb{T}} d(a,t)^p \, d\mu^{\infty}((\omega,t))$. There remains to prove that $\mu^{\alpha} \xrightarrow{\text{W-stably}} \mu^{\infty}$. Let $A \in \mathcal{S}$ and let $f \in \mathrm{BL}_1(d)$, with $f \geq 0$. Let $\varepsilon > 0$. The functions $x \mapsto x^p$ and $x \mapsto x^{1/p}$ are uniformly continuous on $[0,1]$, thus there exists $\eta > 0$ such that

$$\forall x, y \in [0,1] \quad |x-y| \leq \eta \Rightarrow |x^p - y^p| \leq \varepsilon \tag{2.4.13}$$

and there exists $\delta > 0$, with $\delta < \eta/3$, such that

$$\forall x, y \in [0,1] \quad |x-y| \leq \delta \Rightarrow \left|x^{1/p} - y^{1/p}\right| \leq \frac{\eta}{3}. \tag{2.4.14}$$

The measure $\mu^{\infty}(A \times .)$ is inner regular w.r.t. the totally bounded subsets of $\mathbb{T}$ (see Section 1.4), thus there exists a totally bounded subset K of $\mathbb{T}$ (relatively to d) such that $\mu^{\infty}(A \times (\mathbb{T} \setminus K)) < \varepsilon$. For any continuous function g on $\mathbb{T}$ and any $B \subset \mathbb{T}$, let us denote $\|g\|_B := \sup_{x\in B} |g(x)|$. The set of restrictions to K of elements of $\mathrm{BL}_1(d)$ is totally bounded for $\|.\|_K$, thus there exists $g \in \mathrm{BL}_1(d)$, with $g \geq 0$, such that $\left\|f^{1/p} - g\right\|_K \leq \eta/3$. Let $K^{\delta} = \{t \in \mathbb{T};\, d(t,K) < \delta\}$. As $f, g \in \mathrm{BL}_1(\mathbb{T}, d)$, we have , using (2.4.14),

$$\left\|f^{1/p} - g\right\|_{K^{\delta}} \leq \frac{\eta}{3} + \frac{\eta}{3} + \delta \leq \eta$$

and thus, from (2.4.13),

$$\|f - g^p\|_{K^{\delta}} \leq \varepsilon. \tag{2.4.15}$$

Let $h : \mathbb{T} \to [0,1]$ be a continuous function such that $\mathbb{1}_K \leq h \leq \mathbb{1}_{K^{\delta}}$. We can assume that $h^{1/p}$ is d–Lispschitz. From (2.4.15), we have

$$\|(f - g^p)h\|_{\mathbb{T}} \leq \varepsilon. \tag{2.4.16}$$

To shorten notations, for every $\mu \in \mathcal{Y}^1$, we shall denote the measure $\mu(A \times .) \in \mathcal{M}^+(\mathbb{T})$ by μ_A. For all $\alpha \in \mathbb{A}$ we have, from (2.4.16),

$$|(\mu_A^{\infty} - \mu_A^{\alpha})((f - g^p)h)| \leq 2\varepsilon. \tag{2.4.17}$$

On the other hand, from Condition 5, we have

$$\lim_{\alpha} |(\mu_A^{\infty} - \mu_A^{\alpha})(g^p h)| = 0 \tag{2.4.18}$$

because $gh^{1/p}$ is d–Lipschitz. Also, as $\mu_A^\alpha(1) = \mu_A^\infty(1) = \mathrm{P}(A)$ and $\mu_A^\alpha(h) \to \mu_A^\infty(h)$, we have

$$\lim_\alpha \mu_A^\alpha(1-h) = \mu_A^\infty(1-h) \le \mu_A^\infty(\Omega \setminus K) \le \varepsilon. \tag{2.4.19}$$

We thus have, using (2.4.17), (2.4.18) and (2.4.19),

$$\begin{aligned}\limsup_\alpha |(\mu_A^\infty - \mu_A^\alpha)(f)| &\le \limsup_\alpha |(\mu_A^\infty - \mu_A^\alpha)((f-g^p)h)| \\ &\quad + \lim_\alpha |(\mu_A^\infty - \mu_A^\alpha)(g^p h)| \\ &\quad + \limsup_\alpha (\mu_A^\infty + \mu_A^\alpha)(1-h) \\ &\le 4\varepsilon,\end{aligned}$$

and this proves 2. □

Theorem 2.4.3 (The subspace $\mathfrak{X}^p_{(\mathbb{T},\mathcal{D})}$) *Assume that $\mathbb{T}$ is hereditarily Lindelöf and regular (thus completely regular). Assume furthermore that $\mathbb{T}$ is separable (or that $\mathbb{T}$ contains a dense subspace with non–measurable cardinal, see Section 1.4).*

(a) *The restriction to $\mathfrak{X}^p_{(\mathbb{T},\mathcal{D})}$ of the topology $\tau^{\mathrm{W}}_{p,\mathcal{D}}$ is induced by the semidistances $\underline{\Delta}_p^{(d)}$ $(d \in \mathcal{D})$ defined by*

$$\underline{\Delta}_p^{(d)}(X,Y) = \left(\int_\Omega d(X,Y)^p \, d\mathrm{P} \right)^{1/p}.$$

(b) *Furthermore, if $\mathbb{T}$ is Radon and its compact subsets are metrizable (e.g. $\mathbb{T}$ is Suslin) and if P is nonatomic, then $\mathfrak{X}^p_{(\mathbb{T},\mathcal{D})}$ is dense in $\mathcal{Y}^{1,p}_{\mathcal{D}}$.*

Proof.
(a) *Topology induced on $\mathfrak{X}^p_{(\mathbb{T},\mathcal{D})}$.* Let $(X_\alpha)_{\alpha \in \mathbb{A}}$ be a net in $\mathfrak{X}^p_{(\mathbb{T},\mathcal{D})}$ and let $X_\infty \in \mathfrak{X}^p_{(\mathbb{T},\mathcal{D})}$. Assume that $\underline{\Delta}_p^{(d)}(X_\alpha, X_\infty) \to 0$ for each $d \in \mathcal{D}$. We will check Condition 3 of Proposition 2.4.1. Let $d \in \mathcal{D}$. We then have

$$\lim_\alpha E\left(d(X_\alpha, X_\infty)^p\right) = 0, \tag{2.4.20}$$

thus, using Hölder's inequality,

$$\lim_\alpha E\left(d(X_\alpha, X_\infty)\right) = 0.$$

For any $f \in \mathrm{BL}_1(\mathbb{T}, d)$ and any $A \in \mathcal{S}$, we thus have

$$\lim_\alpha E\left(| \mathbb{1}_A f(X_\alpha) - \mathbb{1}_A f(X_\infty)|\right) \le \lim_\alpha E\left(\mathbb{1}_A d(X_\alpha, X_\infty)\right) = 0.$$

From Part E of Theorem 2.1.3, this proves that $(\underline{\delta}_{X_\alpha})_\alpha$ W–stably converges to $\underline{\delta}_{X_\infty}$. Let $a \in \mathbb{T}$. Let $\varepsilon > 0$ and let $R > 0$ such that

$$E\left(\mathbb{1}_{\{t\in\mathbb{T};\, d(a,t)>R\}}(X_\infty)\, d(a,X_\infty)^p \right) < \varepsilon. \tag{2.4.21}$$

Let us define a continuous truncation function θ by

$$\theta(x) = \begin{cases} 0 & \text{if } x \geq R+1, \\ R - R(x-R) & \text{if } R \leq x \leq R+1, \\ x & \text{if } -R \leq x \leq R, \\ -R + R(x+R) & \text{if } -R-1 \leq x \leq -R, \\ 0 & \text{if } x \leq -R-1. \end{cases}$$

From (2.4.21), we have

$$E\left(d(a,X_\infty)^p - \theta^p(d(a,X_\infty))\right) < \varepsilon. \tag{2.4.22}$$

On the other hand, as $(\underline{\delta}_{X_\alpha})_\alpha$ W–stably converges to $\underline{\delta}_{X_\infty}$, we also have

$$\lim_\alpha E\left(\theta^p\left(d(a,X_\alpha)\right)\right) = E\left(\theta^p\left(d(a,X_\infty)\right)\right),$$

thus, using (2.4.20),

$$\lim_\alpha E\left(d(a,X_\alpha)^p - \theta^p\left(d(a,X_\alpha)\right)\right) = E\left(d(a,X_\infty)^p - \theta^p\left(d(a,X_\infty)\right)\right). \tag{2.4.23}$$

From (2.4.22) and (2.4.23), we have

$$\lim_\alpha E\left(d(a,X_\alpha)^p - \theta^p(d(a,X_\alpha))\right) < \varepsilon$$

which proves that $(\underline{\delta}_{X_\alpha})_\alpha$ is asymptotically uniformly p–integrable. Thus, from Proposition 2.4.1, $(\underline{\delta}_{X_\alpha})_\alpha$ converges to $\underline{\delta}_{X_\infty}$ for $\tau^{\mathrm{W}}_{p,\mathcal{D}}$.

Conversely, assume that $\underline{\delta}_{X_\alpha}$ converges to $\underline{\delta}_{X_\infty}$ for $\tau^{\mathrm{W}}_{p,\mathcal{D}}$. For each $d \in \mathcal{D}$, define an integrand $f_d : \Omega \times \mathbb{T} \to \mathbb{R}$ by $f_d(\omega,t) = d(X_\infty(\omega),t)^p$. Using Condition 4 of Proposition 2.4.1, we have

$$\int_\Omega d(X_\alpha,X_\infty)^p \, d\,\mathrm{P} = \underline{\delta}_{X_\alpha}(f_d) = \underline{\delta}_{X_\alpha}(f_d) - \underline{\delta}_{X_\infty}(f_d) \to 0,$$

thus $\underline{\Delta}_p^{(d)}(\underline{\delta}_{X_\alpha}, \underline{\delta}_{X_\infty})$ converges to 0.

(b) Denseness of $\mathfrak{X}^p_{(\mathbb{T},\mathcal{D})}$. Fix an element a in $\mathbb{T}$. Let $\mu \in \mathcal{Y}^{1,p}_{\mathcal{D}}$. Let $\varepsilon > 0$. Let $d_1, \ldots, d_n \in \mathcal{D}$ and, for each $i \in \{1,\ldots,n\}$ let $A_i \in \mathcal{S}$ and $f_i : \mathbb{T} \to \mathbb{R}$ such that $|f_i| \leq 1 + d_i(a,.)^p$. For each $i \in \{1,\ldots,n\}$ and for each $R > 0$, let

$B_R^{(i)} = \{t \in \mathbb{T};\, d_i(a,t) \leq R\}$. As $\mathbb{T}$ is Radon, we can find a compact subset K of $\mathbb{T}$ such that $\mu(\Omega \times K^c) < \varepsilon/2$. We can find $R > 0$ such that $K \subset \cap_{1 \leq i \leq n} B_R^{(i)}$ and

$$\forall i \in \{1, \dots, n\} \quad \int_{\Omega \times \left(\mathbb{T} \setminus B_R^{(i)}\right)} 1 + d_i(a,t)^p \, d\mu(\omega, t) < \varepsilon. \tag{2.4.24}$$

For each $i \in \{1, \dots, n\}$, let $g_i : \mathbb{T} \to [-R, R]$ be defined by

$$g_i(t) = \begin{cases} R & \text{if } f_i(t) \geq R, \\ f_i(t) & \text{if } -R \leq f_i(t) \leq R, \\ -R & \text{if } f_i(t) \leq -R. \end{cases}$$

The functions g_i are bounded continuous thus, from the Denseness Theorem 2.2.3, there exists $X \in \mathfrak{X}$ such that

$$\forall i \in \{1, \dots, n\} \quad \left| \int_{\Omega \times \mathbb{T}} (\mathbb{1}_{A_i} \otimes g_i) \, d(\mu - \underline{\delta}_X) \right| < \varepsilon. \tag{2.4.25}$$

Furthermore, the proof of Theorem 2.2.3 shows that we can assume without loss of generality that X takes its values in K, thus

$$\forall i \in \{1, \dots, n\} \quad \int_\Omega \mathbb{1}_{A_i}(\omega) g_i(X(\omega)) \, d\mathrm{P}(\omega) = \int_\Omega \mathbb{1}_{A_i}(\omega) f_i(X(\omega)) \, d\mathrm{P}(\omega). \tag{2.4.26}$$

We thus have, using (2.4.24), (2.4.25) and (2.4.26),

$$\begin{aligned} \left| \int_{\Omega \times \mathbb{T}} (\mathbb{1}_{A_i} \otimes f_i) \, d(\mu - \underline{\delta}_X) \right| &\leq \left| \int_{\Omega \times \mathbb{T}} (\mathbb{1}_{A_i} \otimes g_i) \, d(\mu - \underline{\delta}_X) \right| \\ &\quad + \int_{\Omega \times \mathbb{T}} (\mathbb{1}_{A_i} \otimes (f_i - g_i)) \, d\mu \\ &\leq 2\varepsilon, \end{aligned}$$

which proves the denseness result. □

Chapter 3

Convergence in probability of Young measures (with some applications to stable convergence)

We assume in this chapter that $\mathbb{T}$ *contains a dense subspace with non–measurable cardinal* (see Section 1.4). This assumption will not be recalled in the hypothesis of the theorems, but we shall mention it every time we use it.

The main results of Sections 3.1, 3.2 and 3.3 appear (in a metric setting) in [CRdF04].

3.1 Stable convergence *versus* convergence in probability

We shall compare here $*$–stable convergence of disintegrable Young measures (mainly W–stable convergence) with convergence in probability of the associated random laws. Actually, this comparison will be continued with some by–products of the results of the other sections.

Convergence in probability of random elements Let us recall Hoffmann–Jørgensen's definition of convergence in (Baire) probability [HJ91, HJ98]. We say that a net $(X_\alpha)_{\alpha\in\mathbb{A}}$ of random elements of $\mathbb{T}$ *converges in probability* to a random element X_∞ if, for any bounded continuous function $f : \mathbb{T} \to \mathbb{R}$, we have

$$\lim_\alpha \int_\Omega |f(X_\alpha) - f(X_\infty)| \, d\mathrm{P} = 0. \tag{3.1.1}$$

It is not difficult to show that, for any bounded real random variable Y, we have

$$\int_\Omega |Y|\, d\mathrm{P} \geq \sup_{A\in\mathcal{S}} \left| \int_A Y\, d\mathrm{P} \right| \geq \frac{1}{2} \int_\Omega |Y|\, d\mathrm{P}, \tag{3.1.2}$$

thus (3.1.1) is equivalent to

$$\lim_\alpha \int_A (f(X_\alpha) - f(X_\infty))\, d\mathrm{P} = 0 \quad \text{uniformly in } A \in \mathcal{S}. \tag{3.1.3}$$

We shall see later (see Theorem 3.1.2) and Corollary 3.2.2), that, if $\mathbb{T}$ is completely regular separable, W–stable convergence coincides on $\mathfrak{X}$ with convergence in probability. This means that we can replace the expression "uniformly in $A \in \mathcal{S}$" in (3.1.3) by "for each $A \in \mathcal{S}$".

One immediately notices that this notion of convergence in probability depends only on the bounded continuous functions on $\mathbb{T}$, not on the whole topology $\tau_\mathbb{T}$. Convergence in probability means convergence with respect to the topology induced by the semidistances $\Delta_{\text{prob}}^{(d_f)}$, where f runs over the space $\mathrm{C}(\mathbb{T}, [0,1])$ of continuous functions from $\mathbb{T}$ to $[0,1]$ and, for all $X, Y \in \mathfrak{X}$, $\Delta_{\text{prob}}^{(d_f)}(X,Y) = E(|f(X) - f(Y)|)$. We denote by $\tau_{\text{prob}}(\mathfrak{X}(\mathbb{T}))$ this topology. The topology induced by $\tau_{\text{prob}}(\mathfrak{X})$ on $\mathbb{T}$ is generated by the semidistances $d_f(t,s) = |f(t) - f(s)|$, thus it is the coarsest topology such that the elements of $\mathrm{C}(\mathbb{T},[0,1])$ be continuous. This topology is coarser than $\tau_\mathbb{T}$ and it coincides with $\tau_\mathbb{T}$ if and only if $\mathbb{T}$ is completely regular (see [Eng89, Example 8.1.19]).

Actually, Hoffmann–Jørgensen also defined in [HJ98] a stronger notion of convergence in probability, which always induces $\tau_\mathbb{T}$, but we shall not consider it here. Note also that Hoffmann–Jørgensen's definitions and theory also apply to non–necessarily measurable random elements of a non–necessarily Hausdorff space.

In the case when the topology of $\mathbb{T}$ is induced by a family $\mathcal{D}$ of semidistances, Hoffmann–Jørgensen's definition coincides with the usual one ([HJ91, Theorem 7.4] or [HJ98, Corollary 4.7]):

Hoffmann–Jørgensen's characterization of convergence in probability *Assume that the topology of $\mathbb{T}$ is induced by a family $\mathcal{D}$ of semidistances (and that $\mathbb{T}$ contains a dense subspace with non–measurable cardinal). The following assertions are equivalent:*

(a) $(X_\alpha)_{\alpha\in\mathbb{A}}$ *converges in probability to* X_∞ *(in the sense given above).*

(b) $\lim_\alpha \mathrm{P}\{d(X_\alpha, X_\infty) > \varepsilon\} = 0$ *for all* $d \in \mathcal{D}$ *and for all* $\varepsilon > 0$.

(c) $\lim_\alpha E(d(X_\alpha, X_\infty) \wedge 1) = 0$ *for all* $d \in \mathcal{D}$.

If d is a semidistance on $\mathbb{T}$, we shall denote by $\Delta_{\text{prob}}^{(d)}$ the semidistance on $\mathfrak{X}$ defined by

$$\Delta_{\text{prob}}^{(d)}(X,Y) = E(d(X_\alpha, X_\infty) \wedge 1).$$

If the topology of $\mathbb{T}$ is induced by a family $\mathcal{D}$ of semidistances, then $\tau_{\text{prob}}(\mathfrak{X})$ is induced by the semidistances $\Delta_{\text{prob}}^{(d)}$ $(d \in \mathcal{D})$. Note that if d is a distance on $\mathbb{T}$, then $\Delta_{\text{prob}}^{(d)}$ is a distance on $\mathfrak{X}(\mathbb{T})$, and if d is compatible with the topology of $\mathbb{T}$, then $\Delta_{\text{prob}}^{(d)}$ is compatible with the topology $\tau_{\text{prob}}(\mathfrak{X})$. If $(X_\alpha)_{\alpha\in\mathbb{A}}$ is a net of random elements which converges in probability to a random element X_∞, we write

$$X_\alpha \xrightarrow{\text{prob}} X_\infty.$$

Proposition 3.1.1 *Assume that $\mathbb{T}$ is a subset of a completely regular space $\mathbb{S}$. Then $\tau_{\text{prob}}(\mathfrak{X}(\mathbb{T}))$ is the restriction on $\mathfrak{X}(\mathbb{T})$ of the topology $\tau_{\text{prob}}(\mathfrak{X}(\mathbb{S}))$.*

If furthermore $\mathbb{T}$ is a closed G_δ–subset of $\mathbb{S}$, or if $\mathbb{T}$ is a closed subset of $\mathbb{S}$ and $\mathbb{S}$ is hereditarily Lindelöf, then $\mathfrak{X}(\mathbb{T})$ is $\tau_{\text{prob}}(\mathfrak{X}(\mathbb{S}))$–closed in $\mathfrak{X}(\mathbb{S})$.

Proof. The first part of Proposition 3.1.1 is clear, in view of the above Hoffmann–Jørgensen characterization of convergence in probability in completely regular spaces.

Assume that $\mathbb{T}$ is a closed subset of $\mathbb{S}$, and let $(X_\alpha)_{\alpha\in\mathbb{A}}$ be a net in $\mathfrak{X}(\mathbb{T})$ which converges in probability to a random element X_∞ of $\mathbb{S}$.

Assume first that $\mathbb{T}$ is a G_δ–subset of $\mathbb{S}$. Let $(G_n)_{n\in\mathbb{N}}$ be a decreasing sequence of open subsets of $\mathbb{S}$ such that $\cap_{n\in\mathbb{N}} G_n = \mathbb{T}$. Take for each $n \in \mathbb{N}$ a continuous function f_n on $\mathbb{S}$ such that $f_n = 1$ on $\mathbb{T}$ and $f_n = 0$ on G_n^c. We can furthermore assume that $(f_n)_n$ is decreasing and $0 \le f_n \le 1$ for every n. We have $f_n \circ X_\alpha = 1$ for every $n \in \mathbb{N}$ and every $\alpha \in \mathbb{A}$, thus, for every $n \in \mathbb{N}$, we have

$$\int_\Omega f_n \circ X_\infty \, d\mathrm{P} = \lim_\alpha \int_\Omega f_n \circ X_\alpha \, d\mathrm{P} = 1,$$

which proves that $X_\infty(\omega) \in \mathbb{T}$ for P–almost every $\omega \in \Omega$.

Assume now that $\mathbb{S}$ is hereditarily Lindelöf. The law $\mathcal{L}(X_\infty)$ is thus τ–regular. Furthermore, the law $\mathcal{L}(X_\alpha)$ of X_α converges in $\mathcal{M}^{+,1}(\mathbb{S})$ to $\mathcal{L}(X_\infty)$. As $\mathbb{S}$ is completely regular and $\mathbb{T}$ is closed we thus have, from the usual Portmanteau Theorem [Top70b, Theorem 8.1],

$$\mathrm{P}\{X_\infty \in \mathbb{T}\} = \mathcal{L}(X_\infty)(\mathbb{T}) \ge \limsup_\alpha \mathcal{L}(X_\alpha)(\mathbb{T}) \ge 1.$$

□

Convergence in probability of Young measures We endow $\mathcal{M}^{+,1}(\mathbb{T})$ with the *weak topology* (see page 13) and we consider the topology $\tau_{\text{prob}}\left(\mathfrak{X}(\mathcal{M}^{+,1}(\mathbb{T}))\right)$ on the space $\mathcal{Y}^1_{\text{dis}}(\mathbb{T}) \sim \mathfrak{X}(\mathcal{M}^{+,1}(\mathbb{T}))$. We denote this topology by $\tau_{\text{prob}}\left(\mathcal{Y}^1_{\text{dis}}(\mathbb{T})\right)$, for short $\tau_{\text{prob}}\left(\mathcal{Y}^1_{\text{dis}}\right)$ if no ambiguity is to fear. The weak topology on $\mathcal{M}^{+,1}(\mathbb{T})$ is induced by the family of semidistances $d_f(\mu, \nu) = |\mu(f) - \nu(f)|$, where f runs over

$\mathrm{C}(\mathbb{T},[0,1])$ thus the topology $\tau_{\text{prob}}(\mathcal{Y}^1_{\text{dis}})$ is induced by the semidistances $\Delta^{(d_f)}_{\text{prob}}$ ($f \in \mathrm{C}(\mathbb{T},[0,1])$), that is, we have

$$\mu^\alpha \xrightarrow{\text{prob}} \mu^\infty \Leftrightarrow \left[\forall f \in \mathrm{C}(\mathbb{T},[0,1]) \ \lim_\alpha \int_\Omega |\mu^\alpha_\omega(f) - \mu^\infty_\omega(f)| \, d\mathrm{P}(\omega) = 0\right].$$

Using (3.1.2), we see that the semidistance $\Delta^{(d_f)}_{\text{prob}}$ is equivalent to the semidistance $\widetilde{\Delta}^{(f)}_{\text{prob}}$ defined, for all $\mu, \nu \in \mathcal{Y}^1_{\text{dis}}$, by

$$\widetilde{\Delta}^{(f)}_{\text{prob}}(\mu,\nu) = \sup_{A \in \mathcal{S}} |\mu(\mathbb{1}_A \otimes f) - \nu(\mathbb{1}_A \otimes f)|. \tag{3.1.4}$$

Now, we see that (3.1.4) still has a meaning if μ or ν is not disintegrable (actually, it has a meaning for any μ and any ν in $\mathcal{M}^+(\Omega \times \mathbb{T})$). Thus *we can extend to* $\mathcal{Y}^1$ *the topology* $\tau_{\text{prob}}(\mathcal{Y}^1_{\text{dis}})$, *using the semidistances* $\widetilde{\Delta}^{(f)}_{\text{prob}}$ *defined by* (3.1.4). This *extended topology of convergence in probability* will be denoted by $\tau_{\text{prob}}(\mathcal{Y}^1)$.

For each $\mu \in \mathcal{Y}^1$ and each $A \in \mathcal{S}$, let us define a finite measure μ_A on $\mathbb{T}$ by

$$\mu_A = \mu(A \times .).$$

From (3.1.4), we see that a net $(\mu^\alpha)_{\alpha \in \mathbb{A}}$ in $\mathcal{Y}^1$ converges in probability to a Young measure μ^∞ if and only if we have

$$\forall f \in \mathrm{C}(\mathbb{T},[0,1]) \ \lim_\alpha \sup_{A \in \mathcal{S}} |\mu^\alpha_A(f) - \mu^\infty_A(f)| = 0, \tag{3.1.5}$$

that is, $(\mu^\alpha_A)_{\alpha \in \mathbb{A}}$ weakly converges to μ^∞ uniformly in $A \in \mathcal{S}$ for the uniformity on $\mathcal{M}^+(\mathbb{T})$ induced by the semidistances $(\mu,\nu) \mapsto |\mu(f) - \nu(f)|$, $f \in \mathrm{C}(\mathbb{T},[0,1])$.

Comparing (3.1.4) or (3.1.5) with Condition 13 in Theorem 2.1.3, we immediately get Part 1 of the following theorem. Part 3 extends a result of Jacod and Mémin [JM81b] for Polish spaces (see also [Let98, FGT00] and, in the case $\mathbb{T} = \mathbb{R}$, the remark of Dellacherie [Del78] which follows the paper of Meyer [Mey78]). The result of Part 3 will be further improved in Corollary 3.2.2.

Theorem 3.1.2 (Comparison of $\tau_{\text{prob}}(\mathcal{Y}^1_{\text{dis}})$ and $\tau^{\mathrm{W}}_{\mathcal{Y}^1}$)

1. *The topology $\tau_{\text{prob}}(\mathcal{Y}^1)$ is finer than $\tau^{\mathrm{W}}_{\mathcal{Y}^1}$.*

2. *$\tau_{\text{prob}}(\mathcal{Y}^1_{\text{dis}})$ is strictly finer than $\tau^{\mathrm{W}}_{\mathcal{Y}^1}$ if and only if $\mathbb{T}$ has more than one element and $(\Omega, \mathcal{S}, \mathrm{P})$ has a nonatomic part.*

3. *If $\mathbb{T}$ is completely regular, both topologies coincide on $\mathfrak{X}(\mathbb{T})$.*

4. *If $\mathbb{T}$ is regular cosmic, then $\mathfrak{X}(\mathbb{T})$ is $\tau_{\text{prob}}(\mathcal{Y}^1_{\text{dis}})$-closed in $\mathfrak{X}(\mathcal{M}^{+,1}(\mathbb{T})) = \mathcal{Y}^1_{\text{dis}}$.*

Remark 3.1.3 Recall that we have proved in Theorem 2.2.3 that, if $\mathbb{T}$ is Radon and its compact subsets are metrizable (e.g. if $\mathbb{T}$ is Suslin), and if P is nonatomic, then $\mathfrak{X}$ is $\tau^{\mathrm{W}}_{\mathcal{Y}^1}$–dense in $\mathcal{Y}^1$. This results contrasts with Part 4 of Theorem 3.1.2.

We shall see with Corollary 3.2.2 that, when $\mathbb{T}$ is regular Suslin, the relation between $\tau_{\mathrm{prob}}\left(\mathcal{Y}^1\right)$ and $\tau^{\mathrm{W}}_{\mathcal{Y}^1}$ is similar to the relation between the topology $\mathcal{U}$ of uniform convergence and the Skorokhod topology J_1 on the Skorokhod space $\mathbb{D}[0,1]$ (see e.g.[Bil68, Page 150]): $\mathcal{U}$ is finer than J_1, but if $(x_n)_n$ J_1–converges to x and $x \in \mathcal{C}[0,1]$, then $(x_n)_n$ $\mathcal{U}$–converges to x. Furthermore, $\mathcal{C}[0,1]$ is a closed subset of $\mathbb{D}$ for $\mathcal{U}$, but it is dense in $\mathbb{D}$ for J_1.

Before we prove Theorem 3.1.2, let us consider the following example, which shows that a sequence (μ^n) of random elements of $\mathcal{M}^{+,1}(\mathbb{T})$ may converge S–stably to a random probability μ^∞ without being convergent in probability.

Example 3.1.4 (Rademacher sequence in a topological space) Assume that $\mathbb{T}$ has at least two distinct elements a and b and assume that $(\Omega, \mathcal{S}, \mathrm{P})$ has no atom. We can build an independent sequence $(\mathfrak{r}_n)_{n\geq 1}$ of $\mathbb{T}$–valued random elements such that $\mathrm{P}\{\mathfrak{r}_n = a\} = \mathrm{P}\{\mathfrak{r}_n = b\} = 1/2$ for every n.

First, the sequence $(\underline{\delta}_{\mathfrak{r}_n})_{n\geq 1}$ S–stably converges to the Young measure $\mu = 1/2(\underline{\delta}_a + \underline{\delta}_b)$. Indeed, let $\mathbb{S}$ be the set $\{a, b\}$ endowed with the discrete topology $\{\emptyset, \{a\}, \{b\}, \mathbb{S}\}$. As $\mathbb{S}$ is a closed subspace of $\mathbb{T}$, it is straightforward to check that $(\mathcal{Y}^1(\mathbb{S}), \tau^{\mathrm{S}}_{\mathcal{Y}^1}(\mathbb{S}))$ is a topological subspace of $(\mathcal{Y}^1(\mathbb{T}), \tau^{\mathrm{S}}_{\mathcal{Y}^1}(\mathbb{T}))$. Thus, we only need to prove that $(\underline{\delta}_{\mathfrak{r}_n})_{n\geq 1}$ S–stably converges to μ in $\mathcal{Y}^1(\mathbb{S})$. The space $\mathbb{S}$ is Polish. From Part G of Portmanteau Theorem for Stable Topology 2.1.3, we thus only need to prove that $(\underline{\delta}_{\mathfrak{r}_n})_{n\geq 1}$ W–stably converges to μ in $\mathcal{Y}^1(\mathbb{S})$. Let $A \in \mathcal{S}$ and let $f : \mathbb{S} \to \mathbb{R}$ be a function (necessarily, f is bounded and continuous!). If $f(a) = f(b)$, we have $\underline{\delta}_{\mathfrak{r}_n}(\mathbb{1}_A \otimes f) = \mathrm{P}(A) f(a) = \mu(\mathbb{1}_A \otimes f)$ for every $n \geq 1$. Assume that $f(a) \neq f(b)$. Using a translation and a rescaling, we assume without loss of generality that $f(a) = -1$ and $f(b) = 1$. For $n \geq 1$, set $X_n = f \circ \mathfrak{r}_n$. We have, from Bessel's inequality,

$$E(\mathbb{1}_A) = E\left(\mathbb{1}_A^2\right) \geq \sum_{n\geq 1} \langle X_n, \mathbb{1}_A \rangle^2,$$

thus $(\langle X_n, \mathbb{1}_A \rangle)_n$ converges to 0 and we have

$$\lim_{n\geq 1} \underline{\delta}_{\mathfrak{r}_n}(\mathbb{1}_A \otimes f) = \lim_{n\geq 1} E(X_n \mathbb{1}_A) = 0 = \frac{1}{2}(f(a) + f(b))\,\mathrm{P}(A) = \mu(\mathbb{1}_A \otimes f).$$

In order to show that $(\mathfrak{r}_n)_{n\geq 1}$ does not converge in probability to μ, let us consider again a bounded function f on $\mathbb{T}$ such that $f(a) = -1$ and $f(b) = 1$. The function $\Phi_f : \mathcal{M}^{+,1}(\mathbb{T}) \to \mathbb{R}$ defined by $\Phi_f(\nu) = \nu(f)$ is bounded continuous, but we have, for any $n \geq 1$ and any $\omega \in \Omega$,

$$\left|\Phi_f\left(\delta_{\mathfrak{r}_n(\omega)}\right) - \Phi_f(\mu_\omega)\right| = \left|f(\mathfrak{r}_n(\omega)) - \frac{1}{2}(f(a) + f(b))\right| = |f(\mathfrak{r}_n(\omega))| = 1.$$

Proof of Theorem 3.1.2.

1. Let $(\mu^\alpha)_{\alpha\in\mathbb{A}}$ be a net in $\mathcal{Y}^1$ which converges in probability to $\mu^\infty \in \mathcal{Y}^1$, that is,

$$\forall f \in \mathrm{C}(\mathbb{T},[0,1]) \quad \lim_\alpha \sup_{A\in\mathcal{S}} |\mu^\alpha(\mathbb{1}_A \otimes f) - \mu^\infty(\mathbb{1}_A \otimes f)| = 0.$$

From Part D of Theorem 2.1.3, we have $\mu^\alpha \xrightarrow{\text{W-stably}} \mu^\infty$.

2. If $\mathbb{T}$ has only one element, $\tau_{\text{prob}}(\mathcal{Y}^1)$ and $\tau^{\mathrm{W}}_{\mathcal{Y}^1}$ coincide obviously. If $(\Omega,\mathcal{S},\mathrm{P})$ has no nonatomic part, let $A_1,\dots,A_n,\dots$ be the atoms of $(\Omega,\mathcal{S},\mathrm{P})$. Let $\varepsilon>0$. There exists an integer N such that $\mathrm{P}(\cup_{n\le N}A_n)\ge 1-\varepsilon$. Let $(\mu^\alpha)_{\alpha\in\mathbb{A}}$ be a net in $\mathcal{Y}^1$ which W–stably converges to $\mu^\infty\in\mathcal{Y}^1$. We then have, for every $f\in\mathrm{C}(\mathbb{T},[0,1])$ and for every $\alpha\in\mathbb{A}$,

$$\begin{aligned}\widetilde{\Delta}^{(f)}_{\text{prob}}(\mu^\alpha,\mu^\infty) &\le \sup_{A\in\mathcal{S},A\subset\cup_{n\le N}A_n} |\mu^\alpha(\mathbb{1}_A\otimes f)-\mu^\infty(\mathbb{1}_A\otimes f)|+2\varepsilon\\ &\le \sum_{n\le N}|\mu^\alpha(\mathbb{1}_{A_n}\otimes f)-\mu^\infty(\mathbb{1}_{A_n}\otimes f)|+2\varepsilon.\end{aligned}$$

Thus $(\mu^\alpha)_{\alpha\in\mathbb{A}}$ converges in probability to μ^∞.

Conversely, assume that $\mathbb{T}$ has at least two elements a and b and that $(\Omega,\mathcal{S},\mathrm{P})$ has an atomic part Ω_0. If $\Omega_0=\Omega$, the proof that $\tau_{\text{prob}}(\mathcal{Y}^1_{\text{dis}})$ is strictly finer than $\tau^{\mathrm{W}}_{\mathcal{Y}^1}$ is provided by Example 3.1.4. Let us show that the general case amounts to this particular case. Let $\mathcal{S}|_{\Omega_0}$ be the σ–algebra $\{A\cap\Omega_0;\ A\in\mathcal{S}\}$ and let $\mathrm{P}|_{\Omega_0}$ be the restriction of P to $\mathcal{S}|_{\Omega_0}$. Let P_0 be the probability on $(\Omega_0,\mathcal{S}|_{\Omega_0})$ defined by $\mathrm{P}_0=\frac{1}{\mathrm{P}(\Omega_0)}\mathrm{P}|_{\Omega_0}$. Then $(\Omega_0,\mathcal{S}|_{\Omega_0},\mathrm{P}_0)$ has no atoms, thus we can construct a $\mathbb{T}$–valued Rademacher sequence $(\mathfrak{r}_n)_n$ on $(\Omega_0,\mathcal{S}|_{\Omega_0},\mathrm{P}_0)$, such that $\mathrm{P}_0\{\mathfrak{r}_n=a\}=\mathrm{P}_0\{\mathfrak{r}_n=b\}=1/2$. We extend $\mathfrak{r}_n$ to Ω by setting $\mathfrak{r}_n(\omega)=a$ for every $\omega\in\Omega\setminus\Omega_0$. The sequence $(\underline{\delta}_{\mathfrak{r}_n})_n$ W–stably converges to the Young measure $\mu\in\mathcal{Y}^1_{\text{dis}}$ defined by

$$\mu_\omega=\begin{cases}1/2(\delta_a+\delta_b) & \text{if } \omega\in\Omega_0\\ \delta_a & \text{if } \omega\in\Omega\setminus\Omega_0.\end{cases}$$

But $(\underline{\delta}_{\mathfrak{r}_n})_n$ does not converge in probability because, otherwise, its restriction to Ω_0 would converge in P_0–probability.

3. We only need to prove that, on $\mathfrak{X}$, W–stable convergence implies convergence in probability. Let $(X_\alpha)_{\alpha\in\mathbb{A}}$ be a net of elements of $\mathfrak{X}$ which W–stably converges to an element X_∞ of $\mathfrak{X}$. Let $\mathcal{D}$ be a set of continuous semidistances which induces the topology of $\mathbb{T}$ and such that each element d of $\mathcal{D}$ satisfies $d\le 1$. Let $d\in\mathcal{D}$. Let $g\in\underline{\mathrm{BL}}_1(\Omega,\mathbb{T},d)$ be defined by

$$g(\omega,t)=d(t,X_\infty(\omega)).$$

Proposition 2.1.10 yields

$$\lim_\alpha\int d(X_\alpha,X_\infty)\,d\mathrm{P}=\lim_\alpha\int g\,d\underline{\delta}_{X_\alpha}=\int g\,d\underline{\delta}_{X_\infty}=0,$$

thus $(X_\alpha)_{\alpha\in\mathbb{A}}$ converges in probability to X_∞.

4. Assume that $\mathbb{T}$ is regular cosmic. Then, from Proposition 1.3.2, $\mathcal{M}^{+,1}(\mathbb{T})$ is regular cosmic, thus perfectly normal. As $\mathbb{T}$ is closed in $\mathcal{M}^{+,1}(\mathbb{T})$, it is thus a G_δ subset of $\mathcal{M}^{+,1}(\mathbb{T})$. We thus only need to apply Proposition 3.1.1. □

Using Part D of Theorem 2.1.13, we immediately get the following link between convergence in probability and narrow convergence of random elements.

Corollary 3.1.5 (Convergence in probability and narrow convergence on the subspace $\mathfrak{X}(\mathbb{T})$) *Assume that $\mathbb{T}$ is cosmic regular (thus completely regular) and $\mathcal{S}$ is the Borel σ-algebra of a cosmic regular topology τ_Ω on Ω, or that $\mathbb{T}$ is Suslin regular and $\mathcal{S}$ is essentially generated by a cosmic regular topology τ_Ω on Ω. Let $(X_n)_n$ ($n \in \mathbb{N}$) and X_∞ be random elements of $\mathbb{T}$. Then $(X_n)_{n\in\mathbb{N}}$ converges in probability to X_∞ if and only if the sequence of probability laws $(\underline{\delta}_{X_n})_n$ converges to $\underline{\delta}_X$ for the narrow convergence on $\mathcal{M}^{+,1}(\Omega \times \mathbb{T})$.*

Corollary 3.1.6 (Random laws seen as degenerate Young measures on $\mathcal{M}^{+,1}(\mathbb{T})$) *Assume that $\mathbb{T}$ is cosmic regular. Then $\tau_{\text{prob}}\left(\mathcal{Y}^1_{\text{dis}}\right)$ is the restriction to $\mathfrak{X}\left(\mathcal{M}^{+,1}(\mathbb{T})\right)$ of $\tau^{\mathrm{W}}_{\mathcal{Y}^1(\mathcal{M}^{+,1}(\mathbb{T}))}$.*

In other words, $\tau_{\text{prob}}\left(\mathcal{Y}^1_{\text{dis}}\right)$ is the coarsest topology on $\mathcal{Y}^1_{\text{dis}}$ such that, for each bounded continuous function Φ on $\mathcal{M}^{+,1}(\mathbb{T})$ and each $A \in \mathcal{S}$, the mapping $\mu \mapsto \int_A \Phi(\mu_\omega)\, d\,\mathrm{P}(\omega)$ is continuous.

Proof. From Proposition 1.3.2, $\mathcal{M}^{+,1}(\mathbb{T})$ is cosmic regular, thus it is separable, hereditarily Lindelöf and completely regular. The result follows by applying Part 3 of Theorem 3.1.2 to the space $\mathcal{M}^{+,1}(\mathbb{T})$. □

Remark 3.1.7 If $\mathbb{T}$ is cosmic regular, we can thus compare $\tau^{\mathrm{W}}_{\mathcal{Y}^1}$ and $\tau_{\text{prob}}\left(\mathcal{Y}^1_{\text{dis}}\right)$ as follows. Let $(\mu^\alpha)_{\alpha\in\mathbb{A}}$ be a net in $\mathcal{Y}^1$ and let $\mu^\infty \in \mathcal{Y}^1$. If $f \in \mathrm{C}(\mathbb{T}, [0,1])$, we denote as usual by Φ_f the bounded continuous function $\nu \mapsto \nu(f)$ on $\mathcal{M}^{+,1}(\mathbb{T})$. We thus have

(3.1.6)

$$\mu^\alpha \xrightarrow{\text{prob}} \mu^\infty \Leftrightarrow \Big[\forall A \in \mathcal{S} \quad \forall \Phi \in \mathrm{C}\left(\mathcal{M}^{+,1}(\mathbb{T}), [0,1]\right) \quad \lim_\alpha \int_A \Phi(\mu^\alpha_\omega)\, d\,\mathrm{P}(\omega) = \int_A \Phi(\mu^\infty_\omega)\, d\,\mathrm{P}(\omega)\Big]$$

(and then, for each $\Phi \in \mathrm{C}\left(\mathcal{M}^{+,1}(\mathbb{T}), [0,1]\right)$, the latter convergence is uniform in $A \in \mathcal{S}$), whereas

(3.1.7)

$$\mu^\alpha \xrightarrow{\text{W-stably}} \mu^\infty \Leftrightarrow \Big[\forall A \in \mathcal{S} \quad \forall f \in \mathrm{C}(\mathbb{T}, [0,1]) \quad \lim_\alpha \int_A \Phi_f(\mu^\alpha_\omega)\, d\,\mathrm{P}(\omega) = \int_A \Phi_f(\mu^\infty_\omega)\, d\,\mathrm{P}(\omega)\Big].$$

The main difference between (3.1.6) and (3.1.7) is that, in (3.1.7), the test functions Φ_f act linearly on $\mathcal{M}^{+,1}(\mathbb{T})$.

To show that the Rademacher functions (or the degenerate Young measures associated with them) do not converge in probability (see Example 3.1.4), we can take $\mathbb{T} = \{-1, 1\}$ and consider a (bounded continuous) function Φ on $\mathcal{M}^{+,1}(\mathbb{T})$ such that $\Phi(\delta_{-1}) = \Phi(\delta_1) = 1$ and $\Phi\left(\frac{1}{2}\delta_{-1} + \frac{1}{2}\delta_1\right) = 0$.

Remark 3.1.8 We can consider $\mathcal{Y}^1_{\text{dis}}(\mathbb{T})$ as a subset of $\mathcal{Y}^1_{\text{dis}}\left(\mathcal{M}^{+,1}(\mathbb{T})\right)$ in two different ways.

1. In the first way, we identify $\mathbb{T}$ with the closed subset $\{\delta_t;\, t \in \mathbb{T}\}$ of $\mathcal{M}^{+,1}(\mathbb{T})$. If $\mu \in \mathcal{Y}^1(\mathbb{T})$, we associate with μ a Young measure $\widetilde{\mu} \in \mathcal{Y}^1\left(\mathcal{M}^{+,1}(\mathbb{T})\right)$, defined by
$$\widetilde{\mu}(A \times B) = \mu(A \times \{t \in \mathbb{T};\, \delta_t \in B\}) \sim \mu(A \times (B \cap \mathbb{T}))$$
for any $A \in \mathcal{S}$ and any $B \in \mathcal{B}_{\mathcal{M}^{+,1}(\mathbb{T})}$. The measure $\widetilde{\mu}$ is thus the image of μ by $(\omega, t) \mapsto (\omega, \delta_t)$. It is easy to check that the topology $\tau^{\mathrm{W}}_{\mathcal{Y}^1(\mathcal{M}^{+,1}(\mathbb{T}))}$ on $\{\widetilde{\mu};\, \mu \in \mathcal{Y}^1(\mathbb{T})\}$ coincides with the topology induced by the mapping $\mu \mapsto \widetilde{\mu}$, that is, the mapping $\mu \mapsto \widetilde{\mu}$ is an embedding of $\mathcal{Y}^1(\mathbb{T})$ in $\mathcal{Y}^1\left(\mathcal{M}^{+,1}(\mathbb{T})\right)$.

2. In the second way, we see each random probability $\mu \in \mathcal{Y}^1_{\text{dis}}(\mathbb{T})$ as a degenerate Young measure on $\mathcal{M}^{+,1}(\mathbb{T})$, that is, we identify $\mathcal{M}^{+,1}(\mathbb{T})$ with the closed subset $\{\delta_\mu;\, \mu \in \mathcal{M}^{+,1}(\mathbb{T})\}$ of $\mathcal{M}^{+,1}(\mathcal{M}^{+,1}(\mathbb{T}))$. Each $\mu \in \mathcal{Y}^1_{\text{dis}}(\mathbb{T})$ is identified with the disintegrable degenerate Young measure $\underline{\mu} = \underline{\delta}_\mu \in \mathcal{Y}^1\left(\mathcal{M}^{+,1}(\mathbb{T})\right)$, defined by
$$\underline{\mu}(A \times B) = \int_A \delta_{\mu_\omega}(B)\, d\,\mathrm{P}(\omega) = \mathrm{P}\left(A \cap \{\omega \in \Omega;\, \mu_\omega \in B\}\right).$$
Thus $\underline{\mu}$ is the image of P by the mapping $\omega \mapsto \mu_\omega$.

In Corollary 3.1.6, we chose the second way, that is, we identified $\mathcal{Y}^1_{\text{dis}}(\mathbb{T})$ with $\{\underline{\mu};\, \mu \in \mathcal{Y}^1(\mathbb{T})\}$. Assume that $\mathbb{T}$ is a cosmic regular space with more than one element and that $(\Omega, \mathcal{S}, \mathrm{P})$ has no atom. From Theorem 3.1.2, it follows that, under this identification, the topology $\tau^{\mathrm{W}}_{\mathcal{Y}^1(\mathbb{T})}$ is strictly coarser than the topology induced on $\mathcal{Y}^1(\mathbb{T})$ by $\tau^{\mathrm{W}}_{\mathcal{Y}^1(\mathcal{M}^{+,1}(\mathbb{T}))}$.

Here is a small complement to Theorem 3.1.2.

Proposition 3.1.9 (*–Stable convergence vs. convergence in probability on the subspace $\mathfrak{X}(\mathbb{T})$) *The topology induced on $\mathfrak{X}$ by $\tau^{\mathrm{M}}_{\mathcal{Y}^1}$ is finer than that of convergence in probability, which is finer than the topology induced by $\tau^{\mathrm{W}}_{\mathcal{Y}^1}$.*

Proof. We have already proved in Theorem 3.1.2 that the topology of convergence in probability on $\mathfrak{X}$ is finer than the topology induced by $\tau^{\mathrm{W}}_{\mathcal{Y}^1}$.

Let $(X_\alpha)_{\alpha\in\mathbb{A}}$ be a net of elements of $\mathfrak{X}$ which M–stably converges to an element X_∞ of $\mathfrak{X}$. Let $f : \mathbb{T} \to \mathbb{R}$ be a bounded continuous function. Let g be the integrand on $\Omega \times \mathbb{T}$ defined by

$$g(\omega, t) = |f(t) - f(X_\infty(\omega))| .$$

For each $\omega \in \Omega$, the function $g(\omega, .)$ is continuous. From the definition of M–stable convergence, we have

$$\lim_\alpha \int_\Omega g(\omega, X_\alpha(\omega))\, d\,\mathrm{P}(\omega) = \int_\Omega g(\omega, X_\infty(\omega))\, d\,\mathrm{P}(\omega) = 0,$$

thus $(X_\alpha)_{\alpha\in\mathbb{A}}$ converges in probability to X_∞. □

3.2 Parametrized Dudley distances

In this section, in the case when $\tau_{\mathbb{T}}$ is induced by an upwards filtering family $\mathcal{D}$ of semidistances, we provide a family of semidistances on $\mathcal{Y}^1$, indexed by $\mathcal{D}$, which induces $\tau_{\mathrm{prob}}\left(\mathcal{Y}^1\right)$ (see Theorems 3.2.1 and 3.2.3). We also continue the comparison of $\tau_{\mathrm{prob}}\left(\mathcal{Y}^1\right)$ with $\tau^{\mathrm{M}}_{\mathcal{Y}^1}$ or $\tau^{\mathrm{W}}_{\mathcal{Y}^1}$ (see Corollary 3.2.2 and Remark 3.2.4).

First, we fix some new notations. Let d be a continuous semidistance on $\mathbb{T}$. Recall (see page 33) that $\underline{\mathrm{BL}}_1(\Omega, \mathbb{T}, d)$ denotes the space of integrands f such that $f(\omega, .) \in \mathrm{BL}_1(\mathbb{T}, d)$ for all $\omega \in \Omega$ and that $\underline{\mathrm{BL}}'_1(\Omega, \mathbb{T}, d)$ denotes the set of elements f of $\underline{\mathrm{BL}}_1(\Omega, \mathbb{T}, d)$ which have the form $f = \sum_{i=1}^n \mathbb{1}_{A_i} \otimes g_i$, where $(A_i)_{1\le i\le n}$ is a measurable partition of Ω (which depends on f). We set, for all $\mu, \nu \in \mathcal{Y}^1_{\mathrm{dis}}$,

$$\underline{\Delta}^{(d)}_{\mathrm{BL}}(\mu, \nu) = \int_\Omega \Delta^{(d)}_{\mathrm{BL}}(\mu_\omega, \nu_\omega)\, d\,\mathrm{P}(\omega).$$

We call $\underline{\Delta}^{(d)}_{\mathrm{BL}}$ the *parametrized Dudley semidistance* associated with d.

Theorem 3.2.1 *Assume that $\mathbb{T}$ is completely regular. Let $\mathcal{D}$ be an upwards filtering set of continuous semidistances which induces the topology of $\mathbb{T}$. The topology $\tau_{\mathrm{prob}}\left(\mathcal{Y}^1_{\mathrm{dis}}\right)$ is induced by the family $\left(\underline{\Delta}^{(d)}_{\mathrm{BL}}\right)_{d\in\mathcal{D}}$ and we have, for each $d \in \mathcal{D}$ and all $\mu, \nu \in \mathcal{Y}^1_{\mathrm{dis}}$,*

$$(3.2.1) \quad \underline{\Delta}^{(d)}_{\mathrm{BL}}(\mu, \nu) = \sup_{f\in\underline{\mathrm{BL}}_1(\Omega,\mathbb{T},d)} (\mu(f) - \nu(f)) = \sup_{f\in\underline{\mathrm{BL}}'_1(\Omega,\mathbb{T},d)} (\mu(f) - \nu(f)) .$$

Furthermore, if d is a distance, then $\underline{\Delta}^{(d)}_{\mathrm{BL}}$ is a distance on $\mathcal{Y}^1_{\mathrm{dis}}$.

Proof. We know from Section 1.3 that the topology of $\mathcal{M}^{+,1}(\mathbb{T})$ is induced by the family $(\Delta^{(d)}_{\mathrm{BL}})_{d\in\mathcal{D}}$ of semidistances, and that, if d is a distance, then $\Delta^{(d)}_{\mathrm{BL}}$ is a distance on $\mathcal{M}^{+,1}(\mathbb{T})$. The corresponding results for $(\underline{\Delta}^{(d)}_{\mathrm{BL}})_{d\in\mathcal{D}}$ come from

Hoffmann–Jørgensen's characterization of convergence in probability (see page 54) and from the fact that we have, for each $d \in \mathcal{D}$ (with horrific notations !)

$$\underline{\underline{\Delta}}_{\mathrm{BL}}^{(d)} = \Delta_{\mathrm{prob}}^{(\Delta_{\mathrm{BL}}^{(d)})}.$$

Now, we have to prove (3.2.1). Let $d \in \mathcal{D}$. Let $\mu, \nu \in \mathcal{Y}_{\mathrm{dis}}^1$. We have

$$\begin{aligned}
\sup_{f \in \underline{\mathrm{BL}}_1'(\Omega,\mathbb{T},d)} (\mu(f) - \nu(f)) &\leq \sup_{f \in \underline{\mathrm{BL}}_1(\Omega,\mathbb{T},d)} (\mu(f) - \nu(f)) \\
&\leq \int_\Omega \sup_{f \in \underline{\mathrm{BL}}_1(\Omega,\mathbb{T},d)} (\mu_\omega(f(\omega,.)) - \nu_\omega(f(\omega,.)))\, d\mathrm{P}(\omega) \\
&\leq \int_\Omega \Delta_{\mathrm{BL}}^{(d)}(\mu_\omega, \nu_\omega)\, d\mathrm{P}(\omega).
\end{aligned}$$

There remains to prove the converse inequalities. By an obvious factorization to a quotient space, we can assume w.l.g. that d is a distance. Let $\varepsilon \in]0,1]$. The measures $\mu(\Omega \times .)$ and $\nu(\Omega \times .)$ are inner regular w.r.t. the totally bounded subsets of $\mathbb{T}$ (see Section 1.4), thus there exists a totally bounded subset K of $\mathbb{T}$ such that

$$\mu(\Omega \times K) \geq 1 - \varepsilon \quad \text{and} \quad \nu(\Omega \times K) \geq 1 - \varepsilon. \tag{3.2.2}$$

For any continuous function f on $\mathbb{T}$ and any $B \subset \mathbb{T}$, let us denote $\|f\|_B := \sup_{t \in B} |f(t)|$. The set of restrictions to K of elements of $\mathrm{BL}_1(\mathbb{T}, d)$ is totally bounded for $\|.\|_K$ (it is a subset of the compact space $\mathrm{BL}_1(\widehat{K}, d)$, where $\widehat{K}$ is the d–completion of K). There exist thus $h_1, \ldots, h_n \in \mathrm{BL}_1(\mathbb{T}, d)$ such that, for each $h \in \mathrm{BL}_1(\mathbb{T}, d)$, we have $\inf_{i=1,\ldots,n} \|h - h_i\|_K \leq \varepsilon$. For every $\omega \in \Omega$, there exists $N(\omega) \in \{1, \ldots, n\}$ such that

$$\mu_\omega(h_{N(\omega)} \mathbb{1}_K) - \nu_\omega(h_{N(\omega)} \mathbb{1}_K) \geq \sup_{h \in \mathrm{BL}_1(\mathbb{T},d)} (\mu_\omega(h \mathbb{1}_K) - \nu_\omega(h \mathbb{1}_K)) - 2\varepsilon.$$

Obviously, we can assume that N is measurable. We have, for every $\omega \in \Omega$,

$$\begin{aligned}
\Delta_{\mathrm{BL}}^{(d)}(\mu_\omega, \nu_\omega) &\leq \mu_\omega K^c + \nu_\omega K^c + \sup_{h \in \mathrm{BL}_1(\mathbb{T},d)} (\mu_\omega(h \mathbb{1}_K) - \nu_\omega(h \mathbb{1}_K)) \\
&\leq \mu_\omega K^c + \nu_\omega K^c + \mu_\omega(h_{N(\omega)} \mathbb{1}_K) - \nu_\omega(h_{N(\omega)} \mathbb{1}_K) + 2\varepsilon \\
&\leq 2\mu_\omega K^c + 2\nu_\omega K^c + \mu_\omega(h_{N(\omega)}) - \nu_\omega(h_{N(\omega)}) + 2\varepsilon. \qquad (3.2.3)
\end{aligned}$$

Using (3.2.2) and (3.2.3), we thus have

$$\begin{aligned}
\int_\Omega \Delta_{\mathrm{BL}}^{(d)}(\mu_\omega, \nu_\omega)\, d\mathrm{P}(\omega) &\leq 2\mu(\Omega \times K^c) + 2\nu(\Omega \times K^c) \\
&\quad + \int_\Omega \mu_\omega(h_{N(\omega)}) - \nu_\omega(h_{N(\omega)})\, d\mathrm{P}(\omega) + 2\varepsilon \\
&\leq \sup_{f \in \underline{\mathrm{BL}}_1'(\Omega,\mathbb{T},d)} (\mu(f) - \nu(f)) + 6\varepsilon
\end{aligned}$$

because the mapping $(\omega, t) \mapsto h_{N(\omega)}(t)$ is in $\underline{\mathrm{BL}}'_1(\Omega, \mathbb{T}, d)$. As ε is arbitrary, this shows that we have

$$\int_\Omega \Delta^{(d)}_{\mathrm{BL}}(\mu_\omega, \nu_\omega)\, d\,\mathrm{P}(\omega) = \sup_{f \in \underline{\mathrm{BL}}'_1(\Omega,\mathbb{T},d)} (\mu(f) - \nu(f))\,.$$

□

The following corollary extends the result of Part 3 of Theorem 3.1.2.

Corollary 3.2.2 (Convergence in probability implied by W–stable convergence) *Assume that $\mathbb{T}$ is completely regular. Let $(\mu^\alpha)_{\alpha \in \mathbb{A}}$ be a net in $\mathcal{Y}^1_{\mathrm{dis}}$ and let $X \in \mathfrak{X}$. Then $(\mu^\alpha)_{\alpha \in \mathbb{A}}$ W–stably converges to $\mu^\infty = \underline{\delta}_X$ if and only if $(\mu^\alpha)_{\alpha \in \mathbb{A}}$ converges in probability to $\underline{\delta}_X$.*

Proof. We know from Part 1 of Theorem 3.1.2 that, on $\mathcal{Y}^1_{\mathrm{dis}}$, convergence in probability implies W–stable convergence.

Assume that $(\mu^\alpha)_{\alpha \in \mathbb{A}}$ W–stably converges to $\mu^\infty = \underline{\delta}_X$. Let $\mathcal{D}$ be an upwards filtering set of continuous semidistances which induces the topology of $\mathbb{T}$. We can assume that each element d of $\mathcal{D}$ satisfies $d \le 1$. Let $d \in \mathcal{D}$.

For each $f \in \underline{\mathrm{BL}}_1(\Omega, \mathbb{T}, d)$, we have

$$\begin{aligned}\left|\int f(\omega,t)\, d(\mu^\alpha_\omega - \mu^\infty_\omega)(t)\, d\,\mathrm{P}(\omega)\right| &= \left|\int \left(\int f(\omega,t) - f(\omega, X(\omega))\, d\mu^\alpha_\omega(t)\right) d\,\mathrm{P}(\omega)\right| \\ &\le \int \left(\int |f(\omega,t) - f(\omega, X(\omega))|\, d\mu^\alpha_\omega(t)\right) d\,\mathrm{P}(\omega) \\ &\le \int d(t, X(\omega))\, d\mu^\alpha(\omega, t).\end{aligned}$$

Let $g \in \underline{\mathrm{BL}}_1(\Omega, \mathbb{T}, d)$ be defined by $g(\omega, t) = d(t, X(\omega))$. As $(\mu^\alpha)_{\alpha \in \mathbb{A}}$ W–stably converges to $\mu^\infty = \underline{\delta}_X$, we have, using Proposition 2.1.10,

$$\lim_\alpha \int g\, d\mu^\alpha = \int g\, d\mu^\infty = \int d(X, X)\, d\,\mathrm{P} = 0.$$

We thus have

$$\begin{aligned}&\sup_{f \in \underline{\mathrm{BL}}_1(\Omega,\mathbb{T},d)} \int f(\omega,t)\, d(\mu^\alpha_\omega - \mu^\infty_\omega)(t)\, d\,\mathrm{P}(\omega) \\ &\qquad\qquad \le \left|\int g(\omega,t)\, d(\mu^\alpha_\omega - \mu^\infty_\omega)(t)\, d\,\mathrm{P}(\omega)\right| \to 0.\end{aligned}$$

□

In the case when $\mathbb{T}$ is completely regular, Theorem 3.2.1 yields a natural extension of the topology $\tau_{\mathrm{prob}}\left(\mathcal{Y}^1_{\mathrm{dis}}\right)$ of convergence in probability to the whole space $\mathcal{Y}^1$, and the results of the Comparison Theorem 3.1.2 continue to apply to this extended topology.

Theorem 3.2.3 (Extended parametrized Dudley distances) *Assume that $\mathbb{T}$ is completely regular and let $\mathcal{D}$ be an upwards filtering set of continuous semidistances which induces the topology of $\mathbb{T}$. For each $d \in \mathcal{D}$ and for all $\mu, \nu \in \mathcal{Y}^1$, the last equality in* (3.2.1) *remains valid if μ or ν are not disintegrable, that is, we have*

$$\sup_{f \in \underline{\mathrm{BL}}_1(\Omega,\mathbb{T},d)} (\mu(f) - \nu(f)) = \sup_{f \in \underline{\mathrm{BL}}'_1(\Omega,\mathbb{T},d)} (\mu(f) - \nu(f)) .$$

Thus we can take (3.2.1) *as a definition of the semidistances $\underline{\Delta}_{\mathrm{BL}}^{(d)}$ on the whole space $\mathcal{Y}^1$. Furthermore, the extended topology of convergence in probability on $\mathcal{Y}^1$ (see page 56) is induced by the extended semidistances $\underline{\Delta}_{\mathrm{BL}}^{(d)}$ $(d \in \mathcal{D})$.*

In particular, the topology induced on $\mathcal{Y}^1$ by the extended semidistances $\underline{\Delta}_{\mathrm{BL}}^{(d)}$ $(d \in \mathcal{D})$ does not depend on $\mathcal{D}$.

Proof. Let $d \in \mathcal{D}$ and let $\mu, \nu \in \mathcal{Y}^1$. We have

$$\sup_{f \in \underline{\mathrm{BL}}_1(\Omega,\mathbb{T},d)} (\mu(f) - \nu(f)) \geq \sup_{f \in \underline{\mathrm{BL}}'_1(\Omega,\mathbb{T},d)} (\mu(f) - \nu(f)) .$$

Let us prove the converse inequality. Considering a quotient space, we can assume w.l.g. that d is a distance. Let $\mathbb{S}$ be the d–completion of $\mathbb{T}$. For simplicity of notations, we identify μ and ν with their unique extensions in $\mathcal{Y}^1(\mathbb{S})$, and we identify each Lipschitz function on $\mathbb{T}$ with its unique extension on $\mathbb{S}$ (otherwise, we could use notations similar to those of the proof of Corollary 2.1.8). Assuming that $\mathbb{T}$ contains a dense subspace with non–measurable cardinal, we can find a Polish subspace $\mathbb{S}_0$ of $\mathbb{S}$ such that $\mu(\Omega \times \mathbb{S}_0) = \nu(\Omega \times \mathbb{S}_0) = 1$. The Young measures μ and ν on $\mathbb{S}_0$ are disintegrable elements of $\mathcal{Y}^1(\mathbb{S}_0)$. We thus have, from Theorem 3.2.1,

$$\begin{aligned} \sup_{f \in \underline{\mathrm{BL}}_1(\Omega,\mathbb{T},d)} \mu(f) - \nu(f) &= \sup_{f \in \underline{\mathrm{BL}}_1(\Omega,\mathbb{S},d)} \mu(f) - \nu(f) \\ &\leq \sup_{f \in \underline{\mathrm{BL}}_1(\Omega,\mathbb{S}_0,d)} \mu(f) - \nu(f) \\ &= \sup_{f \in \underline{\mathrm{BL}}'_1(\Omega,\mathbb{S}_0,d)} \mu(f) - \nu(f). \end{aligned}$$

But, from a theorem of Kirszbraun and McShane, each Lispchitz function on $\mathbb{S}_0$ can be extended into a Lipschitz function on $\mathbb{S}$ with same Lipschitz modulus (see [Dud76, Theorem 7.3] or [Dud02, Theorem 6.1.1]). We thus have

$$\begin{aligned} \sup_{f \in \underline{\mathrm{BL}}_1(\Omega,\mathbb{T},d)} \mu(f) - \nu(f) &\leq \sup_{f \in \underline{\mathrm{BL}}'_1(\Omega,\mathbb{S},d)} \mu(f) - \nu(f) \\ &= \sup_{f \in \underline{\mathrm{BL}}'_1(\Omega,\mathbb{T},d)} \mu(f) - \nu(f). \end{aligned}$$

This proves that the last equality in (3.2.1) remains valid in $\mathcal{Y}^1$.

Now, let $\tau'_{\text{prob}}(\mathcal{Y}^1)$ be the topology induced by the semidistances $\underline{\Delta}_{\text{BL}}^{(d)}$ $(d \in \mathcal{D})$, and let us show that $\tau'_{\text{prob}}(\mathcal{Y}^1) = \tau_{\text{prob}}(\mathcal{Y}^1)$. If Δ is a semidistance on a space $\mathfrak{T}$ and $\mathfrak{t} \in \mathfrak{T}$ and $\varepsilon > 0$, let us denote by $B_\Delta(\mathfrak{t}, \varepsilon)$ the open ball for Δ with center $\mathfrak{t}$ and radius ε. With this notation, we only need to prove that

(*i*) each ball $B_{\underline{\Delta}_{\text{BL}}^{(d)}}(\mu, r)$ contains a finite intersection of balls $B_{\widetilde{\Delta}_{\text{prob}}^{(f_i)}}(\mu, \rho_i)$

(*ii*) each ball $B_{\widetilde{\Delta}_{\text{prob}}^{(f)}}(\mu, \rho)$ contains a finite intersection of balls $B_{\underline{\Delta}_{\text{BL}}^{(d_i)}}(\mu, r_i)$.

To prove (*i*), let $\mu \in \mathcal{Y}^1$, let $d \in \mathcal{D}$, and let $r > 0$. Assuming that $\mathbb{T}$ contains a dense subspace with non–measurable cardinal, we can find a finite subset $K = \{t_1, \ldots, t_n\}$ of $\mathbb{T}$ such that

$$\mu_\Omega\left(\cup_{t\in K} B_d\left(t, \frac{r}{8}\right)\right) \geq 1 - \frac{r}{16},$$

where $\mu_\Omega = \mu(\Omega \times .)$. For $i = 1, \ldots, n$, let

$$U_i = B_d\left(t_i, \frac{r}{8}\right) \quad \text{and} \quad V_i = B_d\left(t_i, \frac{r}{4}\right).$$

Let

$$U = \cup_{1\leq i\leq n} U_i \quad \text{and} \quad V = \cup_{1\leq i\leq n} V_i.$$

Let

$$V_0 = \mathbb{T} \setminus \left(\cup_{1\leq i\leq n} B_d\left(t_i, \frac{r}{8}\right]\right)$$

(where $B_d(t, \rho]$ denotes the closed d–ball with center t and radius ρ), and let $(f_i)_{0\leq i\leq n}$ be a partition of unity subordinated to the open cover $(V_i)_{0\leq i\leq n}$ (such a partition of unity can be constructed e.g. in the quotient space $\mathbb{T}/d$ and then lifted to $\mathbb{T}$). Now, let $\nu \in \mathcal{Y}^1$ satisfying

$$\widetilde{\Delta}_{\text{prob}}^{(f_0)}(\mu, \nu) \leq \frac{r}{16}$$

and

$$\forall i = 1, \ldots, n \quad \widetilde{\Delta}_{\text{prob}}^{(f_i)}(\mu, \nu) \leq \frac{r}{4n}.$$

First, observe that we have

$$\nu_\Omega(f_0) = (\nu_\Omega - \mu_\Omega)(f_0) + \mu_\Omega(f_0) \leq \frac{r}{16} + \mu_\Omega(U^c) \leq \frac{r}{8}. \tag{3.2.4}$$

Let $g \in \underline{\text{BL}}_1(\Omega, \mathbb{T}, d)$. We have, for $i = 1, \ldots, n$,

$$\forall \omega \in \Omega \quad \forall t \in V_i \quad |g(\omega, t) - g(\omega, t_i)| \leq \frac{r}{4}. \tag{3.2.5}$$

Let us make a small technical remark. Let $u : \Omega \to [-1, 1]$ and $h : \mathbb{T} \to [-1, 1]$ be two bounded measurable functions. Let λ be the measure $(\mu - \nu)(. \otimes h)$ on

$(\Omega, \mathcal{S})$. Let (λ^+, λ^-) be the Jordan decomposition of λ. There exist two sets S_1 and S_2 in $\mathcal{S}$ such that $S_1 \cap S_2 = \emptyset$, $S_1 \cup S_2 = \Omega$, and $\lambda^+(S_2) = 0 = \lambda^-(S_1)$. Let

$$A_h = (S_1 \cap \{u \geq 0\}) \cup (S_2 \cap \{u \leq 0\}).$$

We then have

(3.2.6)
$$(\mu - \nu)(u \otimes h) = \lambda(u) \leq \lambda(\mathbb{1}_{A_h}) = (\mu_{A_h} - \nu_{A_h})(h) \leq \sup_{A \in \mathcal{S}} |(\mu_A - \nu_A)(h)|.$$

Now, to avoid heavy notations, let us identify each f_i with the corresponding integrand $\mathbb{1}_\Omega \otimes f_i$. Similarly, let us identify each $g(.,t_i)$ with the integrand $(\omega, t) \mapsto g(\omega, t_i)$. Using (3.2.4), (3.2.5) and (3.2.6), we then get

$$\begin{aligned}
(\mu - \nu)(g) &= \sum_{i=0}^{n} (\mu - \nu)(g f_i) \\
&\leq \sum_{i=1}^{n} (\mu - \nu)(g f_i) + \mu_\Omega(f_0) + \nu_\Omega(f_0) \\
&\leq \sum_{i=1}^{n} (\mu - \nu)(g(.,t_i) f_i) + \sum_{i=1}^{n} (\mu - \nu)((g - g(.,t_i)) f_i) + \frac{r}{8} + \frac{r}{8} \\
&\leq \sum_{i=1}^{n} \sup_{A \in \mathcal{S}} |(\mu_A - \nu_A)(f_i)| + \sum_{i=1}^{n} \mu(|g - g(.,t_i)|\, f_i) \\
&\quad + \sum_{i=1}^{n} \nu(|g - g(.,t_i)|\, f_i) + \frac{r}{4} \\
&\leq n\frac{r}{4n} + \frac{r}{4}\mu(\sum_{i=1}^{n} f_i) + \frac{r}{4}\nu(\sum_{i=1}^{n} f_i) + \frac{r}{4} \\
&\leq \frac{r}{4} + \frac{r}{4} + \frac{r}{4} + \frac{r}{4} = r.
\end{aligned}$$

Setting $\rho_0 = \dfrac{r}{16}$ and $\rho_i = \dfrac{r}{4n}$ for $i = 1, \ldots, n$, this yields

$$\cap_{i=0}^{n} B_{\widetilde{\Delta}^{(f_i)}_{\text{prob}}}(\mu, \rho_i) \subset B_{\underline{\Delta}^{(d)}_{\text{BL}}}(\mu, r).$$

To prove (ii), let $\mu \in \mathcal{Y}^1$, let $f \in \mathrm{C}(\mathbb{T}, [0,1])$ and let $\rho > 0$. We can assume without loss of generality that $\rho < 1$. Some surgery on ρ will be necessary.

First Step For each $t \in \mathbb{T}$, as $\mathcal{D}$ is upwards filtering, we can find $\eta_t > 0$ and an element d_t of $\mathcal{D}$ such that

(3.2.7)
$$\forall s \in \mathbb{T} \quad d_t(s,t) < \eta_t \Rightarrow |f(s) - f(t)| < \rho/8.$$

Let us denote by $\partial^d B$ the boundary of a subset B of $\mathbb{T}$ w.r.t. a semidistance d. As the set of positive numbers r such that

$$\mu_\Omega \partial^{d_t} B_{d_t}(t,r) > 0$$

is at most countable, we can choose η_t such that

$$\mu_\Omega \partial^{d_t} B_{d_t}(t,\eta_t) = 0. \tag{3.2.8}$$

Let

$$U_t = B_{d_t}(t,\eta_t/2) \quad \text{and} \quad V_t = B_{d_t}(t,\eta_t).$$

We have $\mathbb{T} = \cup_{t\in\mathbb{T}} U_t$, thus, from τ–regularity of μ_Ω (assuming that $\mathbb{T}$ contains a dense subspace with non–measurable cardinal), we can find a finite subset $K=\{t_1,\ldots,t_n\}$ of $\mathbb{T}$ such that

$$\mu_\Omega(\cup_{t\in K} U_t) \geq 1 - \frac{\rho}{8}. \tag{3.2.9}$$

Let $\delta \in \mathcal{D}$ such that $\delta \geq d_t$ for every $t \in K$. Set

$$\eta = \min_{t\in K} \eta_t, \quad U = \cup_{t\in K} U_t \quad \text{and} \quad V = \cup_{t\in K} V_t.$$

We can find a function $h \in \mathrm{BL}_1(\mathbb{T},\delta)$ such that $h = \eta/2$ on U and $h = 0$ on V^c, e.g. take $h_\delta(s) = (\eta/2 - \delta(s,U))^+$. We have $\mathbb{1}_U \leq (2/\eta)h \leq \mathbb{1}_V$ thus, for any $\nu \in \mathcal{Y}^1$,

$$\begin{aligned}\nu \in B_{\underline{\Delta}_{\mathrm{BL}}^{(\delta)}}(\mu,\eta\rho/16) \Rightarrow \nu_\Omega(V) &\geq \frac{2}{\eta}\nu_\Omega(h) \geq \frac{2}{\eta}\mu_\Omega(h) - \frac{2}{\eta}\left|\nu_\Omega(h) - \mu_\Omega(h)\right| \\ &\geq \mu_\Omega(U) - \frac{2}{\eta}\underline{\Delta}_{\mathrm{BL}}^{(\delta)}(\mu,\nu) \\ &> 1 - \frac{\rho}{8} - \frac{2}{\eta}\frac{\eta\rho}{16} = 1 - \frac{\rho}{4}. \end{aligned} \tag{3.2.10}$$

Second Step We construct a partition $(V_i)_{1\leq i\leq n}$ of V by setting

$$\begin{aligned} V_1 &= V_{t_1}, \\ V_{i+1} &= V_{t_{i+1}} \setminus \cup_{j\leq i} V_{t_j} \quad (1 \leq i \leq n-1). \end{aligned}$$

For each $\varepsilon > 0$ and each $i = 1,\ldots,n$, let

$$V_i^\varepsilon = B_\delta(V_i,\varepsilon) \quad \text{and} \quad (V_i^c)^\varepsilon = B_\delta(V_i^c,\varepsilon).$$

From (3.2.8), each V_i has μ_Ω–negligible δ–boundary, thus $\mu_\Omega(V_i) = \inf_{\varepsilon>0}\mu_\Omega(V_i^\varepsilon)$. We can thus choose $\varepsilon \in]0,1[$ such that

$$\forall i = 1,\ldots,n \quad \mu_\Omega(V_i^\varepsilon \setminus V_i) \leq \frac{\rho}{8n} \quad \text{and} \quad \mu_\Omega((V_i^c)^\varepsilon \setminus V_i^c) \leq \frac{\rho}{8n}. \tag{3.2.11}$$

Let us now find a proper bound on $\underline{\Delta}_{\mathrm{BL}}^{(\delta)}(\nu,\mu)$ so as to make each $|(\mu_A - \nu_A)(V_i)|$ small enough.

For each $i = 1, \ldots, n$, let g_i and h_i in $\mathrm{BL}_1(\mathbb{T}, \delta)$ be defined by

$$\begin{aligned} g_i(t) &= (\varepsilon - \delta(t, V_i))_+ \\ h_i(t) &= \left(\varepsilon - \delta(t, \partial^\delta V_i)\right)_+ \end{aligned}$$

for all $t \in \mathbb{T}$. We have $0 \le h_i \le g_i \le \varepsilon$, $g_i = \varepsilon$ on V_i, $g_i = 0$ on $(V_i^\varepsilon)^c$ and $h_i = 0$ on $(V_i^\varepsilon)^c \cup ((V_i^c)^\varepsilon)^c$.

If $\nu \in \mathcal{Y}^1$ satisfies $\underline{\Delta}_{\mathrm{BL}}^{(\delta)}(\mu,\nu) < \varepsilon\rho/16n$, we have, for each $i = 1, \ldots, n$ and each $A \in \mathcal{S}$,

$$\left|(\mu_A - \nu_A)\left(\frac{g_i}{\varepsilon}\right)\right| < \frac{1}{\varepsilon}\frac{\varepsilon\rho}{16n} = \frac{\rho}{16n} \tag{3.2.12}$$

$$\left|(\mu_A - \nu_A)\left(\frac{h_i}{\varepsilon}\right)\right| < \frac{1}{\varepsilon}\frac{\varepsilon\rho}{16n} = \frac{\rho}{16n}. \tag{3.2.13}$$

Furthermore, under the same hypothesis, we also have, using (3.2.11) and (3.2.12),

$$\begin{aligned} (\nu_A - \mu_A)(V_i) &\le \nu_A\left(\frac{g_i}{\varepsilon}\right) - \mu_A(V_i) \\ &= (\nu_A - \mu_A)\left(\frac{g_i}{\varepsilon}\right) + \mu_A\left(\frac{g_i}{\varepsilon}\,\mathbb{1}_{V_i^c}\right) \\ &\le (\nu_A - \mu_A)\left(\frac{g_i}{\varepsilon}\right) + \mu_A\left(V_i^\varepsilon \setminus V_i\right) \\ &\le \frac{\rho}{16n} + \frac{\rho}{8n} \le \frac{\rho}{4n} \end{aligned}$$

and, using (3.2.11), (3.2.12) and (3.2.13),

$$\begin{aligned} (\mu_A - \nu_A)(V_i) &= (\mu_A - \nu_A)\left(\frac{g_i}{\varepsilon}\right) - (\mu_A - \nu_A)\left(\frac{g_i}{\varepsilon}\,\mathbb{1}_{V_i^c}\right) \\ &\le (\mu_A - \nu_A)\left(\frac{g_i}{\varepsilon}\right) + \nu_A\left(\frac{h_i}{\varepsilon}\right) - \mu_A\left(\frac{g_i}{\varepsilon}\,\mathbb{1}_{V_i^c}\right) \\ &= (\mu_A - \nu_A)\left(\frac{g_i}{\varepsilon}\right) + (\nu_A - \mu_A)\left(\frac{h_i}{\varepsilon}\right) + \mu_A\left(\frac{h_i}{\varepsilon} - \frac{g_i}{\varepsilon}\,\mathbb{1}_{V_i^c}\right) \\ &\le (\mu_A - \nu_A)\left(\frac{g_i}{\varepsilon}\right) + (\nu_A - \mu_A)\left(\frac{h_i}{\varepsilon}\right) + \mu_A\left((V_i^c)^\varepsilon \setminus V_i^c\right) \\ &\le \frac{\rho}{16n} + \frac{\rho}{16n} + \frac{\rho}{8n} = \frac{\rho}{4n}. \end{aligned}$$

If $\nu \in \mathcal{Y}^1$ satisfies $\underline{\Delta}_{\mathrm{BL}}^{(\delta)}(\mu,\nu) < \varepsilon\rho/16n$, we thus have, for each $i = 1, \ldots, n$ and each $A \in \mathcal{S}$,

$$|(\mu_A - \nu_A)(V_i)| \le \frac{\rho}{4n}. \tag{3.2.14}$$

Third Step We now gather the results of the preceding steps.

Let

$$\alpha = \min\{\eta\rho/16, \varepsilon\rho/16n\}.$$

Let $\nu \in B_{\underline{\Delta}_{\mathrm{BL}}^{(\delta)}}(\mu, \alpha)$. Using (3.2.7), (3.2.9), (3.2.10), and (3.2.14), we thus have, for any $A \in \mathcal{S}$,

$$\begin{aligned}
|\mu_A(f) - \nu_A(f)| &\leq \mu_\Omega(V^c) + \nu_\Omega(V^c) + |\mu_A(f\,\mathbb{1}_V) - \nu_A(f\,\mathbb{1}_V)| \\
&< \frac{\rho}{2} + \sum_{1 \leq i \leq n} |(\mu_A - \nu_A)(f\,\mathbb{1}_{V_i} - f(t_i)\,\mathbb{1}_{V_i})| \\
&\quad + \sum_{1 \leq i \leq n} f(t_i)\,|(\mu_A - \nu_A)(V_i)| \\
&\leq \frac{\rho}{2} + \frac{\rho}{8} \sum_{1 \leq i \leq n} (\mu_A + \nu_A)(V_i) + \sum_{1 \leq i \leq n} |(\mu_A - \nu_A)(V_i)| \\
&\leq \frac{\rho}{2} + \frac{\rho}{4} + \frac{\rho}{4}
\end{aligned}$$

that is,

$$\widetilde{\Delta}_{\mathrm{prob}}^{(f)}(\mu, \nu) \leq \rho.$$

□

Remark 3.2.4 (W–stable topology vs. convergence in probability) Assume that $\mathbb{T}$ is hereditarily Lindelöf and regular (thus completely regular), e.g. $\mathbb{T}$ is a regular cosmic space. Let $\mathcal{D}$ be an upwards filtering set of semidistances which induces the topology of $\mathbb{T}$. From Theorem 3.2.3, $\tau_{\mathrm{prob}}(\mathcal{Y}^1)$ appears to be the topology of uniform convergence on the sets $\underline{\mathrm{BL}}_1(\Omega, \mathbb{T}, d)$, $(d \in \mathcal{D})$, whereas, from Theorem 2.1.3 (see Condition 18) and Proposition 2.1.10, $\tau_{\mathcal{Y}^1}^{\mathrm{W}}$ is the topology of pointwise convergence on $\cup_{d \in \mathcal{D}} \underline{\mathrm{BL}}_1(\Omega, \mathbb{T}, d)$.

3.3 Fiber Product Lemma and applications

The results of this section are consequences of Theorem 3.2.1.

Let $\mathbb{S}$ and $\mathbb{T}$ be topological spaces and let $\mu \in \mathcal{Y}_{\mathrm{dis}}^1(\Omega, \mathbb{S})$ and $\nu \in \mathcal{Y}_{\mathrm{dis}}^1(\Omega, \mathbb{T})$. We call *fiber product* of μ and ν the measure $\mu \underline{\otimes} \nu \in \mathcal{Y}_{\mathrm{dis}}^1(\Omega, \mathbb{S} \times \mathbb{T})$ defined by

$$(\mu \underline{\otimes} \nu)_\omega = \mu_\omega \otimes \nu_\omega$$

for every $\omega \in \Omega$. Note that, in such a general setting, the measure $\mu \underline{\otimes} \nu$ may not be defined on $\mathcal{S} \otimes \mathcal{B}_{\mathbb{S} \times \mathbb{T}}$, because the inclusion $\mathcal{B}_{\mathbb{S}} \otimes \mathcal{B}_{\mathbb{T}} \subset \mathcal{B}_{\mathbb{S} \times \mathbb{T}}$ may be strict. We have already seen that $\mathcal{B}_{\mathbb{S}} \otimes \mathcal{B}_{\mathbb{T}} = \mathcal{B}_{\mathbb{S} \times \mathbb{T}}$ when $\mathbb{S} \times \mathbb{T}$ is hereditarily Lindelöf or when one of the spaces $\mathbb{S}$ and $\mathbb{T}$ has a countable network (see Theorem 2.1.13.C). It is also well-known that, if $\mathbb{S}$ or $\mathbb{T}$ is first countable, then every measure on $\mathcal{B}_{\mathbb{S}} \otimes \mathcal{B}_{\mathbb{T}}$

can be extended to a Borel measure on $\mathbb{S}\times\mathbb{T}$ (see [GP84, Proposition 7.10], slightly generalized in [BC93, Exercice 10.9, with solution]). If μ_ω and ν_ω are Radon, then $\mu_\omega \otimes \nu_\omega$ can be extended in a unique way to a Radon measure on $\mathcal{B}_{\mathbb{S}\times\mathbb{T}}$ [Sch73, Theorem 17 page 63].

The following theorem generalizes a classical result [Fis70, Bal88, Val90b, Val94, Tat02]. In these papers (except in [Tat02]), $(\nu^\alpha)_\alpha$ W–stably converges to a degenerate Young measure, but, from Corollary 3.2.2, this assumption is contained in Hypothesis (1*ii*) below.

Theorem 3.3.1 (Fiber product lemma)

1. *Let $\mathbb{S}$ and $\mathbb{T}$ be topological spaces such that $\mathbb{S}\times\mathbb{T}$ is hereditarily Lindelöf regular (e.g. $\mathbb{S}$ and $\mathbb{T}$ are regular cosmic spaces). Let $(\mu^\alpha)_{\alpha\in\mathbb{A}}$ be a net in $\mathcal{Y}^1_{\text{dis}}(\mathbb{S})$ and $(\nu^\alpha)_{\alpha\in\mathbb{A}}$ be a net in $\mathcal{Y}^1_{\text{dis}}(\mathbb{T})$ (with the same index set). Assume that*

   (*i*) *$(\mu^\alpha)_{\alpha\in\mathbb{A}}$ W–stably converges to $\mu^\infty \in \mathcal{Y}^1_{\text{dis}}(\mathbb{S})$,*

   (*ii*) *$(\nu^\alpha)_{\alpha\in\mathbb{A}}$ converges in probability to $\nu^\infty \in \mathcal{Y}^1_{\text{dis}}(\mathbb{T})$.*

   Then $(\mu^\alpha \underline{\otimes}\, \nu^\alpha)_{\alpha\in\mathbb{A}}$ W–stably converges to $\mu^\infty \underline{\otimes}\, \nu^\infty$.

2. *If furthermore $(\mu^\alpha)_\alpha$ converges in probability, then $(\mu^\alpha \underline{\otimes}\, \nu^\alpha)_\alpha$ converges in probability to $\mu^\infty \underline{\otimes}\, \nu^\infty$.*

It is not very much less general to assume in Theorem 3.3.1 that $\mathbb{S}$ and $\mathbb{T}$ are separable: Recall that it is not known whether there exist nonseparable hereditarily Lindelöf regular spaces (see Remark 1.1.3).

Proof of Theorem 3.3.1.

1. From Part E of the Portmanteau Theorem 2.1.3, the first part of Theorem 3.3.1 only needs to be proved in the case when $\mathbb{S}$ and $\mathbb{T}$ are metrizable spaces (assuming that $\mathbb{S}\times\mathbb{T}$ contains a subspace with non–measurable cardinal). Let $d_\mathbb{S}$ and $d_\mathbb{T}$ be distances which are compatible with the respective topologies of $\mathbb{S}$ and $\mathbb{T}$. For all $(s,t),(s',t') \in \mathbb{S}\times\mathbb{T}$, set

$$d((s,t),(s',t')) = \max\{d_\mathbb{S}(s,s'), d_\mathbb{T}(t,t')\}.$$

Let $A \in \mathcal{S}$ and let $f : \mathbb{S}\times\mathbb{T} \to [0,1]$ be an element of $\mathrm{BL}_1(\mathbb{S}\times\mathbb{T}, d)$. For each $\alpha \in \mathbb{A}\cup\{\infty\}$, each $\omega\in\Omega$ and each $t\in\mathbb{T}$, let

$$g^\alpha(\omega,t) = \mathbb{1}_A(\omega)\int_\mathbb{S} f(s,t)\,d\mu^\alpha_\omega(s).$$

Then $g^\alpha \in \underline{\mathrm{BL}}_1(\Omega,\mathbb{T},d_\mathbb{T})$, thus, from (1*ii*) and Theorem 3.2.1,

$$\limsup_\alpha \sup_{\beta\in\mathbb{A}} \left|\int g^\beta\, d(\nu^\alpha - \nu^\infty)\right| = 0.$$

In particular, we have

$$\lim_\alpha \int \mathbb{1}_A(\omega) f(s,t)\, d\mu^\alpha_\omega(s)\, d(\nu^\alpha_\omega - \nu^\infty_\omega)(t)\, d\,\mathrm{P}(\omega) = \lim_\alpha \int g^\alpha\, d(\nu^\alpha - \nu^\infty) = 0. \tag{3.3.1}$$

Set $h(\omega, s) = \mathbb{1}_A(\omega) \int f(s,t)\, d\nu^\infty_\omega(t)$ for all $(\omega, s) \in \Omega\times\mathbb{S}$. Then h is in $\underline{\mathrm{BL}}_1(\Omega, \mathbb{S}, d_{\mathbb{S}})$, thus, from $(1i)$ and Proposition 2.1.10, we also have

$$\begin{aligned}\lim_\alpha \int \mathbb{1}_A(\omega) f(s,t)\, d\mu^\alpha_\omega(s)\, d\nu^\infty_\omega(t)\, d\,\mathrm{P}(\omega) &= \lim_\alpha \int h\, d\mu^\alpha = \int h\, d\mu^\infty \\ &= \int \mathbb{1}_A(\omega) f(s,t)\, d\,(\mu^\infty \underline{\otimes}\, \nu^\infty)\,(\omega, s, t).\end{aligned} \tag{3.3.2}$$

Using (3.3.1) and (3.3.2), we immediately get

$$\begin{aligned}\lim_\alpha &\int \mathbb{1}_A(\omega) f(s,t)\, d\mu^\alpha \underline{\otimes}\, \nu^\alpha(\omega, s, t) \\ &= \lim_\alpha \int \mathbb{1}_A(\omega) f(s,t)\, d\mu^\alpha_\omega(s)\, d(\nu^\alpha_\omega - \nu^\infty_\omega)(t)\, d\,\mathrm{P}(\omega) \\ &\quad + \lim_\alpha \int \mathbb{1}_A(\omega) f(s,t)\, d\mu^\alpha_\omega(s)\, d\nu^\infty_\omega(t)\, d\,\mathrm{P}(\omega) \\ &= \int \mathbb{1}_A(\omega) f(s,t)\, d(\mu^\infty \underline{\otimes}\, \nu^\infty)(\omega, s, t),\end{aligned}$$

which proves that $(\mu^\alpha \underline{\otimes}\, \nu^\alpha)_{\alpha\in\mathbb{A}}$ W–stably converges to $\mu^\infty \underline{\otimes}\, \nu^\infty$.

2. Assume now that $(\mu^\alpha)_\alpha$ converges in probability to μ^∞. To prove the second part of Theorem 3.3.1, using Theorem 3.2.1, we can again assume without loss of generality that $\mathbb{S}$ and $\mathbb{T}$ are metrizable spaces such that $\mathbb{S}\times\mathbb{T}$ contains a subspace with non–measurable cardinal. Let $f \in \underline{\mathrm{BL}}_1(\Omega, \mathbb{S}\times\mathbb{T}, d)$, where the distance d on $\mathbb{S}\times\mathbb{T}$ is defined in the same way as in the the proof of Part 1. Set

$$g^\alpha(\omega, t) = \int_{\mathbb{S}} f(\omega, s, t)\, d\mu^\alpha_\omega(s)$$

for all $\alpha \in \mathbb{A}\cup\{\infty\}$, $\omega\in\Omega$ and $t \in \mathbb{T}$. We have $g^\alpha \in \underline{\mathrm{BL}}_1(\Omega, \mathbb{T}, d_{\mathbb{T}})$, thus, for any $\alpha \in \mathbb{A}$,

$$\left|(\nu^\alpha - \nu^\infty)\,(g^\alpha)\right| \le \underline{\Delta}^{(d_{\mathbb{T}})}_{\mathrm{BL}}(\nu^\alpha, \nu^\infty). \tag{3.3.3}$$

Set $h(\omega, s) = \int f(\omega, s, t)\, d\nu^\infty_\omega(t)$ for all $(\omega, s) \in \Omega\times\mathbb{S}$. We have $h \in \underline{\mathrm{BL}}_1(\Omega, \mathbb{S}, d_{\mathbb{S}})$ and

$$\left|(\mu^\alpha - \mu^\infty)\,(h)\right| \le \underline{\Delta}^{(d_{\mathbb{S}})}_{\mathrm{BL}}(\mu^\alpha, \mu^\infty). \tag{3.3.4}$$

But we have $(\mu^\alpha \underline{\otimes}\, \nu^\alpha - \mu^\infty \underline{\otimes}\, \nu^\infty)(f) = (\nu^\alpha - \nu^\infty)(g^\alpha) + (\mu^\alpha - \mu^\infty)(h)$. From (3.3.3) and (3.3.4), we thus obtain

$$\begin{aligned}|(\mu^\alpha \underline{\otimes}\, \nu^\alpha - \mu^\infty \underline{\otimes}\, \nu^\infty)(f)| &\leq |(\nu^\alpha - \nu^\infty)(g^\alpha)| + |(\mu^\alpha - \mu^\infty)(h)| \\ &\leq \underline{\Delta}_{\mathrm{BL}}^{(d_{\mathbb{T}})}(\nu^\alpha, \nu^\infty) + \underline{\Delta}_{\mathrm{BL}}^{(d_{\mathbb{S}})}(\mu^\alpha, \mu^\infty).\end{aligned}$$

This is valid for any $f \in \underline{\mathrm{BL}}_1(\Omega, \mathbb{S} \times \mathbb{T}, d)$, thus we have

$$\underline{\Delta}_{\mathrm{BL}}^{(d)}(\mu^\alpha \underline{\otimes}\, \nu^\alpha, \mu^\infty \underline{\otimes}\, \nu^\infty) \leq \underline{\Delta}_{\mathrm{BL}}^{(d_{\mathbb{T}})}(\nu^\alpha, \nu^\infty) + \underline{\Delta}_{\mathrm{BL}}^{(d_{\mathbb{S}})}(\mu^\alpha, \mu^\infty).$$

□

Remark 3.3.2 The hypothesis that $\mathbb{S} \times \mathbb{T}$ be hereditarily Lindelöf regular is used for two reasons: firstly, to ensure that the fiber products $\mu^\alpha \underline{\otimes}\, \nu^\alpha$ are Young measures, secondly, to use Part E of the Portmanteau Theorem 2.1.3, which reduces the proof of W–stable convergence in $\mathbb{S} \times \mathbb{T}$ to the proof of W–stable convergence w.r.t. each product semidistance.

It is possible to replace this hypothesis by assuming that $\mathbb{S}$ and $\mathbb{T}$ are metrizable spaces and that there exist separable subspaces $\mathbb{S}_0$ of $\mathbb{S}$ and $\mathbb{T}_0$ of $\mathbb{T}$ such that $\mu^\infty(\Omega \times \mathbb{S}_0) = \nu^\infty(\Omega \times \mathbb{T}_0) = 1$. The reasoning is the same and we skip it.

Counterexample 3.3.3 ([Val94]) The convergence in probability in the hypothesis $(1ii)$ of Theorem 3.3.1 cannot be weakened to W–stable convergence. For example, let $\mu^n = \underline{\delta}_{\mathfrak{r}_n}$ and $\nu^n = \underline{\delta}_{\mathfrak{r}_n}$ be degenerate Young measures associated with the sequence of Rademacher variables with values in $\{-1, 1\}$. Then

$$\mu^\infty \underline{\otimes}\, \nu^\infty = \frac{1}{4}(\underline{\delta}_{-1} + \underline{\delta}_1) \underline{\otimes} (\underline{\delta}_{-1} + \underline{\delta}_1) = \frac{1}{4}(\underline{\delta}_{(-1,-1)} + \underline{\delta}_{(-1,1)} + \underline{\delta}_{(1,-1)} + \underline{\delta}_{(1,1)}),$$

but $(\mu^n \underline{\otimes}\, \nu^n)_n$ converges to $\frac{1}{2}(\underline{\delta}_{(-1,-1)} + \underline{\delta}_{(1,1)})$.

Counterexample 3.3.4 The hypothesis $(1i)$ and the conclusion of Part 1 of Theorem 3.3.1 cannot be weakened into $\mathcal{U}$–$*$–stable convergence for any sub–σ–algebra $\mathcal{U}$ of $\mathcal{S}$. Consequences of this phenomenon will appear in Chapter 9, see Remark 9.4.9.

For example, assume that $(1i)$ and $(1ii)$ hold for the $\mathcal{U}$–$*$–stable convergence, with $\mathcal{U} = \{\emptyset, \Omega\}$. Assume furthermore that $\Omega = \mathbb{T} = [0, 1]$ and that, for each $\omega \in [0, 1]$, we have

$$\mu_\omega^\infty \otimes \nu_\omega^\infty = \begin{cases} \delta_0 \otimes \delta_0 \text{ if } \omega \leq 1/2, \\ \delta_1 \otimes \delta_1 \text{ if } \omega > 1/2. \end{cases}$$

Take for P the Lebesgue measure on $[0, 1]$ and let μ' be defined by

$$\mu'_\omega = \begin{cases} \delta_1 \text{ if } \omega \leq 1/2, \\ \delta_0 \text{ if } \omega > 1/2. \end{cases}$$

We have $\mu'(\Omega \times .) = \mu^\infty(\Omega \times .)$, thus the net $(\mu^\alpha)_\alpha$ $\mathcal{U}$–stably converges to μ' (the limit in $\mathcal{U}$–stable convergence is not unique). But $\mu' \underline{\otimes} \nu^\infty(\Omega\times.) \neq \mu^\infty \underline{\otimes} \nu^\infty(\Omega\times.)$ and thus $(\mu^\alpha \underline{\otimes} \nu^\alpha)_{\alpha\in\mathbb{A}}$ does not $\mathcal{U}$–stably converge to $\mu^\infty \underline{\otimes} \nu^\infty$. Indeed, let $f : [0,1] \times [0,1] \to [0,1]$ be a continuous function such that $f(0,0) = f(1,1) = 0$ and $f(1,0) = f(0,1) = 1$. We have

$$\begin{aligned}\mu^\infty \underline{\otimes} \nu^\infty(\mathbb{1}_\Omega \otimes f) &= \int f(s,t)\, d\mu^\infty_\omega(s)\, d\nu^\infty_\omega(t)\, d\mathrm{P}(\omega)\\ &= \frac{1}{2} f(0,0) + \frac{1}{2} f(1,1) = 0,\end{aligned}$$

whereas

$$\mu' \underline{\otimes} \nu^\infty(\mathbb{1}_\Omega \otimes f) = \frac{1}{2} f(1,0) + \frac{1}{2} f(0,1) = 1.$$

Typical applications of Theorem 3.3.1 make use of the following corollary. In view of these applications, this corollary is given for nets of functions, but it can be extended without further difficulties to nets of Young measures. The first apparitions of similar results seem to be [Mog66], [Fis70, Théorème 5] and [Fis71] ([Mog66] only considered a special case of stable convergence, called *Rényi–mixing*, see the definition in Chapter 9).

Corollary 3.3.5 *Let $\mathbb{S}_1$ and $\mathbb{S}_2$ be topological spaces such that $\mathbb{S}_1 \times \mathbb{S}_2$ is hereditarily Lindelöf regular (e.g. $\mathbb{S}_1$ and $\mathbb{S}_2$ are regular cosmic spaces). Let Φ be a continuous mapping from $\mathbb{S}_1 \times \mathbb{S}_2$ to the topological space $\mathbb{T}$. Let $(X_\alpha)_{\alpha\in\mathbb{A}}$ be a net in $\mathfrak{X}(\mathbb{S}_1)$ and $(Y_\alpha)_{\alpha\in\mathbb{A}}$ be a net in $\mathfrak{X}(\mathbb{S}_2)$ (with same index set). For every $\alpha \in \mathbb{A}$, let $Z_\alpha = \Phi(X_\alpha, Y_\alpha)$. Let $\underline{\Phi} : \Omega \times \mathbb{S}_1 \times \mathbb{S}_2 \to \Omega \times \mathbb{T}$ be defined by $\underline{\Phi}(\omega, s_1, s_2) = (\omega, \Phi(s_1, s_2))$. Assume that*

(i) $(\underline{\delta}_{X_\alpha})_{\alpha\in\mathbb{A}}$ W–stably converges to $\mu^\infty \in \mathcal{Y}^1_{\text{dis}}(\mathbb{S}_1)$,

(ii) $(Y_\alpha)_{\alpha\in\mathbb{A}}$ converges in probability to $Y_\infty \in \mathcal{Y}^1_{\text{dis}}(\mathbb{S}_2)$.

Let $\lambda^\infty = \underline{\Phi}_\sharp\left(\mu^\infty \underline{\otimes}\, \underline{\delta}_{Y_\infty}\right)$. Then $(\underline{\delta}_{Z_\alpha})_{\alpha\in\mathbb{A}}$ W–stably converges to λ^∞.

The proof of Corollary 3.3.5 is obvious in view of Theorem 3.3.1 and the following very easy lemma.

Lemma 3.3.6 *Let $\mathbb{S}$ and $\mathbb{T}$ be topological spaces. Let $\Phi : \mathbb{S} \to \mathbb{T}$ be continuous. Let $(\mu^\alpha)_{\alpha\in\mathbb{A}}$ be a net in $\mathcal{Y}^1(\mathbb{S})$ which W–stably converges to $\mu^\infty \in \mathcal{Y}^1(\mathbb{S})$. Let $\underline{\Phi} : (\omega, s) \mapsto (\omega, \Phi(s))$ and, for each $\mu \in \mathcal{Y}^1(\mathbb{S})$, let $\widetilde{\mu} = (\underline{\Phi})_\sharp(\mu)$. Then $(\widetilde{\mu}^\alpha)_{\alpha\in\mathbb{A}}$ W–stably converges to $\widetilde{\mu}^\infty$ in $\mathcal{Y}^1(\mathbb{T})$.*

Proof. Let $A \in \mathcal{S}$ and let $g : \mathbb{T} \to [0,1]$ be bounded continuous. We have

$$\begin{aligned}\widetilde{\mu}^\infty(\mathbb{1}_A \otimes g) &= \mu^\infty((\mathbb{1}_A \otimes g) \circ \underline{\Phi}) = \mu^\infty(\mathbb{1}_A \otimes (g \circ \Phi))\\ &= \lim_\alpha \mu^\alpha(\mathbb{1}_A \otimes (g \circ \Phi)) = \lim_\alpha \mu^\alpha((\mathbb{1}_A \otimes g) \circ \underline{\Phi})\\ &= \lim_\alpha \widetilde{\mu}^\alpha(\mathbb{1}_A \otimes g).\end{aligned}$$

□

3.4 Parametrized Lévy–Wasserstein distances and $L^1_{\mathbb{E}}$ spaces

We define and study new semidistances on the space of integrable Young measures, in the same spirit as in the definition of the parametrized Dudley semidistances, that is, uniform convergence on an appropriate set of integrands. The difference between the topology constructed in this way and the topology $\tau^{\mathrm{W}}_{1,\mathcal{D}}$ defined in Section 2.4 is similar to that between $\tau_{\mathrm{prob}}\left(\mathcal{Y}^1_{\mathrm{dis}}\right)$ and $\tau^{\mathrm{W}}_{\mathcal{Y}^1}$.

Assume that $\mathbb{T}$ is regular and let $\mathcal{D}$ be a set of continuous semidistances on $\mathbb{T}$. Let $\mu \in \mathcal{Y}^1$. Recall that μ is *integrable* relatively to $\mathcal{D}$ if, for all $d \in \mathcal{D}$ and for some (equivalently, for all) $t_0 \in \mathbb{T}$, we have $\mu(d(t_0, .)) < +\infty$. We denote by $\mathcal{Y}^{1,1}_{\mathrm{dis},\mathcal{D}}$ the space of integrable (relatively to $\mathcal{D}$) *disintegrable* Young measures.

Let $d \in \mathcal{D}$. Recall that $\mathrm{Lip}_1(d)$ is the set of d–Lipschitz functions on $\mathbb{T}$, with Lipschitz modulus not greater than 1. Let $\underline{\mathrm{Lip}}_1(d)$ be the set of integrands f such that $f(\omega, .) \in \mathrm{Lip}_1(d)$ for each $\omega \in \Omega$ and such that $f(., t_0)$ is integrable for some (or any) $t_0 \in \mathbb{T}$. We define the semidistance $\underline{\Delta}^{(d)}_{\mathrm{LW}}$ on $\mathcal{Y}^{1,1}_{\mathrm{dis},\mathcal{D}}$ by

$$\underline{\Delta}^{(d)}_{\mathrm{LW}}(\mu, \nu) = \sup_{f \in \underline{\mathrm{Lip}}_1(d)} \left(\mu(f) - \nu(f)\right).$$

If d is a distance, then $\underline{\Delta}^{(d)}_{\mathrm{LW}}$ is a distance on $\mathcal{Y}^{1,1}_{\mathrm{dis},\mathcal{D}}$.

We denote by $\tau_{1,\mathcal{D}}$ the topology on $\mathcal{Y}^{1,1}_{\mathrm{dis},\mathcal{D}}$ defined by the semidistances $\underline{\Delta}^{(d)}_{\mathrm{LW}}$, $d \in \mathcal{D}$. As $\underline{\mathrm{BL}}_1(\Omega, \mathbb{T}, d) \subset \underline{\mathrm{Lip}}_1(d)$ for every $d \in \mathcal{D}$, the topology $\tau_{1,\mathcal{D}}$ is finer than the topology induced on $\mathcal{Y}^{1,1}_{\mathrm{dis},\mathcal{D}}$ by the topology of convergence in probability $\tau_{\mathrm{prob}}\left(\mathcal{Y}^1_{\mathrm{dis}}\right)$.

Now, let t_0 be some arbitrarily fixed element of $\mathbb{T}$ and let $\widetilde{\underline{\mathrm{Lip}}}_1(d)$ be the set of elements f of $\underline{\mathrm{Lip}}_1(d)$ such that $f(\omega, t_0) = 0$ for every $\omega \in \Omega$. If $f \in \underline{\mathrm{Lip}}_1(d)$, then the integrand $\widetilde{f}$ defined by $\widetilde{f}(\omega, t) = f(\omega, t) - f(\omega, t_0)$ is in $\widetilde{\underline{\mathrm{Lip}}}_1(d)$, and $\mu(f) - \nu(f) = \mu(\widetilde{f}) - \nu(\widetilde{f})$ for all $\mu, \nu \in \mathcal{Y}^{1,1}_{\mathrm{dis},d}$, because μ and ν have the same margin on Ω. Thus we have also

$$\underline{\Delta}^{(d)}_{\mathrm{LW}}(\mu, \nu) = \sup_{f \in \widetilde{\underline{\mathrm{Lip}}}_1(d)} \left(\mu(f) - \nu(f)\right). \tag{3.4.1}$$

Similarly, let $\widetilde{\underline{\mathrm{BL}}}_1(\Omega, \mathbb{T}, d)$ be the set of elements f of $\underline{\mathrm{BL}}_1(\Omega, \mathbb{T}, d)$ such that $f(\omega, t_0) = 0$ for every $\omega \in \Omega$. We have, for any $\mu, \nu \in \mathcal{Y}^1$,

$$\underline{\Delta}^{(d)}_{\mathrm{BL}}(\mu, \nu) = \sup_{f \in \widetilde{\underline{\mathrm{BL}}}_1(\Omega, \mathbb{T}, d)} \left(\mu(f) - \nu(f)\right). \tag{3.4.2}$$

For each $f \in \widetilde{\underline{\mathrm{Lip}}}_1(d \wedge 1)$, we have $|f(\omega, t)| = |f(\omega, t) - f(\omega, t_0)| \le 1$ for all $(\omega, t) \in \Omega \times \mathbb{T}$, thus $f \in \widetilde{\underline{\mathrm{BL}}}_1(\Omega, \mathbb{T}, d)$. But, for each $f \in \widetilde{\underline{\mathrm{BL}}}_1(\Omega, \mathbb{T}, d)$, we have, for all

$(\omega, t_1, t_2) \in \Omega \times \mathbb{T} \times \mathbb{T}$, $|f(\omega, t_1) - f(\omega, t_2)| \leq 2$, thus $f \in \widetilde{\underline{\mathrm{Lip}}_1}(d \wedge 2)$. We thus have the inclusions

$$\widetilde{\underline{\mathrm{Lip}}_1}(d \wedge 1) \subset \widetilde{\underline{\mathrm{BL}}_1}(\Omega, \mathbb{T}, d) \subset 2\widetilde{\underline{\mathrm{Lip}}_1}(d \wedge 1),$$

which imply, using (3.4.1) and (3.4.2),

$$\underline{\Delta}_{\mathrm{LW}}^{(d \wedge 1)} \leq \underline{\Delta}_{\mathrm{BL}}^{(d)} \leq 2\underline{\Delta}_{\mathrm{LW}}^{(d \wedge 1)}. \tag{3.4.3}$$

Thus the topology $\tau_{\mathrm{prob}}\left(\mathcal{Y}^1_{\mathrm{dis}}\right)$ on $\mathcal{Y}^1$ is induced by the semidistances $\underline{\Delta}_{\mathrm{LW}}^{(d \wedge 1)}$, $d \in \mathcal{D}$.

Kantorovich–Rubinštein Theorem For any $\mu, \nu \in \mathcal{M}^{+,1}(\mathbb{T})$, let $D(\mu, \nu)$ be the set of probability laws π on $\mathbb{T} \times \mathbb{T}$ with margins μ and ν. The set $D(\mu, \nu)$ is closed for the narrow topology on $\mathcal{M}^{+,1}(\mathbb{T} \times \mathbb{T})$. From a general result of Kawabe [Kaw94], if $\mathbb{T}$ is regular and if μ and ν are τ–regular, $D(\mu, \nu)$ is compact (if μ and ν are tight, this is obvious because then $D(\mu, \nu)$ is tight). In particular, if $\mathbb{T}$ is a regular hereditarily Lindelöf space, $D(\mu, \nu)$ is compact for all $\mu, \nu \in \mathcal{M}^{+,1}(\mathbb{T})$.

Recall the Kantorovich–Rubinštein Theorem (see e.g. [Dud02, Rac91, RR98]): If $\mathbb{T}$ is a separable metric space, if d is a distance which is compatible with $\tau_{\mathbb{T}}$, and if $\int d(t_0, .)\, d\mu < +\infty$ and $\int d(t_0, .)\, d\nu < +\infty$, we have

$$\sup_{f \in \mathrm{Lip}_1(d)} (\mu(f) - \nu(f)) = \inf_{\pi \in D(\mu,\nu)} \int_{\mathbb{T} \times \mathbb{T}} d(t, t')\, d\pi(t, t').$$

Furthermore, as $\mathbb{T}$ is regular hereditarily Lindelöf, an easy compactness argument shows that the infimum in the right hand side is attained.

For any $\mu, \nu \in \mathcal{Y}^1$, let $\underline{D}(\mu, \nu)$ be the set of probability laws π on $\Omega \times \mathbb{T} \times \mathbb{T}$ such that $\pi(. \times . \times \mathbb{T}) = \mu$ and $\pi(. \times \mathbb{T} \times .) = \nu$. Set

$$\underline{\Delta}_{\mathrm{KR}}^{(d)}(\mu, \nu) = \inf_{\pi \in \underline{D}(\mu,\nu)} \int_{\Omega \times \mathbb{T} \times \mathbb{T}} d(t, t')\, d\pi(\omega, t, t').$$

Theorem 3.4.1 (Parametrized Kantorovich–Rubinštein Theorem) *Assume that $\mathbb{T}$ is Suslin regular. Let d be a continuous semidistance on $\mathbb{T}$. For any $\mu, \nu \in \mathcal{Y}^{1,1}_{\mathrm{dis},d}$, we have*

$$\underline{\Delta}_{\mathrm{LW}}^{(d)}(\mu, \nu) = \underline{\Delta}_{\mathrm{KR}}^{(d)}(\mu, \nu).$$

Furthermore, the infimum in the definition of $\underline{\Delta}_{\mathrm{KR}}^{(d)}(\mu, \nu)$ is attained, that is, there exists $\pi \in \underline{D}(\mu, \nu)$ such that $\underline{\Delta}_{\mathrm{KR}}^{(d)}(\mu, \nu) = \int_{\Omega \times \mathbb{T} \times \mathbb{T}} d(t, t')\, d\pi(\omega, t, t')$.

Let us first prove the following lemma, which is interesting in itself.

Lemma 3.4.2 *Let d be a continuous distance on $\mathbb{T}$. Assume that $(\mathbb{T}, d)$ is separable. Let $\mathcal{B}^*$ be the universal completion of the σ–algebra $\mathcal{B}_{\mathcal{M}^{+,1}(\mathbb{T}) \times \mathcal{M}^{+,1}(\mathbb{T})}$. For any $\mu, \nu \in \mathcal{M}^{+,1}(\mathbb{T})$, let*

$$r(\mu, \nu) = \inf_{\pi \in D(\mu,\nu)} \int d(t, t')\, d\pi(t, t') \in [0, +\infty].$$

The function r is $\mathcal{B}^*$-measurable. Furthermore, the multifunction

$$K : \begin{cases} \mathcal{M}^{+,1}(\mathbb{T}) \times \mathcal{M}^{+,1}(\mathbb{T}) & \to \mathcal{K}(\mathcal{M}^{+,1}(\mathbb{T} \times \mathbb{T})) \\ (\mu, \nu) & \mapsto \{\pi \in D(\mu, \nu);\ \int d(t, t')\, d\pi(t, t') = r(\mu, \nu)\} \end{cases}$$

has a $\mathcal{B}^*$-measurable selection.

Proof. First, we can assume w.l.g. that $\mathbb{T}$ is Polish. Indeed, let $\mathbb{S}$ be the d–completion of $\mathbb{T}$. For each $\mu \in \mathcal{M}^{+,1}(\mathbb{T})$ (resp. $\mu \in \mathcal{M}^{+,1}(\mathbb{T} \times \mathbb{T})$), let us denote by $\widehat{\mu}$ the law on $\mathbb{S}$ (resp. $\mathbb{S} \times \mathbb{S}$) defined by $\widehat{\mu}(B) = \mu(B \cap \mathbb{T})$ for all $B \in \mathcal{B}_{\mathbb{S}}$ (resp. $\widehat{\mu}(B) = \mu(B \cap (\mathbb{T} \times \mathbb{T}))$ for all $B \in \mathcal{B}_{\mathbb{S}\times\mathbb{S}}$). From Lemma 1.3.1, the mapping $\mu \mapsto \widehat{\mu}$ is an homeomorphism from $\mathcal{M}^{+,1}(\mathbb{T})$ to the space $\mathcal{M}^{+,1}_{\mathbb{T}}(\mathbb{S})$ of laws μ on $\mathbb{S}$ satifying $\mu_*(\mathbb{T}^c) = 0$ (resp. from $\mathcal{M}^{+,1}(\mathbb{T} \times \mathbb{T})$ to $\mathcal{M}^{+,1}_{\mathbb{T}\times\mathbb{T}}(\mathbb{S} \times \mathbb{S})$ satifying $\mu_*((\mathbb{T} \times \mathbb{T})^c) = 0$). For all $(\mu, \nu) \in \mathcal{M}^{+,1}(\mathbb{T})$, let

$$\widehat{D}(\mu, \nu) = \{\widehat{\pi};\ \pi \in D(\mu, \nu)\}.$$

One easily checks that we have, with obvious notations,

$$\widehat{D}(\mu, \nu) = D(\widehat{\mu}, \widehat{\nu}).$$

Furthermore, from Lemma 1.3.1, we have, again with obvious notations,

$$r(\mu, \nu) = \inf_{\pi \in D(\mu,\nu)} \int_{\mathbb{T}\times\mathbb{T}} d(t, t')\, d\pi(t, t') = \inf_{\widehat{\pi} \in D(\widehat{\mu},\widehat{\nu})} \int_{\mathbb{S}\times\mathbb{S}} d(t, t')\, d\widehat{\pi}(t, t') = r(\widehat{\mu}, \widehat{\nu}).$$

Using again Lemma 1.3.1, this yields

$$\begin{aligned} K(\mu, \nu) &= \left\{\pi \in D(\mu, \nu);\ \int_{\mathbb{S}\times\mathbb{S}} d(t, t')\, d\widehat{\pi}(t, t') = r(\widehat{\mu}, \widehat{\nu})\right\} \\ &= \Psi^{-1} K(\widehat{\mu}, \widehat{\nu}), \end{aligned}$$

where Ψ is the homeomorphism $\pi \mapsto \widehat{\pi}$ from $\mathcal{M}^{+,1}(\mathbb{T} \times \mathbb{T})$ to $\mathcal{M}^{+,1}_{\mathbb{T}\times\mathbb{T}}(\mathbb{S} \times \mathbb{S})$.

So, we assume from now on that $\mathbb{T}$ is Polish. We have $D = \Phi^{-1}$, where Φ is the continuous mapping

$$\Phi : \begin{cases} \mathcal{M}^{+,1}(\mathbb{T} \times \mathbb{T}) & \to \mathcal{M}^{+,1}(\mathbb{T}) \times \mathcal{M}^{+,1}(\mathbb{T}) \\ \lambda & \mapsto (\lambda(. \times \mathbb{T}), \lambda(\mathbb{T} \times .)). \end{cases}$$

Therefore, the graph $\mathrm{gph}\,(D)$ of D is a closed subset of the Polish space $\big(\mathcal{M}^{+,1}(\mathbb{T}) \times \mathcal{M}^{+,1}(\mathbb{T})\big) \times \mathcal{M}^{+,1}(\mathbb{T} \times \mathbb{T})$.

Now, the mapping

$$\psi : \begin{cases} \mathcal{M}^{+,1}(\mathbb{T} \times \mathbb{T}) & \to [0, +\infty] \\ \pi & \mapsto \int_{\mathbb{T}\times\mathbb{T}} d(t, t')\, d\pi(t, t') \end{cases}$$

is l.s.c. because it is the supremum of the continuous mappings $\pi \mapsto \pi(d \wedge n)$, $n \in \mathbb{N}$ (if d is bounded, ψ is continuous). From the Projection Theorem, as $\mathbb{T}$ is Suslin and as gph (D) is in the Borel σ–algebra of the Polish space $\left(\mathcal{M}^{+,1}(\mathbb{T}) \times \mathcal{M}^{+,1}(\mathbb{T})\right) \times \mathcal{M}^{+,1}(\mathbb{T} \times \mathbb{T})$, the mapping

$$r : (\mu, \nu) \mapsto \inf \{\psi(\pi);\ \pi \in D(\mu, \nu)\}$$

is $\mathcal{B}^*$–measurable: Indeed, we have

$$\begin{aligned} \forall \alpha \in \mathbb{R} \quad & \{(\mu, \nu);\ r(\mu, \nu) < \alpha\} \\ & = \pi_{\mathcal{M}^{+,1}(\mathbb{T}) \times \mathcal{M}^{+,1}(\mathbb{T})} \left\{((\mu, \nu), \pi) \in \text{gph}\,(D)\,;\ \psi(\pi) < \alpha\right\}. \end{aligned}$$

For each $(\mu, \nu) \in \mathcal{M}^{+,1}(\mathbb{T}) \times \mathcal{M}^{+,1}(\mathbb{T})$, we have

$$K(\mu, \nu) = \{\pi \in D(\mu, \nu);\ \psi(\pi) = r(\mu, \nu)\}\,.$$

The multifunction K has nonempty compact values because D has nonempty compact values and ψ is l.s.c. Let

$$F : \begin{cases} \left(\mathcal{M}^{+,1}(\mathbb{T}) \times \mathcal{M}^{+,1}(\mathbb{T})\right) \times \mathcal{M}^{+,1}(\mathbb{T} \times \mathbb{T}) & \to \quad \mathbb{R} \\ ((\mu, \nu), \pi) & \mapsto \quad \psi(\pi) - r(\mu, \nu). \end{cases}$$

The mapping F is $\mathcal{B}^* \otimes \mathcal{B}_{\mathcal{M}^{+,1}(\mathbb{T}\times\mathbb{T})}$–measurable. Furthermore, the graph of K is

$$\begin{aligned} \text{gph}\,(K) &= \{((\mu, \nu), \pi);\ \mu = \pi(.\times \mathbb{T}),\ \nu = \pi(\mathbb{T} \times .),\ F((\mu, \nu), \pi) = 0\} \\ &= \text{gph}\,(D) \cap F^{-1}(0) \\ &\in \mathcal{B}^* \otimes \mathcal{B}_{\mathcal{M}^{+,1}(\mathbb{T}\times\mathbb{T})}. \end{aligned}$$

As $(\mathbb{T}, d)$ is Suslin, this proves that K is $\mathcal{B}^*$–LV–measurable (see Remark 1.2.1). Thus K has a $\mathcal{B}^*$–measurable selection. □

Proof of Theorem 3.4.1. First, we can reduce the proof to the case when d is a distance. Indeed, let $(\hat{\mathbb{T}}, \hat{d})$ be the quotient metric space and π the canonical projection $\mathbb{T} \to \hat{\mathbb{T}}$. The space $\text{Lip}_1(d)$ can be identified in an obvious way with $\text{Lip}_1(\hat{\mathbb{T}}, \hat{d})$ because, for any $f \in \text{Lip}_1(d)$, if $d(t, t') = 0$, $|f(t) - f(t')| = 0$. We shall denote by $\hat{f}$ the corresponding element of $\text{Lip}_1(\hat{\mathbb{T}}, \hat{d})$. Furthermore, the mapping $\lambda \mapsto \pi_\sharp(\lambda)$, $\mathcal{M}^{+,1}(\mathbb{T}, \tau_\mathbb{T}) \to \mathcal{M}^{+,1}(\hat{\mathbb{T}}, \hat{d})$ is surjective because $\mathbb{T}$ and $\hat{\mathbb{T}}$ are Suslin [Sch73, Theorem 12 page 126]. Let $\underline{\pi}$ be the mapping $(\omega, t) \mapsto (\omega, \pi(t))$. We have

$$\underline{\Delta}_{\text{LW}}^{(d)}(\mu, \nu) = \sup_{f \in \underline{\text{Lip}}_1(\hat{d})} \underline{\pi}_\sharp(\mu - \nu)(\hat{f})$$

and

$$\underline{\Delta}_{\text{KR}}^{(d)}(\mu, \nu) = \inf_{\pi \in \underline{D}(\underline{\pi}_\sharp(\mu), \underline{\pi}_\sharp(\nu))} \int_{\Omega \times \hat{\mathbb{T}} \times \hat{\mathbb{T}}} \hat{d}(t, t')\, d\pi(\omega, t, t').$$

We assume now that $(\mathbb{T}, d)$ is a (Suslin) metric space. The proof will be done in three steps.

Step 1

From Lemma 3.4.2, and with the same notations, the mapping

$$G : \omega \mapsto \inf_{\pi \in D(\mu_\omega, \nu_\omega)} \int_{\mathbb{T}\times\mathbb{T}} d(t,t')\, d\pi(t,t') = r(\mu_\omega, \nu_\omega)$$

is $\mathcal{S}^*$–measurable (indeed, the mapping $\omega \mapsto (\mu_\omega, \nu_\omega)$ is measurable for $\mathcal{S}^*$ and $\mathcal{B}^*$ because it is measurable for $\mathcal{S}$ and $\mathcal{B}_{\mathcal{M}^{+,1}(\mathbb{T})\times\mathcal{M}^{+,1}(\mathbb{T})}$). Since $\mu, \nu \in \mathcal{Y}^{1,1}_{\mathrm{dis},d}$, we have $G(\omega) < +\infty$ almost everywhere. Let us prove that $\underline{\Delta}^{(d)}_{\mathrm{KR}}(\mu,\nu) = \int_\Omega G(\omega)\, d\,\mathrm{P}(\omega)$. First, as $\mathbb{T}$ is Radon, π is disintegrable, and we have, for any $\pi \in \underline{D}(\mu,\nu)$,

$$\int_\Omega \left(\int_{\mathbb{T}} d(t,t')\, d\pi_\omega(t,t') \right) d\,\mathrm{P}(\omega) \geq \int_\Omega G(\omega)\, d\,\mathrm{P}(\omega),$$

thus $\underline{\Delta}^{(d)}_{\mathrm{KR}}(\mu,\nu) \geq \int_\Omega G(\omega)\, d\,\mathrm{P}(\omega)$. But, from Lemma 3.4.2, the multifunction $\omega \mapsto D(\mu_\omega, \nu_\omega)$ has a $\mathcal{B}^*$–measurable selection $\omega \mapsto \lambda_\omega$ such that, for every $\omega \in \Omega$, $G(\omega) = \int_{\mathbb{T}\times\mathbb{T}} d(t,t')\, d\lambda_\omega(t,t')$. We thus have $\underline{\Delta}^{(d)}_{\mathrm{KR}}(\mu,\nu) \leq \int_{\Omega\times\mathbb{T}\times\mathbb{T}} d(t,t')\, d\lambda(\omega,t,t')$ $= \int_\Omega G(\omega)\, d\,\mathrm{P}(\omega)$.

Step 2 This is the shortest and main step. Let Ω_0 be the almost sure set on which $G(\omega) < +\infty$. From the usual Kantorovich–Rubinštein Theorem, we have, for every $\omega \in \Omega_0$,

$$G(\omega) = \sup_{g \in \mathrm{Lip}_1(d)} \left(\mu_\omega(g) - \nu_\omega(g)\right) = \sup_{g \in \mathrm{Lip}_1(d),\ g(t_0)=0} \left(\mu_\omega(g) - \nu_\omega(g)\right). \tag{3.4.4}$$

Step 3 To conclude the proof, we only need to prove the equality $\underline{\Delta}^{(d)}_{\mathrm{LW}}(\mu,\nu) = \int_\Omega G(\omega)\, d\,\mathrm{P}(\omega)$. For every $f \in \underline{\mathrm{Lip}}_1(d)$, we have $\int f\, d(\mu-\nu) \leq \int G\, d\,\mathrm{P}$ by (3.4.4), thus $\underline{\Delta}^{(d)}_{\mathrm{LW}}(\mu,\nu) \leq \int_\Omega G\, d\,\mathrm{P}$. Now, let $\varepsilon > 0$. Let $\widetilde{\mu}$ and $\widetilde{\nu}$ be the finite measures on $\mathbb{T}$ defined by

$$\widetilde{\mu}(B) = \int_{\Omega\times B} d(t_0,t)\, d\mu(\omega,t) \quad \text{and} \quad \widetilde{\nu}(B) = \int_{\Omega\times B} d(t_0,t)\, d\nu(\omega,t)$$

for any $B \in \mathcal{B}_{\mathbb{T}}$. Let $\mathbb{T}_0$ be a totally bounded subset of $\mathbb{T}$ containing t_0 such that $\widetilde{\mu}(\mathbb{T}_0^c) \leq \varepsilon$ and $\widetilde{\nu}(\mathbb{T}_0^c) \leq \varepsilon$. For any $f \in \underline{\mathrm{Lip}}_1(d)$, we have

$$\begin{aligned} &\left| \int_\Omega (\mu_\omega - \nu_\omega)(f(\omega,.))\, d\,\mathrm{P}(\omega) - \int_\Omega (\mu_\omega - \nu_\omega)(f(\omega,.)\, \mathbb{1}_{\mathbb{T}_0})\, d\,\mathrm{P}(\omega) \right| \\ &\qquad = \left| \int_\Omega (\mu_\omega - \nu_\omega)(f(\omega,.)\, \mathbb{1}_{\mathbb{T}_0^c})\, d\,\mathrm{P}(\omega) \right| \leq 2\varepsilon. \end{aligned} \tag{3.4.5}$$

Set, for all $\omega \in \Omega_0$,

$$G'(\omega) = \sup_{g \in \mathrm{Lip}_1(d),\ g(t_0)=0} (\mu_\omega - \nu_\omega)(g\, \mathbb{1}_{\mathbb{T}_0}).$$

We thus have

$$\left| \int_{\Omega_0} G \, d\mathrm{P} - \int_{\Omega_0} G' \, d\mathrm{P} \right| \le 2\varepsilon. \tag{3.4.6}$$

From a theorem of Kirszbraun and McShane, each Lipschitz function on $\mathbb{T}_0$ can be extended into a Lipschitz function on $\mathbb{T}$ with the same Lipschitz modulus (see [Dud76, Theorem 7.3] or [Dud02, Theorem 6.1.1]). Thus we have, for any $\omega \in \Omega_0$,

$$G'(\omega) = \sup_{g \in \mathrm{Lip}_1(\mathbb{T}_0, d),\ g(t_0)=0} (\mu_\omega - \nu_\omega)\!\restriction_{\mathbb{T}_0}(g).$$

Let $(g_n)_n$ be a sequence which is dense for $\|.\|_\infty$ in $\mathrm{C}_b(\mathbb{T}_0) \cap \{g \in \mathrm{Lip}_1(\mathbb{T}_0, d);\ g(t_0) = 0\}$. We have $G'(\omega) = \sup_n (\mu_\omega - \nu_\omega)\!\restriction_{\mathbb{T}_0}(g_n)$. Extend each g_n into a function $g_n \in \mathrm{Lip}_1(\mathbb{T}, d)$ and set, for all $(\omega, t) \in \Omega_0 \times \mathbb{T}$,

$$N(\omega) = \inf\left\{n;\ (\mu_\omega - \nu_\omega)\!\restriction_{\mathbb{T}_0}(g_n) \ge G'(\omega) - \varepsilon\right\} \quad \text{and} \quad f(\omega, t) = g_{N(\omega)}(t).$$

We then have, using (3.4.5) and (3.4.6),

$$\begin{aligned}\underline{\Delta}_{\mathrm{LW}}^{(d)}(\mu, \nu) &\ge \int_{\Omega_0 \times \mathbb{T}} f \, d(\mu - \nu) \ge \int_{\Omega_0 \times \mathbb{T}_0} f \, d(\mu - \nu) - 2\varepsilon \\ &\ge \int_{\Omega_0} G' \, d\mathrm{P} - 3\varepsilon \ge \int_{\Omega_0} G \, d\mathrm{P} - 5\varepsilon.\end{aligned}$$

□

Let us formulate Theorem 3.4.1 in the particular case of degenerate Young measures.

Corollary 3.4.3 *Assume that $\mathbb{T}$ is Suslin regular. Let d be a continuous semidistance on $\mathbb{T}$. For any $X, Y \in \mathfrak{X}^1_{(\mathbb{T},d)}$, we have*

$$\underline{\Delta}_{\mathrm{LW}}^{(d)}(X, Y) := \sup_{f \in \underline{\mathrm{Lip}}_1(d)} \int_\Omega f(\omega, X(\omega)) - f(\omega, Y(\omega)) \, d\mathrm{P}(\omega) = \int_\Omega d(X, Y) \, d\mathrm{P}.$$

In particular, if $\mathcal{D}$ is an upwards filtering set of continuous semidistances which defines the topology of $\mathbb{T}$, Corollary 3.4.3 and Theorem 2.4.3 imply that the topologies $\tau_{1,\mathcal{D}}$ et $\tau^{\mathrm{W}}_{1,\mathcal{D}}$ coincide on $\mathfrak{X}^1_{(\mathbb{T},\mathcal{D})}$.

A Vitali Theorem Let us now go back to the relations between convergence in $\mathcal{Y}^{1,1}_{\mathrm{dis},\mathcal{D}}$ for the semidistances $\underline{\Delta}^{(d)}_{\mathrm{LW}}$ $(d \in \mathcal{D})$ and convergence in probability.

Theorem 3.4.4 (Vitali Convergence Theorem for random laws) *Assume that $\mathbb{T}$ is completely regular. Let $\mathcal{D}$ be an upwards filtering set of continuous semidistances which defines the topology of $\mathbb{T}$. Let $(\mu^\alpha)_{\alpha\in\mathbb{A}}$ be a net in $\mathcal{Y}^{1,1}_{\mathrm{dis},\mathcal{D}}$ and let $\mu^\infty \in \mathcal{Y}^{1,1}_{\mathrm{dis},\mathcal{D}}$. The following assertions are equivalent:*

(a) $(\mu^\alpha)_{\alpha\in\mathbb{A}}$ converges to μ^∞ in $\mathcal{Y}^{1,1}_{\mathrm{dis},\mathcal{D}}$

(b) $\mu^\alpha \xrightarrow{prob} \mu^\infty$ and $(\mu^\alpha)_{\alpha\in\mathbb{A}}$ is asymptotically uniformly integrable relatively to $\mathcal{D}$.

Proof.
$(a)\Rightarrow(b)$. We already know that convergence in $\mathcal{Y}^{1,1}_{\mathrm{dis},\mathcal{D}}$ implies convergence in probability. Furthermore, for any $t_0 \in \mathbb{T}$, the mapping $(\omega, t) \mapsto d(t_0, t)$ is in $\underline{\mathrm{Lip}}_1(d)$. Let $\varepsilon > 0$. Let $a > 0$ such that

$$\int_{\Omega\times\{d(t_0,.)>a\}} d(t_0, t)\, d\mu^\infty(\omega, t) < \varepsilon.$$

We can furthermore choose a such that

$$\mu^\infty \{(\omega, t);\ d(t_0, t) = a\} = 0.$$

We then have, by narrow convergence of $(\mu^\alpha(\Omega\times .))_\alpha$ to $\mu^\infty(\Omega \times .)$,

$$\lim_\alpha \mu^\alpha \{(\omega, t);\ d(t_0, t) > a\} = \mu^\infty \{(\omega, t);\ d(t_0, t) > a\}.$$

Now, we have

$$d(t_0, t)\, \mathbb{1}_{\{d(t_0,.)>a\}}(t) = d(t_0, t) - d(t_0, t) \wedge a + a\, \mathbb{1}_{\{d(t_0,.)>a\}}(t),$$

and the mapping $(\omega, t) \mapsto d(t_0, t) \wedge a$ is also in $\underline{\mathrm{Lip}}_1(d)$. We thus have

$$\begin{aligned}\int_{\Omega\times\{d(t_0,.)>a\}} d(t_0, t)\, d\mu^\alpha(\omega, t) &= \int_{\Omega\times\mathbb{T}} d(t_0, t)\, d\mu^\alpha - \int_{\Omega\times\mathbb{T}} d(t_0, t) \wedge a\, d\mu^\alpha \\ &\quad + a\mu^\alpha \{(\omega, t);\ d(t_0, t) > a\} \\ &\to \int_{\Omega\times\mathbb{T}} d(t_0, t)\, d\mu^\infty - \int_{\Omega\times\mathbb{T}} d(t_0, t) \wedge a\, d\mu^\infty \\ &\quad + a\mu^\infty \{(\omega, t);\ d(t_0, t) > a\} \\ &= \int_{\Omega\times\{d(t_0,.)>a\}} d(t_0, t)\, d\mu^\infty(\omega, t) \\ &< \varepsilon,\end{aligned}$$

thus $(\mu^\alpha)_{\alpha\in\mathbb{A}}$ is asymptotically uniformly integrable.

(b)⇒(a). Let $d \in \mathcal{D}$. Let $\varepsilon > 0$ and $t_0 \in \mathbb{T}$. Let $a > 0$ such that

$$\limsup_\alpha \int_{\Omega\times\{d(t_0,.)>a\}} d(t_0,t)\,d\mu^\alpha(\omega,t) < \varepsilon$$

and

$$\int_{\Omega\times\{d(t_0,.)>a\}} d(t_0,t)\,d\mu^\infty(\omega,t) < \varepsilon.$$

Set $B = \{t \in \mathbb{T};\ d(t_0,t) \le a\}$. For any $f \in \mathrm{Lip}$, let $f^{|a}$ be the bounded Lipschitz integrand $(f \wedge a) \vee -a$. We have, using (3.4.1), and for α large enough,

$$\begin{aligned}
\underline{\Delta}^{(d)}_{\mathrm{LW}}(\mu^\alpha,\mu^\infty) &= \sup_{f\in\widetilde{\underline{\mathrm{Lip}}}_1(d)} |\mu^\alpha(f)-\mu^\infty(f)| \\
&\le \sup_{f\in\widetilde{\underline{\mathrm{Lip}}}_1(d)} \left|\mu^\alpha(f^{|a})-\mu^\infty(f^{|a})\right| \\
&\quad + \int_{\Omega\times\mathbb{T}} \left|f-f^{|a}\right| d\mu^\alpha + \int_{\Omega\times\mathbb{T}} \left|f-f^{|a}\right| d\mu^\infty \\
&\le \sup_{f\in\widetilde{\underline{\mathrm{Lip}}}_1(d)} \left|\mu^\alpha(f^{|a})-\mu^\infty(f^{|a})\right| \\
&\quad + \int_{\Omega\times B^c} d(t_0,t)\,d\mu^\alpha(\omega,t) + \int_{\Omega\times B^c} d(t_0,t)\,d\mu^\infty(\omega,t) \\
&\le \sup_{f\in\widetilde{\underline{\mathrm{Lip}}}_1(d)} \left|\mu^\alpha(f^{|a})-\mu^\infty(f^{|a})\right| + 2\varepsilon.
\end{aligned}$$

The conclusion follows from the fact that, from Theorem 3.2.1, we have

$$\lim_\alpha \sup_{f\in\widetilde{\underline{\mathrm{Lip}}}_1(d)} \left|\mu^\alpha(f^{|a})-\mu^\infty(f^{|a})\right| = 0.$$

□

Remark 3.4.5 Assume that $\mathbb{T}$ is hereditarily Lindelöf and regular. Let $\mathcal{D}$ be a family of semidistances which defines the topology of $\mathbb{T}$. Recall that $\tau^{\mathrm{W}}_{1,\mathcal{D}}$ is the topology on $\mathcal{Y}^{1,1}_{\mathrm{dis},\mathcal{D}}$ defined by the semidistances $|\mu(f)-\nu(f)|$, $f \in \underline{\mathrm{Lip}}_1(d)$, $d \in \mathcal{D}$ (see Proposition 2.4.1). We have the following continuous inclusions, represented by arrows:

$$\begin{array}{ccc}
\left(\mathcal{Y}^{1,1}_{\mathrm{dis},\mathcal{D}}, \tau_{1,\mathcal{D}}\right) & \longrightarrow & \left(\mathcal{Y}^{1}_{\mathrm{dis}}, \tau_{\mathrm{prob}}\left(\mathcal{Y}^{1}_{\mathrm{dis}}\right)\right) \\
\downarrow & & \downarrow \\
\left(\mathcal{Y}^{1,1}_{\mathrm{dis},\mathcal{D}}, \tau^{\mathrm{W}}_{1,\mathcal{D}}\right) & \longrightarrow & \left(\mathcal{Y}^{1}_{\mathrm{dis}}, \tau^{\mathrm{W}}_{\mathcal{Y}^1}\right).
\end{array}$$

The lower horizontal arrow comes from Proposition 2.4.1 and the vertical arrow on the right hand side comes from Theorem 3.1.2. If the elements of $\mathcal{D}$ are bounded, the horizontal arrows represent equalities of topological spaces.

Chapter 4

Compactness

In this chapter, we focus on compactness for the S–stable topology because the $\tau^{S}_{\mathcal{Y}^1}$–compact subsets of $\mathcal{Y}^1$ can be characterized in a rather simple way (see Theorem 4.3.5) and because they are also compact for all coarser Hausdorff topologies on $\mathcal{Y}^1$. In some cases (see Corollary 4.3.7) all stable topologies appear to have the same compact subsets.

4.1 Preliminary remarks and definitions

Recall that a subset $\mathfrak{K}$ of a topological space $\mathbb{T}$ is *net–compact* if every net of elements of $\mathfrak{K}$ admits a subnet which converges in $\mathbb{T}$, or equivalently, if every universal net of elements of $\mathfrak{K}$ is convergent in $\mathbb{T}$ (see [Kel55] about subnets and universal nets). We say that $\mathfrak{K}$ is *relatively compact* if it is contained in a compact subset of $\mathbb{T}$. Thus every relatively compact subset of $\mathbb{T}$ is net–compact. The converse implication is true if $\mathbb{T}$ is regular (see the proof in [PV95] or [OW98]). We say that $\mathfrak{K}$ is *sequentially relatively compact* if every sequence of elements of $\mathfrak{K}$ admits a convergent subsequence. If, furthermore, the limit of the convergent subsequence always lies in $\mathfrak{K}$, we say that $\mathfrak{K}$ is *sequentially compact* (this terminology is not entirely consistent, because, if $\mathfrak{K}$ is sequentially relatively compact, its closure is not necessarily sequentially compact). In the case when $\mathbb{T}$ is metrizable, it is well-known that $\mathfrak{K}$ is sequentially compact if and only if it is compact. Similarly, if $\mathbb{T}$ is metrizable, $\mathfrak{K}$ is sequentially relatively compact if and only if it is relatively compact. Indeed, if $\mathfrak{K}$ is sequentially relatively compact, let $(t_n)_n$ be a sequence in the closure $\mathfrak{C}$ of $\mathfrak{K}$. For each n, let $s_n \in \mathfrak{K}$ such that $d(t_n, s_n) \leq 1/n$ (where d is a distance that metrizes $\mathbb{T}$). Then the convergence of any subsequence of $(t_n)_n$ is equivalent to the convergence (to the same limit) of the subsequence of $(s_n)_n$ which has the same indexes. Thus every subsequence of $(t_n)_n$ has a further subsequence which converges in $\mathfrak{C}$.

Let $\mathfrak{T}$ be a set and let τ and τ_0 be two Hausdorff topologies on $\mathfrak{T}$ such that

$\tau_0 \subset \tau$. Let $\mathfrak{K}$ be a τ–relatively compact subset of $\mathfrak{T}$. Then τ and τ_0 coincide on $\mathfrak{K}$. Indeed, the τ–closure $\mathfrak{C}$ of $\mathfrak{K}$ is τ–compact, thus τ and τ_0 coincide on $\mathfrak{C}$.

When we need to prove sequential compactness of compact subsets of $\mathcal{Y}^1$, the following lemma can be used to dodge the assumption that $\mathcal{S}$ be essentially countably generated.

Lemma 4.1.1 (Sequential relative compactness deduced from relative compactness) *Let $\mathfrak{K}$ be a $\tau^{\mathrm{S}}_{\mathcal{Y}^1}$–relatively compact subset of $\mathcal{Y}^1$. Assume that there exists a separable metrizable topology τ_0 on $\mathbb{T}$ which is coarser than the original topology $\tau_{\mathbb{T}}$ of $\mathbb{T}$, and such that $\tau_{\mathbb{T}}$ and τ_0 have the same Borel sets. Then $\mathfrak{K}$ is sequentially relatively compact.*

Proof. Let $\mathfrak{C}$ be the $\tau^{\mathrm{S}}_{\mathcal{Y}^1}$–closure of $\mathfrak{K}$ in $\mathcal{Y}^1$. Then the topologies $\tau^{\mathrm{S}}_{\mathcal{Y}^1(\tau_{\mathbb{T}})}$ and $\tau^{\mathrm{S}}_{\mathcal{Y}^1(\tau_0)}$ coincide on $\mathfrak{C}$. So, we can assume without loss of generality that $\mathbb{T}$ is metrizable and separable.

Let $\mathbb{S}$ be a Polish space containing $\mathbb{T}$. Then we have

$$\mathcal{S} \otimes \mathcal{B}_{\mathbb{T}} = \{B \cap (\Omega \times \mathbb{T});\ B \in \mathcal{S} \otimes \mathcal{B}_{\mathbb{S}}\}.$$

With each measure $\mu \in \mathcal{Y}^1(\mathbb{T})$, we can associate the measure $\boldsymbol{\mu} \in \mathcal{Y}^1(\mathbb{S})$ defined by $\boldsymbol{\mu}(B) = \mu(B \cap (\Omega \times \mathbb{T}))$ for any $B \in \mathcal{S} \otimes \mathcal{B}_{\mathbb{S}}$, and we have $\mu(A) = \boldsymbol{\mu}^*(A)$ for any $A \in \mathcal{S} \otimes \mathcal{B}_{\mathbb{T}}$, which proves that the mapping $\Phi : \mu \mapsto \boldsymbol{\mu}$ is a bijection from $\mathcal{Y}^1(\mathbb{T})$ to the subspace of elements ν of $\mathcal{Y}^1(\mathbb{S})$ which satisfy $\nu^*(\Omega \times \mathbb{T}) = 1$. Using e.g. Condition 6 of the Portmanteau Theorem 2.1.3, we see that Φ is continuous, thus the restriction of Φ on $\mathfrak{C}$ is a homeomorphism from $\mathfrak{C}$ to the compact subset $\Phi(\mathfrak{C})$ of $\mathcal{Y}^1(\mathbb{S})$. So, we only need to show that $\Phi(\mathfrak{C})$ is sequentially compact, that is, we can assume that $\mathbb{T}$ is Polish.

We can associate with each $\mu \in \mathcal{Y}^1(\mathbb{T})$ a disintegration $\omega \mapsto \mu_\omega$, because $\mathbb{T}$ is Polish. It is well-known that the space $\mathcal{M}^{+,1}(\mathbb{T})$ is metrizable separable (see e.g. [Par67, Theorem 6.2]). Thus the Borel σ–algebra generated by each mapping $\omega \mapsto \mu_\omega$ is countably generated. Let $(\mu^n)_{n\in\mathbb{N}}$ be a sequence of elements of $\mathfrak{K}$. Let $\mathcal{S}_0$ be the σ–algebra generated by the mappings $\omega \mapsto \mu^n_\omega$. Let Q be the restriction of P on $\mathcal{S}_0$. Let π be the canonical projection (that is, the restriction) of $\mathcal{Y}^1(\Omega, \mathcal{S}, \mathrm{P}; \mathbb{T})$ onto $\mathcal{Y}^1(\Omega, \mathcal{S}_0, \mathrm{Q}; \mathbb{T})$. The mapping π is obviously continuous, thus $\pi(\mathfrak{K})$ is compact. From Proposition 2.3.1, the space $\mathcal{Y}^1(\Omega, \mathcal{S}_0, \mathrm{Q})$ (endowed with $\tau^{\mathrm{W}}_{\mathcal{Y}^1} = \tau^{\mathrm{S}}_{\mathcal{Y}^1}$), is metrizable, thus $\pi(\mathfrak{K})$ is sequentially compact. There exists a Young measure $\rho \in \mathcal{Y}^1(\Omega, \mathcal{S}_0, \mathrm{Q}; \mathbb{T})$ and a subsequence $(\lambda^n)_{n\in\mathbb{N}}$ of $(\mu^n)_{n\in\mathbb{N}}$ such that $(\pi(\lambda^n))_{n\in\mathbb{N}}$ converges to ρ.

Let $A \in \mathcal{S}$ and let f be a bounded continuous function on $\mathbb{T}$. We denote by $E^{\mathcal{S}_0}_{\mathrm{P}}$ the conditional expectation w.r.t. $\mathcal{S}_0$ and P. We have

$$\begin{aligned}\int \mathbb{1}_A \otimes f\, d\lambda^n &= \int \mathbb{1}_A(\omega)\lambda^n_\omega(f)\, d\mathrm{P}(\omega) = \int E^{\mathcal{S}_0}_{\mathrm{P}}(\mathbb{1}_A)(\omega)\lambda^n_\omega(f)\, d\mathrm{Q}(\omega)\\ &\to \int E^{\mathcal{S}_0}_{\mathrm{P}}(\mathbb{1}_A)(\omega)\rho_\omega(f)\, d\mathrm{Q}(\omega) = \int \mathbb{1}_A(\omega)\rho_\omega(f)\, d\mathrm{P}(\omega)\end{aligned}$$

In view of Condition 13 in the Portmanteau Theorem 2.1.3, this shows that $(\lambda^n)_{n\in\mathbb{N}}$ converges to the Young measure

$$\lambda : \begin{cases} \mathcal{B} & \to & [0,1] \\ B & \mapsto & \int \rho_\omega(B(\omega))\, d\,\mathrm{P}(\omega). \end{cases}$$

□

We shall prove later, in Proposition 4.5.2, a converse implication.

In this chapter, we do not study directly compactness in spaces of p–integrable Young measures. However, compactness criteria can be easily obtained through the following proposition (as in Section 2.4, we give only the result for $\tau^{\mathrm{W}}_{p,\mathcal{D}}$, but the same method would yield similar results for other topologies $\tau^{*}_{p,\mathcal{D}}$).

Proposition 4.1.2 (Compact subsets of $\mathcal{Y}^{1,p}_{\mathcal{D}}$) *Assume that $\mathbb{T}$ is completely regular and that its topology $\tau_{\mathbb{T}}$ is defined by a set $\mathcal{D}$ of semidistances. Let $p > 0$. Let $\mathfrak{Y} \subset \mathcal{Y}^{1,p}_{\mathcal{D}}$. The following conditions are equivalent:*

(*a*) *The set $\mathfrak{Y}$ is net–compact for $\tau^{\mathrm{W}}_{p,\mathcal{D}}$.*

(*b*) *The set $\mathfrak{Y}$ is net–compact for $\tau^{\mathrm{W}}_{\mathcal{Y}^1}$ and uniformly p–integrable relatively to $\mathcal{D}$.*

Proof. Assume (a). Then $\mathfrak{Y}$ is net–compact for $\tau^{\mathrm{W}}_{\mathcal{Y}^1}$. Assume furthermore that for some $d \in \mathcal{D}$, $\mathfrak{Y}$ is not uniformly p–integrable relatively to d. Let $a \in \mathbb{T}$. There exists $\varepsilon > 0$ such that, for each $R > 0$, we can find $\mu^R \in \mathfrak{Y}$ with

$$\int_{\Omega\times\{t\in\mathbb{T};\, d(a,.)>R\}} d(a,t)^p\, d\mu^R(\omega,t) \geq \varepsilon. \tag{4.1.1}$$

From net–compactness of $\mathfrak{Y}$ for $\tau^{\mathrm{W}}_{p,\mathcal{D}}$, there exists a subnet $(\mu^{R_\alpha})_\alpha$ of $(\mu^R)_R$ which converges to some $\mu \in \mathcal{Y}^{1,p}_{\mathcal{D}}$ for $\tau^{\mathrm{W}}_{p,\mathcal{D}}$. But from Theorem 2.4.1, this implies that $(\mu^{R_\alpha})_\alpha$ is asymptotically uniformly p–integrable, which contradicts (4.1.1). Thus $\mathfrak{Y}$ is uniformly p–integrable relatively to $\mathcal{D}$.

Conversely, if we assume (b), every universal net of elements of $\mathfrak{Y}$ is $\tau^{\mathrm{W}}_{\mathcal{Y}^1}$–convergent and uniformly p–integrable relatively to $\mathcal{D}$, thus, from Theorem 2.4.1, it converges for $\tau^{\mathrm{W}}_{p,\mathcal{D}}$. □

4.2 Necessary and sufficient condition when $\mathbb{T}$ is separably submetrizable: Topsøe Criterion

Under the hypothesis that $\mathbb{T}$ is separably submetrizable and $(\Omega, \mathcal{S}, \mathrm{P})$ is complete, we shall now adapt to Young measures a compactness criterion of Topsøe [Top70a]. The main results from this section are proved in [RdF03], in the more general setting of the space $\mathcal{M}^+(\Omega \times \mathbb{T})$ endowed with the topology $\tau^{\mathrm{S}}_{\mathcal{M}}$ (see Remark and Definition 2.1.5, page 31).

We call *paving* on $\Omega \times \mathbb{T}$ any nonempty set of subsets of $\Omega \times \mathbb{T}$. We now list some properties of the pavings $\underline{\mathcal{G}}$ and $\underline{\mathcal{K}}$ that we shall need later. We present them in a similar way as in [Top70a].

Lemma 4.2.1 (Properties of $\underline{\mathcal{G}}$ and $\underline{\mathcal{K}}$) *Assume that $\mathbb{T}$ is Suslin submetrizable and $(\Omega, \mathcal{S}, \mathrm{P})$ is complete.*

I. $\underline{\mathcal{K}}$ contains $\emptyset$ and is closed under finite unions and countable intersections,

II. $\underline{\mathcal{G}}$ contains $\emptyset$ and is closed under finite unions and finite intersections (actually, $\underline{\mathcal{G}}$ is also closed under countable unions, but we shall not need it),

III. $K \setminus G \in \underline{\mathcal{K}}$ for all $K \in \underline{\mathcal{K}}$ and for all $G \in \underline{\mathcal{G}}$,

IV. $\underline{\mathcal{G}}$ separates the sets in $\underline{\mathcal{K}}$, that is, for any pair K_1, K_2 of disjoint elements of $\underline{\mathcal{K}}$, we can find a pair G_1, G_2 of disjoint elements of $\underline{\mathcal{G}}$ such that $K_1 \subset G_1$ and $K_2 \subset G_2$.

V'. The set $\mathcal{Y}^1$ is uniformly σ–smooth on $\underline{\mathcal{G}}$ at $\emptyset$ w.r.t. $\underline{\mathcal{K}}$, *that is, for any countable family $(K_i)_{i\in I}$ of elements of $\underline{\mathcal{K}}$ which filters downwards to $\emptyset$, we have*

$$\inf_{i\in I} \sup_{\mu\in\mathcal{Y}^1} \inf_{G\in\underline{\mathcal{G}},\, G\supset K_i} \mu(G) = 0$$

(we say that $(K_i)_{i\in I}$ filters downwards to $\emptyset$ if $\cap_{i\in I} K_i = \emptyset$ and if, for any $i \in I$ and any $j \in I$, there exists $k \in I$ such that $K_k \subset K_i \cap K_j$).

Remark 4.2.2 Note that we do not have Property V of [Top70a], that is, semi–compactness of $\underline{\mathcal{K}}$ (a paving $\mathcal{C}$ is said to be *semi-compact* if, for any countable family of elements of $\mathcal{C}$ which has an empty intersection, there exists a finite subfamily which has an empty intersection). We shall see however that the weaker Property V' is sufficient to yield the same conclusion as in Theorem 4 of [Top70a].

Proof of Lemma 4.2.1. Actually, only Properties IV and V' need a proof. Note however that III holds because $\mathbb{T}$ is Hausdorff, which entails that $\mathcal{K} \subset \mathcal{F}$ (recall that, in a non–Hausdorff space, a compact subset need not be closed).

We denote by d a continuous distance on $\mathbb{T}$ and by τ_0 the topology (coarser than $\tau_{\mathbb{T}}$) generated by d.

Proof of IV. Let K_1, K_2 be disjoint elements of $\underline{\mathcal{K}}$.

Assume first that K_1 and K_2 have nonempty values. For each $\omega \in \Omega$, $K_1(\omega)$ and $K_2(\omega)$ are compact for τ_0, thus $d(K_1(\omega), K_2(\omega)) > 0$. Furthermore, as $(\Omega, \mathcal{S}, \mathrm{P})$ is complete, the function $\phi : \omega \mapsto d(K_1(\omega), K_2(\omega))$ is $\mathcal{S}$-measurable. Indeed, let $\widetilde{\mathbb{T}}$ be the d–completion of $\mathbb{T}$. Then, for $i = 1, 2$, the set K_i is an element of $\mathcal{S} \otimes \mathcal{B}_{\widetilde{\mathbb{T}}}$, thus, from (*iii*) of Remark 1.2.1, page 7, we have

$$\{\omega \in \Omega;\ K_i(\omega) \cap U \neq \emptyset\} \in \mathcal{S}$$

for any open subset U of $\mathbb{T}$. But, from Theorem III.9 in [CV77], this is equivalent to each of the following properties:

- for each $t \in \mathbb{T}$, the function $\omega \mapsto d(t, K_i(\omega))$ is $\mathcal{S}$–measurable,

- there exists a sequence $(\varphi_n^i)_{n\in\mathbb{N}}$ of $\mathcal{S}$–measurable mappings $\Omega \to \mathbb{T}$ such that, for every $\omega \in \Omega$, $\varphi_n^i(\omega) \in K_i(\omega)$ and $K_i(\omega)$ is the τ_0–closure of $\{(\varphi_n^i(\omega));\ n \in \mathbb{N}\}$.

Thus $\phi = \inf_{m,n\in\mathbb{N}} d(\varphi_m^1, \varphi_n^2)$ is $\mathcal{S}$–measurable.

Now set, for $i = 1, 2$,

$$G_i = \left\{(\omega, t) \in \Omega \times \mathbb{T};\ d(t, K_i(\omega)) < \frac{\phi(\omega)}{3}\right\}.$$

It is clear that $G_1 \cap G_2 = \emptyset$ and that, for $i = 1, 2$, we have $K_i \subset G_i$ and $G_i(\omega)$ $(i = 1, 2)$ is $\tau_{\mathbb{T}}$–open for each $\omega \in \Omega$. Furthermore, for each $n \in \mathbb{N}$, the function

$$g_n^i : (\omega, t) \mapsto d\left(\varphi_n^i(\omega), t\right) - \frac{\phi(\omega)}{3}$$

is $\mathcal{S} \otimes \mathcal{B}_{\mathbb{T}}$–measurable, thus $G_i = \cup_{n\in\mathbb{N}} (g_n^i)^{-1}([-\infty, 0[)$ belongs to $\mathcal{S} \otimes \mathcal{B}_{\mathbb{T}}$.

Let us now allow each K_i to have empty values. From the Projection Theorem (see page 7), as $(\Omega, \mathcal{S}, \mathrm{P})$ is complete, the sets $\{\omega \in \Omega;\ K_i(\omega) \neq \emptyset\}$ are $\mathcal{S}$-measurable. Consider the $\mathcal{S}$-measurable sets

$$\begin{aligned}
\Omega_0 &= \{\omega \in \Omega;\ K_1(\omega) = \emptyset \text{ and } K_2(\omega) = \emptyset\},\\
\Omega_1 &= \{\omega \in \Omega;\ K_1(\omega) = \emptyset \text{ and } K_2(\omega) \neq \emptyset\},\\
\Omega_2 &= \{\omega \in \Omega;\ K_1(\omega) \neq \emptyset \text{ and } K_2(\omega) = \emptyset\},\\
\Omega_3 &= \{\omega \in \Omega;\ K_1(\omega) \neq \emptyset \text{ and } K_2(\omega) \neq \emptyset\}.
\end{aligned}$$

The same arguments as above show the existence of two disjoint elements G_1' and G_2' of $\underline{\mathcal{G}}$, contained in $\Omega_3 \times \mathbb{T}$, such that $K_i \cap (\Omega_3 \times \mathbb{T}) \subset G_i'$ $(i = 1, 2)$. To prove Property IV, we only need to set

$$G_i(\omega) = \begin{cases} \emptyset & \text{if} \quad \omega \in \Omega_0,\\ \emptyset & \text{if} \quad \omega \in \Omega_i \\ \mathbb{T} & \text{if} \quad \omega \in \Omega_{3-i},\\ G_i'(\omega) & \text{if} \quad \omega \in \Omega_3. \end{cases}$$

Proof of V'. Let $(K_i)_{i\in I}$ be a countable family of elements of $\underline{\mathcal{K}}$ which filters downwards to $\emptyset$. For each $\omega \in \Omega$, $(K_i(\omega))_{i\in I}$ is a family of compact subsets of $\mathbb{T}$ which filters downwards to $\emptyset$, thus there exists an element i of I such that $K_i(\omega) = \emptyset$. We can enumerate the elements of I: $I = \{i_0, i_1, \dots\}$, and we can endow I with the ordering associated with this enumeration: $i_0 \leq i_1 \leq \dots$. For each $\omega \in \Omega$, let us denote by $\alpha(\omega)$ the smallest i such that $K_i(\omega) = \emptyset$. Using the Projection Theorem as in the proof of Property IV, we see that, for each $i \in I$, the set

$$A_i = \{\omega \in \Omega;\ \alpha(\omega) = i\} = (\pi_\Omega(K_i))^c \cap \bigcap_{j\leq i-1} \pi_\Omega(K_j)$$

is measurable. Thus the family $(A_i)_{i\in I}$ is a measurable partition of Ω. For each integer $n \in \mathbb{N}$, let $\Omega_n = A_{i_0} \cup \cdots \cup A_{i_n}$. As $(K_i)_{i\in I}$ filters downwards to $\emptyset$, there exists an element j_1 of I such that $K_{j_1} \subset K_{i_0} \cap K_{i_1}$. We then have $K_{j_1}(\omega) = \emptyset$ on $A_{i_0} \cup A_{i_1} = \Omega_1$. By induction, we can construct a sequence $(K_{j_n})_{n\geq 1}$ such that $K_{j_n}(\omega) = \emptyset$ on Ω_n.

Now, we have $\cup_{n\in\mathbb{N}}\Omega_n = \Omega$. Let $\varepsilon > 0$. Let $n \geq 1$ such that $\mathrm{P}(\Omega_n) \geq 1 - \varepsilon$. Let $G_n = \Omega_n^c \times \mathbb{T}$. We have $K_{j_n} \subset G_n$ and $\mu(G_n) = \mathrm{P}(\Omega_n^c) \leq \varepsilon$ for every $\mu \in \mathcal{Y}^1$. This shows that

$$\inf_{i\in I} \sup_{\mu\in\mathcal{Y}^1} \inf_{G\in\underline{\mathcal{G}},\, G\supset K_i} \mu(G) \leq \varepsilon.$$

As ε is arbitrary, this proves Property V'. □

We can now give an adaptation of Topsøe's compactness criterion ([Top70a, Corollary 2], see also [Top74, OW98]). If $\mathcal{C}$ and $\mathcal{E}$ are two pavings on $\Omega \times \mathbb{T}$, we say that $\mathcal{E}$ *dominates* $\mathcal{C}$ if each element of $\mathcal{C}$ is contained in some element of $\mathcal{E}$.

Theorem 4.2.3 (Topsøe Criterion) *Assume that $\mathbb{T}$ is Suslin submetrizable and $(\Omega, \mathcal{S}, \mathrm{P})$ is complete. Let $\mathfrak{K}$ be a subset of $\mathcal{Y}^1$. Then $\mathfrak{K}$ is $\tau_{\mathcal{Y}^1}^{\mathrm{S}}$-net-compact if and only if, for any subfamily $\underline{\mathcal{G}}'$ of $\underline{\mathcal{G}}$ which dominates $\underline{\mathcal{K}}$ and for each $\varepsilon > 0$, there exists a finite subfamily $\underline{\mathcal{G}}''$ of $\underline{\mathcal{G}}'$ such that, for every $\mu \in \mathfrak{K}$, we can find $G \in \underline{\mathcal{G}}''$ such that $\mu(G^c) < \varepsilon$.*

Proof. It is a simple adaptation of the proof of Topsøe's Theorem 4 and Corollary 2 in [Top70a]. We only need to show that Property V in [Top70a] (see remark 4.2.2) can be replaced by our Property V'. We use here the definitions of [Top70a].

The "only if" part of the proof is exactly as in [Top70a], and does not rely on Properties I to V. For the "if" part, Property V is used in [Top70a] (in the proof of Theorem 4) in the following way. Let $(\mu^\alpha)_{\alpha\in\mathbb{A}}$ be a universal net in $\mathfrak{K}$. We define a set function $\nu : \underline{\mathcal{G}} \to [0, 1]$ by

$$\forall G \in \underline{\mathcal{G}} \quad \nu(G) = \lim_{\alpha} \mu^\alpha(G). \tag{4.2.1}$$

The mapping ν is monotone (that is, $G \subset G' \Rightarrow \nu(G) \leq \nu(G')$), additive (that is, $G \cap G' = \emptyset \Rightarrow \nu(G \cup G') = \nu(G) + \nu(G')$) and subadditive (that is, $\nu(G \cup G') \leq \nu(G) + \nu(G')$). Then Property V is only used (through Theorem 2 of [Top70a]) to ensure that the formula

$$\forall B \in \underline{\mathcal{B}} \quad \mu(B) = \sup_{K\subset B,\, K\in\underline{\mathcal{K}}} \ \inf_{G\in\underline{\mathcal{G}},\, G\supset K} \nu(G) \tag{4.2.2}$$

defines a measure $\mu \in \mathcal{M}^+(\Omega \times \mathbb{T})$. But, from [Top70a, Theorem 2], this result also holds true if ν is σ-smooth at $\emptyset$ w.r.t. $\underline{\mathcal{K}}$, and this last property is an immediate consequence of V'. □

Remark 4.2.4 (Extension to $\mathcal{M}^+(\Omega \times \mathbb{T})$) We can extend Theorem 4.2.3 to the set $\mathcal{M}^+(\Omega \times \mathbb{T})$ and the topology $\tau^{\mathrm{S}}_{\mathcal{M}}$ (see Remark and Definition 2.1.5, page 31) by adding the condition that $(\pi_\Omega)_\sharp(\mathfrak{K})$ must be net–compact in the *strong topology* τ_s on $\mathcal{M}^+(\Omega)$, that is, the coarsest topology such that, for each $A \in \mathcal{S}$, the mapping $\mu \mapsto \mu(A)$, $\mathcal{M}^+(\Omega) \to \mathbb{R}$, is continuous.

Indeed, if a net $(\mu^\alpha)_{\alpha\in\mathbb{A}}$ of elements of $\mathcal{M}^+(\Omega \times \mathbb{T})$ converges for $\tau^{\mathrm{S}}_{\mathcal{M}}$ to some limit μ, then its projection $((\pi_\Omega)_\sharp(\mu^\alpha))_{\alpha\in\mathbb{A}}$ on $\mathcal{M}^+(\Omega)$ converges to $(\pi_\Omega)_\sharp(\mu)$ for τ_s, thus, if $\mathfrak{K} \subset \mathcal{M}^+(\Omega \times \mathbb{T})$ is net–compact for $\tau^{\mathrm{S}}_{\mathcal{M}}$, $(\pi_\Omega)_\sharp(\mathfrak{K})$ is net–compact for τ_s.

Conversely, assume that $(\pi_\Omega)_\sharp(\mathfrak{K})$ is net–compact for τ_s and that $\mathfrak{K}$ satisfies the condition given in Theorem 4.2.3. Let $(\mu^\alpha)_{\alpha\in\mathbb{A}}$ be an universal net in $\mathfrak{K}$. Define the monotone additive and subadditive mapping ν as above, using formula (4.2.1). Let $(K_i)_{i\in I}$ be a countable family of elements of $\underline{\mathcal{K}}$ which filters downwards to $\emptyset$. Define the sets Ω_n and the subsequence $(K_{j_n})_n$ as in the proof of Part V' of Lemma 4.2.1. The net $\left((\pi_\Omega)_\sharp \mu^\alpha\right)_{\alpha\in\mathbb{A}}$ converges for τ_s to a measure $\lambda \in \mathcal{M}^+(\mathbb{T})$. We have

$$\begin{aligned}
\inf_{i\in I} \inf_{G\in\underline{\mathcal{G}},\, G\supset K_i} \nu(G) &= \inf_{i\in I} \inf_{G\in\underline{\mathcal{G}},\, G\supset K_i} \lim_{\alpha\in\mathbb{A}} \mu^\alpha(G) \\
&\leq \inf_{n\in\mathbb{N}} \inf_{G\in\underline{\mathcal{G}},\, G\supset K_{j_n}} \lim_{\alpha\in\mathbb{A}} \mu^\alpha(G) \\
&\leq \inf_{n\in\mathbb{N}} \lim_{\alpha\in\mathbb{A}} \mu^\alpha(G_n) \\
&= \inf_{n\in\mathbb{N}} \lambda(\Omega_n^c) = 0,
\end{aligned}$$

thus ν is σ–smooth at $\emptyset$ w.r.t. $\underline{\mathcal{K}}$. Using Properties I to IV of Lemma 4.2.1, we deduce from [Top70a, Theorem 2] that (4.2.2) defines a measure on $(\Omega \times \mathbb{T}, \underline{\mathcal{B}})$ which satisfies $\mu \leq \nu$. We then prove that $(\mu^\alpha)_{\alpha\in\mathbb{A}}$ converges to μ as in the proof of [Top70a, Theorem 4].

Note that, if furthermore $(\pi_\Omega)_\sharp(\mathfrak{K})$ is relatively compact for τ_s, then a slight modification of the proof of Lemma 4.2.1 shows that Property V' still holds true on $\mathfrak{K}$, that is, for any countable family $(K_i)_{i\in I}$ of elements of $\underline{\mathcal{K}}$ which filters downwards to $\emptyset$, we have

$$\inf_{i\in I} \sup_{\mu\in\mathfrak{K}} \inf_{G\in\underline{\mathcal{G}},\, G\supset K_i} \mu(G) = 0.$$

Indeed, with the notations of the proof of Lemma 4.2.1, we have, by Dini Theorem, $\lim_{n\to\infty} \sup_{\mu\in\mathfrak{K}} \mu(\Omega_n^c \times \mathbb{T}) = 0$, because the mappings $\mu \mapsto \mu(\Omega_n^c \times \mathbb{T})$ are continuous for τ_s. We can then deduce the net–compactness of $\mathfrak{K}$ as in Theorem 4.2.3.

The extension of Theorem 4.2.3 obtained in this way includes the compactness criteria of Schäl [Sch75, Theorem 3.10], Jacod and Mémin [JM81b] and Balder [Bal01] for $\tau^{\mathrm{W}}_{\mathcal{M}}$, see [RdF03].

4.3 Flexible tightness and strict tightness: Prohorov Criterion

Tightness and Prohorov spaces Recall that a subset $\mathfrak{K}$ of $\mathcal{M}^{+,1}(\mathbb{T})$ is said to be *(uniformly) tight* if, for each $\varepsilon > 0$, there exists a compact subset K of $\mathbb{T}$ such that $\sup_{\mu\in\mathfrak{K}} \mu(K^c) \leq \varepsilon$. Every tight subset of $\mathcal{M}_t^{+,1}(\mathbb{T})$ is relatively compact ([Top70a, Top70b], Theorem 9.1). In the case where each compact subset of $\mathcal{M}_R^{+,1}(\mathbb{T})$ is tight, $\mathbb{T}$ is said to be a *Prohorov space.* J. Hoffmann–Jørgensen [HJ72] has shown that every Čech complete space (that is, every space which is a G_δ subset of some compact space) is Prohorov (in particular every completely metrizable space and every locally compact space are Prohorov), and furthermore that every locally convex topological space which is the inductive limit of a sequence of closed Prohorov linear subspaces is Prohorov. The most general sufficient conditions for a Hausdorff space to be Prohorov are given in [Bou96, Bou98, Bou02]. Other references on Prohorov spaces are the surveys [Top74], [Whe83] and [Bog98b, Section 8.3]. The result of Małgorzata Wójcicka [Wój87] is particularly interesting in our context and we shall use it later: If $\mathbb{T}$ is completely regular and Prohorov, the space $\mathcal{M}^{+,1}(\mathbb{T})$ is also Prohorov. If $\mathbb{T}$ is a completely regular Čech complete space, then $\mathcal{K}(\mathbb{T})$ is also completely regular Čech complete (see [Eng89, page 285]), thus $\mathcal{K}(\mathbb{T})$ is Prohorov. If $\mathbb{E}$ is an infinite dimensional Banach space and $\mathbb{E}^*$ its topological dual, the weak topology $\sigma(\mathbb{E}, \mathbb{E}^*)$ and the weak* topology $\sigma(\mathbb{E}^*, \mathbb{E})$ are never Prohorov [Fer94].

Flexible tightness and strict tightness Let $\mathfrak{K} \subset \mathcal{Y}^1$.

- We say that $\mathfrak{K}$ is *flexibly tight* if, for each $\varepsilon > 0$, there exists $K \in \underline{\mathcal{K}}$ such that
$$\sup_{\mu\in\mathfrak{K}} \mu(K^c) < \varepsilon.$$
- We say that $\mathfrak{K}$ is *strictly tight* if, for each $\varepsilon > 0$, there exists a compact subset K of $\mathbb{T}$ such that
$$\sup_{\mu\in\mathfrak{K}} \mu(\Omega \times K^c) < \varepsilon.$$
In other words, $\mathfrak{K}$ is strictly tight if and only if the set $(\pi_{\mathbb{T}})_\sharp \mathfrak{K}$ of margins on $\mathbb{T}$ of elements of $\mathfrak{K}$ is a tight subset of $\mathcal{M}^{+,1}(\mathbb{T})$.

It is clear that, if $\mathfrak{K}$ is strictly tight, it is flexibly tight. We shall prove later that the converse implication holds in some important cases.

Let us put in a lemma a simple observation.

Lemma 4.3.1 *Assume that the compact subsets of* $\mathbb{T}$ *are metrizable. Let* $\mu \in \mathcal{Y}^1$ *such that* $\{\mu\}$ *is strictly tight. Then* $\mu \in \mathcal{Y}^1_{\text{dis}}$.

Proof. For each integer $n \geq 1$, let $K_n \in \mathcal{K}$ such that $\mu(\Omega \times K_n) \geq 1 - 1/n$. Let $\mathbb{T}_0 = \cup_{n \geq 1} K_n$. We can identify μ with a measure on $\Omega \times \mathbb{T}_0$, and the space $\mathbb{T}_0$ is a countable union of Suslin spaces, thus it is Suslin, thus it is Radon. □

We give below some characterizations of strict tightness and flexible tightness.

We start with strict tightness. Recall that a function $f : \mathbb{T} \to [0, +\infty]$ is *inf-compact* if, for any $M \in [0, +\infty[$, the set $\{t \in \mathbb{T};\ f(t) \leq M\}$ is compact. Every inf-compact function is l.s.c., thus Borel.

Theorem 4.3.2 (Equivalence theorem for strict tightness) *Let $\mathfrak{Y} \subset \mathcal{Y}^1$. The following conditions are equivalent:*

(*a*) *$\mathfrak{Y}$ is strictly tight.*

(*b*) *There exists an inf-compact function $h : \mathbb{T} \to [0, +\infty]$ such that*

$$\sup_{\mu \in \mathfrak{Y}} \mu(\mathbb{1}_\Omega \otimes h) < +\infty.$$

Furthermore, if $\mathfrak{Y} \subset \mathcal{Y}^1_{\text{dis}}$, then these conditions are equivalent to the following one:

(*c*) *For each $\varepsilon > 0$, there exists a tight subset $\mathfrak{K}_\varepsilon$ of $\mathcal{M}^{+,1}(\mathbb{T})$ such that, for any $\mu \in \mathfrak{Y}$,*

$$\mathrm{P}_* \{\omega \in \Omega;\ \mu_\omega \in \mathfrak{K}_\varepsilon\} \geq 1 - \varepsilon.$$

Remarks 4.3.3

1. If the compact subsets of $\mathbb{T}$ are metrizable and if $\mathfrak{Y}$ is a strictly tight subset of $\mathcal{Y}^1$, then, from Lemma 4.3.1, each element of $\mathfrak{Y}$ is disintegrable and Condition (*c*) holds true.

2. If $\mathbb{T}$ is Prohorov, Condition (*c*) means that the set of random measures $\{\mu_{.}\,;\ \mu \in \mathfrak{K}\}$ is tight in the usual sense for random elements of a topological space. This condition is taken as a definition in e.g. [Jak88].

Proof of Theorem 4.3.2. The equivalence of (*a*) and (*b*) is well-known: see Exercice 10 of §5 in [Bou69]. The proof may be found in [Jaw84, Val90b, Bal95] (in fact a more general form is proved there, see Remark 4.3.4 below).

(*b*) ⇒ (*c*). Let $\varepsilon > 0$. Let h as in (*b*), and let $M = \sup_{\mu \in \mathfrak{Y}} \mu(\mathbb{1}_\Omega \otimes h)$. Let

$$\mathfrak{K}_\varepsilon = \left\{\nu \in \mathcal{M}^{+,1}(\mathbb{T});\ \nu(h) \leq \frac{M}{\varepsilon}\right\}.$$

From [Bou69, Exercice 10 of §5], $\mathfrak{K}_\varepsilon$ is a tight subset of $\mathcal{M}^{+,1}(\mathbb{T})$. For each $\mu \in \mathfrak{Y}$, let

$$\Omega_\varepsilon^\mu = \{\omega \in \Omega;\ \mu_\omega \notin \mathfrak{K}_\varepsilon\}.$$

We have

$$\mathrm{P}(\Omega_\varepsilon^\mu) \leq \frac{\varepsilon}{M}\int \mu_\omega(h)\, d\,\mathrm{P}(\omega) \leq \varepsilon.$$

$(c) \Rightarrow (a)$. Let $\varepsilon > 0$. There exists a tight subset $\mathfrak{K}_{\varepsilon/2}$ of $\mathcal{M}^{+,1}(\mathbb{T})$ and, for each $\mu \in \mathfrak{Y}$, a measurable set $\Omega_{\mu,\varepsilon} \in \mathcal{S}$ such that $\mathrm{P}(\Omega_{\mu,\varepsilon}) \geq 1 - \varepsilon/2$ and, for each $\omega \in \Omega_{\mu,\varepsilon}$, $\mu_\omega \in \mathfrak{K}_{\varepsilon/2}$. Let $K \in \mathcal{K}$ such that $\nu(K) \geq 1 - \varepsilon/2$ for every $\nu \in \mathfrak{K}_{\varepsilon/2}$. We then have, for every $\mu \in \mathfrak{Y}$,

$$\mu(\Omega \times K) \geq \int_{\Omega_{\mu,\varepsilon}} \mu_\omega(K)\, d\,\mathrm{P}(\omega) \geq (1 - \varepsilon/2)^2 \geq 1 - \varepsilon.$$

□

Remark 4.3.4 (Equivalence theorem for flexible tightness) Let $\mathfrak{Y} \subset \mathcal{Y}^1$. The following conditions are equivalent (see [Bal84a, Jaw84, Val90b, Bal95]):

(a) $\mathfrak{Y}$ is flexibly tight.

(b) There exists an inf–compact integrand $h : \Omega \times \mathbb{T} \to [0, +\infty]$ such that

$$\sup_{\mu \in \mathfrak{Y}} \mu(h) < +\infty.$$

Tightness and relative compactness in $(\mathcal{Y}^1, \tau^{\mathrm{S}}_{\mathcal{Y}^1})$ As a consequence of Topsøe Criterion (Theorem 4.2.3), we easily get Part A of the following theorem. This result was proved by Balder [Bal89a], in the case when $\mathbb{T}$ is a Suslin regular space. The converse Part B follows easily from the definition.

Theorem 4.3.5 (Prohorov Criterion)

A) **(Direct Prohorov Criterion)**

1. *Assume that the compact subsets of $\mathbb{T}$ are metrizable. Any strictly tight subset of $\mathcal{Y}^1$ is $\tau^{\mathrm{S}}_{\mathcal{Y}^1}$–relatively compact and $\tau^{\mathrm{S}}_{\mathcal{Y}^1}$–sequentially relatively compact.*

2. *If furthermore $\mathbb{T}$ is Suslin submetrizable, any flexibly tight subset of $\mathcal{Y}^1$ is $\tau^{\mathrm{S}}_{\mathcal{Y}^1}$–relatively compact and $\tau^{\mathrm{S}}_{\mathcal{Y}^1}$–sequentially relatively compact.*

B) **(Converse Prohorov Criterion)** *Assume that $\mathbb{T}$ is Prohorov and that furthermore $\mathcal{M}^{+,1}(\mathbb{T}) = \mathcal{M}^{+,1}_t(\mathbb{T})$ (e.g. $\mathbb{T}$ is Suslin Prohorov). Then every $\tau^{\mathrm{N}}_{\mathcal{Y}^1}$–relatively compact subset of $\mathcal{Y}^1$ is strictly tight.*

Proof. Let $\mathfrak{K} \subset \mathcal{Y}^1$.

A) 1 For each integer $n \geq 1$, let $K_n \in \mathcal{K}(\mathbb{T})$ such that, for every $\mu \in \mathfrak{K}$, $\mu(\Omega \times K_n) \geq 1 - 1/n$. Let $\mathbb{T}_0 = \cup_n K_n$. The space $\mathbb{T}_0$ is a countable union of Lusin spaces, thus it is Lusin. Furthermore, we have $\mu(\Omega \times \mathbb{T}_0) = 1$ for every $\mu \in \mathfrak{K}$, thus we can consider $\mathfrak{K}$ as a strictly tight subset of $\mathcal{Y}^1(\mathbb{T}_0)$. But, for any subset $\mathbb{T}'$ of $\mathbb{T}$, the topology $\tau^{\mathrm{S}}_{\mathcal{Y}^1(\mathbb{T}')}$ is finer than the topology induced by $\tau^{\mathrm{S}}_{\mathcal{Y}^1(\mathbb{T})}$. Thus we only need to prove that $\mathfrak{K}$ is $\tau^{\mathrm{S}}_{\mathcal{Y}^1(\mathbb{T}_0)}$–relatively compact and $\tau^{\mathrm{S}}_{\mathcal{Y}^1(\mathbb{T}_0)}$–sequentially relatively compact. Furthermore, $\mathbb{T}_0$ is a countable union of second countable Suslin spaces which are Borel subsets of $\mathbb{T}_0$, thus, from Lemma 2.1.2, we have $\tau^{\mathrm{S}}_{\mathcal{Y}^1(\mathbb{T}_0)} = \tau^{\mathrm{S}}_{\mathcal{Y}^1(\Omega, \mathcal{S}^*_{\mathrm{P}}, \mathrm{P}^*, \mathbb{T}_0)}$. Thus 1 is a consequence of 2.

2 Denote as usual by $(\Omega, \mathcal{S}^*_{\mathrm{P}}, \mathrm{P}^*)$ the P–completion of $(\Omega, \mathcal{S}, \mathrm{P})$. The topology $\tau^{\mathrm{S}}_{\mathcal{Y}^1(\Omega, \mathcal{S}^*_{\mathrm{P}}, \mathrm{P}^*)}$ is finer than $\tau^{\mathrm{S}}_{\mathcal{Y}^1(\Omega, \mathcal{S}, \mathrm{P})}$, thus we only need to prove the result when $(\Omega, \mathcal{S}, \mathrm{P})$ is complete. To do that, observe first that the $\tau^{\mathrm{S}}_{\mathcal{Y}^1}$–closure $\overline{\mathfrak{K}}$ of $\mathfrak{K}$ is also tight. From Theorem 4.2.3, $\overline{\mathfrak{K}}$ is $\tau^{\mathrm{S}}_{\mathcal{Y}^1}$–net–compact. Indeed, let $\underline{\mathcal{G}}'$ be a subfamily of $\underline{\mathcal{G}}$ which dominates $\underline{\mathcal{K}}$ and let $\varepsilon > 0$. There exists $K_\varepsilon \in \underline{\mathcal{K}}$ such that $\mu(K^c_\varepsilon) < \varepsilon$ for each $\mu \in \overline{\mathfrak{K}}$. Choose an element G_ε of $\underline{\mathcal{G}}'$ such that $K_\varepsilon \subset G_\varepsilon$ and let $\underline{\mathcal{G}}'' = \{G_\varepsilon\} \subset \underline{\mathcal{G}}'$. We have $\mu(G^c_\varepsilon) < \varepsilon$ for each $\mu \in \overline{\mathfrak{K}}$, thus Topsøe's criterion applies. Thus $\overline{\mathfrak{K}}$ is compact and $\mathfrak{K}$ is relatively compact. From Lemma 4.1.1, $\mathfrak{K}$ is also sequentially relatively compact.

B) Let $\mathfrak{K}$ be a $\tau^{\mathrm{N}}_{\mathcal{Y}^1}$–relatively compact subset of $\mathcal{Y}^1$. For each $\mu \in \mathcal{Y}^1$, let $\mu_\Omega = \mu(\Omega \times .)$ be the marginal of μ on $\mathbb{T}$, defined by $\mu_\Omega(B) = \mu(\Omega \times B)$ for any $B \in \mathcal{B}_{\mathbb{T}}$. It is clear that the mapping $\phi : \mu \mapsto \mu_\Omega$, $\mathcal{Y}^1 \to \mathcal{M}^{+,1}(\mathbb{T})$ is continuous for $\tau^{\mathrm{N}}_{\mathcal{Y}^1}$, thus $\phi(\mathfrak{K})$ is relatively compact, thus it is tight, which means that, for any $\varepsilon > 0$, there exists a compact subset K of $\mathbb{T}$ such that

$$\sup_{\mu \in \mathfrak{K}} \mu\left((\Omega \times K)^c\right) < \varepsilon.$$

□

Remarks 4.3.6 (Extension to $(\mathcal{M}^{+,1}(\Omega \times \mathbb{T}), \tau^{\mathrm{N}}_{\mathcal{M}})$)

1. The generalization to $(\mathcal{M}^{+,1}(\Omega \times \mathbb{T}), \tau^{\mathrm{N}}_{\mathcal{M}})$ of Theorem 4.3.5 (except the results on relative sequential compactness) is immediate.

2. If Ω is standard and $\mathbb{T}$ is Suslin regular, we can complete this result in the following way. Let τ_Ω be a Polish topology on Ω such that $\mathcal{S} = \mathcal{B}_{(\Omega, \tau_\Omega)}$. Then, from Theorem 2.1.13, $\tau^{\mathrm{N}}_{\mathcal{M}}$ coincides with the narrow topology. If $\mathfrak{K}$ is strictly tight, it is tight w.r.t. the set $\mathcal{K}(\Omega \times \mathbb{T})$. Conversely, if $\mathbb{T}$ is Prohorov, as the product of two Prohorov spaces is (trivially) Prohorov, the space $\Omega \times \mathbb{T}$ endowed with the product topology $\mathcal{T}_s$ is Prohorov. Thus any relatively compact subset of $\mathcal{M}^{+,1}(\Omega \times \mathbb{T})$ is tight w.r.t. $\mathcal{K}(\Omega \times \mathbb{T})$.

The following result can be seen as a semicontinuity theorem.

Corollary 4.3.7 (When $\tau^{\mathrm{W}}_{\mathcal{Y}^1}$ and $\tau^{\mathrm{S}}_{\mathcal{Y}^1}$ have the same compact subsets) *Assume that $\mathbb{T}$ is Prohorov and completely regular, and that its compact subsets are metrizable. Assume furthermore that $\mathcal{M}^{+,1}(\mathbb{T}) = \mathcal{M}^{+,1}_{\tau}(\mathbb{T})$ (all these conditions are statisfied if $\mathbb{T}$ is Suslin regular Prohorov). Let $\mathfrak{K} \subset \mathcal{Y}^1$. The following conditions are equivalent:*

(a) $\mathfrak{K}$ is $\tau^{\mathrm{S}}_{\mathcal{Y}^1}$–relatively compact.

(b) $\mathfrak{K}$ is $\tau^{\mathrm{W}}_{\mathcal{Y}^1}$–relatively compact.

(c) $\mathfrak{K}$ is strictly tight.

Proof. As $\mathbb{T}$ is completely regular, the topology $\tau^{\mathrm{W}}_{\mathcal{Y}^1}$ is Hausdorff. The implication $(a) \Rightarrow (b)$ is obvious. Assume (b). Then the set $(\pi_{\mathbb{T}})_\sharp(\mathfrak{K}) = \{\mu(\Omega \times .);\, \mu \in \mathfrak{K}\}$ of margins on $\mathbb{T}$ of elements of $\mathfrak{K}$ is relatively compact for the weak topology on $\mathcal{M}^{+,1}(\mathbb{T}) = \mathcal{M}^{+,1}_{\tau}(\mathbb{T})$, which coincides with the narrow topology because $\mathbb{T}$ is completely regular. As $\mathbb{T}$ is Prohorov, $(\pi_{\mathbb{T}})_\sharp(\mathfrak{K})$ is tight, thus we have (c). The implication $(c) \Rightarrow (a)$ comes from the Direct Prohorov Criterion. □

Theorem 4.3.5 yields the following complement to the Portmanteau Theorem 2.1.3, which extends by a different method [CV98, Proposition 3.1]. This theorem is also a semicontinuity theorem. Note that the limit of the net does not need to be explicited.

Theorem 4.3.8 (Portmanteau Theorem continued) *Assume that $\mathcal{M}^{+,1}_{\tau}(\mathbb{T}) = \mathcal{M}^{+,1}(\mathbb{T})$ (e.g. $\mathbb{T}$ is Radon or hereditarily Lindelöf). Assume furthermore that $\mathbb{T}$ is completely regular and that each compact subset of $\mathbb{T}$ is metrizable. Let $(\mu^\alpha)_{\alpha\in\mathbb{A}}$ be a net in $\mathcal{Y}^1$. Let $\mathcal{C}$ be a set of nonnegative $\mathcal{S}$–measurable bounded functions which is stable under multiplication of two elements, which contains the constant function 1 and which generates $\mathcal{S}$ (e.g. $\mathcal{C}$ consists in the indicator functions of the elements of a subset of $\mathcal{S}$ which is stable under finite intersection, which contains Ω and which generates $\mathcal{S}$).*

Assume that one of the following conditions is satisfied:

(i) $(\mu^\alpha)_\alpha$ is strictly tight

(ii) or $(\mu^\alpha)_\alpha$ is flexibly tight and $\mathbb{T}$ is Suslin regular.

The following are equivalent:

1. *$(\mu^\alpha)_\alpha$ is S–stably convergent.*
2. *$(\mu^\alpha)_\alpha$ is W–stably convergent.*
3. *For each $f \in \mathcal{C}$, the net $(\mu^\alpha(f \otimes .))_\alpha$ has a limit in $\mathcal{M}^+(\mathbb{T})$.*
4. *For each $f \in \mathcal{C}$ and each $g \in \mathrm{C}_b(\mathbb{T})$, the net $(\mu^\alpha(f \otimes g))_\alpha$ is convergent.*

Remark 4.3.9 Under Assumption 3, if $(\mu^\alpha)_\alpha$ is a sequence and if $\mathbb{T}$ is Prohorov, the set of margins $\{\mu^\alpha(\Omega \times .);\ \alpha \in \mathbb{A}\}$ is a relatively compact subset of $\mathcal{M}^{+,1}(\mathbb{T})$, thus Condition (i) is satisfied. We shall give a more general result for sequences in Theorem 4.5.1.

Remark 4.3.10 The hypothesis that $\mathcal{M}_\tau^{+,1}(\mathbb{T}) = \mathcal{M}^{+,1}(\mathbb{T})$ can be skipped in Theorem 4.3.8, if we assume that $(\mu^\alpha)_{\alpha\in\mathbb{A}}$ is a net of τ–regular Young measures, that is, $\mu^\alpha(A \times .) \in \mathcal{M}_\tau^+(\mathbb{T})$ for each $\alpha \in \mathbb{A}$ and each $A \in \mathcal{S}$.

Proof of Theorem 4.3.8. The implications $1 \Rightarrow 2 \Rightarrow 4$ and $1 \Rightarrow 3 \Rightarrow 4$ are obvious. So, we only need to prove $4 \Rightarrow 3$ and $3 \Rightarrow 1$.

Note that, as $\mathbb{T}$ is completely regular, the weak and narrow topologies coincide on $\mathcal{M}_\tau^{+,1}(\mathbb{T}) = \mathcal{M}^{+,1}(\mathbb{T})$.

Now, if (i) or (ii) holds true, then, by Theorem 4.3.5.(A), the net $(\mu^\alpha)_\alpha$ is $\tau_{\mathcal{Y}^1}^{\mathrm{S}}$–relatively compact. Therefore, in any case, each subnet of $(\mu^\alpha)_\alpha$ has a convergent subnet $(\mu^{\alpha_\beta})_\beta$ which converges to a limit in $\mathcal{Y}^1$.

Assume 4. For each limit λ of a subnet of $(\mu^\alpha)_\alpha$ and for every $f \in \mathcal{C}$, we have

$$\forall g \in \mathrm{C}_b(\mathbb{T}) \quad \lambda(f \otimes g) = \lim_\alpha \mu^\alpha(f \otimes g),$$

that is, for every $f \in \mathcal{C}$ and for all such limits λ, the measures $\lambda(f \otimes .)$ coincide on $\mathrm{C}_b(\mathbb{T})$. As $\mathbb{T}$ is completely regular, $\mathrm{C}_b(\mathbb{T})$ separates the elements of $\mathcal{M}^+(\mathbb{T})$, thus, for each $f \in \mathcal{C}$, there exists a measure $\nu \in \mathcal{M}^+(\mathbb{T})$ such that each subnet of $(\mu^\alpha(f \otimes .))_\alpha$ has a further subnet which converges to ν. Therefore, $(\mu^\alpha(f \otimes .))_\alpha$ converges to ν, that is, Condition 3 is satisfied.

Assume now 3. Let $\mathcal{E}$ be the set of bounded measurable functions f on Ω such that $(\mu^\alpha(f \otimes .))_\alpha$ has a limit in $\mathcal{M}^+(\mathbb{T})$. As in the proof of Part C in Theorem 2.1.3, the set $\mathcal{E}$ is a monotone vector space which contains $\mathcal{C}$, thus it contains all bounded measurable functions on Ω. This proves that 3 does not depend on the choice of $\mathcal{C}$, thus we can assume without loss of generality that $\mathcal{C} = \{\,\mathbb{1}_A;\ A \in \mathcal{S}\}$.

For each limit λ of a subnet of $(\mu^\alpha)_\alpha$ and for every $A \in \mathcal{S}$, we have

$$\lambda(A \times .) = \lim_\alpha \mu^\alpha(A \times .).$$

Thus all such limits λ coincide on the product $\mathcal{S} \times \mathcal{B}_\mathbb{T}$. As $\mathcal{S} \times \mathcal{B}_\mathbb{T}$ is stable by finite intersections and generates $\underline{\mathcal{B}}$, all thoses limits coincide on $\underline{\mathcal{B}}$. We have proved that there exists a Young measure λ such that, for each subnet of $(\mu^\alpha)_\alpha$, there exists a further subnet which S–stably converges to λ. This proves that $(\mu^\alpha)_\alpha$ S–stably converges to λ. □

Remark 4.3.11 In the cases (i) and (ii), if $(\mu^\alpha)_\alpha$ is a *sequence*, we can reproduce the reasoning of the proof $3 \Rightarrow 1$ using subsequences instead of subnets, because, from Theorem 4.3.5, $(\mu^\alpha)_\alpha$ is relatively sequentially compact.

Support Theorem for strictly tight sequences We now give a result on the support of the limit of a sequence of Young measures. This result will be used in Section 6.1. Further developments on the same topic will be made in Section 6.5. If $\mu \in \mathcal{M}^{+,1}(\mathbb{T})$, we denote by $\operatorname{Supp}\mu$ the support of μ (if it exists), that is, the intersection of all closed subsets F such that $\mu(F) = 1$. It is easy to check that the support of μ exists when μ is τ–regular, in particular when $\mathbb{T}$ is Radon or when $\mathbb{T}$ is hereditarily Lindelöf. If μ is a disintegrable Young measure and $\operatorname{Supp}\mu_\omega$ exists for all $\omega \in \Omega$, we shall denote by $\operatorname{Supp}\mu$ the multifunction $\omega \mapsto \operatorname{Supp}\mu_\omega$.

We also have to introduce the *superior limit in the sense of Kuratowski* of a sequence $(B_n)_n$ of subsets of $\mathbb{T}$: this is the set

$$\operatorname{ls}(B_n) = \bigcap_n \overline{\bigcup_{m \geq n} B_m}.$$

The following theorem is a generalization of [Val90b, Proposition 5 page 159].

Theorem 4.3.12 (Support Theorem) *Assume that $\mathbb{T}$ has metrizable compact subsets. Let $(\mu^n)_{n\in\mathbb{N}}$ be a strictly tight sequence in $\mathcal{Y}^1$ which W–stably converges to some $\mu^\infty \in \mathcal{Y}^1$. For each $n \in \mathbb{N}$ and each $\omega \in \Omega$, let $\operatorname{Supp}\mu^n_\omega$ be the support of the measure μ^n_ω. Let L be the graph of the multifunction $\operatorname{ls}(\operatorname{Supp}\mu^n_.)$, that is,*

$$L = \{(\omega, t) \in \Omega \times \mathbb{T};\, t \in \operatorname{ls}(\operatorname{Supp}\mu^n_\omega)\}.$$

We have

$$\mu^\infty_*(L) = 1,$$

where μ^∞_* is the interior measure associated with μ^∞.

Proof. We can assume w.l.g. that $\mathbb{T}$ is Suslin, because there exists a K_σ subset $\mathbb{T}_0$ of $\mathbb{T}$ such that $\mu^n(\Omega \times \mathbb{T}_0) = 1$ for every $n \in \mathbb{N} \cup \{\infty\}$. As $\mathbb{T}$ is Radon, there exists a sequence $(K_j)_{j\geq 1}$ of elements of $\mathcal{K}$ such that, for each $j \geq 1$ and for each $n \in \mathbb{N} \cup \{\infty\}$, $\mu^n(\Omega \times K_j) \geq 1/j$. For each $n \in \mathbb{N}$ and each $j \geq 1$, let $\Gamma_{n,i}$ be the compact valued multifunction $\omega \mapsto \operatorname{Supp}\mu^n_\omega(.\cap K_j)$. For each $U \in \mathcal{G}(\mathbb{T})$, we have

$$\Gamma^-_{n,i}U = \{\omega \in \Omega;\, \mu^n_\omega(U \cap K_j) > 0\} \in \mathcal{S},$$

thus $\Gamma_{n,j}$ is LV–measurable. For every $n \in \mathbb{N}$, the multifunction

$$\Delta_{n,j} : \omega \mapsto \overline{\cup_{m\geq n}\Gamma_{m,j}(\omega)}$$

is thus LV–measurable, with values contained in K_j. Since the space K_j is separable metrizable, we have $\operatorname{gph}(\Delta_{n,j}) \in \underline{\mathcal{K}}(K_j)$ (see (*vi*) page 7), which implies $\operatorname{gph}(\Delta_{n,j}) \in \mathcal{S} \otimes \mathcal{B}_\mathbb{T}$. We thus have $\cap_n \Delta_{n,j} \in \mathcal{S} \otimes \mathcal{B}_\mathbb{T}$, that is,

$$L_j := \cap_n \overline{\cup_{m\geq n} \operatorname{gph}(\Gamma_{m,j})} \in \mathcal{S} \otimes \mathcal{B}_\mathbb{T}.$$

Clearly, we also have $L_j \subset L$. Thus, we only need to prove that we have

$$\mu^\infty(\cup_{j\geq 1} L_j) = 1 \tag{4.3.1}$$

But we have, for each $j \geq 1$,

$$\forall n_0 \in \mathbb{N} \quad \forall n \geq n_0 \quad \mu^n \left(\text{gph}\left(\overline{\cup_{m \geq n_0} \Gamma_{m,j}}\right)\right) \geq 1 - \frac{1}{j},$$

thus

$$\limsup_n \mu^n(L_j) \geq 1 - \frac{1}{j}.$$

Now, from Theorem 4.3.8, the sequence $(\mu^n)_{n\in\mathbb{N}}$ S–stably converges to μ^∞. We thus have

$$\mu^\infty(L_j) \geq \limsup_n \mu^n(L_j) \geq 1 - \frac{1}{j},$$

which proves (4.3.1). □

Comparison of flexible tightness and strict tightness The following theorem is a generalization (with a similar proof) of a result of the third author, which was given in [Jaw84].

Theorem 4.3.13 (Equivalence of tightness notions) *Assume that $\mathbb{T}$ is Suslin and that the set $\mathcal{K}$ of compact subsets of $\mathbb{T}$, endowed with the Vietoris topology, is Radon. A set $\mathfrak{Y} \subset \mathcal{Y}^1$ is flexibly tight if and only if it is strictly tight.*

Such a theorem is not very useful if we do not know when $\mathcal{K}$ is Radon. This is why we now exhibit two particular cases.

Let us say that a sequence $(K_n)_n$ of compact subsets of $\mathbb{T}$ is a *cofinal sequence* if each compact subset of $\mathbb{T}$ is contained in some K_n (in this case, it is said that $\mathbb{T}$ is *hemicompact*). Note that, if the compact subspaces K_n are metrizable, $\mathbb{T}$ is a countable union of Polish spaces, thus it is Lusin.

Theorem 4.3.14 (Cases of equivalence of tightness notions) *Assume that one of the following conditions is satisfied:*

(*i*) *$\mathbb{T}$ is a regular Suslin Prohorov space,*

(*ii*) *$\mathbb{T}$ has a cofinal sequence of metrizable compact subsets.*

Then the space $\mathcal{K}$ is Radon, thus any flexibly tight subset of $\mathcal{Y}^1$ is strictly tight.

Polish spaces are regular Suslin Prohorov spaces. Another useful example of Case (*i*) is provided by submetrizable k_ω–spaces, that we shall investigate later in Section 4.4.

Submetrizable k_ω–spaces also belong to Case (*ii*). Here is another example, which is not included in Case (*i*).

Example 4.3.15 (Topological vector space with a cofinal sequence of compact subsets) Condition (*ii*) of Theorem 4.3.14 is satisfied if $\mathbb{T}$ is the weak dual $\mathbb{E}^*_\sigma$ of a separable (locally convex) Fréchet space $\mathbb{E}$. Indeed, as $\mathbb{E}$ is barrelled, each bounded subset of $\mathbb{E}^*_\sigma$ is relatively compact, and, from [Bou81, Proposition 2 page IV.21], there exists a sequence $(K_n)_{n\geq 1}$ of closed bounded subsets of $\mathbb{E}^*_\sigma$ such that each bounded subset of $\mathbb{E}^*_\sigma$ is contained in some K_n.

Furthermore, from the separability of $\mathbb{E}$, there exists a countable family of continuous functions on $\mathbb{E}^*_\sigma$ which separates the points of $\mathbb{E}^*_\sigma$, thus $\mathbb{E}^*_\sigma$ is submetrizable, thus its compact subsets are metrizable.

But, if $\mathbb{E}$ is not nuclear, $\mathbb{E}^*_\sigma$ does not belong to Case (*i*), because $\sigma(\mathbb{E}^*, \mathbb{E})$ is not Prohorov [Fer94].

Proof of Theorem 4.3.13. Assume that $\mathfrak{Y}$ is flexibly tight and let $\varepsilon > 0$. Let $K \in \underline{\mathcal{K}}$ be such that

$$\sup_{\mu\in\mathfrak{Y}} \mu\left(K^c\right) = \sup_{\mu\in\mathfrak{Y}} \int_\Omega \mu_\omega(\mathbb{T}\setminus K(\omega))\, d\,\mathrm{P}(\omega) < \frac{\varepsilon}{2}.$$

From Remark 1.2.1, as $\mathbb{T}$ is Suslin, K can be seen as a random element of $\mathcal{K}$ (defined on the universal completion $(\Omega, \mathcal{S}^*, \mathrm{P}^*)$ of $(\Omega, \mathcal{S}, \mathrm{P})$), because, for any open subset U of $\mathbb{T}$, the sets

$$\{\omega\in\Omega;\ K(\omega)\cap U \neq \emptyset\} = \pi_\Omega\left(K\cap(\Omega\times U)\right)$$

and

$$\{\omega\in\Omega;\ K(\omega)\subset U\} = \pi_\Omega\left(K\cap(\Omega\times U^c)\right)^c$$

belong to $\mathcal{S}^*$, where π_Ω denotes the canonical projection $\Omega\times\mathbb{T}\to\Omega$. To simplify notations, we assume now w.l.g. that $(\Omega, \mathcal{S}, \mathrm{P}) = (\Omega, \mathcal{S}^*, \mathrm{P}^*)$.

As $\mathcal{K}$ is radon, the law $\mathcal{L}\left(K\right)$ of the random element K is Radon, thus there exists a compact subset $\mathfrak{K}$ of $\mathcal{K}$ such that

$$\mathrm{P}\left\{\omega\in\Omega;\ K(\omega)\in\mathfrak{K}\right\} \geq 1-\frac{\varepsilon}{2}.$$

From Michael's characterization of compact subsets of $\mathcal{K}$ [Mic51, Theorem 2.5.2] (see also [Chr74, Theorem 3.1] for the converse implication), the set

$$H = \cup_{L\in\mathfrak{K}} L$$

is compact.

Now, let $\Omega' = \{\omega\in\Omega;\ K(\omega)\in\mathfrak{K}\}$. We have $\mathrm{P}(\Omega\setminus\Omega') \leq \varepsilon/2$ thus, for any $\mu\in\mathfrak{Y}$,

$$\begin{aligned}\mu(\Omega\times H^c) &= \int_{\Omega'} \mu_\omega(H^c)\, d\,\mathrm{P}(\omega) + \int_{\Omega\setminus\Omega'} \mu_\omega(H^c)\, d\,\mathrm{P}(\omega)\\ &\leq \int_{\Omega'} \mu_\omega(\mathbb{T}\setminus K(\omega))\, d\,\mathrm{P}(\omega) + \frac{\varepsilon}{2}\\ &< \varepsilon.\end{aligned}$$

□

Remark 4.3.16 It is easy to prove directly that, under the hypothesis (i) or (ii) of Theorem 4.3.14, any flexibly tight subset of $\mathcal{Y}^1$ is strictly tight.

Indeed, let $\mathfrak{Y} \subset \mathcal{Y}^1$ be flexibly tight. If (i) is satisfied, from Prohorov Criterion, $\mathfrak{Y}$ is relatively compact in $\mathcal{Y}^1$. From the Converse Prohorov Criterion, it is thus strictly tight.

Assume now (ii). Then $\mathbb{T}$ is Lusin. Let $(L_n)_{n\in\mathbb{N}}$ be a nondecreasing cofinal sequence of metrizable compact subsets of $\mathbb{T}$. Let $\varepsilon > 0$ and let $K \in \underline{\mathcal{K}}$ such that $\mu(K) \geq 1 - \varepsilon/2$ for each $\mu \in \mathfrak{Y}$. By the Projection Theorem, for each $n \in \mathbb{N}$, the set $A_n = \{\omega \in \Omega;\ K(\omega) \subset L_n\}$ is in $\mathcal{S}^*$, and we have $\cup_n A_n = \Omega$ and $(A_n)_n$ is nondecreasing. There exists thus $n_0 \in \mathbb{N}$ such that $\mathrm{P}^*(A_{n_0}) \geq 1 - \varepsilon/2$. We then have, for every $\mu \in \mathfrak{Y}$,

$$\mu(\Omega \times L_{n_0}) \geq \mu(K \setminus (A_{n_0}^c \times \mathbb{T})) \geq 1 - \varepsilon/2 - \varepsilon/2 = 1 - \varepsilon.$$

A part of the proof of Theorem 4.3.14 lies in the following lemma. We recall that, if A is a random subset of $\mathbb{T}$ and $f : \Omega \to \mathbb{T}$ a mapping, we say that f is a selection of A if $f(\omega) \in A(\omega)$ for each $\omega \in \Omega$.

Lemma 4.3.17 *Assume that $\mathbb{T}$ is Suslin submetrizable. Let K be a random element of $\mathcal{K}$. Let $\mathcal{Z}$ be the set of measurable selections of K and set $\mathfrak{Y} = \{\mathcal{L}(f);\ f \in \mathcal{Z}\}$. Then $\mathfrak{Y}$ is a relatively compact subset of $\mathcal{M}^{+,1}(\mathbb{T})$.*

Proof. From Remark 1.2.1 (vi), K is graph–measurable, that is, $K \in \underline{\mathcal{K}}$. Let $\mathfrak{Y}' = \{\underline{\delta}_f;\ f \in \mathcal{Z}\}$. The set $\mathfrak{Y}'$ is flexibly tight thus, from the Direct Prohorov Criterion (Theorem 4.3.5), it is relatively compact in $\mathcal{Y}^1$. Thus $\mathfrak{Y} = (\pi_{\mathbb{T}})_\sharp(\mathfrak{Y}')$ is relatively compact in $\mathcal{M}^{+,1}(\mathbb{T})$. □

Proof of Theorem 4.3.14.

(i) We denote as usual by $\tau_{\mathbb{T}}$ the topology of $\mathbb{T}$ and by τ_0 a separable metrizable topology which is coarser than $\tau_{\mathbb{T}}$. Necessarily, τ_0 has the same Borel sets as $\tau_{\mathbb{T}}$.

Let us denote by $\tau_{\mathrm{V}}(\tau_{\mathbb{T}})$ the Vietoris topology on $\mathcal{K}$ and by $\tau_{\mathrm{V}}(\tau_0)$ the Vietoris topology on the bigger set $\mathcal{K}(\mathbb{T}, \tau_0)$. The topology $\tau_{\mathrm{V}}(\tau_{\mathbb{T}})$ is submetrizable, because it is finer than the trace on $\mathcal{K}$ of $\tau_{\mathrm{V}}(\tau_0)$, which is metrizable. Thus, to prove that $\mathcal{K}(\mathbb{T}, \tau_{\mathbb{T}})$ is Radon, we only need to prove that every element of $\mathcal{M}^{+,1}(\mathcal{K}(\mathbb{T}, \tau_{\mathbb{T}}))$ is tight (see page 12).

Let K be a random element of $\mathcal{K}$. With the notations of Lemma 4.3.17, the set $\mathfrak{Y}$ is relatively compact. As $\mathbb{T}$ is a completely regular Prohorov space, the space $\mathcal{M}^{+,1}(\mathbb{T})$ is Prohorov [Wój87]. There exists thus a nondecreasing sequence $(L_m)_{m\geq 1}$ of compact subsets of $\mathbb{T}$ such that, for each element f of $\mathcal{Z}$ and each m, we have $\mathrm{P}\{\omega \in \Omega;\ f(\omega) \notin L_m\} \leq 1/m$.

Assume that the law μ of K is not tight. From Michael's characterization of compact subsets of $\mathcal{K}$ [Mic51, Theorem 2.5.2], there exists $\varepsilon > 0$ such that, for each $m \geq 1$, the set $A_m := \{\omega \in \Omega;\ K(\omega) \not\subset L_m\}$ satisfies $\mathrm{P}(A_m) > \varepsilon$. From [CV77, Theorem III.22], for each $m \geq 1$, there exists on A_m a measurable selection f_m of

$\omega \mapsto K(\omega) \setminus L_m$. Let σ_0 be an arbitrary measurable selection of K. We extend f_m on Ω by setting $f_m(\omega) = \sigma_0(\omega)$ for $\omega \notin A_m$. We have $f_m \in \mathcal{Z}$, thus

$$\mathrm{P}(A_m) = \mathrm{P}\{\omega \in \Omega;\ f_m(\omega) \notin L_m\} \leq 1/m$$

which, for $m > 1/\varepsilon$, contradicts $\mathrm{P}(A_m) > \varepsilon$.

(*ii*) Let $(L_n)_{n\in\mathbb{N}}$ be a cofinal sequence of metrizable compact subsets of $\mathbb{T}$. From [Chr74, Theorem 3.1], if $\mathfrak{C}$ is a compact subset of $\mathcal{K}(\mathbb{T})$, there exists $n \in \mathbb{N}$ such that every element of $\mathfrak{C}$ is a subset of L_n. Thus we have

$$\mathcal{K}(\mathbb{T}) = \cup_{n\in\mathbb{N}}\mathcal{K}(L_n).$$

Furthermore, the Vietoris topology on each $\mathcal{K}(L_n)$ is induced by the Vietoris topology on $\mathbb{T}$. But each $\mathcal{K}(L_n)$ is Polish (e.g. [CV77, Corollary II.9]), thus $\mathcal{K}(\mathbb{T})$ is a countable union of Lusin spaces, thus it is Lusin, thus it is Radon. □

4.4 Submetrizable k_ω–spaces, Change of Topology Lemma

Submetrizable k_ω–spaces Submetrizable k_ω–spaces are Suslin spaces which share many properties with Polish spaces. They are interesting not only because of these nice properties, but also because, in the study of tight sets of Young measures, it is sometimes possible to replace the topology of $\mathbb{T}$ by a stronger one which makes $\mathbb{T}$ a submetrizable k_ω–space (see the "Change of Topology Lemma" 4.4.3, and some of its applications in Section 4.5).

We say that a Hausdorff topological space $\mathbb{T}$ is a *k_ω–space* if there exists a countable family $\mathcal{K}$ of compact subsets of $\mathbb{T}$ such that $\mathbb{T} = \cup_{K\in\mathcal{K}}K$ and any subset U of $\mathbb{T}$ is open if and only if, for each $K \in \mathcal{K}$, $K \cap U$ is an open subset of K (in the relative topology). We then say that $\mathcal{K}$ *determines* the topology of $\mathbb{T}$. The k_ω–spaces are the same spaces as the *hemicompact k_R–spaces* considered in [Whe83].

It is easy to check that, if $\mathcal{K} = \{K_0, K_1, \ldots\}$ determines the topology of $\mathbb{T}$, then $\{K_0, K_0 \cup K_1, \ldots, \cup_{j\leq n}K_n, \ldots\}$ also determines the topology of $\mathbb{T}$, thus we can always assume that $\mathcal{K} = \{K_0, K_1, \ldots\}$ for a nondecreasing sequence $(K_n)_n$. In other words, a k_ω–space is the inductive limit of a nondecreasing sequence of compact subsets.

Here are some properties of k_ω–spaces which will be useful in the sequel.

1. Assume that $(K_n)_n$ is a nondecreasing sequence of compact subsets of $\mathbb{T}$ which determines the topology of $\mathbb{T}$. Then $(K_n)_n$ is a cofinal sequence of compact subsets of $\mathbb{T}$. Indeed, assume the contrary and let K be a compact subset of $\mathbb{T}$ which is not contained in any K_n. There exist a sequence $(x_k)_k$ of elements of K and a subsequence $(K_{n_k})_k$ of $(K_n)_n$ such that $x_k \in K_{n_{k+1}} \setminus K_{n_k}$ for each k. Let $L = \{x_k;\ k \in \mathbb{N}\}$. For each $n \in \mathbb{N}$, the set $L \cap K_n$ is finite,

thus it is closed. As $\mathbb{T}$ is determined by the sets K_n, this means that L is closed, thus L is compact because it is contained in K. But the relative topology of L is determined by the sets $\{K_n \cap L;\ n \in \mathbb{N}\}$, which are finite. Thus L is discrete. This leads to a contradiction, because every discrete compact space is finite.

2. Every Hausdorff k_ω–space is normal, as was proved by Morita [Mor56] (see also [All96, Proposition 2.7.1]).

3. Every k_ω–space whose compact subsets are metrizable is a (normal, thus regular) Lusin space, because it is a countable union of Polish spaces.

4. Thus a k_ω–space is submetrizable if and only if its compact subsets are metrizable.

5. Every k_ω–space is Prohorov: This is a consequence of Corollary 6 in [HJ72].

Examples 4.4.1 (k_ω–spaces: a recipe) To build a k_ω–space, one can take a topological space $(\mathbb{T}, \tau_0)$ which admits a cofinal sequence of compact subsets, and then endow $\mathbb{T}$ with the finest topology τ which has the same compact sets as τ_0.

The following example comes from [HJ72]: If $\mathbb{E}$ is a Fréchet space, we can take for $(\mathbb{T}, \tau_0)$ the weak dual $\mathbb{E}^*_\sigma$ as in Example 4.3.15. Then, from Banach–Dieudonné Theorem (e.g. [Bou81, Sch66]), τ is the topology τ_c of uniform convergence on compact subsets of $\mathbb{E}$. If, furthermore, $\mathbb{E}$ is separable, then $\mathbb{T} = \mathbb{E}^*_c$ is a submetrizable k_ω–space.

In particular, if $\mathbb{E}$ is a Fréchet–Montel space, then $\tau = \tau_c$ is also the topology τ_b of uniform convergence on bounded subsets of $\mathbb{E}$, that is, $\mathbb{T}$ is the strong dual $\mathbb{E}^*_b$ of $\mathbb{E}$. Thus the space $\mathcal{S}'$ of tempered distributions is a submetrizable k_ω–space.

The following lemma shows how close to Polish spaces are submetrizable k_ω–spaces, with respect to properties of tight sets of Young measures. It will be used in particular in the proof of Proposition 4.5.2, in the case when $\mathbb{T}$ is a submetrizable k_ω–space.

Lemma 4.4.2 (Lifting Lemma) *Assume that $\mathbb{T}$ is a submetrizable k_ω–space and let $(K_n)_n$ be a nondecreasing sequence of compact subsets of $\mathbb{T}$ which determines the topology of $\mathbb{T}$. There exist a Polish locally compact k_ω–space $\mathbb{S}$, a continuous surjection $\pi : \mathbb{S} \to \mathbb{T}$ and a Borel injection $\psi : \mathbb{T} \to \mathbb{S}$ such that*

1. *$\pi \circ \psi$ is the identity mapping of $\mathbb{T}$ (in other words, ψ is a lifting of π),*

2. *For each compact subset K of $\mathbb{T}$, $\psi(K)$ is a relatively compact subset of $\mathbb{S}$.*

3. *A subset $\mathfrak{K}$ of $\mathcal{M}^{+,1}(\mathbb{T})$ is tight if and only if its image $\psi_\sharp(\mathfrak{K})$ by ψ in $\mathcal{M}^{+,1}(\mathbb{S})$ is tight.*

Furthermore, the mappings

$$\underline{\pi} : \begin{cases} \Omega \times \mathbb{S} & \rightarrow & \Omega \times \mathbb{T} \\ (\omega, s) & \mapsto & (\omega, \pi(s)) \end{cases} \quad \text{and } \underline{\psi} : \begin{cases} \Omega \times \mathbb{T} & \rightarrow & \Omega \times \mathbb{S} \\ (\omega, t) & \mapsto & (\omega, \psi(t)) \end{cases}$$

are measurable and satisfy

4. $\underline{\pi} \circ \underline{\psi}$ *is the identity mapping of* $\Omega \times \mathbb{T}$,

5. *For each* $K \in \underline{\mathcal{K}}$, $\overline{\underline{\psi}(K)}$ *is a random compact subset of* $\mathbb{S}$, *and, for each random compact subset* K *of* $\mathbb{S}$, *we have* $\underline{\pi}(K) \in \underline{\mathcal{K}}$.

6. *For* $* = S, M, N, W$, *the mapping* $\underline{\pi}_\sharp$ *maps continuously* $(\mathcal{Y}^1(\mathbb{S}), \tau^{\mathrm{W}}_{\mathcal{Y}^1(\mathbb{S})})$ *onto* $(\mathcal{Y}^1(\mathbb{T}), \tau^{*}_{\mathcal{Y}^1})$ *(note that, from Part G of Portmanteau Theorem 2.1.3, we know that the topologies* $\tau^{*}_{\mathcal{Y}^1(\mathbb{S})}$ *coincide). Furthermore, a subset* $\mathfrak{K}$ *of* $\mathcal{Y}^1(\mathbb{T})$ *is strictly tight if and only if its image* $\underline{\psi}_\sharp(\mathfrak{K})$ *in* $\mathcal{Y}^1(\mathbb{S})$ *by* $\underline{\psi}$ *is strictly tight.*

Proof. For each integer $m \geq 1$, let d_m be a distance on K_m which is compatible with the topology induced by τ and such that $d_m \leq 1$.

For each $m \geq 1$, let $\widetilde{K}_m = \{m\} \times K_m$. We then set

$$\mathbb{S} = \cup_{m \geq 1} \{m\} \times K_m$$

and we endow $\mathbb{S}$ with the distance $\widetilde{d}$ defined by $\widetilde{d}((m,x),(n,y)) = d_m(x,y)$ if $m = n$ and $\widetilde{d}((m,x),(n,y)) = 1$ otherwise. The space $\mathbb{S}$ is a K_σ metric space (it is even locally compact), thus it is separable, and it is easy to check that it is complete (see [Bou74], Proposition 1 page IX.57), that the canonical surjection $\pi : \mathbb{S} \rightarrow \mathbb{T}$, $(m, x) \mapsto x$ is continuous and that, for each $m \geq 1$, π induces a homeomorphism of $\widetilde{K}_m$ onto K_m. The topology of $\mathbb{S}$ is determined by the sets $C_n = \cup_{j \leq n} \widetilde{K}_j$.

For each $m \geq 1$, let π_m be the restriction of π to $\widetilde{K}_m$. Let $\psi : \mathbb{T} \rightarrow \mathbb{S}$ defined by $\psi(x) = \pi_m^{-1}(x)$ if $x \in K_m \setminus K_{m-1}$ (we set $K_0 = \emptyset$). Thus $\pi \circ \psi(x) = x$ for every $x \in \mathbb{T}$, and ψ is a measurable injection.

The proof of 3 follows immediately from the fact that ψ and π map compact subsets into relatively compact subsets.

The verification of 4 and 5 is trivial.

Let us prove 6. To prove the continuity property, we only need to prove that $\underline{\pi}_\sharp$ maps continuously $(\mathcal{Y}^1(\mathbb{S}), \tau^{\mathrm{S}}_{\mathcal{Y}^1(\mathbb{S})})$ onto $(\mathcal{Y}^1(\mathbb{T}), \tau^{\mathrm{S}}_{\mathcal{Y}^1})$.

Let G be a random open subset of $\mathbb{T}$. The set $\underline{\pi}^{-1}(G)$ is measurable and, for each $\omega \in \Omega$, the set $\pi^{-1}(G(\omega))$ is open, thus $\underline{\pi}^{-1}(G)$ is a random open subset of $\mathbb{S}$. Let $(\mu^\alpha)_\alpha$ be a net in $\mathcal{Y}^1(\mathbb{S})$ which S–stably converges to an element μ^∞ of $\mathcal{Y}^1(\mathbb{S})$. We thus have

$$\begin{aligned} \liminf_\alpha (\underline{\pi}_\sharp \mu^\alpha)(G) &= \liminf_\alpha \mu^\alpha(\underline{\pi}^{-1}(G)) \\ &\geq \mu^\infty(\underline{\pi}^{-1}(G)) \\ &= (\underline{\pi}_\sharp \mu^\infty)(G), \end{aligned}$$

which shows the continuity of $\underline{\pi}_\sharp$ for $\tau^{\mathrm{S}}_{\mathcal{Y}^1(\mathbb{S})}$ and $\tau^{\mathrm{S}}_{\mathcal{Y}^1}$. As there is only one stable topology on $\mathcal{Y}^1(\mathbb{S})$, this entails the continuity for $* = $ M, N and W on $\mathcal{Y}^1(\mathbb{T})$.

For every $\mu \in \mathcal{Y}^1(\mathbb{T})$ and for every $B \in \mathcal{B}_{\mathbb{S}}$, we have

$$(\underline{\psi}_\sharp \mu)(\Omega \times B) = \mu(\underline{\psi}^{-1}(\Omega \times B)) = \mu(\Omega \times \psi^{-1}(B)),$$

thus

$$(\underline{\psi}_\sharp \mu)(\Omega \times .) = \psi_\sharp(\mu(\Omega \times .)). \tag{4.4.1}$$

Let $\mathfrak{K} \subset \mathcal{Y}^1(\mathbb{T})$. The set $\mathfrak{K}$ is strictly tight if and only if $\{\mu(\Omega \times .);\ \mu \in \mathfrak{K}\}$ is a tight subset of $\mathcal{M}^{+,1}(\mathbb{T})$, But, from Property 3, the set $\{\mu(\Omega \times .);\ \mu \in \mathfrak{K}\}$ is tight if and only if its image by ψ is a tight subset of $\mathcal{M}^{+,1}(\mathbb{S})$, which, by (4.4.1), means that $\underline{\psi}_\sharp(\mathfrak{K})$ is strictly tight. □

Change of Topology Lemma Let $(K_n)_{n\in\mathbb{N}}$ be a sequence of compact subsets of $\mathbb{T}$. Let $\mathfrak{K} \subset \mathcal{M}^{+,1}(\mathbb{T})$. Let us say that $(K_n)_{n\in\mathbb{N}}$ *tightens* $\mathfrak{K}$ if, for each $\varepsilon > 0$, there exists an $n \in \mathbb{N}$ such that $\sup_{\mu\in\mathfrak{K}} \mu(K_n^c) \le \varepsilon$.

If each compact subset of $\mathbb{T}$ is metrizable, and if $\mathfrak{K}$ is a subset of $\mathcal{Y}^1$ which is tightened by a nondecreasing sequence $(K_n)_{n\in\mathbb{N}}$ of metrizable compact subsets of $\mathbb{T}$, the topology of $\mathfrak{K}$ remains unchanged if we replace the topology of $\mathbb{T}$ by a stronger k_ω–topology. This is shown in the following lemma, the idea of which is due to A. Bouziad.

Lemma 4.4.3 (Change of topology [Bou]) *Assume that each compact subset of $\mathbb{T}$ is metrizable. Let $\mathfrak{K}$ be a strictly tight subset of $\mathcal{Y}^1$. Let $(K_n)_{n\in\mathbb{N}}$ be a non-decreasing sequence of compact subsets of $\mathbb{T}$ which tightens $\mathfrak{K}$. Let $\mathbb{T}_0 = \cup_{n\in\mathbb{N}} K_n$. We thus have $\mu(\Omega \times \mathbb{T}_0) = 1$ for every $\mu \in \mathfrak{K}$, that is, $\mathfrak{K} \subset \mathcal{Y}^1(\mathbb{T}_0)$. Let τ_0' be the topology on $\mathbb{T}_0$ determined by the sets K_n. Then*

(i) $(\mathbb{T}_0, \tau_0')$ is a submetrizable k_ω–space,

(ii) τ_0' is finer than the topology τ_0 induced on $\mathbb{T}_0$ by $\tau_{\mathbb{T}}$, and τ_0' has the same Borel sets as τ_0,

(iii) $(K_n)_n$ is a sequence of τ_0'–compact subsets of $\mathbb{T}_0$ which tightens $\mathfrak{K}$,

(iv) The topologies $\tau^{\mathrm{S}}_{\mathcal{Y}^1(\mathbb{T}_0,\tau_0')}$, $\tau^{\mathrm{W}}_{\mathcal{Y}^1(\mathbb{T}_0,\tau_0')}$, $\tau^{\mathrm{S}}_{\mathcal{Y}^1(\mathbb{T},\tau_{\mathbb{T}})}$ and $\tau^{\mathrm{W}}_{\mathcal{Y}^1(\mathbb{T},\tau_{\mathbb{T}})}$ coincide on $\mathfrak{K}$.

Proof. The only point which needs a proof is (iv). First, for $* = \mathrm{S}, \mathrm{M}, \mathrm{N}, \mathrm{W}$, let $\tau^{*}_{\mathcal{Y}^1(\mathbb{T})}\big|_{\mathcal{Y}^1(\mathbb{T}_0)}$ be the restriction of $\tau^{*}_{\mathcal{Y}^1(\mathbb{T})}$ to the subspace $\mathcal{Y}^1\,(\mathbb{T}_0)$. We have

$$\tau^{\mathrm{W}}_{\mathcal{Y}^1(\mathbb{T})}\big|_{\mathcal{Y}^1(\mathbb{T}_0)} \subset \tau^{\mathrm{S}}_{\mathcal{Y}^1(\mathbb{T})}\big|_{\mathcal{Y}^1(\mathbb{T}_0)} \subset \tau^{\mathrm{S}}_{\mathcal{Y}^1(\mathbb{T}_0,\tau_0)} \subset \tau^{\mathrm{S}}_{\mathcal{Y}^1(\mathbb{T}_0,\tau_0')}.$$

But the $\tau^{\mathrm{S}}_{\mathcal{Y}^1(\mathbb{T}_0,\tau_0')}$–closure $\overline{\mathfrak{K}}$ of $\mathfrak{K}$ is strictly tight in $\mathcal{Y}^1(\mathbb{T}_0, \tau_0')$, thus it is $\tau^{\mathrm{S}}_{\mathcal{Y}^1(\mathbb{T}_0,\tau_0')}$–compact, therefore $\tau^{\mathrm{S}}_{\mathcal{Y}^1(\mathbb{T}_0,\tau_0')}$ and $\tau^{\mathrm{W}}_{\mathcal{Y}^1(\mathbb{T})}\big|_{\mathcal{Y}^1(\mathbb{T}_0)}$ coincide on $\overline{\mathfrak{K}}$. □

Remark 4.4.4 Taking $\mathcal{S} = \{\emptyset, \Omega\}$, the Change of Topology Lemma implies that the narrow topologies on $\mathcal{M}^{+,1}(\mathbb{T}, \tau_{\mathbb{T}})$ and $\mathcal{M}^{+,1}(\mathbb{T}_0, \tau_0')$ coincide on any subset $\mathfrak{K}$ of $\mathcal{M}^{+,1}(\mathbb{T})$ which is tightened by $(K_n)_{n\in\mathbb{N}}$.

4.5 Sequential properties: sequential Prohorov property, Komlós convergence, Mazur–compactness

Sequential Prohorov property Let us say that $\mathbb{T}$ is *sequentially Prohorov* if every convergent sequence in $\mathcal{M}_t^{+,1}(\mathbb{T})$ is tight. Clearly, Prohorov spaces are sequentially Prohorov. Topsøoe [Top70b, Theorem 9.3] provides another sufficient criterion, generalizing a result of Le Cam: If, for any $K \in \mathcal{K}$, there exists $K_0 \in \mathcal{K}$ such that $K \subset K_0$ and K_0 has a countable base of neighbourhoods, then $\mathbb{T}$ is sequentially Prohorov. In particular, all metric spaces and all locally compact spaces are sequentially Prohorov. Fernique [Fer67, Exemple 1.6.4] gives a counterexample showing that the weak topology $\sigma(\mathbb{E}, \mathbb{E})$ of an infinite dimensional Hilbert space is never sequentially Prohorov. This counterexample is generalized in [FGH72, page 126] to prove that, if $\mathbb{E}$ is an infinite dimensional Banach space with dual $\mathbb{E}^*$, neither $(\mathbb{E}, \sigma(\mathbb{E}, \mathbb{E}^*))$ nor $(\mathbb{E}^*, \sigma(\mathbb{E}^*, \mathbb{E}))$ is sequentially Prohorov.

The following result is almost obvious, but very important!

Theorem 4.5.1 (Sequential Semicontinuity Theorem) *Assume that $\mathbb{T}$ is sequentially Prohorov and that its compact subsets are metrizable. Let $(\mu^n)_{n\in\mathbb{N}}$ be a sequence of elements of $\mathcal{Y}^1$ and let $\mu^\infty \in \mathcal{Y}^1$. Assume that, for each $n \in \mathbb{N} \cup \{\infty\}$, $\mu^n(\Omega \times .)$ is in $\mathcal{M}_t^+(\mathbb{T})$ (this is the case if $\mathbb{T}$ is metrizable or if $\mathbb{T}$ is Radon). The following implication holds true:*

$$\mu^n \xrightarrow{N\text{-stably}} \mu^\infty \Rightarrow \mu^n \xrightarrow{S\text{-stably}} \mu^\infty.$$

In particular, if $\mathbb{T}$ is completely regular and Radon or if $\mathbb{T}$ is metrizable, we have

$$\mu^n \xrightarrow{W\text{-stably}} \mu^\infty \Rightarrow \mu^n \xrightarrow{S\text{-stably}} \mu^\infty.$$

Proof. Indeed, from the Converse Prohorov Criterion (Theorem 4.3.5), the set $\mathfrak{K} = \{\mu^n;\, n \in \mathbb{N} \cup \{\infty\}\}$ is strictly tight, because the margins $\mu^n(\Omega \times .)$ narrowly converge to $\mu^\infty(\Omega \times .)$. Thus, from the Direct Prohorov Criterion (Theorem 4.3.5), $\mathfrak{K}$ is $\tau_{\mathcal{Y}^1}^{\mathrm{S}}$–compact, and the topologies $\tau_{\mathcal{Y}^1}^{\mathrm{S}}$ and $\tau_{\mathcal{Y}^1}^{\mathrm{N}}$ coincide on $\mathfrak{K}$.

If $\mathbb{T}$ is is completely regular and Radon or if $\mathbb{T}$ is metrizable (provided that $\mathbb{T}$ contains a dense subspace with non–measurable cardinal), we have $\mathcal{M}_\tau^+(\mathbb{T}) = \mathcal{M}^+(\mathbb{T})$, thus $\tau_{\mathcal{Y}^1}^{\mathrm{N}} = \tau_{\mathcal{Y}^1}^{\mathrm{W}}$ (see Part D of Theorem 2.1.3). □

Proposition 4.5.2 (Relative compactness deduced from sequential relative compactness) *Assume that $\mathbb{T}$ is a Polish space or a sequentially Prohorov space with a cofinal sequence of metrizable compact subsets. Let $\mathfrak{K}$ be a $\tau^{\mathrm{S}}_{\mathcal{Y}^1}$-sequentially relatively compact subset of $\mathcal{Y}^1$. Then $\mathfrak{K}$ is strictly tight.*

In particular, any sequentially Prohorov space with a cofinal sequence of metrizable compact subsets is Prohorov.

Remark 4.5.3 That $\tau^{\mathrm{S}}_{\mathcal{Y}^1}$-sequential relative compactness implies $\tau^{\mathrm{S}}_{\mathcal{Y}^1}$-relative compactness is obvious if $\mathcal{S}$ is essentially countably generated and $\mathbb{T}$ is a Suslin metrizable space, because then, from Proposition 2.3.1, $\tau^{\mathrm{S}}_{\mathcal{Y}^1}$ is metrizable.

Proof of Proposition 4.5.2. Assume first that $\mathbb{T}$ is Polish. From continuity of the mapping $\mu \mapsto \mu(\Omega \times .)$, the set $\{\mu(\Omega \times .);\, \mu \in \mathfrak{K}\}$ is sequentially relatively compact. From metrizability of $\mathcal{M}^{+,1}(\mathbb{T})$, it is thus relatively compact, thus it is tight. This means that $\mathfrak{K}$ is strictly tight. Thus, from the Direct Prohorov Criterion (Theorem 4.3.5), $\mathfrak{K}$ is relatively compact.

Assume now that $\mathbb{T}$ is a sequentially Prohorov space with a cofinal sequence $(K_m)_{m\geq 1}$ of metrizable compact subsets. We can assume that $(K_m)_{m\geq 1}$ is nondecreasing. Let τ' be the k_ω-topology determined by $(K_m)_{m\geq 1}$. Let $(\mu^n)_n$ be a sequence in $\mathfrak{K}$. Let $(\nu^n)_n$ be a subsequence of $(\mu^n)_n$. There exists a $\tau^{\mathrm{S}}_{\mathcal{Y}^1}$-convergent subsequence $(\lambda^n)_n$ of $(\nu^n)_n$. From the sequential Prohorov property, the set $\{\lambda^n;\, n \in \mathbb{N}\cup\{\infty\}\}$ is strictly tight. Thus $\{\lambda^n;\, n \in \mathbb{N}\cup\{\infty\}\}$ is strictly tight for τ'. Thus $\mathfrak{K}$ is sequentially relatively compact for $\tau^{\mathrm{S}}_{\mathcal{Y}^1(\tau')}$. As τ' is finer than $\tau_{\mathbb{T}}$, we can thus assume without loss of generality that $\mathbb{T}$ is a submetrizable k_ω-space. From Lemma 4.4.2, and with the same notations, the set $\underline{\psi}_\sharp(\{\lambda^n;\, n \in \mathbb{N}\cup\{\infty\}\})$ is strictly tight. From Prohorov Criterion (Theorem 4.3.5), there exists a subsequence $(\rho^n)_n$ of $(\underline{\psi}_\sharp \lambda^n)_n$ which is convergent for $\tau^{\mathrm{S}}_{\mathcal{Y}^1(\mathbb{S})}$. This shows that the set $\underline{\psi}_\sharp \mathfrak{K}$ is $\tau^{\mathrm{S}}_{\mathcal{Y}^1(\mathbb{S})}$-sequentially relatively compact. Therefore, from continuity of the mapping $\underline{\pi}_\sharp : \mathcal{Y}^1(\mathbb{S}) \to \mathcal{Y}^1(\mathbb{T})$, we can deduce the result from the Polish case.

□

Komlós convergence Let $(u_n)_n$ be a sequence of random elements of a topological vector space $\mathfrak{T}$, defined on $(\Omega, \mathcal{S}, \mathrm{P})$, and let u be a random element of $\mathfrak{T}$ defined on $(\Omega, \mathcal{S}, \mathrm{P})$. Following Balder [Bal91], we say that $(u_n)_n$ *Komlós converges*, or *K–converges* to u if, for every subsequence $(v_n)_n$ of $(u_n)_n$, the sequence $\left(\frac{1}{n}\sum_{j=1}^n v_j\right)_n$ almost everywhere converges to u, and, when we consider random laws (i.e. disintegrable Young measures), we adapt naturally this definition to the convex set $\mathcal{M}^{+,1}(\mathbb{T})$ endowed with the topology of narrow convergence.

We recall the celebrated Komlós Theorem [Kom67]: If $(u_n)_n$ is a bounded sequence of elements of $\mathrm{L}^1(\Omega, \mathcal{S}, \mathrm{P})$, then there exists a subsequence $(v_n)_n$ of $(u_n)_n$ which K–converges to an element v of $\mathrm{L}^1(\Omega, \mathcal{S}, \mathrm{P})$.

This theorem has been extended to random elements of a Banach space (with additional conditions) in [Bou79, Gar79, BBR79, Bal89b, Gue97, Saa98]. Multi-valued versions of Komlós's theorem are given in [BH96, CE98].

E. Balder [Bal90] was the first who applied this theorem to Young measures (but the notion of K–convergence is already at work in [Fis67, Section 3] or in [Fis71]).

In particular, he used it as a tool to prove sequential compactness in $\mathcal{Y}^1_{\text{dis}}$. On the contrary, we deduce here K–convergence of Young measures from sequential compactness. The following lemma is very powerful, because it allows to pass from S–stable convergence to almost everywhere convergence. It is essentially due to Balder [Bal90], who proved it in the Suslin regular case.

Lemma 4.5.4 (K–convergence from S–stable convergence) *Assume that the compact subsets of $\mathbb{T}$ are metrizable. Let $(\mu^n)_{n\in\mathbb{N}}$ be a strictly tight sequence in $\mathcal{Y}^1 = \mathcal{Y}^1_{\text{dis}}$ (see Lemma 4.3.1), which S-stably converges to a limit $\mu^\infty \in \mathcal{Y}^1_{\text{dis}}$. Then there exists a subsequence of $(\mu^n)_{n\in\mathbb{N}}$ which K–converges to μ^∞.*

Proof.

First step Assume first that $\mathbb{T}$ is a separable metric space. There exists thus a countable set $\{f_k;\ k \in \mathbb{N}\}$ of nonnegative bounded continuous functions on $\mathbb{T}$ such that the topology of $\mathcal{M}^{+,1}(\mathbb{T})$ is the coarsest topology such that the mappings $\nu \mapsto \nu(f_k)$ $(k \in \mathbb{N})$ are continuous.

For each $k \in \mathbb{N}$ and each $n \in (\mathbb{N} \cup \{\infty\})$, let φ^n_k be the random variable $\omega \mapsto \mu^n_\omega(f_k)$. We have $\lim_{n\to\infty} \int_A \varphi^n_k \, d\mathrm{P} = \int_A \varphi^\infty_k \, d\mathrm{P}$ for every $A \in \mathcal{S}$ and every $k \in \mathbb{N}$. In particular, $(\varphi^n_k)_{n\in\mathbb{N}}$ is bounded in L^1. From Komlós Theorem, there exists a subsequence of $(\varphi^n_k)_{n\in\mathbb{N}}$ which K–converges to a limit g_k. From Lebesgue's dominated convergence theorem, we have necessarily $g_k = \varphi^\infty_k$. By an obvious diagonal process, we can extract a subsequence $(\widetilde{\mu}^n)_{n\in\mathbb{N}}$ of $(\mu^n)_{n\in\mathbb{N}}$ such that, for each $k \in \mathbb{N}$, the associated sequence $(\widetilde{\varphi}^n_k)_{n\in\mathbb{N}}$ K–converges to φ^∞_k. This proves that $(\widetilde{\mu}^n)_{n\in\mathbb{N}}$ K–converges to μ^∞.

Second step Assume now that $\mathbb{T}$ is a submetrizable k_ω–space. We use the notations of the Lifting Lemma 4.4.2. From the Converse Prohorov Criterion (Theorem 4.3.5), the set $\mathfrak{Y} := \{\mu^n;\ n \in (\mathbb{N} \cup \{\infty\})\}$ is strictly tight, thus, from Lemma 4.4.2, the image $\mathfrak{Y}'$ of $\mathfrak{Y}$ by the lifting $\underline{\psi}$ is a strictly tight subset of $\mathcal{Y}^1_{\text{dis}}(\mathbb{S})$. From the Direct Prohorov Criterion (Theorem 4.3.5), there exists a subsequence $(\nu^n)_n$ of $(\underline{\psi}(\mu^n))_n$ which S–stably converges to a limit ν, and from continuity of $\mu \mapsto \underline{\pi}(\mu)$, we have $\underline{\pi}(\nu) = \mu^\infty$. From the first step, there exists a subsequence $(\widetilde{\nu}^n)_n$ of $(\nu^n)_n$ which K–converges to ν, and from continuity of π, the sequence $(\underline{\pi}(\widetilde{\nu}^n))_n$ K–converges to μ^∞.

Third step We now consider the general case. As $(\mu^n)_{n\in\mathbb{N}}$ is tight, we can assume that $\mathbb{T}$ is a K_σ–space. From the Change of Topology Lemma 4.4.3, there exists thus a topology τ' on $\mathbb{T}$ which is finer than the original topology $\tau_{\mathbb{T}}$ of $\mathbb{T}$, such that $(\mathbb{T}, \tau')$ is a submetrizable k_ω–space and such that $(\mu^n)_{n\in\mathbb{N}}$ converges to μ^∞ for $\tau^{\mathrm{S}}_{\mathcal{Y}^1(\tau')}$. From the second step, there exists a subsequence $(\widetilde{\mu}^n)_{n\in\mathbb{N}}$ of $(\mu^n)_{n\in\mathbb{N}}$ which K–converges to μ^∞ in $\mathcal{M}^{+,1}(\tau')$, thus in $\mathcal{M}^{+,1}(\tau_{\mathbb{T}})$. □

Let us introduce in the next lemma a condition of "sequential almost everywhere tightness" for disintegrable Young measures.

Lemma 4.5.5 *Assume that $\mathbb{T}$ is a Suslin regular Prohorov space or that $\mathbb{T}$ has a cofinal sequence of metrizable compact subsets. Let $\mathfrak{Y} \subset \mathcal{Y}^1$ be such that, for each sequence $(\mu^n)_{n\in\mathbb{N}}$ of elements of $\mathfrak{Y}$, the set $\{\mu^n_\omega;\ n\in\mathbb{N}\}$ is almost everywhere tight. Then $\mathfrak{Y}$ is $\tau^{\mathrm{S}}_{\mathcal{Y}^1}$–sequentially relatively compact.*

Proof. Assume first that $\mathbb{T}$ is a regular Suslin Prohorov space. Let $(\mu^n)_{n\in\mathbb{N}}$ be a sequence in $\mathfrak{Y}$. For each $\omega \in \Omega$, let

$$H(\omega) = \overline{\bigcup_{n\in\mathbb{N}} \{\mu^n_\omega\}}.$$

For almost every $\omega\in\Omega$, the set $H(\omega)$ is closed and tight, thus compact.

For any open subset U of $\mathcal{M}^{+,1}(\mathbb{T})$, we have

$$H^- U = \{\omega\in\Omega;\ H(\omega)\cap U \neq \emptyset\} = \bigcup_{n\in\mathbb{N}} \{\omega\in\Omega;\ \mu^n_\omega \in U\} \in \mathcal{S},$$

thus H is LV–measurable. Now, as $\mathcal{M}^{+,1}(\mathbb{T})$ is Suslin regular, there exists from Lemma 1.1.2 a continuous distance δ on $\mathcal{M}^{+,1}(\mathbb{T})$ such that U is δ–open. From [CV77, Theorem III.2], as H is δ–LV–measurable, it is also δ–V–measurable, thus we have also $H^+U \in \mathcal{S}$. Therefore H is a Borel mapping from Ω to $\mathcal{K}(\mathcal{M}^{+,1}(\mathbb{T}))$.

Let $\varepsilon > 0$. The space $\mathcal{M}^{+,1}(\mathbb{T})$ is Suslin regular. Furthermore, from [Wój87], $\mathcal{M}^{+,1}(\mathbb{T})$ is Prohorov. Thus, from Theorem 4.3.14, the space $\mathcal{K}(\mathcal{M}^{+,1}(\mathbb{T}))$ is Radon. There exist thus $\Omega_1 \in \mathcal{S}$ and a compact subset $\widehat{\mathfrak{C}}$ of $\mathcal{K}(\mathcal{M}^{+,1}(\mathbb{T}))$ such that

$$\mathrm{P}(\Omega_1) \geq 1 - \varepsilon/2 \text{ and } \forall\omega\in\Omega_1 \quad H(\omega)\in\widehat{\mathfrak{C}}.$$

Now, from [Mic51, Theorem 2.5.2], the set

$$\mathfrak{C} = \cup_{L\in\widehat{\mathfrak{C}}} L$$

is a compact subset of $\mathcal{M}^{+,1}(\mathbb{T})$. We thus have

$$\forall\omega\in\Omega_1 \quad H(\omega)\subset\mathfrak{C}.$$

As $\mathcal{M}^{+,1}(\mathbb{T})$ is Prohorov, there exists a compact subset L of $\mathbb{T}$ such that, for any $\nu\in\mathfrak{C}$, we have $\nu(L)\geq 1-\varepsilon/2$. We thus have

$$\sup_{n\in\mathbb{N}} \mu^n(\Omega\times L) \geq \sup_{n\in\mathbb{N}} \mu^n(\Omega_1\times L) \geq (1-\varepsilon/2)^2 \geq 1-\varepsilon.$$

This shows that the set $\{\mu^n; n \in \mathbb{N}\}$ is strictly tight, thus, from the Direct Prohorov Criterion (Theorem 4.3.5), it admits a convergent subsequence. Thus $\mathfrak{Y}$ is $\tau^{\mathrm{S}}_{\mathcal{Y}^1}$–sequentially relatively compact.

Assume now that $\mathbb{T}$ admits a cofinal sequence $(K_n)_n$ of metrizable compact subsets. We assume w.l.g. that $(K_n)_n$ is nondecreasing. Let τ' be the topology on $\mathbb{T}$ determined by $(K_n)_n$. For each sequence $(\mu^n)_{n\in\mathbb{N}}$ of elements of $\mathfrak{Y}$, the sequence $(\mu^n_\omega)_{n\in\mathbb{N}}$ is almost everywhere tight in $\mathcal{M}^{+,1}(\mathbb{T},\tau')$. As $(\mathbb{T},\tau')$ is a submetrizable k_ω–space, it is regular Suslin and Prohorov, thus, from the preceding study, $\mathfrak{Y}$ is $\tau^{\mathrm{S}}_{\mathcal{Y}^1(\mathbb{T},\tau')}$–sequentially relatively compact, and therefore it is also $\tau^{\mathrm{S}}_{\mathcal{Y}^1(\mathbb{T},\tau_{\mathbb{T}})}$–sequentially relatively compact. □

We deduce from Lemma 4.5.5 the following converse to Lemma 4.5.4. We shall use again Lemma 4.5.5 in the proof of Theorem 4.5.8.

Proposition 4.5.6 (S–stable convergence from K–convergence) *Let $(\mu^n)_n$ be a sequence in $\mathcal{Y}^1_{\mathrm{dis}}$ which K–converges to a limit $\mu^\infty \in \mathcal{Y}^1_{\mathrm{dis}}$. Assume that $\mathbb{T}$ is Suslin regular. Assume furthermore that $(\mu^n)_n$ is strictly tight, or that $\mathbb{T}$ is Prohorov, or that $\mathbb{T}$ has a cofinal sequence of metrizable compact subsets. Then $(\mu^n)_n$ S–stably converges to μ^∞.*

Proof. From the Direct Prohorov Criterion (Theorem 4.3.5) if $(\mu^n)_n$ is strictly tight, or else from Lemma 4.5.5, $(\mu^n)_n$ is $\tau^{\mathrm{S}}_{\mathcal{Y}^1}$–sequentially relatively compact. For each subsequence of $(\mu^n)_n$, there exists a further subsequence $(\widetilde{\mu}^n)_n$ of $(\mu^n)_n$ which S–stably converges to a limit ν. But, as $(\mu^n)_n$ K–converges to μ^∞, we have $\mu^\infty = \nu$. Indeed, for each $A \in \mathcal{S}$ and each $f \in \mathrm{C}_b(\mathbb{T})$, we have, using Lebesgue's dominated convergence theorem,

$$\nu(\mathbb{1}_A \otimes f) = \lim_n \widetilde{\mu}^n(\mathbb{1}_A \otimes f) = \lim_n \frac{1}{n}\sum_{i=1}^{n} \widetilde{\mu}^i(\mathbb{1}_A \otimes f) = \mu(\mathbb{1}_A \otimes f).$$

We have proved that each subsequence of $(\mu^n)_n$ has a further subsequence which S–stably converges to μ^∞. Thus $(\mu^n)_n$ S–stably converges to μ^∞. □

Mazur–compactness For any subset A of a vector space, we denote by $\mathrm{co}\,A$ the convex hull of A.

Let us recall the following theorem of Mazur (see [Maz33, Section 2]): If $(x_n)_n$ is a weakly convergent sequence of a Banach space $\mathbb{E}$, then there exists a sequence $(y_n)_n$ of elements of $\mathbb{E}$ which converges in $\mathbb{E}$ to the same limit, with $y_n \in \mathrm{co}\{x_n, x_{n+1}, \dots\}$. Indeed, let x be the weak limit of $(x_n)_n$. Let us denote respectively by w–cl $[A]$ and cl $[A]$ the weak and strong closures of a subset A of $\mathbb{E}$. We then have, for every n,

$$x \in \text{w–cl}\left[\{x_n, \dots\}\right] \subset \text{w–cl}\left[\mathrm{co}\{x_n, \dots\}\right] = \mathrm{cl}\left[\mathrm{co}\{x_n, \dots\}\right].$$

Let $\mathfrak{Y}$ be a set of random elements of a topological vector space. We assume that all elements of $\mathfrak{Y}$ are defined on $(\Omega, \mathcal{S}, \mathrm{P})$. We say that $\mathfrak{Y}$ is *Mazur–compact* if, for each sequence $(u_n)_n$ of elements of $\mathfrak{Y}$, there exists a sequence $(v_n)_n$ which is almost everywhere convergent and such that, for each $n \in \mathbb{N}$, $v_n \in \operatorname{co}\{u_m;\, m \geq n\}$.

Of course, we can apply this definition to disintegrable Young measures by embedding the convex set $\mathcal{M}^{+,1}(\mathbb{T})$ in the vector space of signed measures on $\mathbb{T}$.

Yet another characterization of strict tightness and flexible tightness (in the case of a Polish space or a submetrizable k_ω–space) This characterization will be given in Theorem 4.5.8, using K–convergence and Mazur compactness.

The following lemma gives a necessary and sufficient condition of relative compactness in $\mathcal{M}^{+,1}(\mathbb{T})$. It was proved in [CV98, Proposition 3.3] in the case when $\mathbb{T}$ is Polish (see also Proposition 6.4.4 page 158). We shall use this lemma in the proof of Theorem 4.5.8, which gives necessary and sufficient conditions of relative compactness for $\tau^{\mathrm{S}}_{\mathcal{Y}^1}$.

Lemma 4.5.7 *Assume that $\mathbb{T}$ is Polish or that $\mathbb{T}$ has a cofinal sequence of metrizable compact subsets (e.g. $\mathbb{T}$ is a submetrizable k_ω–space). Let $\mathfrak{Y} \subset \mathcal{M}^+(\mathbb{T})$. The following conditions are equivalent:*

1. *$\mathfrak{Y}$ is tight.*
2. *For each sequence $(\mu_n)_n$ of elements of $\mathfrak{Y}$, there exists a tight sequence $(\nu_n)_n$ in $\mathcal{M}^+(\mathbb{T})$ such that, for every $n \in \mathbb{N}$, $\nu_n \in \operatorname{co}\{\mu_m;\, m \geq n\}$.*

Proof.

The proof in the Polish case is given in [CV98, Proposition 3.3]. We reproduce it in Chapter 6 (Proposition 6.4.4 page 158), instead of giving it here, because it uses duality arguments which are best integrated in this chapter. But there is no logical circle !

Assume that $\mathbb{T}$ has a cofinal sequence of metrizable compact subsets. The implication 1⇒2 is obvious, because, for every n, we have $\mu_n \in \operatorname{co}\{\mu_m;\, m \geq n\}$.

Conversely, assume 2. Let $(\mu_n)_{n\in\mathbb{N}}$ be a sequence of elements of $\mathfrak{Y}$. Let $(K_j)_{j\in\mathbb{N}}$ be a cofinal sequence of metrizable compact subsets of $\mathbb{T}$. Each tight subset of $\mathcal{M}^{+,1}(\mathbb{T})$ is tightened by $(K_j)_{j\in\mathbb{N}}$. Thus, if τ' denotes the k_ω topology determined by $(K_j)_{j\in\mathbb{N}}$, each tight subset of $\mathcal{M}^{+,1}(\mathbb{T})$ is tight for $\tau_{\mathbb{T}}$ if and only if it is tight for τ'. We can thus assume without loss of generality that $\mathbb{T}$ is a submetrizable k_ω–space. With the notations of Lemma 4.4.2, let $\mathfrak{C} = \psi(\mathfrak{Y})$. Let $(\mu^n)_n$ be a sequence of elements of $\mathfrak{C}$. Then there exists a tight sequence $(\nu^n)_n$ in $\mathcal{M}^+(\mathbb{T})$ such that, for every $n \in \mathbb{N}$, $\nu^n \in \operatorname{co}\{\pi_\sharp\, \mu^m;\, m \geq n\}$. We thus have $\psi_\sharp\, \nu^n \in \operatorname{co}\{\mu^m;\, m \geq n\}$. From Proposition 3.3 in [CV98], the set $\mathfrak{C}$ is tight, thus, from Lemma 4.4.2, $\mathfrak{Y} = \pi_\sharp(\mathfrak{C})$ is tight. □

Theorem 4.5.8 *Assume that $\mathbb{T}$ is Polish or that $\mathbb{T}$ is a Prohorov space with a cofinal sequence of metrizable compact subsets (e.g. $\mathbb{T}$ is a submetrizable k_ω–space). Let $\mathfrak{Y} \subset \mathcal{Y}^1$. The following conditions are equivalent:*

1. $\mathfrak{Y}$ *is (strictly or flexibly) tight.*
2. *For each sequence* $(\mu^n)_n$ *of elements of* $\mathfrak{Y}$, *there exists a subsequence* $(\lambda^n)_n$ *of* $(\mu^n)_n$ *which is K–convergent.*
3. *For each sequence* $(\mu^n)_n$ *of elements of* $\mathfrak{Y}$, *there exists a subsequence* $(\lambda^n)_n$ *of* $(\mu^n)_n$ *such that the sequence*

$$\left(1/n\sum_{i=1}^{n}\lambda_\omega^i\right)_n$$

is almost everywhere convergent.
4. $\mathfrak{K}$ *is Mazur–compact.*
5. *For each sequence* $(\mu^n)_n$ *of elements of* $\mathfrak{Y}$, *there exists a sequence* $(\nu^n)_n$ *in* $\mathcal{Y}^1$ *such that, for every* $n\in\mathbb{N}$, $\nu^n\in\operatorname{co}\{\mu^m;\, m\geq n\}$ *and, for almost every* $\omega\in\Omega$, $(\nu_\omega^n)_n$ *is tight.*
6. *For each sequence* $(\mu^n)_n$ *of elements of* $\mathfrak{Y}$, *there exists a (strictly or flexibly) tight sequence* $(\nu^n)_n$ *in* $\mathcal{Y}^1$ *such that, for every* $n\in\mathbb{N}$, $\nu^n\in\operatorname{co}\{\mu^m;\, m\geq n\}$.

Proof.

$1\Rightarrow 2$ comes from Lemma 4.5.4.

$2\Rightarrow 3$ is obvious.

$3\Rightarrow 4$ Assume 3. Let $(\mu^n)_n$ be a sequence of elements of $\mathfrak{Y}$ and let $(\lambda^n)_n$ be a subsequence of $(\mu^n)_n$ such that the sequence $(1/n\sum_{i=1}^{n}\lambda_\omega^i)_n$ is almost everywhere convergent. For almost every $\omega\in\Omega$, the sequence

$$\begin{aligned}(\nu_\omega^n)_n &:= \left(\frac{1}{n(n-1)}\sum_{i=n+1}^{n^2}\lambda_\omega^i\right)_n\\ &= \left(\frac{n^2}{n^2-n}\left(\frac{1}{n^2}\sum_{i=1}^{n^2}\lambda_\omega^i\right)\right)_n - \left(\frac{1}{n-1}\left(\frac{1}{n}\sum_{i=1}^{n}\lambda_\omega^i\right)\right)_n\end{aligned}$$

is almost everywhere convergent, and we have $\nu^n\in\operatorname{co}\{\mu^m;\, m\geq n\}$ for every $n\geq 1$.

$4\Rightarrow 1$ Assume 4. Let $(\mu^n)_n$ be a sequence of elements of $\mathfrak{Y}$. For each subsequence $(\lambda^n)_n$ of $(\mu^n)_n$, there exists a subsequence $(\nu^n)_n$ in $\mathcal{Y}^1$ such that, for every $n\in\mathbb{N}$, $\nu^n\in\operatorname{co}\{\lambda^m;\, m\geq n\}$ and such that $(\nu_\omega^n)_n$ almost everywhere converges. The sequence $(\nu^n(\Omega\times .))$ is narrowly convergent, thus $(\nu^n)_n$ is strictly tight. Thus, for each subsequence $(\lambda^n(\Omega\times .))_n$ of $(\mu^n(\Omega\times .))_n$, there exists a narrowly convergent sequence $(\widetilde{\nu}^n)_n$ in $\mathcal{M}^{+,1}(\mathbb{T})$ such that $\widetilde{\nu}^n\in\operatorname{co}\{\lambda^m(\Omega\times .);\, m\geq n\}$ for every $n\in\mathbb{N}$. From [CV98, Proposition 3.3], this implies that $(\mu^n)_n$ is strictly tight.

$4\Rightarrow 5$ is obvious from the (sequential) Prohorov property.

$5 \Rightarrow 6$ comes from Lemma 4.5.5 and the fact that $\mathbb{T}$ is sequentially Prohorov.

$6 \Rightarrow 4$ Assume 6. Let $(\mu^n)_n$ be a sequence of elements of $\mathfrak{Y}$ and let $(\nu^n)_n$ be a flexibly tight sequence $(\nu^n)_n$ in $\mathcal{Y}^1$ such that, for every $n \in \mathbb{N}$, $\nu^n \in \operatorname{co}\{\mu^m;\, m \geq n\}$. Using again Balder's result (Lemma 4.5.4), we see that there exists a subsequence of $(\nu^n)_n$ which is K–convergent. We denote this subsequence again by $(\nu^n)_n$. We then obtain 4 by the same technique as for the implication $3 \Rightarrow 4$. □

Corollary 4.5.9 *Assume that $\mathbb{T}$ is Polish or that $\mathbb{T}$ is a Prohorov space with a cofinal sequence of metrizable compact subsets (e.g. $\mathbb{T}$ is a submetrizable k_ω-space). Let $\mathfrak{Y}$ be a* convex *subset of $\mathcal{Y}^1$. The following conditions are equivalent:*

1. *$\mathfrak{Y}$ is sequentially closed for $\tau^{\mathrm{W}}_{\mathcal{Y}^1}$.*

2. *$\mathfrak{Y}$ is sequentially closed for $\tau_{\mathrm{prob}}\left(\mathcal{Y}^1\right)$.*

In particular, if $\mathbb{T}$ is Polish and $\mathcal{S}$ is essentially countably generated, then every closed convex subset of $(\mathcal{Y}^1, \tau^{\mathrm{W}}_{\mathcal{Y}^1}) = (\mathcal{Y}^1, \tau^{\mathrm{S}}_{\mathcal{Y}^1})$ is closed for $\tau_{\mathrm{prob}}\left(\mathcal{Y}^1\right)$.

Proof. From Theorem 3.1.2, we have $1 \Rightarrow 2$. Assume 2 and let $(\mu^n)_{n\in\mathbb{N}}$ be a sequence in $\mathcal{Y}^1$ which W–stably converges to some $\mu \in \mathcal{Y}^1$. Then $\{\mu^n;\, n \in \mathbb{N}\}$ is strictly tight, thus, from Theorem 4.5.8, there exists a subsequence $(\nu_n)_n$ of $(\mu^n)_n$ such that

$$\lim_n \frac{1}{n} \sum_{i=1}^{n} \nu^i_\omega = \mu_\omega \text{ a.e.}$$

Thus the sequence $(1/n \sum_{i=1}^n \nu^i)_n$ converges in probability to μ. As $\mathfrak{Y}$ is convex and sequentially closed in probability, we thus have $\mu \in \mathfrak{Y}$, which proves 1.

If $\mathbb{T}$ is Polish and $\mathcal{S}$ is essentially countably generated, we know from Proposition 2.3.1 that $\tau^{\mathrm{S}}_{\mathcal{Y}^1} = \tau^{\mathrm{W}}_{\mathcal{Y}^1}$ is metrizable. Furthermore, from e.g. Theorem 3.2.1, $\tau_{\mathrm{prob}}\left(\mathcal{Y}^1\right)$ is also metrizable. Thus sequential closedness amounts to closedness in any of the topologies $\tau^{*}_{\mathcal{Y}^1}$ and $\tau_{\mathrm{prob}}\left(\mathcal{Y}^1\right)$, which entails the second part of Corollary 4.5.9. □

4.6 Tables

In this section, we gather some important results which are spread in Chapters 4 and 2.

TABLE 1: SEMICONTINUITY THEOREM

$\mu^\alpha \xrightarrow{\text{W-stably}} \mu^\infty \Rightarrow \mu^\alpha \xrightarrow{\text{S-stably}} \mu^\infty$	
Sufficient condition	**Proof**
$\mathbb{T}$ is Suslin metrizable	Theorem 2.1.3 (Part G) page 25
$(\mu^\alpha)_\alpha$ is strictly tight and $\mathcal{M}_\tau^{+,1}(\mathbb{T}) = \mathcal{M}^{+,1}(\mathbb{T})$ (e.g. $\mathbb{T}$ is Radon or hereditarily Lindelöf) and $\mathbb{T}$ is completely regular and the compact subsets of $\mathbb{T}$ are metrizable or $(\mu^\alpha)_\alpha$ is flexibly tight and $\mathbb{T}$ is Suslin regular	Theorem 4.3.8 page 94
$(\mu^\alpha)_\alpha$ is a sequence and $\mathbb{T}$ is sequentially Prohorov such that $\mathcal{M}_t^+(\mathbb{T}) = \mathcal{M}_\tau^+(\mathbb{T}) = \mathcal{M}^+(\mathbb{T})$ (e.g. $\mathbb{T}$ is a metric space or a submetrizable k_ω–space)	Theorem 4.5.1 page 104

TABLE 2: RELATIVE SEQUENTIAL COMPACTNESS CRITERIA

$\mathfrak{K} \subset \mathcal{Y}^1$ is $\tau^{\mathrm{s}}_{\mathcal{Y}^1}$–sequentially relatively compact	
Sufficient condition	**Proof**
$\mathfrak{K}$ is strictly tight and the compact subsets of $\mathbb{T}$ are metrizable or $\mathfrak{K}$ is flexibly tight and $\mathbb{T}$ is Suslin submetrizable	Theorem 4.3.5 (Prohorov Criterion) page 92
$\mathfrak{K}$ is flexibly tight and $\mathbb{T}$ is Suslin and $\mathcal{K}(\mathbb{T})$ is Radon (e.g. $\mathbb{T}$ has a cofinal sequence of metrizable compact subsets)	Theorem 4.3.5 (Prohorov Criterion) page 92 and Theorem 4.3.13 page 97 (see also Theorem 4.3.14)

$\mathfrak{K}$ is $\tau^{\mathrm{s}}_{\mathcal{Y}^1}$–sequentially relatively compact$\Rightarrow$ $\mathfrak{K}$ is $\tau^{\mathrm{s}}_{\mathcal{Y}^1}$–relatively compact	
Sufficient condition	**Proof**
$\mathbb{T}$ is Suslin metrizable and $\mathcal{S}$ is essentially countably generated (thus $\tau^{\mathrm{s}}_{\mathcal{Y}^1}$ is metrizable)	Remark 4.5.3 page 105
$\mathbb{T}$ is a Polish space or a sequentially Prohorov space with a cofinal sequence of compact metrizable subsets (e.g. $\mathbb{T}$ is a submetrizable k_ω–space)	Proposition 4.5.2 page 105

$\mathfrak{K}$ is $\tau^{\mathrm{s}}_{\mathcal{Y}^1}$–relatively compact $\Rightarrow$ $\mathfrak{K}$ is $\tau^{\mathrm{s}}_{\mathcal{Y}^1}$–sequentially relatively compact	
Sufficient condition	**Proof**
$\mathbb{T}$ is Suslin submetrizable	Lemma 4.1.1 page 84

Chapter 5

Strong tightness

5.1 Equivalent definitions

The following definition is a generalization to Young measures of the definition of strong tightness given by Kruk and Zięba (see [KZ94, KZ95, KZ96]) for random elements of a Polish space. We denote by P_* the inner probability associated with P.

Let $\mathfrak{Y} \subset \mathcal{Y}^1_{\mathrm{dis}}$. We say that $\mathfrak{Y}$ is *strongly tight* if $\mathfrak{Y}$ admits a disintegration $(\mu_\omega)_{\mu\in\mathfrak{Y},\,\omega\in\Omega}$ such that Condition *(i)* below is satisfied.

(i) For each $\varepsilon > 0$, there exists a tight subset $\mathfrak{K}_\varepsilon$ of $\mathcal{M}^{+,1}(\mathbb{T})$ such that

$$\mathrm{P}_* \{\omega \in \Omega;\ \forall \mu \in \mathfrak{Y} \quad \mu_\omega \in \mathfrak{K}_\varepsilon\} > 1 - \varepsilon.$$

Thus, for disintegrable Young measures, strong tightness is a stronger property than strict tightness (compare with (c) of Theorem 4.3.2). In the case when there is no Ω (that is, $\mathcal{S} = \{\emptyset, \Omega\}$), these three notions of tightness (including flexible tightness) coincide with the usual tightness notion in $\mathcal{M}^{+,1}(\mathbb{T})$. We give in Theorem 5.1.1 below the main features of strong tightness.

Theorem 5.1.1 (Equivalence theorem for strong tightness) *Let $\mathfrak{Y} \subset \mathcal{Y}^1_{\mathrm{dis}}$. Condition (i) is equivalent to each of the following conditions:*

(ii) There exists a sequence $(K_n)_{n\in\mathbb{N}}$ of compact subsets of $\mathbb{T}$ and, for every $\varepsilon > 0$, there exists $\Omega_\varepsilon \in \mathcal{S}$, with $\mathrm{P}(\Omega_\varepsilon) > 1 - \varepsilon$, such that

$$\forall \delta > 0 \quad \exists n \in \mathbb{N} \quad \forall \omega \in \Omega_\varepsilon \quad \forall \mu \in \mathfrak{Y} \quad \mu_\omega(K_n) \geq 1 - \delta.$$

(iii) For each $\varepsilon > 0$, there exists a compact subset K_ε of $\mathbb{T}$ such that

$$\mathrm{P}_* \{\omega \in \Omega;\ \forall \mu \in \mathfrak{Y} \quad \mu_\omega(K_\varepsilon) \geq 1 - \varepsilon\} > 1 - \varepsilon.$$

(iv) *There exists a sequence $(K_n)_{n\in\mathbb{N}}$ of compact subsets of $\mathbb{T}$ and a measurable subset Ω^* of Ω, with $\mathrm{P}(\Omega^*) = 1$, such that*

$$\forall\omega\in\Omega^*\quad \forall\delta>0\quad \exists n\in\mathbb{N}\quad \forall\mu\in\mathfrak{Y}\quad \mu_\omega(K_n)\geq 1-\delta.$$

(v) *There exists an inf-compact function $h : \mathbb{T} \to [0, +\infty]$ such that*

$$\int_\Omega^* \left(\sup_{\mu\in\mathfrak{Y}} \int_\mathbb{T} h(t)\, d\mu_\omega(t)\right) d\,\mathrm{P}(\omega) < +\infty.$$

Furthermore, if $\mathbb{T}$ is Suslin, then $\mathfrak{Y} \subset \mathcal{Y}^1_{\mathrm{dis}}$ is strongly tight if and only if one of the following conditions is satisfied:

(iii)* *For each $\varepsilon > 0$, there exists a compact random set K_ε such that*

$$\mathrm{P}_* \{\omega\in\Omega;\ \forall\mu\in\mathfrak{Y}\quad \mu_\omega(K_\varepsilon(\omega)) \geq 1-\varepsilon\} > 1-\varepsilon.$$

(iii)** *For each $\varepsilon > 0$, there exists a compact random set K_ε such that*

$$\mathrm{P}_* \{\omega\in\Omega;\ \forall\mu\in\mathfrak{Y}\quad \mu_\omega(K_\varepsilon(\omega)) \geq 1-\varepsilon\} = 1.$$

Remark and Definition 5.1.2 Let $(K_n)_{n\in\mathbb{N}}$ be a sequence of compact subsets of $\mathbb{T}$. Let $\mathfrak{K} \subset \mathcal{M}^{+,1}(\mathbb{T})$. Recall that $(K_n)_{n\in\mathbb{N}}$ *tightens* $\mathfrak{K}$ if, for each $\varepsilon > 0$, there exists an $n \in \mathbb{N}$ such that $\sup_{\mu\in\mathfrak{K}} \mu(K_n^c) \leq \varepsilon$. In particular, if $(K_n)_{n\in\mathbb{N}}$ tightens $\mathfrak{K}$, then $\mathfrak{K}$ is tight and all $\mu \in \mathfrak{K}$ satisfy $\mu(\mathbb{T}_0) = 1$, where $\mathbb{T}_0$ is the K_σ subspace $\cup_{n\in\mathbb{N}} K_n$, but this information tells nothing about the "rate" of tightness, that is, the way the sequence $(\sup_{\mu\in\mathfrak{K}} \mu(K_n^c))_{n\in\mathbb{N}}$ converges to 0.

Let $\mathfrak{Y} \subset \mathcal{Y}^1_{\mathrm{dis}}$. We say that $(K_n)_{n\in\mathbb{N}}$ *strongly tightens* $\mathfrak{Y}$ if $(K_n)_{n\in\mathbb{N}}$ satisfies *(ii)*. This implies that, for each $\varepsilon > 0$, there exists a subset $\mathfrak{K}_\varepsilon$ of $\mathcal{M}^{+,1}(\mathbb{T})$ *which is tightened by* $(K_n)_{n\in\mathbb{N}}$, such that

$$\mathrm{P}_* \{\omega\in\Omega;\ \forall\mu\in\mathfrak{Y}\quad \mu_\omega \in \mathfrak{K}_\varepsilon\} > 1-\varepsilon.$$

Thus we can take all $\mathfrak{K}_\varepsilon$ in Condition *(i)* in such a way that they be tightened by the same sequence $(K_n)_{n\in\mathbb{N}}$, but not necessarily with the same "rate" of tightness. In the non–Suslin case, this is probably the main difference between strong tightness and Condition *(iii)**, where the choice of a sequence $(K_n)_{n\in\mathbb{N}}$ which tightens $\{\mu_\omega;\ \mu\in\mathfrak{Y}\}$ depends *a priori* on ω.

We could have called the strong tighness property expressed by *(i)* "*strong strict tightness*", whereas *(iii)** would express "*strong flexible tightness*".

Proof of Theorem 5.1.1.

First, let us note that *(i)*⇔*(j)*, where *(j)* is the following condition:

(j) For each $\varepsilon > 0$, there exists $\Omega_\varepsilon \in \mathcal{S}$ such that $\mathrm{P}(\Omega_\varepsilon) > 1-\varepsilon$ and such that, for each $\delta > 0$, there exists a compact subset $K^{\varepsilon,\delta}$ of $\mathcal{M}^{+,1}(\mathbb{T})$ satisfying

$$\forall\omega\in\Omega_\varepsilon\quad \forall\mu\in\mathfrak{Y}\quad \mu_\omega(K^{\varepsilon,\delta}) \geq 1-\delta.$$

We shall prove the implications *(j)⇒(ii)⇒(i)*, *(j)⇔(iii)*, *(iii)⇒(iv)*, *(iv)⇒(iii)*, *(iii)⇒(v)*, *(v)⇒(iii)*, *(iii)⇔(iii)** and *(iii)*⇔(iii)***.

(j)⇒(ii)
If *(j)* holds true, let $(\eta_n)_{n\in\mathbb{N}}$ be any decreasing sequence of positive numbers with limit 0, and set $K_n = K^{\eta_n,\eta_n}$ for any $n \in \mathbb{N}$. This gives Condition *(ii)*.

(ii)⇒(i)
Assume *(ii)*. For each $\varepsilon > 0$ and each $\delta > 0$, let $n(\varepsilon, \delta) \in \mathbb{N}$ be such that $\mu_\omega(K_{n(\varepsilon,\delta)}) \geq 1 - \delta$ for each $\omega \in \Omega_\varepsilon$ and each $\mu \in \mathfrak{Y}$. To prove *(i)*, we only need to set

$$\mathfrak{K}_\varepsilon = \left\{\nu \in \mathcal{M}^{+,1}(\mathbb{T});\ \forall\delta > 0 \quad \nu(K_{n(\varepsilon,\delta)}) \geq 1 - \delta\right\}.$$

(j)⇔(iii)
Condition *(j)* can be written in the form

(jj) For each $\varepsilon > 0$ and each $\delta > 0$, there exists a compact subset $K^{\varepsilon,\delta}$ of $\mathcal{M}^{+,1}(\mathbb{T})$ such that

$$\mathrm{P}_*\left\{\omega \in \Omega;\ \forall\delta > 0 \quad \forall\mu \in \mathfrak{Y} \quad \mu_\omega(K^{\varepsilon,\delta}) \geq 1 - \delta\right\} > 1 - \varepsilon.$$

Now, *(jj)* clearly implies *(iii)*. Conversely, assume *(iii)*, and let $\varepsilon > 0$. For each $n \in \mathbb{N}$, let $K^n \in \mathcal{K}(\mathbb{T})$ and $\Omega^n_\varepsilon \in \mathcal{S}$ such that $\mathrm{P}(\Omega^n) \geq 1 - 2^{-n}\varepsilon$ and

$$\Omega^n_\varepsilon \subset \left\{\omega \in \Omega;\ \forall\mu \in \mathfrak{Y} \quad \mu_\omega(K^n) \geq 1 - 2^{-n}\varepsilon\right\}.$$

and set

$$\Omega_\varepsilon = \cap_{n\geq 2}\Omega^n_\varepsilon.$$

We then have $\Omega_\varepsilon \in \mathcal{S}$ and $\mathrm{P}(\Omega_\varepsilon) \geq 1 - \varepsilon/2 > 1 - \varepsilon$. Let $\delta > 0$. There exists $n \geq 2$ such that $\delta \geq 2^{-n}\varepsilon$. We thus have, for any $\omega \in \Omega_\varepsilon$ and for any $\mu \in \mathfrak{Y}$, $\mu_\omega(K^n) \geq 1 - 2^{-n}\varepsilon \geq 1 - \delta$, which proves *(jj)*.

(iii)⇒(iv)
Assume *(iii)*. Changing notations, let us denote by K_n the compact subset $\cup_{j\leq n} K_{1/(j+1)}$. The sequence $(K_n)_{n\in\mathbb{N}}$ is increasing and there exists for each $n \in \mathbb{N}$ a measurable subset Ω_n of Ω, with $\mathrm{P}(\Omega_n) > 1 - 1/(n+1)$ such that, for each $\omega \in \Omega_n$ and each $\mu \in \mathfrak{Y}$, $\mu_\omega(K_n) \geq 1 - 1/(n+1)$. Let $\Omega^* = \cap_{m\in\mathbb{N}} \cup_{n\geq m} \Omega_n$. Then Ω^* is measurable and $\mathrm{P}(\Omega^*) = \lim_{m\to\infty} \mathrm{P}(\cup_{n\geq m}\Omega_n) \geq \lim_{m\to\infty} \mathrm{P}(\Omega_m) = 1$. Furthermore, let $\delta > 0$ and let $\omega \in \Omega^*$. For each $m \in \mathbb{N}$, there exists an $n \geq m$ such that $\omega \in \Omega_n$, in particular one can choose n such that $1/(n+1) < \delta$. We then have $\mu_\omega(K_n) \geq 1 - 1/(n+1) > 1 - \delta$ for each $\mu \in \mathfrak{Y}$.

(iv)⇒(iii)
Let $\varepsilon > 0$. For each $n \in \mathbb{N}$, set

$$\Omega^*_{n,\varepsilon} = \{\omega \in \Omega;\ \forall \mu \in \mathfrak{Y}\quad \mu_\omega(K_n) \geq 1 - \varepsilon\}$$

and let $\Omega_{n,\varepsilon}$ be a measurable subset of $\Omega^*_{n,\varepsilon}$ such that $\mathrm{P}(\Omega_{n,\varepsilon}) \geq \mathrm{P}_*(\Omega^*_{n,\varepsilon}) - 1/(n+1)$. We assume w.l.g. that $(\Omega_{n,\varepsilon})_{n\in\mathbb{N}}$ is increasing. For each $\omega \in \Omega^*$, there exists an $n \in \mathbb{N}$ such that $\omega \in \Omega^*_{n,\varepsilon}$. Thus $\mathrm{P}\left(\cup_{n\in\mathbb{N}}\Omega_{n,\varepsilon}\right) = 1$, and, as the sequence $(\Omega_{n,\varepsilon})_{n\in\mathbb{N}}$ is increasing, there exists an $n \in \mathbb{N}$ such that $\mathrm{P}(\Omega_{n,\varepsilon}) > 1 - \varepsilon$. Then, we only need to take $K_\varepsilon = K_n$.

(iii)⇒(v)
For each integer $n \geq 1$, let $K'_n = K_{3^{-n}}$. We may (and shall) assume that the sequence $(K'_n)_n$ is increasing. For each $n \geq 1$, let Ω'_n be a measurable subset of Ω such that $\mathrm{P}(\Omega'_n) > 1 - 3^{-n}$ and, for every $\omega \in \Omega'_n$ and every $\mu \in \mathfrak{Y}$, $\mu_\omega(K'_n) \geq 1 - 3^{-n}$. Furthermore, we choose the sets Ω'_n such that the sequence $(\Omega'_n)_{n\geq 1}$ is increasing. Define the inf–compact function h by $h(t) = 2^n$ on $K'_n \setminus K'_{n-1}$ (with the convention $K'_0 = \emptyset$) and $h(t) = +\infty$ on $\mathbb{T} \setminus \cup_{n\geq 1} K'_n$. We have

$$\begin{aligned}
&\int^*_\Omega \sup_{\mu\in\mathfrak{Y}} \int_{\mathbb{T}} h(t)\, d\mu_\omega(t)\, d\mathrm{P}(\omega) \\
&\quad = \sum_{n\geq 0} \int^*_{\Omega'_{n+1}\setminus\Omega'_n} \sup_{\mu\in\mathfrak{Y}} \int_{\mathbb{T}} h(t)\, d\mu_\omega(t)\, d\mathrm{P}(\omega) \\
&\quad = \sum_{n\geq 0} \int^*_{\Omega'_{n+1}\setminus\Omega'_n} \sup_{\mu\in\mathfrak{Y}} \sum_{m\geq 0} 2^{m+1} \mu_\omega\left(K'_{m+1} \setminus K'_m\right) d\mathrm{P}(\omega) \\
&\quad \leq \sum_{n\geq 0} \int^*_{\Omega'_{n+1}\setminus\Omega'_n} \sup_{\mu\in\mathfrak{Y}} \Bigg(\sum_{0\leq m\leq n} 2^{m+1} \mu_\omega\left(K'_{m+1} \setminus K'_m\right) \\
&\qquad\qquad + 3 \sum_{m>n} \left(\frac{2}{3}\right)^{m+1} \Bigg) d\mathrm{P}(\omega)
\end{aligned}$$

(because, for $\omega \in \Omega'_{n+1} \setminus \Omega'_n$ and $m \geq n+1$, we have $\omega \in \Omega'_m$, thus $\mu_\omega(K'_{m+1} \setminus K'_m) \leq 3^{-m}$)

$$\begin{aligned}
&\quad \leq \sum_{n\geq 0} 3^{-n} \left(\sum_{0\leq m\leq n} 2^{m+1} + 3 \sum_{m>n} \left(\frac{2}{3}\right)^{m+1} \right) \\
&\quad \leq \sum_{n\geq 0} 3^{-n} \left(2^{n+2} - 2\right) + \sum_{n\geq 0} 3^{-n-1} \sum_{m\geq 0} \left(\frac{2}{3}\right)^{m+1} \\
&\quad < +\infty.
\end{aligned}$$

(v)⇒(iii)
For each $\varepsilon > 0$, there exists a real number $M_\varepsilon > 0$ such that

$$\mathrm{P}^* \left\{ \sup_{\mu \in \mathfrak{Y}} \int_{\mathbb{T}} h(t)\, d\mu_\omega(t) \geq M_\varepsilon \right\} \leq \varepsilon.$$

For each $\varepsilon > 0$ and each $\omega \in \Omega$, let

$$K_\varepsilon = \{t \in \mathbb{T};\ h(t) \leq M_\varepsilon/\varepsilon\}.$$

From inf–compactness of h, the set K_ε is compact. Moreover,

$$\begin{aligned}\mathrm{P}^* \{\omega \in \Omega;\ \forall \mu \in \mathfrak{Y} \quad \mu_\omega(K_\varepsilon) \geq 1 - \varepsilon\} &= \mathrm{P}^*\{\omega \in \Omega;\ \sup_{\mu \in \mathfrak{Y}} \int_{\mathbb{T}} \mathbb{1}_{\mathbb{T}\setminus K_\varepsilon}(t)\, d\mu_\omega(t) \leq \varepsilon\} \\ &\geq \mathrm{P}^*\{\omega \in \Omega;\ \sup_{\mu \in \mathfrak{Y}} \int_{\mathbb{T}} h(t)\, d\mu_\omega(t) \leq \varepsilon M_\varepsilon/\varepsilon\} \\ &\geq 1 - \varepsilon.\end{aligned}$$

Assume now that $\mathbb{T}$ is Suslin. Clearly, *(iii)* implies *(iii)**. To prove the converse implication, we use the same method as in Theorem 4.3.13. Assume that *(iii)** is satisfied. Let $\varepsilon > 0$ and $K \in \underline{\mathcal{K}}$ such that $\mathrm{P}_* \{\omega \in \Omega;\ \forall \mu \in \mathfrak{Y} \quad \mu_\omega(K(\omega)) \geq 1 - \varepsilon/2\} > 1 - \varepsilon/2$. Define the compact subset H of $\mathbb{T}$ and $\Omega' \in \mathcal{S}$ as in the proof of Theorem 4.3.13: We have $\mathrm{P}(\Omega') \geq 1 - \varepsilon/2$ and $K(\omega) \subset H$ for every $\omega \in \Omega'$. This yields

$$\begin{aligned}&\mathrm{P}_* \{\omega \in \Omega;\ \forall \mu \in \mathfrak{Y} \quad \mu_\omega(H) \geq 1 - \varepsilon\} \\ &\qquad\geq \mathrm{P}_* \{\omega \in \Omega';\ \forall \mu \in \mathfrak{Y} \quad \mu_\omega(K(\omega)) \geq 1 - \varepsilon\} \\ &\qquad\geq \mathrm{P}_* \{\omega \in \Omega;\ \forall \mu \in \mathfrak{Y} \quad \mu_\omega(K(\omega)) \geq 1 - \varepsilon\} - \varepsilon/2 \\ &\qquad\geq 1 - \varepsilon/2 - \varepsilon/2 = 1 - \varepsilon.\end{aligned}$$

(iii)*⇔*(iii)***
The implication *(iii)***⇒*(iii)** is clear. For the converse, assume *(iii)**. Let $\varepsilon > 0$. For each integer $n \geq 1$, let K^n be a random compact set and $\Omega_n \in \mathcal{S}$ such that $\mathrm{P}(\Omega_n) \geq 1 - 1/n$ and, for each $\omega \in \Omega_n$ and each $\mu \in \mathfrak{Y}$, $\mu_\omega(K^n(\omega)) \geq 1 - \varepsilon$. Set $\Omega_0 = \emptyset$ and, for each $n \geq 1$, $K_\varepsilon(\omega) = K^n(\omega)$ for $\omega \in \Omega_n \setminus \Omega_{n-1}$ (we define arbitrarily K_ε on the negligible set $\Omega \setminus \cup_n \Omega_n$). □

It is worth writing down Theorem 5.1.1 in the case of random elements of $\mathbb{T}$. Note that the adaptation of Conditions *(i)*, *(ii)* and *(iii)* gives the single Condition *(i')* below: This comes from the fact that, for Dirac measures, there is only one possible "rate" of tightness (see Remark and Definition 5.1.2). This Condition *(i')* is in fact the condition originally given in [KZ94] by Kruk and Zięba to define strong tightness. Moreover, Condition *(iii')** (adapted in *(iii')***) has lost its parameter ε and simply expresses the fact that all elements of $\mathfrak{Y}$ are selections of some random compact set.

Corollary 5.1.3 *Let $\mathfrak{Y} \subset \mathfrak{X}$. The following conditions are equivalent:*

(i') *For each $\varepsilon > 0$, there exists a compact subset K_ε of $\mathbb{T}$ such that*

$$\mathrm{P}_* \{\omega \in \Omega;\ \forall X \in \mathfrak{Y} \quad X(\omega) \in K_\varepsilon\} > 1 - \varepsilon.$$

(iv') *There exists a sequence $(K_n)_{n\in\mathbb{N}}$ of compact subsets of $\mathbb{T}$ and a measurable subset Ω^* of Ω, with $\mathrm{P}(\Omega^*) = 1$, such that, for each $\omega \in \Omega^*$, there exists an $n \in \mathbb{N}$ such that*

$$\forall X \in \mathfrak{Y} \quad X(\omega) \in K_n.$$

(v') *There exists an inf-compact function $h : \mathbb{T} \to [0, +\infty]$ such that*

$$\int_\Omega^* \left(\sup_{X\in\mathfrak{Y}} h \circ X \right) d\mathrm{P} < +\infty.$$

If $\mathbb{T}$ is Suslin, these conditions are equivalent to any one of the following two conditions:

(iii')* *For each $\varepsilon > 0$, there exists a random compact set K_ε such that*

$$\mathrm{P}_* \{\omega \in \Omega;\ \forall X \in \mathfrak{Y} \quad X(\omega) \in K_\varepsilon(\omega)\} > 1 - \varepsilon.$$

(iii')** *There exists a random compact set K such that, for almost every $\omega \in \Omega$ and for every $X \in \mathfrak{Y}$, $X(\omega) \in K(\omega)$.*

5.2 Strong tightness of almost everywhere convergent sequences

We start with some results for random elements, which generalize by new methods a result of Kruk and Zięba [KZ95] (see also [KZ96]).

The idea of the proof of the following theorem is the same as in the proof of Theorem 4.3.13.

Theorem 5.2.1 *Assume that the space of countable compact subsets of $\mathbb{T}$ (with the Vietoris topology) is Radon (e.g. $\mathbb{T}$ is a regular Suslin Prohorov space, or a Lusin space with a cofinal sequence of compact subsets, see Theorem 4.3.14). Let $(X_n)_{n\in\mathbb{N}}$ be a sequence of random elements of $\mathbb{T}$ which almost everywhere converges to a random element X_∞. Then $(X_n)_n$ is strongly tight.*

Proof. For each $\omega \in \Omega$, let $K(\omega) = \{X_n(\omega);\ n \in \mathbb{N} \cup \{\infty\}\}$. Let $G \in \mathcal{G}$. We have

$$\{\omega \in \Omega;\ K(\omega) \subset G\} = \cap_{n\in\mathbb{N}\cup\{\infty\}} \{\omega \in \Omega;\ X_n(\omega) \in G\} \in \mathcal{S}$$

and

$$\{\omega \in \Omega;\ K(\omega) \cap G \neq \emptyset\} = \cup_{n \in \mathbb{N} \cup \{\infty\}} \{\omega \in \Omega;\ X_n(\omega) \in G\} \in \mathcal{S}.$$

Thus the mapping $K : \Omega \to \mathcal{K}$ is measurable. Let $\varepsilon > 0$. As the space of countable compact subsets of $\mathbb{T}$ is Radon, there exists a compact subset $\mathfrak{K} \in \mathcal{K}$ such that

$$\mathrm{P}\{\omega \in \Omega;\ K(\omega) \in \mathfrak{K}\} \geq 1 - \varepsilon.$$

Let $H = \cup_{L \in \mathfrak{K}} L$. We thus have

$$\mathrm{P}\{\omega \in \Omega;\ K(\omega) \subset H\} \geq 1 - \varepsilon.$$

but, from a result of Michael ([Mic51], Theorem 2.5.2), the set H is compact.

□

Corollary 5.2.2 *Assume that $\mathbb{T}$ is Suslin regular Prohorov and let $(\mu^n)_{n \in \mathbb{N}}$ be a sequence of elements of $\mathcal{Y}^1_{\text{dis}}$ which almost everywhere converges to a Young measure μ^∞. Then $(\mu^n)_n$ is strongly tight.*

Proof. The space $\mathcal{M}^{+,1}(\mathbb{T})$ is Suslin regular and, from Wójcicka's result [Wój87], it is Prohorov. Thus, from Theorem 5.2.1, $(\mu^n)_{n \in \mathbb{N}}$ is a strongly tight sequence in $\mathfrak{X}\left(\mathcal{M}^{+,1}(\mathbb{T})\right)$, that is, for each $\varepsilon > 0$, there exists a compact subset $\mathfrak{K}_\varepsilon$ of $\mathcal{M}^{+,1}(\mathbb{T})$ such that

$$\mathrm{P}\{\omega \in \Omega;\ \forall n \in \mathbb{N}\ \ \mu^n_\omega \in \mathfrak{K}_\varepsilon\} > 1 - \varepsilon.$$

But, as $\mathbb{T}$ is Prohorov, each $\mathfrak{K}_\varepsilon$ is tight, thus $(\mu^n)_{n \in \mathbb{N}}$ is a strongly tight sequence in $\mathcal{Y}^1_{\text{dis}}$. □

Corollary 5.2.3 *With the hypothesis and notations of Theorem 4.5.8, Conditions 1 to 6 are equivalent to*

4′. *For each sequence $(\mu^n)_n$ of elements of $\mathfrak{Y}$, there exists a strongly tight sequence $(\nu^n)_n$ in $\mathcal{Y}^1$ such that, for every $n \in \mathbb{N}$, $\nu^n \in \operatorname{co}\{\mu_m;\ m \geq n\}$.*

Proof. From Corollary 5.2.2, we have $4 \Rightarrow 4'$. But $4' \Rightarrow 5$ is obvious, and 4 and 5 are equivalent to 1, 2, 3 and 6. □

Remark 5.2.4 The Radon property is not necessary for results of the type of Theorem 5.2.1. For example, there exists a compact space which is not Radon (see [Sch73, page 45], for an example of Dieudonné). For such a space $\mathbb{T}$, $\mathcal{K}(\mathbb{T})$ cannot be Radon. But any sequence of random elements of $\mathbb{T}$ is obviously strongly tight.

Let us give another example. Assume that $\mathbb{T}$ is a (non–necessarily submetrizable) k_ω–space and let $(X_n)_{n \in \mathbb{N}}$ be a sequence of random elements of $\mathbb{T}$ which almost everywhere converges to a random element X_∞. Then $(X_n)_n$ is strongly tight.

Indeed, let $(K_m)_{m\in\mathbb{N}}$ be an increasing sequence of compact subsets of $\mathbb{T}$ which determines the topology of $\mathbb{T}$. Let $\Omega_1 \in \mathcal{S}$ be an almost sure set on which $(X_n(\omega))_n$ converges to $X_\infty(\omega)$. Let $\hat{\mathbb{N}} = \mathbb{N} \cup \{\infty\}$. For every $\omega \in \Omega_1$, the set $Y(\omega) = \{X_n(\omega);\ n \in \mathbb{N} \cup \{\infty\}\}$ is compact, thus it is contained in some K_m. Set

$$M(\omega) = \inf\{m \in \mathbb{N};\ Y(\omega) \subset K_m\}.$$

The function M has finite values on $\omega \in \Omega_1$. Furthermore, we have, for every $m \in \mathbb{N}$,

$$M^{-1}(m) = \left\{\omega \in \Omega;\ \forall n \in \overline{\mathbb{N}}\quad X_n(\omega) \in K_m\right\} \setminus \cup_{k\le m} M^{-1}(k),$$

Thus M is measurable. Let $\varepsilon > 0$. We have $\Omega_1 \subset \cup_{N\in\mathbb{N}} \{\omega \in \Omega;\ M(\omega) \le N\}$, thus there exists $N \in \mathbb{N}$ such that $\mathrm{P}\{\omega \in \Omega;\ M(\omega) \le N\} \ge 1 - \varepsilon$.

Chapter 6

Young measures on Banach spaces. Applications

6.1 Preliminaries, Biting Lemma and some basic results

In this chapter, the probability space $(\Omega, \mathcal{S}, \mathrm{P})$ is assumed to be complete. We use the following notations:

- $\mathbb{E}$ is a separable Banach space, $\mathbb{E}^*$ is its dual, $\mathbb{E}_\sigma$ is the vector space $\mathbb{E}$ equipped with the $\sigma(\mathbb{E}, \mathbb{E}^*)$ topology, $\mathbb{E}^*_\sigma$ is the vector space $\mathbb{E}^*$ equipped with the $\sigma(\mathbb{E}^*, \mathbb{E})$ topology.

- For each $x \in \mathbb{E}$ and each $r > 0$, the closed ball with center x and radius r is the closure of the open ball $B_{\mathbb{E}}(0, r)$, so we shall denote it by $\overline{B}_{\mathbb{E}}(0, r)$. The closed unit ball of $\mathbb{E}$ is simply denoted by $\overline{B}_{\mathbb{E}}$.

- Recall that the convex hull of a subset A of $\mathbb{E}$ is denoted by $\operatorname{co} A$. We shall denote by $\overline{\operatorname{co}}\, A$ the closed convex hull of A.

- Recall that $\mathrm{L}^1_{\mathbb{E}} = \mathrm{L}^1_{\mathbb{E}}(\Omega, \mathcal{S}, \mathrm{P})$ is the space of Bochner integrable mappings defined on $(\Omega, \mathcal{S}, \mathrm{P})$ with values in $\mathbb{E}$. The vector space $\mathfrak{X}(\mathbb{E})$ will be denoted by $\mathrm{L}^0_{\mathbb{E}}$.

- $\mathrm{c}\mathcal{K}(\mathbb{E})$ is the collection of nonempty convex compact subsets of $\mathbb{E}$,

  $\mathrm{cw}\mathcal{K}(\mathbb{E})$ is the collection of nonempty convex weakly compact subsets of $\mathbb{E}$,

  $\mathcal{L}\mathrm{cw}\mathcal{K}(\mathbb{E})$ is the collection of nonempty closed convex locally weakly compact subsets of $\mathbb{E}$ containing no line [CV77, Theorem I.14] (some authors say: linefree subsets),

$\mathcal{R}\mathrm{cw}\mathcal{K}(\mathbb{E})$ (resp. $\mathcal{R}\mathrm{cw}\mathcal{K}(\mathbb{E})$) is the collection of closed (resp. closed convex) convex subsets of $\mathbb{E}$ such that their intersection with any closed ball in $\mathbb{E}$ is weakly compact.

- If $(x_n)_n$ is a sequence which converges in $\mathbb{E}_\sigma$ to an element x of $\mathbb{E}$, we write $x = \text{w-}\lim x_n$.

  If $(C_n)_n$ is a sequence of nonempty closed convex subsets of $\mathbb{E}$, the *weak sequential upper limit* w-ls C_n of $(C_n)_n$ is the set of elements x of $\mathbb{E}$ such that there exist a subsequence $(C_{n_j})_j$ of $(C_n)_n$ and, for each j, an element x_{n_j} of C_{n_j} such that $x = \text{w-}\lim_j x_{n_j}$.

- A sequence (u_n) in $\mathrm{L}^0_{\mathbb{E}}(\Omega, \mathcal{S}, \mathrm{P})$ is *norm tight* (resp. *weakly strictly tight*) if there is an inf-compact (resp. inf-weakly compact) function $h : \mathbb{E} \to [0, +\infty]$ such that
$$\sup_n \int h(u_n(\omega))\, d\,\mathrm{P}(\omega) < +\infty.$$
  If (u_n) is norm tight (resp. weakly tight), then the sequence $(\underline{\delta}_{u_n})$ of Young measures associated with (u_n) is strictly tight (cf. Theorem 4.3.5) in the space $\mathcal{Y}^1(\Omega, \mathcal{S}, \mathrm{P}; \mathbb{E})$ (resp. $\mathcal{Y}^1(\Omega, \mathcal{S}, \mathrm{P}; \mathbb{E}_\sigma)$) of Young measures corresponding to $\mathbb{E}$ and $\mathbb{E}_\sigma$ respectively. We will consider also the space of Young measures $\mathcal{Y}^1(\Omega, \mathcal{S}, \mathrm{P}; \mathbb{E}^*_\sigma)$.

- A sequence (u_n) in $\mathrm{L}^1_{\mathbb{E}}(\mathrm{P})$ is said *weakly flexibly tight* if $(\underline{\delta}_{u_n})$ is flexibly tight in $\mathcal{Y}^1(\Omega, \mathrm{P}; \mathbb{E}_\sigma)$. This is equivalent to: For any $\varepsilon > 0$, there exists a measurable multifunction $\Gamma_\varepsilon : \Omega \to \mathrm{cw}\mathcal{K}(\mathbb{E})$ such that
$$\sup_n \mathrm{P}(\{\omega; u_n(\omega) \notin \Gamma_\varepsilon(\omega)\}) < \varepsilon.$$
  A sequence (u_n) in $\mathrm{L}^1_{\mathbb{E}}(\mathrm{P})$ is $\mathcal{R}\mathrm{cw}\mathcal{K}(\mathbb{E})$*-flexibly tight* if the foregoing condition holds with Γ_ε valued in $\mathcal{R}\mathrm{cw}\mathcal{K}(\mathbb{E})$. Note that if furthermore (u_n) is bounded in $\mathrm{L}^1_{\mathbb{E}}$, then it is weakly-flexibly tight. Indeed set $\Omega'_\varepsilon := \{\omega; u_n(\omega) \notin \Gamma_\varepsilon(\omega)\}$ and $\Omega''_\varepsilon := \{\omega; \|u_n(\omega)\| > \frac{M}{\varepsilon}\}$ where $M := \sup_n \|u_n\|_{\mathrm{L}^1}$. Then by Markov's inequality $\mathrm{P}((\Omega'_\varepsilon \cup \Omega''_\varepsilon)^c) < 2\varepsilon$ and since $\Gamma_\varepsilon(\omega) \cap \overline{B}(0, M/\varepsilon)$ belongs to $\mathrm{cw}\mathcal{K}(\mathbb{E})$, we have proved that [$\mathcal{R}\mathrm{cw}\mathcal{K}(\mathbb{E})$-flexibly tight and bounded in $\mathrm{L}^1_{\mathbb{E}}$] imply weakly-flexibly tight.

  One can wonder if weakly flexibly tight implies weakly strictly tight. By (*ii*) of Theorem 4.3.14 this holds when $\mathbb{E}$ is reflexive but for $\mathbb{E}$ not reflexive neither part (*i*) of the same theorem nor parts A.2 and B of Theorem 4.3.5 apply since $\mathbb{E}_\sigma$ is not Prohorov (see reference to Fernique [Fer94] above page 90).

- Taking up the definition of Section 4.5, we say that a sequence (u_n) in $\mathrm{L}^1_{\mathbb{E}}(\mathrm{P})$ is *weakly* (resp. *strongly*) Mazur–compact if there is a sequence (v_n) with $v_n(\omega) \in \mathrm{co}\{u_m(\omega); m \geq n\}$ such that $(v_n(\omega))$ converges weakly (resp.

strongly) a.e. to $u(\omega)$ where u is a function in $\mathrm{L}^1_{\mathbb{E}}(\mathrm{P})$. More generally, the sequence (u_n) is cw$\mathcal{K}(\mathbb{E})$-*Mazur–tight* (resp. $\mathcal{R}$cw$\mathcal{K}(\mathbb{E})$-*Mazur–tight*) if there is a sequence (v_n) with $v_n(\omega) \in \operatorname{co}\{u_m(\omega);\ m \geq n\}$ such that (v_n) is cw$\mathcal{K}(\mathbb{E})$-tight (resp. $\mathcal{R}$cw$\mathcal{K}(\mathbb{E})$-tight). These notions are useful in the study of weak compactness in $\mathrm{L}^1_{\mathbb{E}}(\mathrm{P})$. We refer to [BC97, CG99] for a complete study of the Mazur-tightness properties in $\mathrm{L}^1_{\mathbb{E}}(\mathrm{P})$. In particular, if $\mathbb{E}$ is reflexive, then any bounded sequence in $\mathrm{L}^1_{\mathbb{E}}(\mathrm{P})$ is strongly Mazur–compact. These tightness notions have been extended in [CS00, page 51] to the space of convex weakly compact valued integrably bounded multifunctions.

It is worth to compare some aspects of the preceding tightness notions with the ones developed for Young measures in the foregoing chapters. The two following results will precise these differences.

Lemma 6.1.1 *Let $\mathbb{S}$ be a topological Suslin space. Let $\mathcal{R}$ be a class of Borel subsets of $\mathbb{S}$ such that $\emptyset \in \mathcal{R}$; $A, B \in \mathcal{R} \Longrightarrow A \cup B \in \mathcal{R}$. Let $\mathcal{H}$ be a subset of $\mathrm{L}^0_{\mathbb{S}}(\Omega, \mathcal{S}, \mathrm{P})$. Then the following are equivalent:*

(a) *For any $\varepsilon > 0$, there exists an $\mathcal{R}$-valued measurable multifunction L_ε on Ω such that*

$$\forall u \in \mathcal{H} \quad \mathrm{P}[\{\omega \in \Omega;\ u(\omega) \notin L_\varepsilon(\omega)\}] < \varepsilon.$$

(b) *There exists an $\mathcal{S} \otimes \mathcal{B}_{\mathbb{S}}$-measurable integrand $\varphi : \Omega \times \mathbb{S} \longrightarrow [0, +\infty]$ such that for all $\omega \in \Omega$ and all $r \in [0, +\infty[$, $\{x \in \mathbb{S};\ \varphi(\omega, x) \leq r\} \in \mathcal{R}$ and such that*

$$\sup_{u \in \mathcal{H}} \int_\Omega \varphi(\omega, u(\omega))\, d\,\mathrm{P}(\omega) < +\infty;$$

(c) *There exists an $\mathcal{S} \otimes \mathcal{B}_{\mathbb{S}}$-measurable integrand $\varphi : \Omega \times \mathbb{S} \longrightarrow [0, +\infty]$ such that for all $\omega \in \Omega$ and all $r \in [0, +\infty[$, $\{x \in \mathbb{S};\ \varphi(\omega, x) \leq r\} \in \mathcal{R}$ and such that*

$$\lim_{\lambda \to +\infty} \sup_{u \in \mathcal{H}} \mathrm{P}[\{\omega \in \Omega;\ \varphi(\omega, u(\omega)) \geq \lambda\}] = 0.$$

Proof. $(a) \Longrightarrow (b)$. Let $\varepsilon_p = 3^{-p}$ $(p \in \mathbb{N})$. By (a) there exists a $\mathcal{R}$-valued measurable multifunction $L_p : \Omega \longrightarrow \mathbb{S}$ such that

$$\forall u \in \mathcal{H} \quad \mathrm{P}[\{\omega \in \Omega;\ u(\omega) \notin L_p(\omega)\}] < \varepsilon_p\,.$$

Let us consider the $\mathcal{B}_{\mathbb{S}}$-valued multifunctions $K_n : \Omega \longrightarrow \mathbb{S}$ $(n \in \mathbb{N} \cup \{\infty\})$ given by

$$\forall \omega \in \Omega \quad \forall n \geq 1 \quad K_0(\omega) = L_0(\omega), K_n(\omega) = L_n(\omega) \setminus K_{n-1}(\omega)$$

and

$$K_\infty(\omega) = \mathbb{S} \setminus \bigcup_{n \in \mathbb{N}} K_n(\omega) = \mathbb{S} \setminus \bigcup_{n \in \mathbb{N}} L_n(\omega).$$

Then it is obvious that each K_n ($n \in \mathbb{N} \cup \{\infty\}$) is measurable and the sequence $(\mathrm{gph}\,(K_n))_{n\in\mathbb{N}\cup\{\infty\}}$ is an $\mathcal{S}\otimes\mathcal{B}_{\mathbb{S}}$-measurable partition of $\Omega\times\mathbb{S}$. Set

$$\varphi(\omega,x)=\begin{cases}2^n & \text{if } (\omega,x)\in \mathrm{gph}\,(K_n)\,,\ n\in\mathbb{N},\\ +\infty & \text{if } (\omega,x)\in\mathrm{gph}\,(K_\infty)\,.\end{cases}$$

We claim that φ is an $\mathcal{S}\otimes\mathcal{B}_{\mathbb{S}}$-measurable integrand which satisfies condition (b). Indeed, let $r\geq 0$. If $r<1$, $\{x\in\mathbb{S};\ \varphi(\omega,x)\leq r\}$ is empty; if $r\geq 1$, let m be the unique integer such that $m\leq \log r/\log 2<m+1$. Then

$$\{(\omega,x)\in\Omega\times S;\ \varphi(\omega,x)\leq r\}=\bigcup_{n=0}^{m}\mathrm{gph}\,(K_n)\in\mathcal{S}\otimes\mathcal{B}_{\mathbb{S}}.$$

Similarly for all $\omega\in\Omega$, we have

$$\{x\in\mathbb{S};\ \varphi(\omega,x)\leq r\}=\bigcup_{n=0}^{m}K_n(\omega)=\bigcup_{n=0}^{m}L_n(\omega)\in\mathcal{R}.$$

It remains to check that $\sup_{u\in\mathcal{H}}\int_\Omega\varphi(\omega,u(\omega))\,d\mathrm{P}(\omega)<+\infty$.

For each $u\in\mathcal{H}$ and each $n\in\mathbb{N}\cup\{\infty\}$, set

$$\Omega_n^u=\{\omega\in\Omega;\ u(\omega)\in K_n(\omega)\}.$$

Then $(\Omega_n^u)_{n\in\mathbb{N}\cup\{\infty\}}$ is an $\mathcal{S}$-measurable partition of Ω with $\mathrm{P}(\Omega_n^u)<\varepsilon_{n-1}$ for every $n\geq 1$ and $\mathrm{P}(\Omega_\infty^u)=0$. Consequently we have

$$\begin{aligned}\int_\Omega\varphi(\omega,u(\omega))\,d\mathrm{P}(\omega)&=\sum_{n=0}^{\infty}\int_{\Omega_n^u}\varphi(\omega,u(\omega))\,d\mathrm{P}(\omega)=\sum_{n=0}^{\infty}2^n\,\mathrm{P}(\Omega_n^u)\\&\leq 1+\sum_{n=1}^{\infty}\frac{2^n}{3^{n-1}}<+\infty\end{aligned}$$

thus proving the implication $(a)\Longrightarrow(b)$.

$(b)\Longrightarrow(c)$ follows immediately from Markov's inequality.

Let us prove the implication $(c)\Longrightarrow(a)$. For every $\varepsilon>0$, there exists $\lambda_\varepsilon>0$ such that $\sup_{u\in\mathcal{H}}\mathrm{P}[\{\omega\in\Omega;\ \varphi(\omega,u(\omega))>\lambda_\varepsilon\}]<\varepsilon$. Since φ is $\mathcal{S}\otimes\mathcal{B}_{\mathbb{S}}$-measurable, the multifunction $L_\varepsilon(\omega):=\{x\in\mathbb{S};\ \varphi(\omega,x)\leq\lambda_\varepsilon\}$, $\forall\omega\in\Omega$, is measurable and takes its values in $\mathcal{R}$ by (c). As

$$\forall u\in\mathcal{H}\quad,\mathrm{P}[\{\omega\in\Omega;\ u(\omega)\notin L_\varepsilon(\omega)\}]=\mathrm{P}[\{\omega\in\Omega;\ \varphi(\omega,u(\omega))>\lambda_\varepsilon\}]<\varepsilon,$$

$(c)\Longrightarrow(a)$ follows. □

References: [ACV92, Théorème E page 175], [Jaw84, Proposition 2.2 page 13.17], [BC97, Lemma 2.5]. The above proof is borrowed from [BC97].

Let (u_n) be a sequence in $\mathrm{L}^1_{\mathbb{E}}(\Omega, \mathcal{S}, \mathrm{P})$. It turns out that the following condition:

There exists a measurable integrand $\varphi : \Omega \times \mathbb{E} \to [0, +\infty]$ satisfying
(6.1.1)
$$\begin{cases} \forall \omega \in \Omega \quad \forall r \in [0, +\infty[\quad \{x \in \mathbb{E};\ \varphi(\omega, x) \le r\} \in \mathrm{cw}\mathcal{K}(\mathbb{E}) \ (\text{resp. } \mathcal{R}\mathrm{cw}\mathcal{K}(\mathbb{E})), \\ \sup_n \int_\Omega \varphi(\omega, u_n(\omega))\, d\,\mathrm{P}(\omega) < +\infty, \end{cases}$$

implies the weak flexible tightness (resp. $\mathcal{R}\mathrm{cw}\mathcal{K}(\mathbb{E})$-flexible tightness) condition, whereas the last tightness condition implies Condition (6.1.1) provided some additional assumption.

The following result focuses the difference between tightness condition for Young measures and the one given for sequences in $\mathrm{L}^1_{\mathbb{E}}$. Indeed it is known that the parametric Prohorov tightness condition for a sequence (λ^n) of Young measures on $\Omega \times \mathbb{S}$, $\mathbb{S}$ being a completely regular Suslin space: For every $\varepsilon > 0$ there exists a compact-valued measurable multifunction $\Gamma_\varepsilon : \Omega \to \mathcal{K}(\mathbb{S})$ such that

$$\sup_n \int_\Omega \lambda^n_\omega(S \setminus \Gamma_\varepsilon(\omega))\, d\,\mathrm{P}(\omega) \le \varepsilon$$

is equivalent to the existence of a nonnegative measurable integrand h defined on $\Omega \times \mathbb{S}$ which is inf-compact in $s \in \mathbb{S}$ such that

$$\sup_n \int_\Omega \Big[\int_{\mathbb{S}} h(\omega, x)\, d\lambda^n_\omega(s)\Big]\, d\,\mathrm{P}(\omega) < +\infty.$$

Set $\varphi(\omega, \nu) = \int_{\mathbb{S}} h(\omega, s)\, d\nu(s)$ for $(\omega, \nu) \in \Omega \times \mathcal{M}^{+,1}(\mathbb{S})$, then the preceding condition is written as

$$\sup_n \int_\Omega \varphi(\omega, \lambda^n_\omega)\, d\,\mathrm{P}(\omega) < +\infty.$$

Since $\varphi(\omega, \nu)$ is convex (even affine) in the second variable, one could think that the $\mathrm{cw}\mathcal{K}(\mathbb{E})$-tightness for a sequence (u_n) in $\mathrm{L}^1_{\mathbb{E}}(\mathrm{P})$ is equivalent to:

There is a measurable integrand $\varphi : \Omega \times \mathbb{E} \to [0, +\infty]$ which is convex and inf-weakly compact in x such that

$$\sup_n \int_\Omega \varphi(\omega, u_n(\omega))\, d\,\mathrm{P}(\omega) < +\infty.$$

An example [Saa98, Proposition p.115] due to the third author shows that this equivalence fails. In the following lemma $\mathrm{P} = dt$ is the Lebesgue measure on $[0, 1]$. Recall the *James (or James–Pryce) Theorem:* If $\mathbb{E}$ is a Banach space and C is a $\sigma(\mathbb{E}, \mathbb{E}^*)$–closed subset of $\mathbb{E}$, then $\mathbb{E}$ is $\sigma(\mathbb{E}, \mathbb{E}^*)$–compact if and only if, for every $x' \in \mathbb{E}^*$, the exists $x \in C$ such that $\langle x', x\rangle = \max_{y \in C} \langle x', y\rangle$. This result was proved by James [Jam64] and extended by Pryce [Pry66] to locally convex spaces with a simpler proof.

Lemma 6.1.2 *Let $\mathbb{E}$ be a non reflexive Banach space. Then there exists a bounded set $\mathcal{H} \subset L^1_{\mathbb{E}}([0,1], dt)$ which is norm-tight and such that there exists no measurable integrand $\varphi : [0,1] \times \mathbb{E} \to [0, +\infty]$ which is convex and inf-weakly compact in x and which satisfies*

$$\sup_{u \in \mathcal{H}} \int_{[0,1]} \varphi(t, u(t))\, dt < +\infty.$$

Proof. Firstly by James' theorem, $\mathbb{E}$ contains a bounded sequence (e_n) which does not admit any weakly convergent subsequence. Indeed there exists $x' \in \mathbb{E}^*$ which does not achieve its supremum on $\overline{B}_{\mathbb{E}}$. Let (e_n) be a maximizing sequence. It has not any weakly convergent subsequence. We may suppose without loss of generality that $\forall n$, $\|e_n\| = 1$, hence $m \neq n \Longrightarrow me_n \neq ne_n \neq 0$.

Let us consider $\mathcal{H} := \{ne_n \mathbb{1}_A;\ n \in \mathbb{N}^*, P(A) = 1/n\} \cup \{0\}$. Then $\mathcal{H}$ is bounded in $L^1_{\mathbb{E}}$. Let h_0 be defined by $h_0(0) = 0$, $h_0(ne_n) = n$ for $n \geq 1$ and $h_0(x) = +\infty$ elsewhere. Thus h_0 is inf-strongly compact and satisfies

$$\sup_{u \in \mathcal{H}} \int_{[0,1]} (h_0 \circ u)(t)\, dt \leq 1,$$

hence $\mathcal{H}$ is norm-tight.

Now suppose that there exists some convex inf-weakly compact integrand $h : [0,1] \times \mathbb{E} \to [0, +\infty]$ satisfying

$$\sup_{u \in \mathcal{H}} \int_{[0,1]} h(t, u(t))\, dt < +\infty. \tag{6.1.2}$$

Then $h(t, e_n) \to +\infty$ because otherwise there would exist $M < +\infty$ and a subsequence $(e_{n_k})_k$ such that $\sup_k h(\omega, e_{n_k}) \leq M$ and the sequence $(e_{n_k})_k$ would admit a weakly convergent further subsequence. For $u_n := ne_n \mathbb{1}_{A_n} \in \mathcal{H}$, we have

$$\int_{[0,1]} h(t, u_n(t))\, dt \geq \int_{A_n} h(t, ne_n)\, dt.$$

It is possible to choose A_n such that $\int_{[0,1]} h(t, u_n(t))\, dt \to +\infty$. Indeed, since $0 \in \mathcal{H}$, one has $h(t, 0) < +\infty$ a.e. By convexity $h(t, ne_n) \geq n[h(t, e_n) - (1 - 1/n)h(t, 0)] \geq n[h(t, e_n) - h(t, 0)]$. Then by Egorov's theorem, there exists a Borel subset B of $[0,1]$ such that $P(B) \geq 1/2$ and such that the sequence $(h(., e_n) - h(., 0))_n$ converges uniformly to $+\infty$ on B. It remains to choose, for $n \geq 2$, A_n contained in B, to obtain a contradiction with (6.1.2). □

Recall that a sequence (u_n) in $L^1_{\mathbb{E}}(\Omega, \mathcal{S}, P)$ *Komlós converges to* $u_\infty \in L^1_{\mathbb{E}}(\Omega, \mathcal{S}, P)$, if, for any subsequence (v_n) of (u_n), one has

$$\lim_{n \to \infty} \frac{1}{n} \sum_{j=1}^{n} v_j(\omega) = u_\infty(\omega) \text{ a.e.}$$

The space $L^1_{\mathbb{E}}(\Omega, \mathcal{S}, P)$ has the *strong Komlós property* if, for any bounded sequence (u_n) in $L^1_{\mathbb{E}}(\Omega, \mathcal{S}, P)$, there is $u_\infty \in L^1_{\mathbb{E}}(\Omega, \mathcal{S}, P)$, such that some subsequence of (u_n) Komlós converges to u_∞. Recall that the Komlós Theorem [Kom67] states that $L^1_{\mathbb{R}}(\Omega, \mathcal{S}, P)$ has the strong Komlós property. This theorem also holds for $L^1_{\mathbb{R}}(\Omega, \mathcal{S}, P)$ when $\mathbb{E}$ is super-reflexive, or more generally when $\mathbb{E}$ is B–convex reflexive, see ([Gar79, Bou79]). An elementary proof of Komlós theorem in $L^1_{\mathbb{E}}(\Omega, \mathcal{S}, P)$ where $\mathbb{E}$ is a Hilbert space is given by [Gue97].

We recall the Biting Lemma. Here we follow the proof given by Słaby [Sła85] reproduced in [Egg84, VIII.1.17 pp.303–305] and [Val90b, Theorem 23 pp.173–174][1]. For further reference, see Kadec–Pelzynski [KP62], Gapoškin [Gap72]. See also the proof given by H.P. Rosenthal [Ros79] reproduced in [CM97, Lemma 2.1.3 pp.54–56].

The *modulus of uniform integrability* of H.P. Rosenthal, $\eta((u_n))$, of the sequence $(u_n)_n$ in $L^1_{\mathbb{E}}(\Omega, \mathcal{S}, P)$, is

$$\eta((u_n)) := \lim_{\substack{\varepsilon \to 0 \\ \varepsilon > 0}} \left[\sup\{ \int_A \|u_n(\omega)\| \, d\,P(\omega);\ n \in \mathbb{N},\ P(A) \leq \varepsilon \} \right].$$

The sequence $(u_n)_n$ is uniformly integrable iff $\eta((u_n)) = 0$.

Lemma 6.1.3 *Let $(u_n)_n$ be a bounded sequence in $L^1_{\mathbb{E}}(\Omega, \mathcal{S}, P)$. Then*

$$\eta((u_n)) = \lim_{t \to +\infty} \left[\sup_{n \in \mathbb{N}} \int_{\{\|u_n(.)\| > t\}} \|u_n(\omega)\| \, d\,P(\omega) \right].$$

Proof. The two limits are nonincreasing limits, hence infima. Let us denote η_1 the modulus of the statement and $M := \sup_n \|u_n\|_{L^1_{\mathbb{E}}}$. Here $\|u_n\|$ denotes the scalar function $\|u_n(.)\|$. By definition of η_1, for all $\delta > 0$ one has

$$\forall t \quad \exists n \quad \int_{\{\|u_n(.)\| > t\}} \|u_n\| \, d\,P > \eta_1 - \delta.$$

Let $\varepsilon > 0$. There exists t large enough such that $\frac{M}{t}$ is $\leq \varepsilon$. Then if n corresponds to such a t, $A := \{\|u_n(.)\| > t\}$ satisfies $P(A) \leq \varepsilon$ and $\int_A \|u_n\| \, d\,P \geq \eta_1 - \delta$. Hence $\eta((u_n)) \geq \eta_1 - \delta$, and this holds for all $\delta > 0$, so $\eta((u_n)) \geq \eta_1$.

Let us prove the converse inequality. Let $t > 0$ and $\delta > 0$ be given. Take $\varepsilon > 0$ small enough such that $t\,\varepsilon < \delta$. By definition of $\eta((u_n))$ there exist A and n such

[1] The statements and proofs of Lemma 6.1.3 and Theorem 6.1.4 reproduce corresponding parts of [SV95c].

that $\mathrm{P}(A) \le \varepsilon$ and $\int_A \|u_n\| \, d\mathrm{P} > \eta((u_n)) - \delta$. Then

$$\begin{aligned}
\int_{\{\|u_n\|>t\}} \|u_n\| \, d\mathrm{P} &= \int_A \|u_n\| \, d\mathrm{P} - \int_{A\cap\{\|u_n\|\le t\}} \|u_n\| \, d\mathrm{P} + \int_{\{\|u_n\|>t\}\setminus A} \|u_n\| \, d\mathrm{P} \\
&\ge \int_A \|u_n\| \, d\mathrm{P} - \int_{A\cap\{\|u_n\|\le t\}} \|u_n\| \, d\mathrm{P} \\
&\ge (\eta((u_n)) - \delta) - t\,\varepsilon \\
&\ge \eta((u_n)) - 2\,\delta.
\end{aligned}$$

So one has $\eta_1 \ge \eta((u_n)) - 2\,\delta$ for all $\delta > 0$, hence $\eta_1 \ge \eta((u_n))$. □

The expression of $\eta((u_n))$ given in Lemma 6.1.3 is used in Słaby's proof of the Biting Lemma. We reproduce this proof here in order to introduce its notations.

In the sequel, UI abbreviates "uniformly integrable".

Theorem 6.1.4 (Biting Lemma) *Let $\mathbb{E}$ be a separable Banach space and let $(u_n)_{n\in\mathbb{N}}$ be a bounded sequence in $\mathrm{L}^1_{\mathbb{E}}(\Omega, \mathcal{S}, \mathrm{P})$. There exist a subsequence $(u_{n_k})_{k\in\mathbb{N}}$ and a sequence $(B_p)_{p\in\mathbb{N}}$ in $\mathcal{S}$ decreasing to $\emptyset$ such that the sequence $(\mathbb{1}_{B_k^c}\, u_{n_k})_{k\in\mathbb{N}}$ is* UI.

Proof. Denote, for $i \in \mathbb{N}$,

$$g_i(t) := \sup_{n\ge i} \int_{\{\|u_n(.)\|>t\}} \|u_n(\omega)\| \; d\mathrm{P}(\omega).$$

The functions g_i are nonincreasing with values in $[0, +\infty[$. Hence, $\lim_{t\to+\infty} g_i(t)$ exists. By Lemma 6.1.3, $\eta((u_n)) = \lim_{t\to+\infty} g_0(t)$. There exists a sequence increasing to $+\infty$, $(t_q)_{q\ge1}$ such that for every q, $g_0(t_q) \le \eta((u_n)) + \frac{1}{q}$. Since for each i, $\{u_0, \ldots, u_{i-1}\}$ is UI, it is easy to check that $\lim_{t\to+\infty} g_i(t) = \eta((u_n))$. Hence for every i and for every t, $g_i(t) \ge \eta((u_n))$, so there exists an increasing sequence $(m_q)_q$ such that

$$\int_{\{\|u_{m_q}(.)\|>t_q\}} \|u_{m_q}(\omega)\| \, d\mathrm{P}(\omega) \ge \eta((u_n)) - \frac{1}{q}. \tag{6.1.3}$$

Let $A_p = \{\|u_{m_p}(.)\| > t_p\}$. Then $t_p\ \mathrm{P}(A_p) \le \sup_{n\in\mathbb{N}} \|u_n\|_{\mathrm{L}^1}$ implies $\mathrm{P}(A_p) \to 0$. Now we show that $(\mathbb{1}_{A_q^c}\, u_{m_q})_{q\in\mathbb{N}}$ is UI. Let

$$g(t) := \sup_{q\in\mathbb{N}} \int_{A_q^c\cap\{\|u_{m_q}(.)\|>t\}} \|u_{m_q}(\omega)\| \, d\mathrm{P}(\omega).$$

We have to prove that $g(t) \to 0$ when $t \to +\infty$. One has

$$\begin{aligned}
g(t_j) &= \sup_{q>j} \int_{\{t_j < \|u_{m_q}(.)\| \le t_q\}} \|u_{m_q}(\omega)\| \, d\mathrm{P}(\omega) \\
&= \sup_{q>j} \Big[\int_{\{\|u_{m_q}(.)\| > t_j\}} \|u_{m_q}(\omega)\| \, d\mathrm{P}(\omega) - \int_{\{\|u_{m_q}(.)\| > t_q\}} \|u_{m_q}(\omega)\| \, d\mathrm{P}(\omega) \Big] \\
&\le \sup_{q>j} \Big[g_0(t_j) - \int_{\{\|u_{m_q}(.)\| > t_q\}} \|u_{m_q}(\omega)\| \, d\mathrm{P}(\omega) \Big] \\
&\le \sup_{q>j} \big[(\eta((u_n)) + \frac{1}{j}) - (\eta((u_n)) - \frac{1}{q}) \big] \\
&\le \frac{2}{j}.
\end{aligned}$$

This proves: $t \to +\infty \Rightarrow g(t) \to 0$. It remains to replace $(A_p)_p$ by a sequence decreasing to $\emptyset$ (or to a negligible set). There exists an increasing sequence $(l_k)_k$ such that $\mathrm{P}(A_{l_k}) \le 2^{-k}$. Then $B_p := \cup_{k \ge p} A_{l_k}$ decreases to a negligible set. The subsequence $u_{n_k} := u_{m_{l_k}}$ (or $u_{m_{(l_k)}}$) has the required properties. The uniform integrability of $(\mathbb{1}_{B_k^c} u_{n_k})_{k \in \mathbb{N}}$ follows from the inclusion $B_k^c \subset A_{l_k}^c$. □

Remark. It is worth to mention that the sequence $(\mathbb{1}_{B_k} \|u_{n_k}(.)\|)$ converges to 0 a.e.

Taking the Biting Lemma into account, we say that a bounded sequence $(u_n)_{n \in \mathbb{N}}$ in $\mathrm{L}^1_{\mathbb{E}}(\Omega, \mathcal{S}, \mathrm{P})$ *has a subsequence which weakly biting converges to* a function $u \in \mathrm{L}^1_{\mathbb{E}}(\Omega, \mathcal{S}, \mathrm{P})$ if there exist an increasing sequence $(A_q)_{q \in \mathbb{N}}$ in $\mathcal{S}$ with $\lim_q \mathrm{P}(A_q) = 1$ and a subsequence (u'_q) of (u_n) such that $(\mathbb{1}_{A_q} u'_q)$ weakly converges to u.

Remark 6.1.5 If $\mathbb{E}$ is reflexive, then any bounded sequence in $\mathrm{L}^1_{\mathbb{E}}(\Omega, \mathcal{S}, \mathrm{P})$ weakly biting converges because any uniformly integrable sequence in $\mathrm{L}^1_{\mathbb{E}}(\Omega, \mathcal{S}, \mathrm{P})$ is relatively sequentially weakly compact.

In this vein the following lemma can be considered as a substitute of the Biting Lemma. These two results are useful in the study of bounded sequences in $\mathrm{L}^1_{\mathbb{E}}(\Omega, \mathcal{S}, \mathrm{P})$. If (f_n) is a sequence in $\mathrm{L}^1_{\mathbb{E}}(\Omega, \mathcal{S}, \mathrm{P})$, if $N \in \mathbb{N}$, we denote by $(f_n^N)_n$ the truncated sequence associated with (f_n) and N, where $f_n^N := \mathbb{1}_{\{\|f_n\| < N\}} f_n$.

Lemma 6.1.6 *Let (f_n) be a bounded sequence in $\mathrm{L}^1_{\mathbb{E}}(\Omega, \mathcal{S}, \mathrm{P})$. Then there exists a subsequence (g_n) of (f_n) such that for each subsequence (h_n) of (g_n)*

(a) the sequence $(\mathbb{1}_{\{\|h_n\| < n\}} h_n)$ is uniformly integrable;

(b) the sequence $(h_n - \mathbb{1}_{\{\|h_n\| < n\}} h_n)$ converges strongly a.e. to 0 in $\mathbb{E}$.

Proof. Put $\mathrm{M} = \sup_n \|f_n\|_1$. For each integer $k \ge 1$ the real valued sequence

$$(\mathrm{P}(\{k - 1 \le \|f_n\| < k\}))_n$$

is bounded. Then there exist subsequences $(f_n^1), (f_n^2), ..., (f_n^k), ...$ of (f_n), where (f_n^{k+1}) is a subsequence of (f_n^k), and a sequence $(p_k)_k$ in $[0,1]$ such that

$$\forall k \geq 1 \quad \lim_n \mathrm{P}(\{k-1 \leq \|f_n^k\| < k\}) = p_k \tag{6.1.4}$$

and

$$\forall k \geq 1 \quad \forall n \geq 1 \quad \mathrm{P}(\{k-1 \leq \|f_n^k\| < k\}) < p_k + \frac{1}{k^3}. \tag{6.1.5}$$

Indeed we can extract from $(f_n)_n$ a subsequence $(f_{0,n}^1)_n$ satisfying (6.1.4) for $k = 1$. Since $\mathrm{P}(\{0 \leq \|f_{0,n}^1\| < 1\}) \to p_1$, there exists n_0 such that

$$\forall n \geq n_0 \quad \mathrm{P}(\{0 \leq \|f_{0,n}^1\| < 1\}) < p_1 + 1.$$

So it suffices to set $f_n^1 = f_{0,n+n_0-1}^1$ in order to have (6.1.5) for all n. Suppose that (f_n^{k-1}) has been obtained, then using the similar arguments we can construct from (f_n^{k-1}) a subsequence (f_n^k) with the required properties.

Put $g_n = f_n^{n^2}$ and let (h_n) be a subsequence of (g_n). Then

$$\forall k \geq 1 \quad \lim_n \mathrm{P}(\{k-1 \leq \|h_n\| < k\}) = p_k, \tag{6.1.6}$$

and

$$\forall n \in \mathbb{N} \quad \forall k \in [1, n^2] \quad \mathrm{P}(\{k-1 \leq \|h_n\| < k\}) < p_k + \frac{1}{k^3}. \tag{6.1.7}$$

We only need to check (6.1.7). Since (h_n) is extracted from $(g_n)_n = (f_n^{n^2})_n$, each h_n is of the form $h_n = f_m^{m^2}$ with $m \geq n$. Let $k \in \mathbb{N}$ be fixed. Let us notice that $(f_n^{m^2})_n$ is a subsequence of $(f_n^k)_n$ when $m^2 \geq k$. So let us suppose that $n \geq \sqrt{k}$. Then $m^2 \geq k$, and so $f_m^{m^2}$ is a term f_l^k of $(f_n^k)_n$ with $l \geq m$, hence by (6.1.5) and what has been said we deduce (6.1.7). In this part of proof the symbol $(f_n^k)_n$ has been used for notational convenience and there was no risk of confusion with the truncated sequences $(f_n^N)_n$ $(N \in \mathbb{N})$ above mentioned. We have $\sum_{k=1}^N p_k = \lim_n \sum_{k=1}^N \mathrm{P}(\{k-1 \leq \|h_n\| < k\}) \leq \mathrm{P}(\Omega)$, and $\sum_{k=1}^N (k-1)p_k = \lim_n \sum_{k=1}^N (k-1)\,\mathrm{P}(\{k-1 \leq \|h_n\| < k\}) \leq M$. Then $\sum_{k=1}^{+\infty} p_k \leq 1$ and $\sum_{k=1}^{+\infty} (k-1)p_k \leq M$.

We are now ready to obtain (a) and (b).

(a) Let $\varepsilon > 0$. As $\sum_{k=1}^{+\infty} kp_k$ converges, so is $\sum_{k=1}^{+\infty} k(p_k + \frac{1}{k^3})$, hence there exists an integer $N \in \geq 1$ such that

$$\sum_{k \geq N} k(p_k + \frac{1}{k^3}) < \varepsilon.$$

If $n \in \{1,..,N\}$, then (now the superscript n means truncation) $\|h_n^n\| < n \le N$, so $\{\|h_n^n\| \ge N\} = \emptyset$. If $n > N$, then

$$\{\|h_n^n\| \ge N\} = \bigcup_{N+1 \le k \le n} \{k-1 \le \|h_n^n\| < k\},$$

and by (6.1.7)

$$\int_{\{\|h_n^n\| \ge N\}} \|h_n^n\| \, d\mathrm{P} \le \sum_{N+1 \le k \le n} k(p_k + \frac{1}{k^3}) < \varepsilon$$

so (h_n^n) is uniformly integrable.

(b) We have by (6.1.7)

$$\begin{aligned}
\mathrm{P}(\{\|h_n\| \ge n\}) &= \sum_{k=n+1}^{n^2} \mathrm{P}(\{k-1 \le \|h_n\| < k\}) + \mathrm{P}(\{\|h_n\| \ge n^2\}) \\
&< \sum_{k=n+1}^{n^2} (p_k + \frac{1}{k^3}) + \frac{1}{n^2} \|h_n\|_1 \\
&\le \sum_{k \ge n+1} p_k + \frac{1}{2n^2} + \frac{M}{n^2}.
\end{aligned}$$

Hence by summing we get

$$\begin{aligned}
\sum_{n \ge 1} \mathrm{P}(\{\|h_n\| \ge n\}) &< \sum_{n \ge 1} [\sum_{k \ge n+1} p_k + \frac{2M+1}{2n^2}] \\
&= \sum_{k \ge 2} (k-1) p_k + \sum_{n \ge 1} \frac{2M+1}{2n^2} < +\infty.
\end{aligned}$$

Then (b) follows from Borel–Cantelli Lemma. □

Reference: [CG99, Lemma 4.1].

We recall now a famous result due to Talagrand [Tal84].

Theorem 6.1.7 (Talagrand) *Let (u_n) be a bounded sequence in $\mathrm{L}^1_{\mathbb{E}}(\mathrm{P})$. Then there exists a sequence $(\tilde{u}_n)$ with $\tilde{u}_n \in \mathrm{co}\{u_m;\, m \ge n\}$ and two sets A and B in $\mathcal{S}$ with $\mathrm{P}(A \cup B) = 1$ such that*

(a) *for each ω in A, the sequence $(\tilde{u}_n(\omega))$ is weakly Cauchy,*

(b) *for each ω in B, there exists some integer k such that the sequence $(\tilde{u}_n(\omega))_{n \ge k}$ is equivalent to the vector unit basis of l^1.*

Analogous results in $L^1_{\mathbb{E}}(\Omega, \mathcal{S}, P)$ are in [BC97]. See also [BC01] dealing with the space $L^1_{\mathbb{E}^*}[\mathbb{E}](\Omega, \mathcal{S}, P)$.

Here we present some structure-type theorems for two classes of bounded sequences in $L^1_{\mathbb{E}}(\Omega, \mathcal{S}, P)$.

The following theorem gives a characterization of the Mazur-tightness property for bounded subsets in $L^1_{\mathbb{E}}(\Omega, \mathcal{S}, P)$.

Theorem 6.1.8 *Let $\mathcal{H}$ a bounded subset of $L^1_{\mathbb{E}}(\Omega, \mathcal{S}, P)$. Then the following are equivalent:*

(1) *Given any sequence (f_n) in $\mathcal{H}$, there exists a sequence (g_n) with $g_n \in \operatorname{co}\{f_k;\, k \geq n\}$ such that (g_n) strongly converges a.e. in $\mathbb{E}$, shortly (f_n) is strongly Mazur–compact.*

(2) *Given any sequence (f_n) in $\mathcal{H}$, there exists a sequence (g_n) with $g_n \in \operatorname{co}\{f_k;\, k \geq n\}$ such that (g_n) weakly converges a.e. in $\mathbb{E}$, shortly (f_n) is weakly Mazur–compact.*

(3) *Given any sequence (f_n) in $\mathcal{H}$, there exists a subsequence (g_n) of (f_n) such that $(\mathbb{1}_{\{\|h_n\|<n\}} h_n)$ is $\sigma(L^1, L^\infty)$ relatively compact for each subsequence (h_n) of (g_n).*

(4) *Given any sequence (f_n) in $\mathcal{H}$, there exists a subsequence (g_n) of (f_n) such that $(\mathbb{1}_{\{\|g_n\|<n\}} g_n)$ is $\sigma(L^1, L^\infty)$ relatively compact.*

(5) *Given any sequence (f_n) in $\mathcal{H}$, there exists a subsequence (f_{n_k}) of (f_n) such that $(\mathbb{1}_{\{\|f_{n_k}\|<n_k\}} f_{n_k})$ converges $\sigma(L^1, L^\infty)$. Consequently the sequence $(f^k_{n_k}) = (\mathbb{1}_{\{\|f_{n_k}\|<k\}} f_{n_k})$ converges $\sigma(L^1, L^\infty)$ too.*

(6) *Given any sequence (f_n) in $\mathcal{H}$, there exists a subsequence (f_{n_k}) of (f_n) such that $f_{n_k} = u_{n_k} + v_{n_k}$, where (u_{n_k}) converges $\sigma(L^1, L^\infty)$ and (v_{n_k}) converges a.e. to 0.*

Proof.

(1)$\Longrightarrow$(2) is clear.

(2)$\Longrightarrow$(3). Let (f_n) be a sequence in $\mathcal{H}$. By Lemma 6.1.6 (a) there is a subsequence (g_n) of (f_n) such that $(\mathbb{1}_{\{\|h_n\|<n\}} h_n)$ is uniformly integrable, for each subsequence (h_n) of (g_n). If $(\mathbb{1}_{\{\|h_{n_p}\|<n_p\}} h_{n_p})$ is a subsequence of $(\mathbb{1}_{\{\|h_n\|<n\}} h_n)$ there is, by (2), a sequence (u_p) with $u_p \in \operatorname{co}\{h_{n_k};\, k \geq p\}$ such that (u_p) weakly converges a.e. in $\mathbb{E}$. We have

$$u_p = \sum_{i=p}^{k_p} \lambda_i^p h_{n_i} = \sum_{i=p}^{k_p} \lambda_i^p \mathbb{1}_{\{\|h_{n_i}\|<n_i\}} h_{n_i} + \sum_{i=p}^{k_p} \lambda_i^p (h_{n_i} - \mathbb{1}_{\{\|h_{n_i}\|<n_i\}} h_{n_i})$$

with $0 \le \lambda_i^p \le 1$ and $\sum_{i=p}^{k_p} \lambda_i^p = 1$. By Lemma 6.1.6 (b), we have

$$\sum_{i=p}^{k_p} \lambda_i^p (h_{n_i} - \mathbb{1}_{\{\|h_{n_i}\|<n_i\}} h_{n_i}) \to 0$$

a.e. as $p \to \infty$. Hence $\sum_{i=p}^{k_p} \lambda_i^p \mathbb{1}_{\{\|h_{n_i}\|<n_i\}} h_{n_i}$ weakly converges a.e. as $p \to \infty$. By the Ülger–Diestel–Ruess–Schachermayer theorem [Ülg91, DRS93], we conclude that $(\mathbb{1}_{\{\|h_n\|<n\}} h_n)$ is $\sigma(\mathrm{L}^1, \mathrm{L}^\infty)$ relatively compact.

(3) $\Longrightarrow$ (4) is obvious.

Let us prove that (4)$\Longrightarrow$(5). By the Eberlein–Šmulian theorem there is a subsequence $(f_{n_k})_k$ of (f_n) such that $(f_{n_k}^{n_k})_k$ converges $\sigma(\mathrm{L}^1, \mathrm{L}^\infty)$ and so is the sequence $(f_{n_k}^k)_k$. Indeed we have

$$f_{n_k}^{n_k} - f_{n_k}^k = \mathbb{1}_{\{k \le \|f_{n_k}\| < n_k\}} f_{n_k} = \mathbb{1}_{\{k \le \|f_{n_k}^{n_k}\|\}} f_{n_k}^{n_k}.$$

Since $(f_{n_k}^{n_k})_k$ is uniformly integrable, $f_{n_k}^{n_k} - f_{n_k}^k \to 0$ strongly in $\mathrm{L}^1_{\mathbb{E}}(\mathrm{P})$ as $k \to \infty$. It follows that $(f_{n_k}^k)_k$ converges $\sigma(\mathrm{L}^1, \mathrm{L}^\infty)$.

(5)$\Longrightarrow$(6). Let (f_n) be a sequence in $\mathcal{H}$. By (5) there exists a subsequence (f_{n_k}) of (f_n) such that $(f_{n_k}^{n_k})_k$ converges $\sigma(\mathrm{L}^1, \mathrm{L}^\infty)$ in $\mathrm{L}^1_{\mathbb{E}}(\mathrm{P})$. Put $u_{n_k} = f_{n_k}^{n_k}$ and $v_{n_k} = f_{n_k} - u_{n_k}$. Then (5)$\Longrightarrow$(6) follows immediately.

(6)$\Longrightarrow$(1). Let us apply the notations in (5). Since (u_{n_k}) converges $\sigma(\mathrm{L}^1, \mathrm{L}^\infty)$ in $\mathrm{L}^1_{\mathbb{E}}(\mathrm{P})$, there is a sequence $(\widetilde{u}_m)$ of the form $\widetilde{u}_m = \sum_{i=m}^{\nu_m} \lambda_i^m u_{n_i}$ with $\lambda_i^m \ge 0$, $\sum_{i=m}^{\nu_m} \lambda_i^m = 1$ such that $(\widetilde{u}_m)$ converges strongly in $\mathrm{L}^1_{\mathbb{E}}(\mathrm{P})$. There is a subsequence $(\widetilde{u}_{m_p})$ which strongly converges a.e. in $\mathbb{E}$ and so is $(\sum_{i=m_p}^{\nu_{m_p}} \lambda_i^{m_p} f_{n_i})_p$.

□

Reference: [CG99, Theorem 6.1].

In particular, if $\mathbb{E}$ is reflexive, any bounded sequence in $\mathrm{L}^1_{\mathbb{E}}(\Omega, \mathcal{S}, \mathrm{P})$ is strongly Mazur–compact. See [BCG99a, Proposition 6.7] for details. To finish this section we give another structure theorem for bounded norm tight (flexibly and strictly are equivalent, cf. Theorem 4.3.14 (i))) sequences in $\mathrm{L}^1_{\mathbb{E}}(\Omega, \mathcal{S}, \mathrm{P})$.

We use the following definitions[2]:

Let u_n ($n \in \mathbb{N} \cup \{+\infty\}$) be functions in $\mathrm{L}^1_{\mathbb{E}}(\Omega, \mathcal{S}, \mathrm{P})$. One says that u_n converges *purely weakly to* u_∞ *on* $W \in \mathbb{S}$ if the restrictions $u_n|_W$ converge weakly to $u_\infty|_W$ and, for any non negligible measurable subset A of W, not any subsequence of $(u_n|_A)_n$ converges strongly to $u_\infty|_A$.

Remark It is equivalent to say: No subsequence of $(u_n|_A)_n$ converges in measure (to any function). For more about convergence in measure, see [SV95a, SV95b]

[2] The next Theorems 6.1.9 and 6.1.11 are straightforward extensions to Banach space of corresponding theorems of [SV95c]. Example 6.1.13 also comes from this paper.

where compactness in measure with applications to $\mathrm{L}^1_{\mathbb{E}}$ is characterized in the line of an old criterion of Fréchet.

For clarity we give first the simplified version of the general result (Theorem 6.1.11) in the case when the sequence is UI. All assertions are consequences of the Prohorov theorem (Theorem 4.3.5), of the fact that convergence in measure coincide with the topology induced by the stable one (cf. Theorem 3.1.2) and of the "characterizing continuity property" extended to Carathéodory integrands with linear growth.

Theorem 6.1.9 *Let $(u_n)_n$ be a sequence in $\mathrm{L}^1_{\mathbb{E}}(\Omega, \mathcal{S}, \mathrm{P})$ which is uniformly integrable and tight relatively to the norm of $\mathbb{E}$ (again note that flexibly and strictly are equivalent). There exist a subsequence $(u'_n)_n$, a function $u_\infty \in \mathrm{L}^1_{\mathbb{E}}(\Omega, \mathcal{S}, \mathrm{P})$ and $M \in \mathcal{S}$ satisfying*[3]*:*

1) 1b) *u'_n converges weakly to u_∞,*

   1c) *The restrictions $u'_n|_M$ converge strongly and in measure to $u_\infty|_M$.*

   1d) *The restrictions $u'_n|_{M^c}$ converge purely weakly to $u_\infty|_{M^c}$.*

2) *There exists a Young measure τ, such that*

$$M^c = \{\omega \in \Omega; \int_{\mathbb{E}} \|\xi - u_\infty(\omega)\| \, d\tau_\omega(\xi) > 0\}$$

and such that for $\omega \in M$, $\tau_\omega = \delta_{u_\infty(\omega)}$ and, for any Carathéodory integrand ψ with linear growth (i.e. $|\psi(\omega, \xi)| \le \alpha(\omega) + K\,|\xi|$, where $\alpha \in \mathrm{L}^1$)

$$\int_\Omega \psi(\omega, u'_n(\omega)) \, d\,\mathrm{P}(\omega) \longrightarrow \int_{\Omega\times\mathbb{E}} \psi \, d\tau. \qquad (6.1.8)$$

In particular for any $u \in \mathrm{L}^1_{\mathbb{E}}(\Omega, \mathcal{S}, \mathrm{P})$, $\|u'_n - u\|_{\mathrm{L}^1}$ converges to

$$\int_{M^c\times\mathbb{E}} \|\xi - u(\omega)\| \, d\tau(\omega, \xi) + \| \mathbb{1}_M \, (u_\infty - u)\|_{\mathrm{L}^1}.$$

Remark 6.1.10 1) We will not give the proof since the statement, except (6.1.8), is a particular case of Theorem 6.1.11. First versions of this statement appear in [Val90b, Theorem 19 page 169] and [Val94, Theorem 9]. The novelty is the introduction of the partition (M, M^c).

2) The norm $\|u'_n - u\|_{\mathrm{L}^1}$ equals $\int_{\Omega\times\mathbb{E}} \psi \, d\underline{\delta}_{u'_n}$, where ψ is the integrand $\psi(\omega, \xi) := \|\xi - u(\omega)\|$. This is a Carathéodory integrand with linear growth (for some applications of these integrands see [PV95]).

[3] Here 1a) of Theorem 6.1.11 is useless.

Theorem 6.1.11 *Let $(u_n)_n$ be a sequence in $\mathrm{L}^1_{\mathbb{E}}(\Omega, \mathcal{S}, \mathrm{P})$ which is bounded and tight relatively to the norm of $\mathbb{E}$ (again note that flexibly and strictly are equivalent). There exist a subsequence $(u'_n)_n$, a function $u_\infty \in \mathrm{L}^1_{\mathbb{E}}(\Omega, \mathcal{S}, \mathrm{P})$, a sequence $(B_p)_p$ in $\mathcal{S}$ which decreases to $\emptyset$ (if $(u_n)_n$ is* UI*, one may take $B_p = \emptyset$ for every p) and $M \in \mathcal{S}$ satisfying*

1) 1a) $(\mathbb{1}_{B^c_n}\, u'_n)_n$ *is* UI,
   1b) *for every p, the restrictions $u'_n|_{B^c_p}$ converge weakly to $u_\infty|_{B^c_p}$,*
   1c) *the restrictions $u'_n|_M$ converge in measure to $u_\infty|_M$ and for every p, the restrictions $u'_n|_{(M\setminus B_p)}$ converge strongly to $u_\infty|_{(M\setminus B_p)}$,*
   1d) *for any non negligible $A \in \mathcal{S}$ contained in $W := \Omega \setminus M$, no subsequence of $(u'_n|_A)_n$ converges in measure to $u_\infty|_A$ (nor to any other function),*
   1e) *for any subsequence $(u''_n)_n$ of $(u'_n)_n$, $\eta((u''_n)_n) = \eta((u_n)_n)$.*

2) *There exists a Young measure τ whose disintegration is a family of first order probabilities on $\mathbb{E}$, $(\tau_\omega)_{\omega\in\Omega}$, such that τ_ω is carried by the set $\mathrm{ls}(u_n(\omega))$ of limit points of the sequence $(u_n(\omega))_n$, such that for $\omega \in W$, τ_ω is not a Dirac mass and $\mathrm{bar}\,(\tau_\omega) = u_\infty(\omega)$ (where $\mathrm{bar}\,(\tau_\omega)$ denotes the barycenter of the measure τ_ω) such that for $\omega \in M$, $\tau_\omega = \delta_{u_\infty(\omega)}$ and such that, for any $u \in \mathrm{L}^1_{\mathbb{E}}(\Omega, \mathcal{S}, \mathrm{P})$, $\|u'_n - u\|_{\mathrm{L}^1}$ converges to*

$$\eta((u_n)_n) + \int_{\Omega\times\mathbb{E}} |\xi - u(\omega)|\, d\tau(\omega,\xi)$$
$$= \eta((u_n)_n) + \int_{W\times\mathbb{E}} |\xi - u(\omega)|\, d\tau(\omega,\xi) + \|\,\mathbb{1}_M\,(u_\infty - u)\|_{\mathrm{L}^1}.$$

Moreover for any bounded Carathéodory integrand ψ

$$\int_\Omega \psi(\omega, u_n(\omega))\, d\,\mathrm{P}(\omega) \longrightarrow \int_{\Omega\times\mathbb{E}} \psi\, d\tau.$$

Remark 6.1.12 Assertion 1e) means that, for the subsequence $(u'_n)_n$, the concentration of mass (non null iff $\eta((u_n)_n)$ is > 0) equals the maximum of mass concentrations among all the subsequences of $(u_n)_n$. As a preliminary but less precise result in this line recall: If $(u_n)_n$ converges weakly (hence is UI) and does not converge strongly, there exists a subsequence whose associated Young measures converge to a Young measure τ which is not associated with a function [Val94, Theorem 20 page 169], [Val94, Theorem 9].

Example 6.1.13 Here is an example where B_p necessarily bites both M and W. Moreover this example illustrates the general situation. Let $\Omega = [0,1]^2$ with P as the Lebesgue measure on Ω, $\mathbb{E} = \mathbb{R}$ and

$$u_n(x,y) = n\ \mathbb{1}_{[0,\frac{1}{n}]}(x) + \mathbb{1}_{[0,\frac{1}{2}]}(y)\ \sin(n\,x).$$

Then $u_\infty = 0$, $W = [0,1] \times [0, \frac{1}{2}]$ and[4] $M = [0,1] \times]\frac{1}{2}, 1]$. Roughly speaking, mass concentration appears around $\{0\} \times [0,1]$. In any way the B_p are chosen, they meet M and W: Since $(\mathbb{1}_{B_k^c} u_{n_k})_k$ is UI, $\| \mathbb{1}_{[0,1/n_k]\times[0,1]} \mathbb{1}_{B_k^c} u_{n_k} \|_{\mathrm{L}^1}$ tends to 0 when $k \to +\infty$, hence

$$\iint_{[0,1/n_k]\times[0,1]} \mathbb{1}_{B_k^c}(x,y)\,[n_k + \mathbb{1}_{[0,\frac{1}{2}]}(y)\,\sin(n_k\,x)]\,dx\,dy \longrightarrow 0.$$

Thus $n_k\,\mathrm{P}(B_k^c \cap ([0,1/n_k] \times [0,1])) \to 0$ which is equivalent to

$$\frac{\mathrm{P}(B_k \cap ([0,1/n_k] \times [0,1]))}{\mathrm{P}([0,1/n_k] \times [0,1])} \longrightarrow 1.$$

Consequently

$$\frac{\mathrm{P}(B_k \cap W)}{\mathrm{P}([0,1/n_k] \times [0,1/2])} \longrightarrow 1 \quad \text{and} \quad \frac{\mathrm{P}(B_k \cap M)}{\mathrm{P}([0,1/n_k] \times]1/2,1])} \longrightarrow 1.$$

Hence B_k covers a big part of $W \cap ([0,1/n_k] \times [0,1])$ and a big part of $M \cap ([0,1/n_k] \times [0,1])$.

Note that $\sin(n\,y)$ in place of $\sin(n\,x)$ would give the same W and τ despite the change of directions of the "waves." When Ω is an open subset of $\mathbb{R}^N$, there exist more powerful tools than Young measures: see G. Allaire [All92], L. Tartar [Tar90], and J.J. Alibert and G. Bouchitté [AB97].

Proof of Theorem 6.1.11. 1) Let $(B_p)_p$ denote the sequence of the proof of the Biting Lemma and $(u'_n)_n$ the subsequence denoted by $(u_{n_k})_k$ in that proof. Let $(u''_n)_n$ be a further subsequence. A priori $\eta((u''_n)) \le \eta((u_n)_n)$. Using the notations of the proof of Theorem 6.1.4, u''_n is some u_{m_q}. For all t, there exists q large enough such that $t_q \ge t$ and such that the function u_{m_q} is some u''_n. By (6.1.3) $\eta((u_n)_n)$ is approximated with a gap less than $1/q$ by

$$\int_{\{\|u''_n(.)\|>t_q\}} \|u''_n(\omega)\|\,d\,\mathrm{P}(\omega)$$

hence a fortiori by

$$\int_{\{\|u''_n(.)\|>t\}} \|u''_n(\omega)\|\,d\,\mathrm{P}(\omega).$$

This proves $\eta((u''_n)) = \eta((u_n)_n)$.

2) Since $(u_n)_n$ is tight, thanks to the Prohorov theorem, one may assume that the Young measures $\underline{\delta}_{u'_n}$ converge to a Young measure τ. Thanks to Proposition 2.1.12,

$$\text{(6.1.9)} \qquad \int_{\Omega\times\mathbb{E}} \|\xi\|\,d\tau(\omega,\xi) \le \sup_n \|u_n\|_{\mathrm{L}^1}.$$

[4] One can check (see [Val94, Theorem 4]) that for $(x,y) \in W$, $\tau_{(x,y)}$ is the probability on $\mathbb{R}$ with the density $\xi \mapsto (\pi\sqrt{1-\xi^2})^{-1}$ on $]-1,1[$.

Thus, for P-almost every ω, τ_ω has a barycenter bar (τ_ω) and, setting $u_\infty(\omega) :=$ bar (τ_ω), one gets the integrable function u_∞. Moreover τ_ω is carried by $\mathrm{ls}(u_n(\omega))$ (see Theorem 4.3.12 or [Val90b, Proposition 5 page 159]).

For fixed $z \in \mathrm{L}^\infty_{\mathbb{E}^*_\sigma}$ and p, since $u'_n|_{B_p^c}$ is UI, $\left\langle z, u'_n|_{B_p^c}\right\rangle \to \left\langle z, u_\infty|_{B_p^c}\right\rangle$. Indeed

$$\begin{aligned}\lim_{n\to+\infty} \int_\Omega \langle z(\omega), [u'_n|_{B_p^c}](\omega)\rangle \, d\mathrm{P}(\omega) &= \int_{\Omega\times\mathbb{E}} \mathbb{1}_{B_p^c}(\omega)\langle z(\omega), \xi\rangle \, d\tau^\infty(\omega,\xi)\\ &= \int_\Omega \mathbb{1}_{B_p^c}(\omega)\Big\langle z(\omega), \int_{\mathbb{E}} \xi \, d\tau^\infty_\omega(\xi)\Big\rangle \, d\mathrm{P}(\omega)\\ &= \int_\Omega \mathbb{1}_{B_p^c}(\omega)\langle z(\omega), u_\infty(\omega)\rangle \, d\mathrm{P}(\omega).\end{aligned}$$

For a reference see [Val90b, Theorem 19 page 169]; see also the proof of Theorem 9 in [Val94]. The set W of all ω where τ_ω is not a Dirac mass, is

$$W = \Big\{\omega \in \Omega;\ \int_{\mathbb{E}} |\xi - u_\infty(\omega)| \, d\tau_\omega(\xi) > 0\Big\}.$$

By the Fubini theorem it is measurable. Let $M := \Omega \setminus W$. Note that P-almost everywhere on M, $\tau_\omega = \delta_{u_\infty(\omega)}$. By Part 3 of Theorem 3.1.2, $u'_n|_M \to u_\infty|_M$ in measure and, for any non negligible A contained in W, not any subsequence of $(u'_n)_n$ converges in measure on A. Thanks to UI and to the Lebesgue–Vitali theorem, strong convergence holds on $M \setminus B_p$.

3) Let $u \in \mathrm{L}^1_{\mathbb{E}}(\Omega, \mathcal{S}, \mathrm{P})$ and $\varepsilon > 0$. Thanks to (6.1.9), for p large enough one has

$$\begin{aligned}\int_{B_p\times\mathbb{E}} \|\xi - u(\omega)\| \, d\tau(\omega,\xi) &\le \int_{B_p\times\mathbb{E}} \|\xi\| \, d\tau(\omega,\xi) + \int_{B_p\times\mathbb{E}} \|u(\omega)\| \, d\tau(\omega,\xi)\\ &= \int_{B_p\times\mathbb{E}} \|\xi\| \, d\tau(\omega,\xi) + \int_{B_p} \|u(\omega)\| \, d\mathrm{P}(\omega)\\ &\le \frac{\varepsilon}{4}.\end{aligned}$$

On $B_p^c \times \mathbb{E}$ the integrand ψ defined by $\psi(\omega,\xi) := \|\xi - u(\omega)\|$ is Carathéodory with linear growth and the u'_n are UI on B_p^c, so by Theorem 2.4.1,

$$\| \mathbb{1}_{B_p^c}(u'_n - u)\|_{\mathrm{L}^1} = \int_{B_p^c\times\mathbb{E}} \|\xi - u(\omega)\| \, d\underline{\delta}_{u'_n}(\omega,\xi)$$

converges as $n \to +\infty$ to

$$\int_{B_p^c\times\mathbb{E}} \|\xi - u(\omega)\| \, d\tau(\omega,\xi).$$

Hence, p being fixed, for n large enough,

$$\left| \| \mathbb{1}_{B_p^c}(u'_n - u)\|_{\mathrm{L}^1} - \int_{\Omega\times\mathbb{E}} \|\xi - u(\omega)\| \, d\tau(\omega,\xi) \right| \le \frac{\varepsilon}{2}.$$

As for $\| \mathbb{1}_{B_p}(u'_n - u)\|_{L^1}$, firstly, if p is large enough,

$$\big| \| \mathbb{1}_{B_p}(u'_n - u)\|_{L^1} - \| \mathbb{1}_{B_p} u'_n\|_{L^1} \big| \leq \| \mathbb{1}_{B_p} u\|_{L^1} \leq \frac{\varepsilon}{4}.$$

It remains to show that, for n large enough, one has

$$\big| \| \mathbb{1}_{B_p} u'_n\|_{L^1} - \eta((u_n)_n) \big| \leq \frac{\varepsilon}{4}.$$

Recall (see the notations of the proof of Theorem 6.1.4) that $B_p = A_{l_p} \cup A_{l_{p+1}} \cup \ldots$ and that, from (6.1.3),

$$\int_{A_{l_k}} \|u_{m_{l_k}}(\omega)\| \, d\mathrm{P}(\omega) \geq \eta((u_n)_n) - \frac{1}{l_k},$$

hence

$$\forall k \geq p \quad \int_{B_p} \|u_{m_{l_k}}(\omega)\| \, d\mathrm{P}(\omega) \geq \eta((u_n)_n) - \frac{1}{l_p}.$$

If p has been chosen such that $l_p \geq 4/\varepsilon$, then, for n large enough, $\| \mathbb{1}_{B_p} u'_n\|_{L^1} \geq \eta((u_n)_n) - \frac{\varepsilon}{4}$. Finally, setting

$$\eta((u_n)_n; \delta) := \sup\Big\{ \int_A \|u_n(\omega)\| \, d\mathrm{P}(\omega);\ n \in \mathbb{N},\ \mu(A) \leq \delta \Big\},$$

one has, as soon as $\mathrm{P}(B_p) \leq \delta$,

$$\sup_n \int_{B_p} \|u'_n\| \, d\mu \leq \eta((u_n)_n; \delta).$$

Consequently, if one has chosen δ small enough such that $\eta((u_n)_n; \delta) \leq \eta((u_n)_n) + \frac{\varepsilon}{4}$ and then p large enough such that $\mu(B_p) \leq \delta$, one has

$$\forall n \quad \int_{B_p} \|u'_n\| \, d\mu \leq \eta((u_n)_n) + \frac{\varepsilon}{4}.$$

Finally, for a given ε, a good choice of p is possible and one gets, for n large enough,

$$\left| \|u'_n - u\|_{L^1} - \left[\eta((u_n)_n) + \int_{\Omega \times F} \|\xi - u(\omega)\| \, d\tau(\omega, \xi) \right] \right| \leq \varepsilon.$$

□

6.2 Weak convergence in $\mathrm{L}^1_{\mathbb{E}}(\Omega, \mathcal{S}, \mathrm{P})$ using Young measures

We will show in this section some applications of Young measures and truncations techniques to weak compactness in $\mathrm{L}^1_{\mathbb{E}}(\Omega, \mathcal{S}, \mathrm{P})$ and the space $\mathrm{P}^1_{\mathbb{E}}(\Omega, \mathcal{S}, \mathrm{P})$ of Pettis-integrable $\mathbb{E}$-valued functions. In the sequel, $\mathcal{C}th^b(\Omega, \mathbb{E})$ denotes the set of bounded Carathéodory integrands on $\Omega \times \mathbb{E}$ and $\mathcal{C}th^1(\Omega, \mathbb{E})$ denotes the set of Carathéodory integrands on $\Omega \times \mathbb{E}$ such that there exists some $C \in [0, \infty[$ satisfying $|h(\omega, x)| \leq C(1 + \|x\|)$ for all $(\omega, x) \in \Omega \times \mathbb{E}$. The elements of $\mathcal{C}th^1(\Omega, \mathbb{E})$ are called *Carathéodory integrands of first order.* Here is a useful lemma.

Lemma 6.2.1 *Suppose that (u_n) is a uniformly integrable sequence in $\mathrm{L}^1_{\mathbb{E}}(\Omega, \mathcal{S}, \mathrm{P})$ such that the sequence $(\underline{\delta}_{u_n})$ of Young measures associated with (u_n) stably converges to $\lambda \in \mathcal{Y}^1(\Omega, \mathcal{S}, \mathrm{P}; \mathbb{E})$, then for any Carathéodory integrand of first-order $h \in \mathcal{C}th^1(\Omega, \mathbb{E})$, we have*

$$\lim_{n \to \infty} \int_\Omega h(\omega, u_n(\omega))\, d\mathrm{P}(\omega) = \int_\Omega \Big[\int_{\mathbb{E}} h(\omega, x)\, d\lambda_\omega(x)\Big]\, d\mathrm{P}(\omega).$$

Remark 6.2.2 Lemma 6.2.1 is a particular case of the implication 3 $\Rightarrow$ 1 in Proposition 2.4.1, but replacing the topology $\tau^{\mathrm{s}}_{p,\mathcal{D}}$ by $\tau^{\mathrm{s}}_{p,\mathcal{D}}$. This result was annouced in Remark 2.4.2. Note that, in the metrizable case considered here, $\tau^{\mathrm{S}}_{\mathcal{Y}^1}$ coincides with $\tau^{\mathrm{W}}_{\mathcal{Y}^1}$.

Proof of Lemma 6.2.1. We follow the arguments in [PV95]. Thanks to a lower semicontinuity theorem (cf. Portmanteau Theorem 2.1.3) for Young measures, λ is of first order, more precisely

$$\int_{\Omega \times \mathbb{E}} \|x\|\, d\lambda(\omega, x) \leq M := \sup_n \int_\Omega \|u_n(\omega)\|\, d\mathrm{P}(\omega).$$

Denote $C := \sup\{|h(\omega, x)|/(1 + \|x\|);\ (\omega, x) \in \Omega \times \mathbb{E}\}$ and, for $K \in [0, +\infty[$,

$$h^K(\omega, x) = \max\{-C(1 + K), \min(C(1 + K), h(\omega, x))\}.$$

Then $h^K \in \mathcal{C}th^b(\Omega, \mathcal{S}; \mathbb{E})$ and $\|x\| \leq K$ implies $h(\omega, x) = h^K(\omega, x)$. Fix $\varepsilon > 0$. One has

$$\begin{aligned}
&\left|\int_\Omega h(\omega, u_n(\omega))\, d\mathrm{P}(\omega) - \int_\Omega \Big[\int_{\mathbb{E}} h(\omega, x)\, d\lambda_\omega(x)\Big]\, d\mathrm{P}(\omega)\right| \\
&\quad \leq \left|\int_\Omega h(\omega, u_n(\omega))\, d\mathrm{P}(\omega) - \int_\Omega h^K(\omega, u_n(\omega))\, d\mathrm{P}(\omega)\right| \\
&\qquad + \left|\int_\Omega h^K(\omega, u_n(\omega))\, d\mathrm{P}(\omega) - \int_\Omega \Big[\int_{\mathbb{E}} h^K(\omega, x)\, d\lambda_\omega(x)\Big]\, d\mathrm{P}(\omega)\right| \\
&\qquad + \left|\int_\Omega \Big[\int_{\mathbb{E}} h^K(\omega, x)\, d\lambda_\omega(x)\Big]\, d\mathrm{P}(\omega) - \int_\Omega \Big[\int_{\mathbb{E}} h(\omega, x)\, d\lambda_\omega(x)\Big]\, d\mathrm{P}(\omega)\right|.
\end{aligned}$$

The first term of the second member is bounded from above by

$$\begin{aligned}
&\int_\Omega |h(\omega, u_n(\omega)) - h^K(\omega, u_n(\omega))| \, d\,\mathrm{P}(\omega) \\
&\qquad = \int_{\{\omega;\, h(\omega,u_n(\omega)) \neq h^K(\omega,u_n(\omega))\}} |h(\omega, u_n(\omega)) - h^K(\omega, u_n(\omega))| \, d\,\mathrm{P}(\omega) \\
&\qquad \leq \int_{\{\omega;\, \|u_n(\omega)\| > K\}} 2C(1 + \|u_n(\omega)\|) \, d\,\mathrm{P}(\omega).
\end{aligned}$$

Since (u_n) is uniformly integrable this expression is, for K large enough, $\leq \varepsilon/3$, this uniformly in n. As λ is of first order, by Lebesgue Theorem the third term of the second member is also $\leq \varepsilon/3$, for K large enough. Indeed, it is less than

$$\int_{\Omega \times \{\|x\| > K\}} 2C(1 + \|x\|) \, d\lambda(\omega, x).$$

Finally, after K has been fixed, the second term is $\leq \varepsilon/3$ for n large enough.

□

Proposition 6.2.3 (Compactness of the set of integrable selections) *Let $\Gamma : \Omega \to c\mathcal{K}(\mathbb{E})$ be a measurable and integrably bounded multifunction (i.e. there exists $g \in \mathrm{L}^1_{\mathbb{R}^+}(\Omega, \mathcal{S}, \mathrm{P})$ such that $\Gamma(\omega) \subset g(\omega)\overline{B}_{\mathbb{E}}$ for all $\omega \in \Omega$). Then the set S^1_Γ of all integrable selections of Γ is convex and $\sigma(\mathrm{L}^1_{\mathbb{E}}, \mathrm{L}^\infty_{\mathbb{E}^*})$-compact. Moreover the multivalued integral*

$$\int_\Omega \Gamma(\omega) \, d\,\mathrm{P}(\omega) := \{\int_\Omega f(\omega) \, d\,\mathrm{P}(\omega);\ f \in \mathrm{S}^1_\Gamma\}$$

is convex and norm compact in $\mathbb{E}$.

First proof. It is obvious that S^1_Γ is bounded convex in $\mathrm{L}^1_{\mathbb{E}}(\Omega, \mathcal{S}, \mathrm{P})$. It is closed in the norm topology of $\mathrm{L}^1_{\mathbb{E}}$ because, if $(f_n)_n$ is a sequence in S^1_Γ which converges to a limit f in the norm topology of $\mathrm{L}^1_{\mathbb{E}}$, there is a subsequence of $(f_n)_n$ which converges almost everywhere to f, thus f still belongs to S^1_Γ. To prove the first part of Proposition 6.2.3, in view of James' theorem, it is enough to show that, for every $g \in \mathrm{L}^\infty_{\mathbb{E}^*}(\Omega, \mathcal{S}, \mathrm{P})$, there is $f \in \mathrm{S}^1_\Gamma$ such that

$$\delta^*(g, \mathrm{S}^1_\Gamma) = \langle g, f \rangle = \int_\Omega \langle g(\omega), f(\omega) \rangle \, d\,\mathrm{P}(\omega),$$

where, for any $A \subset \mathbb{E}$, $\delta^*(., A)$ is the *support function of* A, that is, for every $x' \in \mathbb{E}^*$, $\delta^*(x', A) := \sup_{x \in A} \langle x', x \rangle$. By a measurable selection theorem [CV77, Theorem III.22], there is a measurable and integrable selection f of Γ such that

$$\delta^*(g(\omega), \Gamma(\omega)) = \langle g(\omega), f(\omega) \rangle,$$

for all $\omega \in \Omega$. By Strassen's theorem [CV77, Theorem V.14] this entails

$$\delta^*(g, S^1_\Gamma) = \int_\Omega \delta^*(g(\omega), \Gamma(\omega))\, d\,P(\omega) = \int_\Omega \langle g(\omega), f(\omega)\rangle\, d\,P(\omega).$$

Now the norm compactness of $\int_\Omega \Gamma(\omega)\, d\,P(\omega)$ is equivalent to the continuity of its support function on $\mathbb{E}^*$ with respect to the compact convergence topology (continuity at 0 is sufficient): This is a consequence of a general result of duality about the bipolar of a set (for a more general result see [CV77, Corollary I.15 pages 16–17]. Then thanks to Banach–Dieudonné's theorem we only have to check that this support function is continuous on the unit ball $\overline{B}_{\mathbb{E}^*}$ of the dual $\mathbb{E}^*$ for the compact convergence. Since $\overline{B}_{\mathbb{E}^*}$ is compact metrizable for both the weak*-topology and the topology of compact convergence, it remains to show that this function is sequentially continuous for the topologies under consideration. Let $x'_n \to x'$ in $\overline{B}_{\mathbb{E}^*}$ for the compact convergence. Then

$$\delta^*(x'_n, \Gamma(\omega)) \to \delta^*(x', \Gamma(\omega))$$

for each $\omega \in \Omega$. Using Lebesgue's theorem and Strassen's theorem, we easily get

$$\delta^*(x'_n, \int_\Omega \Gamma(\omega)\, d\,P(\omega)) = \int_\Omega \delta^*(x'_n, \Gamma(\omega))\, d\,P(\omega)$$
$$\to \int_\Omega \delta^*(x', \Gamma(\omega))\, d\,P(\omega) = \delta^*(x', \int_\Omega \Gamma(\omega)\, d\,P(\omega)).$$

□

Second proof. By Eberlein–Šmulian's theorem, it is enough to show that S^1_Γ is sequentially compact for the topology $\sigma(L^1_{\mathbb{E}}, L^\infty_{\mathbb{E}^*})$. Let (u_n) be a sequence in S^1_Γ. By the hypothesis of integrable boundedness, the sequence of Young measures $(\underline{\delta}_{u_n})$ is tight in $\mathcal{Y}^1(\Omega, \mathcal{S}, P; \mathbb{E})$. Hence there is a subsequence $(\underline{\delta}_{v_n})$ which stably converges to $\lambda \in \mathcal{Y}^1(\Omega, \mathcal{S}, P; \mathbb{E})$, that is, $(\underline{\delta}_{v_n})$ converges $\sigma(\mathcal{Y}^1(\Omega, \mathcal{S}, P; \mathbb{E}), \mathcal{C}th^b(\Omega, \mathbb{E}))$ to λ. As (v_n) is uniformly integrable, Lemma 6.2.1 entails that $(\underline{\delta}_{v_n})$ converges $\sigma(\mathcal{Y}^1(\Omega, \mathcal{S}, P; \mathbb{E}), \mathcal{C}th^1(\Omega, \mathbb{E}))$ to λ. Let $g \in L^\infty_{\mathbb{E}^*_\sigma}(\Omega, \mathcal{S}, P)$. Then the integrand h defined by

$$h(\omega, x) = \langle g(\omega), x\rangle, \ (\omega, x) \in \Omega \times \mathbb{E}$$

belongs to $\mathcal{C}th^1(\Omega, \mathbb{E})$. It follows that

$$\lim_{n\to\infty} \int_\Omega \langle g(\omega), v_n(\omega)\rangle\, d\,P(\omega) = \int_\Omega \Big[\int_{\mathbb{E}} \langle g(\omega), x\rangle\, d\lambda_\omega(x)\Big]\, d\,P(\omega).$$

In view of the Portmanteau Theorem 2.1.3, it is easily seen that $\lambda_\omega(\Gamma(\omega)) = 1$ a.e. and the mapping $\omega \mapsto u(\omega) := \mathrm{bar}\,(\lambda_\omega)$ where the barycenter $\mathrm{bar}\,(\lambda_\omega)$ of λ_ω is integrable and belongs to $\Gamma(\omega)$ a.e., thus proving the weak compactness of S^1_Γ in $L^1_{\mathbb{E}}(\Omega, \mathcal{S}, P)$. The norm compactness of the multivalued integral $\int_\Omega \Gamma(\omega)\, d\,P(\omega)$ can be proved by applying Banach–Dieudonné's theorem as in the end of the first proof. □

Remark 6.2.4 The $\sigma(\mathrm{L}^1_{\mathbb{E}}, \mathrm{L}^\infty_{\mathbb{E}^*})$ compactness property of S^1_Γ holds true if we suppose that Γ is convex weakly compact valued. Indeed the first proof is the same using James' Theorem while the second proof needs a careful look if we want to use again the compactness of Young measures in the S–stable topology. Note that $\mathbb{E}_\sigma$ is a completely regular Lusin space, and that the sequence $(\underline{\delta}_{u_n})$ associated with (u_n) is flexibly tight in $\mathcal{Y}^1(\Omega, \mathcal{S}, \mathrm{P}; \mathbb{E}_\sigma)$, hence, from Theorem 4.3.5, there is a subsequence $(\underline{\delta}_{u_{\varphi(n)}})$ which S–stably converges to $\lambda \in \mathcal{Y}^1(\Omega, \mathcal{S}, \mathrm{P}; \mathbb{E}_\sigma)$. Repeating the truncation techniques given in Lemma 6.2.1 shows that $(\underline{\delta}_{u_{\varphi(n)}})$ converges $\sigma(\mathcal{Y}^1(\Omega, \mathcal{S}, \mathrm{P}; \mathbb{E}_\sigma), \mathcal{C}th^1(\Omega, \mathbb{E}_\sigma))$ to λ, where $\mathcal{C}th^1(\Omega, \mathbb{E}_\sigma)$ denotes the set of all first order Carathéodory integrand defined on on $\Omega \times \mathbb{E}_\sigma$. So we can finish the proof in the same way as in the second proof of Proposition 6.2.3 by observing that the integrand $h : \Omega \times \mathbb{E}_\sigma \mapsto \langle g(\omega), x\rangle$ belongs to $\mathcal{C}th^1(\Omega, \mathbb{E}_\sigma)$. In this example it turns out that the second proof using Young measures is rather long and less direct than the first one. In this context other proofs not involving Young measures are available, essentially when we deal with weak sequential compactness in Pettis integration in the next paragraph.

The following result is stated in [ACV92, Théorème 6 pp.174–175]. We provide the proof since it can be applied to other situations.

Theorem 6.2.5 *If (u_n) is a bounded uniformly integrable* cw$\mathcal{K}(\mathbb{E})$*-tight sequence in $\mathrm{L}^1_{\mathbb{E}}(\Omega, \mathcal{S}, \mathrm{P})$, then (u_n) is relatively weakly compact in $\mathrm{L}^1_{\mathbb{E}}(\Omega, \mathcal{S}, \mathrm{P})$.*

Proof. We will follow [ACV92, Lemme 5 and Théorème 6]. Let $\varepsilon > 0$. There exist $\alpha > 0$ and $\eta > 0$ such that

$$\sup_n \int_{\{\|u_n(\cdot)\| > \alpha\}} \|u_n(\omega)\| \, d\mathrm{P}(\omega) \le \frac{\varepsilon}{2},$$

and

$$\forall A \in \mathcal{S} \quad \mathrm{P}(A) < \eta \Longrightarrow \sup_n \int_A \|u_n(\omega)\| \, d\mathrm{P}(\omega) \le \frac{\varepsilon}{2}.$$

By hypothesis there exists a cw$\mathcal{K}(\mathbb{E})$-valued measurable multifunction L_η such that

$$\sup_n \mathrm{P}(\{\omega \in \Omega;\, u_n(\omega) \notin L_\eta(\omega)\}) \le \eta.$$

We have

$$\begin{aligned} u_n(\omega) = {} & \mathbb{1}_{\{u_n(\omega) \in L_\eta(\omega)\} \cap \{\|u_n(\omega)\| \le \alpha\}} u_n(\omega) + \mathbb{1}_{\{u_n(\omega) \notin L_\eta(\omega)\} \cap \{\|u_n(\omega)\| \le \alpha\}} u_n(\omega) \\ & + \mathbb{1}_{\{\|u_n(\omega)\| > \alpha\}} u_n(\omega). \end{aligned}$$

So $u_n = v_n + w_n$ with

$$v_n(\omega) = \mathbb{1}_{\{u_n(\omega) \in L_\eta(\omega)\} \cap \{\|u_n(\omega)\| \le \alpha\}} u_n(\omega),$$

and

$$\sup_n \int_\Omega \|w_n(\omega)\| \, d\,\mathrm{P}(\omega) \le \varepsilon.$$

As

$$v_n(\omega) \in \Gamma(\omega) := \overline{\mathrm{co}}[\overline{B}\,(0,\alpha) \cap L_\eta(\omega) \cup \{0\}]$$

and the multifunction Γ is measurable cw$\mathcal{K}(\mathbb{E})$-valued and integrably bounded, by Proposition 6.2.3, (v_n) is relatively weakly compact. Since $\int_\Omega \|w_n\| \, d\,\mathrm{P}(\omega) \le \varepsilon$ for all n, by Grothendieck's lemma [Gro64, Chapitre 5 page 296] we deduce that (u_n) is relatively weakly compact in $L^1_{\mathbb{E}}(\Omega, \mathcal{S}, \mathrm{P})$. □

The following is an easy consequence of the preceding theorem and the Biting Lemma.

Theorem 6.2.6 *If (u_n) is a bounded* cw$\mathcal{K}(\mathbb{E})$*-tight sequence in $L^1_{\mathbb{E}}(\Omega, \mathcal{S}, \mathrm{P})$, then there is a sequence $(\widetilde{u}_n)$ with $\widetilde{u}_n \in$ co$\{u_m;\ m \ge n\}$ and $u_\infty \in L^1_{\mathbb{E}}(\Omega, \mathcal{S}, \mathrm{P})$ such that $(\widetilde{u}_n(\omega))$ strongly converges a.e. to u_∞. Consequently, we have*

$$u_\infty(\omega) \in \bigcap_n \overline{\mathrm{co}}\{u_m(\omega);\ m \ge n\} \ \text{ a.e.}$$

Proof. Applying the Biting Lemma to the bounded sequence $(\|u_n(.)\|)$ provides a subsequence still denoted by $(\|u_n(.)\|)$ and an increasing sequence (A_n) with $\mathrm{P}(A_n) \uparrow \mathrm{P}(\Omega)$ such that $(\mathbb{1}_{A_n} \|u_n(.)\|)$ is uniformly integrable and $(\mathbb{1}_{A_n^c} \|u_n(.)\|) \to 0$ a.e. As the sequence $(v_n) = (\mathbb{1}_{A_n} u_n)$ is uniformly integrable and cw$\mathcal{K}(\mathbb{E})$-tight, (v_n) is relatively weakly compact in $L^1_{\mathbb{E}}(\Omega, \mathcal{S}, \mathrm{P})$ in view of Theorem 6.2.5. By extracting a subsequence we can suppose that this sequence converges weakly to $u_\infty \in L^1_{\mathbb{E}}(\Omega, \mathcal{S}, \mathrm{P})$. Hence there exists a sequence $(\tilde{v}_n)$ with $\tilde{v}_n \in$ co$\{v_m(\omega);\ m \ge n\}$ which converges strongly a.e. to u_∞. It follows that $u_\infty(\omega) \in \cap_n \overline{\mathrm{co}}\{u_m(\omega);\ m \ge n\}$ a.e. □

There are some useful consequences of the preceding theorem that we summarize as follows.

Proposition 6.2.7 *If $\mathcal{H}$ is a bounded convex* cw$\mathcal{K}(\mathbb{E})$*-tight subset in $L^1_{\mathbb{E}}(\Omega, \mathcal{S}, \mathrm{P})$ and closed for the convergence in measure, and if $J : \mathcal{H} \to [0, \infty[$ is a convex lower semicontinuous on $\mathcal{H}$ for the convergence in measure, then J reaches its mimimum on $\mathcal{H}$.*

In particular, if $\mathbb{E}$ is a reflexive separable Banach space, $\mathcal{H}$ is bounded convex in $L^1_{\mathbb{E}}$, closed in measure, then any positive convex lower semicontinuous on $\mathcal{H}$ for the convergence in measure reaches its mimimum on $\mathcal{H}$. Theorem 6.2.6 shows that any bounded cw$\mathcal{K}(\mathbb{E})$-tight sequence in $L^1_{\mathbb{E}}(\Omega, \mathcal{S}, \mathrm{P})$ is strongly Mazur–compact.

Proposition 6.2.8 *Let $\mathcal{H}$ be a bounded uniformly integrable set in $L^1_{\mathbb{E}}(\Omega, \mathcal{S}, \mathrm{P})$ which satisfies the following condition: For any sequence (u_n) in $\mathcal{H}$, there is a* cw$\mathcal{K}(\mathbb{E})$*-tight sequence (v_n) such that $v_n \in$ co$\{u_m;\ m \ge n\}$ for each n, shortly (u_n) is Mazur* cw$\mathcal{K}(\mathbb{E})$*-tight, then $\mathcal{H}$ is relatively weakly compact in $L^1_{\mathbb{E}}(\Omega, \mathcal{S}, \mathrm{P})$.*

The proof is straightforward since it follows easily from the above weak compactness theorem. A multivalued version of the preceding result is available [CS00, Theorem 3.4].

We end this section with a useful lemma.

Lemma 6.2.9 *Let $\mathbb{E}$ be a separable Banach space and $(u_n)_{n\in\mathbb{N}}$ a sequence of scalarly integrable $\mathbb{E}$-valued functions satisfying*

(i) $\{\langle x', u_n\rangle;\ x' \in \overline{B}_{\mathbb{E}^*},\ n \in \mathbb{N}\}$ is uniformly integrable,

(ii) for every $A \in \mathcal{S}$, the set $\mathcal{H}_A := \{\int_A u_n \, d\mathrm{P};\ n \in \mathbb{N}\}$ is relatively weakly compact.

Then there is a subsequence $(u_{n_k})_{k\in\mathbb{N}}$ such that

$$\forall A \in \mathcal{S} \quad \forall x' \in \mathbb{E}^* \quad \lim_{k\to\infty} \int_A \langle x', u_{n_k}\rangle \, d\mathrm{P}$$

exists in $\mathbb{R}$.

Proof. We may suppose that for every $A \in \mathcal{S}$, there exists a convex weakly compact subset $\mathcal{K}_A$ such that $\mathcal{H}_A \subset \mathcal{K}_A$. Let D^* be a countable dense sequence in $\mathbb{E}^*$ for the Mackey topology (see [CV77, Lemma III.32]) and let $\mathcal{A} = \sigma(A_i\,, i \in \mathbb{N})$ be the σ-algebra generated by $(u_n)_{n\in\mathbb{N}}$. Then by (*ii*) and by extracting diagonal subsequences, we find a subsequence $(u_{n_k})_{k\in\mathbb{N}}$ such that for any fixed $i \in \mathbb{N}$, $(\int_{A_i} u_{n_k} \, d\mathrm{P})$ weakly converges to an element $c_i \in \mathcal{K}_{A_i}$. It follows that

$$\lim_{k\to\infty} \langle x', \int_{A_i} u_{n_k} \, d\mathrm{P}\rangle) = \lim_{k\to\infty} \int_{A_i} \langle x', u_{n_k}\rangle \, d\mathrm{P} = \langle x', c_i\rangle$$

for all $x' \in D^*$ and for all $i \in \mathbb{N}$. Now since D^* is dense for the Mackey topology, the preceding equalities are valid for every $x' \in \mathbb{E}^*$. Let $A \in \mathcal{A}$ and $\varepsilon > 0$. Since the set $\{\langle x', u_{n_k}\rangle;\ x' \in \overline{B}_{\mathbb{E}^*},\ k \in \mathbb{N}\}$ is uniformly integrable by (*i*), there is a measurable set A_i such that

$$\int_{A_i \Delta A} |\langle x', u_{n_k}\rangle| \, d\mathrm{P} \leq \varepsilon$$

for all $x' \in \overline{B}_{\mathbb{E}^*}$ and for all $k \in \mathbb{N}$ so that

$$|\int_A \langle x', u_{n_k}\rangle \, d\mathrm{P} - \int_{A_i} \langle x', u_{n_k}\rangle \, d\mathrm{P}| \leq \int_{A_i \Delta A} |\langle x', u_{n_k}\rangle| \, d\mathrm{P} \leq \varepsilon$$

for all $x' \in \overline{B}_{\mathbb{E}^*}$ and for all $k \in \mathbb{N}$. It follows that $\lim_{k\to\infty} \int_A \langle x', u_{n_k}\rangle \, d\mathrm{P}$ exists in $\mathbb{R}$. Consequently, for any $x' \in \mathbb{E}^*$ and for any positive $\mathcal{A}$-measurable and bounded function h

$$\lim_{k\to\infty} \int_\Omega h \, \langle x', u_{n_k}\rangle \, d\mathrm{P}$$

exists in $\mathbb{R}$. Now let h be any positive $\mathcal{S}$-measurable and bounded function and let $E^{\mathcal{A}}h$ the conditional expectation of h, then we have

$$\lim_{k\to\infty}\int_{\Omega} h\,\langle x', u_{n_k}\rangle\, d\,\mathrm{P} = \lim_{k\to\infty}\int_{\Omega} E^{\mathcal{A}}h\,\langle x', u_{n_k}\rangle\, d\,\mathrm{P}\,.$$

□

6.3 Weak compactness and convergences in Pettis integration

Most results in this section section are borrowed from [AC97].

We first give the following basic result. See also [Gei81, Huf86, Mus91].

Theorem 6.3.1 *Let $\mathbb{E}$ be a Fréchet space, $(f_n)_{n\in\mathbb{N}}$ a sequence of $\mathbb{E}$-valued Pettis integrable functions and $f:\Omega\to\mathbb{E}$ a scalarly integrable function such that*

(*i*) *for every convex weakly compact subset $B\subset\mathbb{E}^*$, the set $\{\langle x', f\rangle\,;\, x'\in B\}$ is uniformly integrable,*

(*ii*) *for every $x'\in\mathbb{E}^*$, $\langle x', f_n\rangle$ converges $\sigma(\mathrm{L}^1,\mathrm{L}^\infty)$ to $\langle x', f\rangle$.*

Then f is Pettis-integrable.

Proof. By hypothesis, for every $x'\in\mathbb{E}^*$ and for every $A\in\mathcal{S}$, we have

$$\lim_{n\to\infty}\langle x', \int_A f_n\, d\,\mathrm{P}\rangle = \lim_{n\to\infty}\int_A \langle x', f_n\rangle\, d\,\mathrm{P} = \int_A \langle x', f\rangle\, d\,\mathrm{P}\,.$$

So, in order to prove the theorem, it is sufficient to show that for every $A\in\mathcal{S}$, the sequence $(\int_A f_n\, d\,\mathrm{P})_{n\in\mathbb{N}}$ is relatively weakly compact in $\mathbb{E}$. By the Eberlein–Šmulian–Grothendieck Theorem [Gro52, Corollaire 1 of Théorème 7] it is equivalent to prove: For every convex weakly compact subset $B\subset\mathbb{E}^*$, for every sequence $(x'_k)_{k\in\mathbb{N}}$ in B and for every subsequence $(f_{n_m})_{m\in\mathbb{N}}$ of $(f_n)_{n\in\mathbb{N}}$, we have

$$\alpha := \lim_{k\to\infty}\lim_{m\to\infty}\langle x'_k, \int_A f_{n_m}\, d\,\mathrm{P}\rangle = \beta := \lim_{m\to\infty}\lim_{k\to\infty}\langle x'_k, \int_A f_{n_m}\, d\,\mathrm{P}\rangle \tag{6.3.1}$$

provided these limits exist. First, by (*ii*), we have

$$\lim_{m\to\infty}\langle x'_k, \int_A f_{n_m}\, d\,\mathrm{P}\rangle = \lim_{m\to\infty}\int_A \langle x'_k, f_{n_m}\rangle\, d\,\mathrm{P} = \int_A \langle x'_k, f\rangle\, d\,\mathrm{P}\,. \tag{6.3.2}$$

By Komlós Theorem [Kom67] applied to the sequence $(\langle x'_k, f\rangle)_{k\in\mathbb{N}}$, there exist a sequence $(y'_n)_{n\in\mathbb{N}}$ with $y'_n = 1/n\sum_{i=1}^n x'_{k_i}$ and a real valued integrable function h

such that $\langle y'_n, f\rangle$ converges to h almost everywhere. So, by (6.3.1), (6.3.2) and (i), we have

$$\alpha = \lim_{k\to\infty} \int_A \langle x'_k, f\rangle \, d\mathrm{P} = \lim_{n\to\infty} \int_A \langle y'_n, f\rangle \, d\mathrm{P} = \int_A h \, d\mathrm{P}. \tag{6.3.3}$$

Let y'_0 be a weak* cluster point of $(y'_n)_{n\in\mathbb{N}}$, then for every $m \in \mathbb{N}$, we have

$$\begin{aligned} \lim_{k\to\infty} \langle x'_k, \int_A f_{n_m} \, d\mathrm{P}\rangle &= \lim_{n\to\infty} \langle y'_n, \int_A f_{n_m} \, d\mathrm{P}\rangle \\ &= \langle y'_0, \int_A f_{n_m} \, d\mathrm{P}\rangle \\ &= \int_A \langle y'_0, f_{n_m}\rangle \, d\mathrm{P}. \end{aligned} \tag{6.3.4}$$

Taking the limit when $m \to \infty$ in the last integral in (6.3.4) and using (6.3.1) and (6.3.2), we obtain

$$\beta = \lim_{m\to\infty} \int_A \langle y'_0, f_{n_m}\rangle \, d\mathrm{P} = \int_A \langle y'_0, f\rangle \, d\mathrm{P}. \tag{6.3.5}$$

Since $\langle y'_n, f\rangle$ converges to h almost everywhere and y'_0 is a weak* cluster point of $(y'_n)_{n\in\mathbb{N}}$, $h = \langle y'_0, f\rangle$ almost everywhere. Returning to (6.3.1) and using (6.3.3), (6.3.4) and (6.3.5), we get $\alpha = \beta$. □

Theorem 6.3.2 *Let $\mathbb{E}$ be a separable Banach space and $\mathcal{H}$ a subset of $\mathrm{P}^1_{\mathbb{E}}(\Omega, \mathcal{S}, \mathrm{P})$ satisfying:*

(1) *$\{\langle x', f\rangle;\, x' \in \overline{B}_{\mathbb{E}^*},\, f \in \mathcal{H}\}$ is uniformly integrable.*

(2) *Given any sequence (f_n) in $\mathcal{H}$, there are a sequence $(\widetilde{f}_n)$ with $\widetilde{f}_n \in \mathrm{co}\{f_k;\, k \geq n\}$ and $\widetilde{f}_\infty \in \mathrm{P}^1_{\mathbb{E}}(\Omega, \mathcal{S}, \mathrm{P})$ such that, $\forall x' \in \mathbb{E}^*$, $\langle x', \widetilde{f}_n\rangle$ converges $\sigma(\mathrm{L}^1, \mathrm{L}^\infty)$ to $\langle x', \widetilde{f}_\infty\rangle$.*

Then $\mathcal{H}$ is relatively sequentially compact for the topology of pointwise convergence on $\mathrm{L}^\infty_{\mathbb{R}} \otimes \mathbb{E}^*$.

Proof. *Step 1.* Let $(f_n)_{n\in\mathbb{N}}$ in $\mathcal{H}$. For every measurable set $A \in \mathcal{A}$, let $\mathcal{H}_A := \{\int_A f_n \, d\mathrm{P};\, n \in \mathbb{N}\}$. By (1) $\mathcal{H}_A$ is bounded for every $A \in \mathcal{A}$. Now we claim that, $\forall A \in \mathcal{A}$, $\mathcal{H}_A$ is relatively weakly compact or equivalently $\mathcal{K}_A := \overline{\mathrm{co}}\,\mathcal{H}_A$ is weakly compact. By James' theorem it is enough to prove that for every $x' \in \mathbb{E}^*$, there exists $\zeta \in \mathcal{K}_A$ such that

$$\langle x', \zeta\rangle = \sup_{x\in\mathcal{K}_A} \langle x', x\rangle = \delta^*(x', \mathcal{K}_A) = \delta^*(x', \mathcal{H}_A).$$

Let $(f_{n_k})_{k\in\mathbb{N}}$ be a subsequence of $(f_n)_{n\in\mathbb{N}}$ such that

$$\lim_{k\to\infty} \langle x', \int_A f_{n_k}\, d\mathrm{P}\rangle = \delta^*(x', \mathcal{H}_A).$$

Let $(\widetilde{f}_n)_{n\in\mathbb{N}}$ and $\widetilde{f}_\infty \in \mathrm{P}^1_{\mathbb{E}}(\Omega, \mathcal{S}, \mathrm{P})$ associated with $(f_{n_k})_{k\in\mathbb{N}}$ by (2). Since each $\widetilde{f}_n$ has the form $\widetilde{f}_n = \sum_{i=n}^{\nu_n} \lambda_i^n\, f_{n_i}$ with $0 \leq \lambda_i^n \leq 1$ and $\sum_{i=n}^{\nu_n} \lambda_i^n = 1$, then we have

$$\begin{aligned}
\delta^*(x', \mathcal{K}_A) &= \lim_{k\to\infty} \langle x', \int_A f_{n_k}\, d\mathrm{P}\rangle \\
&= \lim_{n\to\infty} \langle x', \sum_{i=n}^{\nu_n} \lambda_i^n \int_A f_{n_i}\, d\mathrm{P}\rangle \\
&= \langle x', \int_A \widetilde{f}_\infty\, d\mathrm{P}\ \rangle \\
&\leq \delta^*(x', \mathcal{K}_A).
\end{aligned}$$

So the claim is true. Note that in this step, it is not necessary to suppose that $\mathbb{E}$ is separable.

Step 2. Since $(\int_A f_n\, d\mathrm{P})_{n\in\mathbb{N}}$ is relatively weakly compact, we may apply Lemma 6.2.9 which provides a subsequence still denoted by $(f_{n_k})_{k\in\mathbb{N}}$, such that for every measurable set A and every $x' \in \mathbb{E}^*$, $\lim_{k\to\infty} \int_A \langle x', f_{n_k}\rangle\, d\mathrm{P}$ exists in $\mathbb{R}$. Let $(\widetilde{f}_n)_{n\in\mathbb{N}}$ and $\widetilde{f}_\infty \in \mathrm{P}^1_{\mathbb{E}}(\Omega, \mathcal{S}, \mathrm{P})$ associated with $(f_{n_k})_{k\in\mathbb{N}}$ by (2). Then we have

$$\lim_{k\to\infty} \int_A \langle x', f_{n_k}\rangle\, d\mathrm{P} = \lim_{n\to\infty} \int_A \langle x', \widetilde{f}_n\rangle\, d\mathrm{P} = \int_A \langle x',\ \widetilde{f}_\infty\rangle\, d\mathrm{P}$$

so that by standard arguments we get

$$\lim_{k\to\infty} \int h\, \langle x', f_{n_k}\rangle\, d\mathrm{P} = \int h\, \langle x', f_\infty\rangle\, d\mathrm{P}$$

for all $h \in \mathrm{L}^\infty_{\mathbb{R}}$ and $x' \in \mathbb{E}^*$. □

If Γ is a multifunction defined on a measurable space, with values in the nonempty subsets of a measurable space $\mathbb{E}$, we denote by S_Γ the set of its measurable selections. If furthermore $\mathbb{E}$ is a Banach space, we denote by $\mathrm{S}^{\mathrm{Pe}}_\Gamma$ the set of Pettis integrable elements of S_Γ.

Corollary 6.3.3 *Let $\mathbb{E}$ be a separable Banach space. Let $\Gamma : \Omega \to \mathrm{cw}\mathcal{K}(\mathbb{E})$ be a $\mathrm{cw}\mathcal{K}(\mathbb{E})$-valued* scalarly integrable multifunction, *that is, for any $x' \in \mathbb{E}^*$, $\delta^*(x', \Gamma(.))$ is integrable. If $\{\langle x', f\rangle;\, x' \in \overline{B}_{\mathbb{E}^*}, f \in \mathrm{S}_\Gamma\}$ is uniformly integrable, then the set $\mathrm{S}^{\mathrm{Pe}}_\Gamma$ is nonempty and sequentially compact for the topology of pointwise convergence on $\mathrm{L}^\infty_{\mathbb{R}} \otimes \mathbb{E}^*$.*

Proof. By hypothesis and [Gei81, Huf86] $\mathrm{S}_\Gamma^{\mathrm{Pe}}$ is nonempty. Now let $(f_n)_{n\in\mathbb{N}} \subset \mathrm{S}_\Gamma^{\mathrm{Pe}}$ and let $(e'_p)_{p\in\mathbb{N}}$ be a dense sequence in $\overline{B}_{\mathbb{E}^*}$ for the Mackey topology. Since for each $p \in \mathbb{N}$, the sequence $(\langle e'_p, f_n\rangle)_{n\in\mathbb{N}}$ is uniformly integrable, using Komlós theorem [Kom67] and an appropriate diagonal process, we find a subsequence $(f_{n_k})_{k\in\mathbb{N}}$ and a sequence $(\varphi_p)_{p\in\mathbb{N}}$ in $\mathrm{L}^1_{\mathbb{R}}(\Omega, \mathcal{S}, \mathrm{P})$ such that

$$\lim_{n\to\infty} \frac{1}{n}\sum_{k=1}^{n} \langle e'_p, f_{n_k}(\omega)\rangle = \varphi_p(\omega) \quad \text{a.e.}$$

Since $\frac{1}{n}\sum_{k=1}^n f_{n_k}(\omega) \in \Gamma(\omega)$ for all $n \in \mathbb{N}$ and for all $\omega \in \Omega$, and $\Gamma(\omega)$ is convex weakly compact, $(\frac{1}{n}\sum_{k=1}^n f_{n_k}(\omega))_{n\in\mathbb{N}}$ weakly converges a.e. So by Theorem 6.3.2 and, using the fact that Γ is scalarly integrable with convex weakly compact values, we conclude that $\mathrm{S}_\Gamma^{\mathrm{Pe}}$ is sequentially compact for the topology of pointwise convergence on $\mathrm{L}^\infty_{\mathbb{R}} \otimes \mathbb{E}^*$. □

The following result is a Pettis analog of Proposition 6.2.3 and the arguments developed therein.

Proposition 6.3.4 *Suppose that $\mathbb{E}$ is a separable Banach space and $\Gamma : \Omega \to c\mathcal{K}(\mathbb{E})$ is a Pettis integrable multifunction, that is, $\{\delta^*(x', \Gamma(.)); \|x'\| \le 1\}$ is uniformly integrable, then the set $\mathrm{S}_\Gamma^{\mathrm{Pe}}$ is nonempty and sequentially compact for the topology of pointwise convergence on $\mathrm{L}^\infty_{\mathbb{R}} \otimes \mathbb{E}^*$ and, the multivalued integral*

$$\int_\Omega \Gamma\, d\mathrm{P} := \{\int_\Omega f\, d\mathrm{P};\ f \in \mathrm{S}_\Gamma^{\mathrm{Pe}}\}$$

is convex and compact in $\mathbb{E}$.

Proof. Sequential compactness of $\mathrm{S}_\Gamma^{\mathrm{Pe}}$ is an immediate consequence of Corollary 6.3.3. The norm compactness of the multivalued integral $\int_\Omega \Gamma(\omega)\, d\mathrm{P}(\omega)$ can be proved by applying Strassen's theorem and Banach–Dieudonné's theorem as in the end of the first proof of Proposition 6.2.3. □

Now we want to show that Proposition 6.3.4 can be deduced from a general convergence result for Young measures.

Theorem 6.3.5 *Suppose that $\mathbb{S}$ is a Lusin topological space and let (u_n) be a sequence of measurable mappings from $(\Omega, \mathcal{S}, \mathrm{P})$ into $\mathbb{S}$ such that the sequence $(\underline{\delta}_{u_n})$ of Young measures associated with (u_n) stably converges to a Young measure $\sigma^\infty \in \mathcal{Y}^1(\Omega, \mathcal{S}, \mathrm{P}; \mathbb{S})$. And suppose h is a Carathéodory integrand defined on $\Omega \times \mathbb{S}$ such that the sequence $(h(u_n)) := (h(., u_n(.)))$ is uniformly integrable in $\mathrm{L}^1_{\mathbb{R}}(\Omega, \mathcal{S}, \mathrm{P})$. Then*

$$(a) \qquad \lim_{n\to\infty} \int_\Omega h(u_n)\, d\mathrm{P} = \int_\Omega [\int_{\mathbb{S}} h(\omega, s)\, d\sigma^\infty_\omega(s)]\, d\mathrm{P}(\omega).$$

Suppose further that $(\mathbb{S}, d)$ *is Polish space and* $\mathcal{H}$ *is a set of continuous functions defined on* $\mathbb{S}$ *such that* $|g(x)-g(y)| \leq d(x,y)$ *for all* $g \in \mathcal{H}$ *and for all* $(x,y) \in \mathbb{S}\times\mathbb{S}$, *then the following holds:*

$$(b) \qquad \sup_{g\in\mathcal{H}} \left| \int_\Omega g(u_n(\omega))\, d\,\mathrm{P}(\omega) - \int_\Omega \Big[\int_{\mathbb{S}} g(s)\, d\sigma_\omega^\infty(s)\Big]\, d\,\mathrm{P}(\omega) \right| \to 0$$

as $n \to \infty$.

Proof. (a) Suppose that h is a Carathéodory integrand on $\Omega \times \mathbb{S}$ such that $h(u_n)$ is uniformly integrable. Since $h = h^+ - h^-$ and both $h^+(u_n)$ and $h^-(u_n)$ are uniformly integrable, we may suppose that $h \geq 0$. For every $k \in \mathbb{N}$, let us define a continuous function $\alpha_k : \mathbb{R}^+ \to \mathbb{R}^+$ as follows: $\alpha_k(x) \leq x$ for all $x \geq 0$, $\alpha_k(x) = x$ if $x \leq k$, $\alpha_k(x) = 0$ if $x \geq k+1$. We claim that

$$A := \int_\Omega \Big[\int_{\mathbb{S}} h(\omega, s)\, d\sigma_\omega^\infty(s)\Big]\, d\,\mathrm{P}(\omega) < +\infty.$$

Note that if h is bounded, the result follows by hypothesis. By Beppo Levi's theorem we have

$$\begin{aligned} A &= \sup_k \int_\Omega \Big[\int_{\mathbb{S}} \alpha_k(h(\omega,s))\, d\sigma_\omega^\infty(s)\Big]\, d\,\mathrm{P}(\omega) = \sup_k \lim_n \int_\Omega \alpha_k(h(\omega, u_n(\omega)))\, d\,\mathrm{P}(\omega) \\ &\leq \limsup_n \sup_k \int_\Omega \alpha_k(h(\omega, u_n(\omega)))\, d\,\mathrm{P}(\omega) \leq \limsup_n \int_\Omega h(\omega, u_n(\omega))\, d\,\mathrm{P}(\omega) < +\infty, \end{aligned}$$

because $(h(u_n))$ is uniformly integrable. Let $\varepsilon > 0$. We need to prove that there is an integer N such that

$$\left| \int_\Omega h(\omega, u_n(\omega))\, d\,\mathrm{P}(\omega) - \int_\Omega \Big[\int_{\mathbb{S}} h(\omega,s)\, d\sigma_\omega^\infty(s)\Big]\, d\,\mathrm{P}(\omega) \right| \leq 3\varepsilon$$

for all $n \geq N$. Notice that

$$\left| \int_\Omega h(\omega, u_n(\omega))\, d\,\mathrm{P}(\omega) - \int_\Omega \Big[\int_{\mathbb{S}} h(\omega,s)\, d\sigma_\omega^\infty(s)\Big]\, d\,\mathrm{P}(\omega) \right|,$$

is $\leq L_1(k,n) + L_2(k,n) + L_3(k)$ where, for every $k \in \mathbb{N}$,

$$\begin{aligned} L_1(k,n) &= \left| \int_\Omega h(\omega, u_n(\omega)) - \alpha_k(h(\omega, u_n(\omega)))\, d\,\mathrm{P}(\omega) \right|, \\ L_2(k,n) &= \left| \int_\Omega \alpha_k(h(\omega, u_n(\omega)))\, d\,\mathrm{P}(\omega) - \int_\Omega \Big[\int_{\mathbb{S}} \alpha_k(h(\omega,s))\, d\sigma_\omega^\infty(s)\Big]\, d\,\mathrm{P}(\omega) \right|, \\ L_3(k) &= \left| \int_\Omega \Big[\int_{\mathbb{S}} \alpha_k(h(\omega,s))\, d\sigma_\omega^\infty(s)\Big]\, d\,\mathrm{P}(\omega) - \int_\Omega \Big[\int_{\mathbb{S}} (h(\omega,s)\, d\sigma_\omega^\infty(s)\Big]\, d\,\mathrm{P}(\omega) \right|. \end{aligned}$$

Pick $K \in \mathbb{N}$ such that

$$\sup_n \int_{\{h(u_n)>K+1\}} h(\omega, u_n(\omega))\, d\,\mathrm{P}(\omega) \le \varepsilon$$

because $(h(u_n))$ is uniformly integrable, and $L_3(K) \le \varepsilon$, using Beppo Levi's theorem. Since (δ_{u_n}) converges to σ^∞ in $\mathcal{Y}^1(\Omega, \mathcal{S}, \mathrm{P}; \mathbb{S})$ and $\alpha_K \circ h$ is a bounded Carathéodory integrand, we have that $L_2(K, n) \to 0$ when $n \to \infty$. Hence there is some integer N such that $n \ge N$ implies $L_2(K, n) \le \varepsilon$. Notice that

$$L_1(K, n) \le \sup_n \int_{\{h(u_n)>K+1\}} h(\omega, u_n(\omega))\, d\,\mathrm{P}(\omega)$$

for all $n \in \mathbb{N}$. So, for $n \ge N$, we have

$$\left| \int_\Omega h(\omega, u_n(\omega))\, d\,\mathrm{P}(\omega) - \int_\Omega \Big[\int_{\mathbb{S}} h(\omega, s)\, d\sigma_\omega^\infty(s) \Big]\, d\,\mathrm{P}(\omega) \right| \\ \le L_1(K, n) + L_2(K, n) + L_3(K) \le 3\varepsilon.$$

(b) Let us set

$$\mu_n := \int_\Omega \delta_{u_n(\omega)}\, d\,\mathrm{P}(\omega) \quad \text{and} \quad \sigma := \int_\Omega \sigma_\omega^\infty\, d\,\mathrm{P}(\omega).$$

Since $(\underline{\delta}_{u_n})$ stably converges to σ^∞ in $\mathcal{Y}^1(\Omega, \mathcal{S}, \mathrm{P}; \mathbb{S})$, the sequence (μ_n) narrowly converges to σ in $\mathcal{M}^{+,1}(\mathbb{S})$. According to Skorokhod's theorem, there exist a probability space $(\Omega', \mathcal{S}', \mathrm{Q})$ and random elements $X_m, X_\infty : \Omega' \to \mathbb{S}$ such that the law of X_n (resp. X_∞) is μ_n (resp. σ) with $X_n \to X_\infty$ Q-a.e. First we claim that $\{g(X_n);\ n \in \mathbb{N},\ g \in \mathcal{H}\}$ is uniformly integrable in $\mathrm{L}^1_{\mathbb{R}}(\Omega', \mathcal{S}', \mathrm{Q})$. As $(g(u_n))$ is uniformly integrable in $\mathrm{L}^1_{\mathbb{R}}(\Omega, \mathcal{S}, \mathrm{P})$, by de La Vallée Poussin's theorem, there is a continuous convex even function $\varphi : \mathbb{R} \to \mathbb{R}^+$ with $\varphi(t)/|t| \to \infty$ when $|t| \to +\infty$ such that

$$\sup_{g \in \mathcal{H}} \sup_{n \in \mathbb{N}} \int_\Omega \varphi(g(u_n))\, d\,\mathrm{P} < +\infty,$$

so that

$$\sup_{g \in \mathcal{H}} \sup_{n \in \mathbb{N}} \int_\Omega \varphi(g(X_n))\, d\,\mathrm{Q} < +\infty,$$

thus proving the claim.

Secondly, we claim that $\mathcal{H}$ is uniformly integrable with respect to $\sigma \in \mathcal{M}^{+,1}(\mathbb{S})$. For each $x \in \mathbb{R}^+$, let $\beta_k(x) := x - \alpha_k(x)$ $(k \in \mathbb{N})$. Let $g \in \mathcal{H}$. We have the estimate

$$x\, \mathbb{1}_{\{y \in \mathbb{R}^+;\, y \ge k+1\}}(x) \le \beta_k(x) \le x\, \mathbb{1}_{\{y \in \mathbb{R}^+;\, y \ge k\}}(x),$$

for all $x \in \mathbb{R}^+$. So, for every $k \in \mathbb{N}$, we have the estimate

$$\begin{aligned}\int_{\{g(.)>k+1\}} g(s)\,d\sigma(s) &\leq \int_{\mathbb{S}} \beta_k(g)\,d\sigma \\ &= \lim_{n\to\infty} \int_\Omega \beta_k(g(u_n))\,d\mathrm{P} \\ &\leq \sup_{g'\in\mathcal{H}} \sup_n \int_{\{g'(u_n)>k\}} g'(u_n)\,d\mathrm{P}\,. \qquad (6.3.6)\end{aligned}$$

Let $\varepsilon > 0$. As $\{g'(u_n);\ g' \in \mathcal{H},\ n \in \mathbb{N}\}$ is uniformly integrable by hypothesis, there is an integer $K > 0$ such that $\sup_{g'\in\mathcal{H}} \sup_n \int_{\{g'(u_n)>K\}} g'(u_n)\,d\mathrm{P} < \varepsilon$. Returning to the estimate (6.3.6) we get

$$\int_{\{g(.)>K+1\}} g(s)\,d\sigma(s) < \varepsilon,$$

thus proving the claim.

Now we can finish the proof as follows. For every $A \in \mathcal{S}'$, we have

$$\begin{aligned}&\left|\int_\Omega g(u_n(\omega))\,d\mathrm{P}(\omega) - \int_\Omega \Big[\int_{\mathbb{S}} g(s)\,d\sigma^\infty_\omega(s)\Big]\,d\mathrm{P}(\omega)\right| \\ &\quad= \left|\int_{\mathbb{S}} g\,d\mu_n - \int_{\mathbb{S}} g\,d\sigma\right| \\ &\quad\leq \left|\int_A g(X_n(\omega'))\,d\mathrm{Q}(\omega') - \int_A g(X_\infty(\omega'))\,d\mathrm{Q}(\omega')\right| \\ &\qquad + \int_{A^c} g(X_n(\omega'))\,d\mathrm{Q}(\omega') + \int_{A^c} g(X_\infty(\omega'))\,d\mathrm{Q}(\omega').\end{aligned}$$

Let $\varepsilon > 0$. Since $(h(X_n))$ is uniformly integrable in $\mathrm{L}^1_{\mathbb{R}}(\Omega', \mathcal{S}', \mathrm{Q})$, and $\mathcal{H}$ is uniformly integrable with respect to $\sigma \in \mathcal{M}^{+,1}(\mathbb{S})$, equivalently $\{h(X_\infty);\ h \in \mathcal{H}\}$ is uniformly integrable in $\mathrm{L}^1_{\mathbb{R}}(\Omega', \mathcal{S}', \mathrm{Q})$, there exists $\eta > 0$ such that $\mathrm{Q}(B) < \eta$ implies

$$\sup_{g\in\mathcal{H}} \sup_{n\in\mathbb{N}} \int_B g(X_n(\omega'))\,d\mathrm{Q}(\omega') < \varepsilon,$$

and

$$\sup_{g\in\mathcal{H}} \int_B g(X_\infty(\omega'))\,d\mathrm{Q}(\omega') < \varepsilon.$$

As $X_n \to X_\infty$ Q-a.e., by Egorov's theorem there exists a Q-measurable set A with $\mathrm{Q}(A^c) < \eta$ such that $X_n \to X_\infty$ uniformly on A. Taking account of the above estimate and the choice of η, we get

$$\sup_{g\in\mathcal{H}} \sup_{n\in\mathbb{N}} \int_{A^c} g(X_n(\omega'))\,d\mathrm{Q}(\omega') < \varepsilon$$

and

$$\sup_{g\in\mathcal{H}}\int_{A^c} g(X_\infty(\omega'))\,d\,\mathrm{Q}(\omega') < \varepsilon.$$

So it remains to check that

$$\sup_{g\in\mathcal{H}}\int_{A} |g(X_n(\omega')) - g(X_\infty(\omega'))|\,d\,\mathrm{Q}(\omega') < \varepsilon.$$

There exists $N \in \mathbb{N}$ such that $n \geq N$ implies

$$d(X_n(\omega'), X_\infty(\omega')) < \varepsilon,$$

for all $\omega' \in A$. It follows that

$$|g(X_n(\omega') - g(X_\infty(\omega')| \leq d(X_n(\omega'), X_\infty(\omega')) < \varepsilon,$$

for all $\omega' \in A$ so that

$$\sup_{g\in\mathcal{H}}\int_{A} |g(X_n(\omega')) - g(X_\infty(\omega'))|\,d\,\mathrm{Q}(\omega') < \varepsilon.$$

The proof is therefore complete. □

There is a direct application of the preceding theorem to Pettis integration.

Proposition 6.3.6 *Suppose that $\mathbb{E}$ is a separable Banach space, (u_n) is a Pettis integrable sequence in the space $\mathrm{P}^1_{\mathbb{E}}(\Omega, \mathcal{S}, \mathrm{P})$ of Pettis integrable $\mathbb{E}$-valued functions such that (u_n) is norm-tight and that $\{\langle x', u_n\rangle;\ \|x'\| \leq 1, n \in \mathbb{N}\}$ is uniformly integrable in $\mathrm{L}^1_{\mathbb{R}}(\Omega, \mathcal{S}, \mathrm{P})$, then there exist a subsequence (v_m) of (u_n) and a Young measure σ^∞ in $\mathcal{Y}^1(\Omega, \mathcal{S}, \mathrm{P}; \mathbb{E})$ such that*

$$\sup_{\|x'\|\leq 1}\left|\int_\Omega \langle x', v_m(\omega)\rangle\,d\,\mathrm{P}(\omega) - \int_\Omega \Big[\int_{\mathbb{E}} \langle x', x\rangle\,d\sigma^\infty_\omega(x)\Big]\,d\,\mathrm{P}(\omega)\right| \to 0$$

when $m \to +\infty$.

Proof. Applying Theorem 6.3.5 by taking $\mathbb{S} = \mathbb{E}$ and $\mathcal{H} = \{x' \in \mathbb{E}^*;\ \|x'\| \leq 1\}$ gives the result. In particular, if Γ is Pettis integrable convex compact-valued multifunction from Ω into $\mathbb{E}$, that is, $\{\delta^*(x', \Gamma(.));\ \|x'\| \leq 1\}$ is uniformly integrable in $\mathrm{L}^1_{\mathbb{R}}(\Omega, \mathcal{S}, \mathrm{P})$, then the set-valued integral

$$\int_\Omega \Gamma(\omega)\,d\,\mathrm{P}(\omega) := \{\int_\Omega u(\omega)\,d\,\mathrm{P}(\omega);\ u \in \mathrm{S}^{\mathrm{Pe}}_\Gamma\}$$

is compact in $\mathbb{E}$, $\mathrm{S}^{\mathrm{Pe}}_\Gamma$ being the set of all Pettis integrable selections of Γ. Remembering that $\mathrm{S}^{\mathrm{Pe}}_\Gamma$ is nonempty, and any sequence (u_n) in $\mathrm{S}^{\mathrm{Pe}}_\Gamma$ is norm tight, we can extract a subsequence (v_m) which converges stably to σ^∞ in $\mathcal{Y}^1(\Omega, \mathcal{S}, \mathrm{P}; \mathbb{E})$ whose barycenter bar (σ^∞_ω) belongs to $\Gamma(\omega)$ and satisfies the required property in the theorem, thus proving the norm-compactness of $\int_\Omega \Gamma(\omega)\,d\,\mathrm{P}(\omega)$. □

Remark 6.3.7 In the present context, Theorem 6.3.5 provides a subtitute of Banach–Dieudonné's theorem (cf. Proposition 6.2.3).

Let us mention a useful fact.

Lemma 6.3.8 *Suppose that $L : \Omega \to \mathcal{L}\mathrm{cw}\mathcal{K}(\mathbb{E})$ is a measurable multifunction, $(u_n)_{n\in\mathbb{N}}$ is a sequence of scalarly integrable $\mathbb{E}$-valued selections of L and $u : \Omega \to \mathbb{E}$ is a scalarly integrable function such that*

$$\lim_{n\to\infty} \int_A \langle x', u_n\rangle \, d\,\mathrm{P} = \int_A \langle x', u\rangle \, d\,\mathrm{P}$$

for every $A \in \mathcal{S}$ and every $x' \in \mathbb{E}^*$, then $u(\omega) \in L(\omega)$-a.e.

Proof. Suppose that the conclusion is not true. Then by [CV77, Lemma III.34] there exist $x' \in \mathbb{E}^*$, $A \in \mathcal{S}$ with $\mathrm{P}(A) > 0$ such that

$$\langle x', u(\omega)\rangle > \delta^*(x', L(\omega)) := \sup\{\langle x', x\rangle;\ x \in L(\omega)\} \tag{6.3.7}$$

for all $\omega \in A$. By integrating we get

$$\int_A \delta^*(x', L) \, d\,\mathrm{P} < \int_A \langle x', u\rangle \, d\,\mathrm{P}\,. \tag{6.3.8}$$

Since $\langle x', u_n\rangle$ converges $\sigma(\mathrm{L}^1, \mathrm{L}^\infty)$ to $\langle x', u\rangle$ and the u_n are integrable Pettis selections of L, we deduce that

$$\int_A \delta^*(x', L) \, d\,\mathrm{P} \geq \lim_{n\to\infty} \int_A \langle x', u_n\rangle \, d\,\mathrm{P} = \int_A \langle x', u\rangle \, d\,\mathrm{P} \tag{6.3.9}$$

which contradicts (6.3.7). □

6.4 Narrow compactness of Young measures via the Dudley embedding theorem

This special section is a continuation of the two preceding ones. It is concerned with stable convergence in Young measures and applications to weak compactness in Bochner integration. Here sophisticated techniques are used. Firstly, thanks to a theorem of Dudley, the set of probability measures on a Polish space embeds in a dual Banach space: There the James (or James–Pryce) Theorem can be used and the Eberlein–Šmulian theorem allows the use of ordinary sequences (with index set $\mathbb{N}$). Secondly Young measures sometimes give a rather weak limit object which can be connected to some other more classical limit objects. Thirdly, some sequences are weakly Cauchy and the foregoing techniques give a candidate to be the limit. A condition in the line of Ülger [Ülg91, DRS93] is often used — as in Propositions

6.4.4–6.4.5 where it takes the form "there exists a subsequence $(\nu_n)_n$ satisfying $\nu_n \in \mathrm{co}\{\theta_m;\, m \geq n\}$ for every n, which is stably convergent".

Let $(\mathbb{S}, d)$ be a complete separable metric space. We could assume that $\mathbb{S}$ is a Polish space, but then we would have to precise that d is a complete metric. Recall that $\mathcal{M}^{+,1}(\mathbb{S})$ is set of probability measures on $\mathbb{S}$. We endow it with the narrow topology $\sigma(\mathcal{M}^{+,1}(\mathbb{S}), \mathrm{C}_b(\mathbb{S}))$ where $\mathrm{C}_b(\mathbb{S})$ is the set of all real-valued bounded continuous functions defined on $\mathbb{S}$. This topology is metrizable: see [Bou69, Proposition 10 page 62], [Par67, Theorem 6.2 page 43]. Recall also that $\mathrm{BL}(\mathbb{S}, d)$ denotes the vector space of real-valued bounded Lipschitz functions defined on $\mathbb{S}$. It is a Banach space with the norm

$$\|f\|_{\mathrm{BL}(\mathbb{S},d)} := \|f\|_\infty + \sup\{\frac{|f(x) - f(y)|}{d(x, y)};\, x \neq y\}.$$

Let $\mathrm{BL}(\mathbb{S}, d)^*$ denote its topological dual. The space $\mathrm{BL}(\mathbb{S}, d)$ has been introduced and studied by Dudley [Dud66, Dud89]. It is a (usually strict) subspace of $\mathrm{C}_b(\mathbb{S})$. For strict inclusion think of $\mathbb{S} = \mathbb{R}$. Moreover the $\|.\|_\infty$-closure of $\mathrm{BL}(\mathbb{S}, d)$ in $\mathrm{C}_b(\mathbb{S})$ is the set of all bounded uniformly continuous functions: [Dud66, Lemma 8 page 255], on $\mathbb{S}$. There is a natural embedding $\mathcal{M}^+(\mathbb{S}) \to \mathrm{BL}(\mathbb{S}, d)^*$ defined by $\nu \mapsto [f \mapsto \int_{\mathbb{S}} f\, d\nu]$. We will consider $\mathcal{M}^{+,1}(\mathbb{S})$ as a subset of $\mathrm{BL}(\mathbb{S}, d)^*$. The narrow topology on $\mathcal{M}^{+,1}(\mathbb{S})$ coincides with the topology defined by the dual norm on $\mathrm{BL}(\mathbb{S}, d)^*$: see [Dud66, Theorem 6 page 258 and Theorem 8 page 259], [Dud89, Theorem 11.3.3 page 310]. This is mainly due to the fact that bounded sets in $\mathrm{BL}(\mathbb{S}, d)$ are equi-continuous sets (of functions). Moreover one can easily check that $\mathcal{M}^{+,1}(\mathbb{S})$ is a convex subset of $\mathrm{BL}(\mathbb{S}, d)^*$ contained in the unit sphere (i.e. elements of norm 1) of $\mathrm{BL}(\mathbb{S}, d)^*$. But there is more:

1) $\mathcal{M}^{+,1}(\mathbb{S})$ is a closed subset in the Banach space $\mathrm{BL}(\mathbb{S}, d)^*$. This relies on the fact that any sequence in $\mathcal{M}^{+,1}(\mathbb{S})$ which is Cauchy in the $\|.\|_{\mathrm{BL}(\mathbb{S},d)^*}$-norm is narrowly convergent (see [Dud66, Theorem 9 page 260], the main point is that the sequence is tight; see also the proof in [Bou83, pages 191–192]).

2) We have $\nu_n \to \nu_\infty$ in the weak* topology of $\mathrm{BL}(\mathbb{S}, d)^*$, that is, $[\forall f \in \mathrm{BL}(\mathbb{S}, d)$, $\int_{\mathbb{S}} f\, d\nu_n \to \int_{\mathbb{S}} f\, d\nu_\infty]$, iff $\nu_n \to \nu_\infty$ narrowly. This is a particular case of Corollary 2.1.11 (see also Lemma 6.4.2 given below) and rather classical: see [Bou83, pages 189–190] and, for related results, [Par67, Theorem 6.6 page 47].

Besides these properties, R.D. Bourgin [Bou83, Theorem 6.3.8 page 193] proves that the subset $\mathcal{M}^{+,1}(\mathbb{S})$ of the Banach dual space $(\mathrm{BL}(\mathbb{S}, d)^*, \|.\|_{\mathrm{BL}(\mathbb{S},d)^*})$ has the Radon–Nikodým property, which means that any $\mathrm{BL}(\mathbb{S}, d)^*$-valued measure m defined on $\mathcal{S}$ which is absolutely continuous with respect to P and whose *average range*

$$AR(m) := \{m(A)/\,\mathrm{P}(A);\ A \in \mathcal{S},\ \mathrm{P}(A) > 0\}$$

is contained in $\mathcal{M}^{+,1}(\mathbb{S})$ and admits a density f which belongs to the space of Bochner P-integrable functions $\mathrm{L}^1(\Omega, \mathcal{S}, \mathrm{P}; \mathrm{BL}(\mathbb{S}, d)^*)$. Necessarily $f(\omega) \in \mathcal{M}^{+,1}(\mathbb{S})$ P-a.e.

Proposition 6.4.5 *Let $\mathcal{H}$ be a Mazur–compact subset of $\mathcal{Y}^1(\Omega, \mathcal{S}, \mathrm{P}; \mathbb{S})$, that is, for any sequence $(\lambda^n)_n$ in $\mathcal{H}$, there exists a sequence $(\nu^n)_n$ with $\nu^n \in \mathrm{co}\{\lambda^m;\, m \geq n\}$ such that, for* P*-almost every ω, $(\nu^n_\omega)_n$ converges narrowly in $\mathcal{M}^{+,1}(\mathbb{S})$. Then $\mathcal{H}$ is sequentially relatively $\sigma(\mathcal{Y}^1(\Omega, \mathcal{S}, \mathrm{P}; \mathbb{S}), \mathrm{L}^\infty(\Omega, \mathcal{S}, \mathrm{P}) \otimes \mathrm{BL}(\mathbb{S}, d)^{**})$-compact i.e. for any sequence $(\lambda^n)_n$ in $\mathcal{H}$, there exists a subsequence $(\lambda^{\alpha(n)})_n$ and $\lambda^\infty \in \mathcal{Y}^1(\Omega, \mathcal{S}, \mathrm{P}; \mathbb{S})$ such that*

$$\forall h \in \mathrm{L}^\infty(\Omega, \mathcal{S}, \mathrm{P}) \otimes \mathrm{BL}(\mathbb{S}, d)^{**} \quad \int_\Omega \langle h, \lambda^{\alpha(n)} \rangle \, d\mathrm{P} \to \int_\Omega \langle h, \lambda^\infty \rangle \, d\mathrm{P}. \tag{6.4.2}$$

Proof. We proceed as in Theorem 6.3.2. Let $\mathbb{E}$ denote the Banach space $\mathrm{BL}(\mathbb{S}, d)^*$.

1) For $A \in \mathcal{S}$, let us define $\mathcal{H}_A := \{\int_A \lambda_\omega \, d\mathrm{P}(\omega);\, \lambda \in \mathcal{H}\} \subset \mathbb{E}$. We are going to prove, using James' theorem, that $\overline{\mathrm{co}}(\mathcal{H}_A)$ is $\sigma(\mathbb{E}, \mathbb{E}^*)$ compact. We have to check that any $\zeta \in \mathbb{E}^*$ attains its supremum on $\overline{\mathrm{co}}(\mathcal{H}_A)$. Let $(\lambda^n)_n$ be a sequence in $\mathcal{H}$ such that $(\theta^n)_n$ with $\theta^n := \int_A \lambda^n \, d\mathrm{P}$ is a maximizing sequence, that is, $\langle \zeta, \theta^n \rangle \nearrow \delta^*(\zeta, \overline{\mathrm{co}}(\mathcal{H}_A))$. From the hypothesis there exists $\nu^n \in \mathrm{co}\{\lambda^m;\, m \geq n\}$ which P-a.e. converges narrowly in $\mathcal{M}^{+,1}(\mathbb{S})$. The limit λ^∞_ω is scalarly measurable (i.e for every $f \in \mathrm{C}_b(\mathbb{S})$, $\omega \mapsto \langle f, \lambda^\infty_\omega \rangle$ is measurable), hence is a Young measure. Each ν^n has the form: $\nu^n = \sum_{i=0}^{k_n} \alpha^n_i \lambda^{n+i}$ with $\alpha^n_i \geq 0$, $\sum_i \alpha^n_i = 1$. Since $\langle \zeta, \nu^n_\omega \rangle$ converges a.e. to $\langle \zeta, \lambda^\infty_\omega \rangle$, by Lebesgue's theorem

$$\langle \zeta, \int_A \lambda^\infty \, d\mathrm{P} \rangle = \lim_{n \to \infty} \langle \zeta, \int_A \nu^n \, d\mathrm{P} \rangle = \lim_{n \to \infty} \sum_{i=0}^{k_n} \alpha^n_i \langle \zeta, \int_A \lambda^{n+i} \, d\mathrm{P} \rangle \tag{6.4.3}$$

and $\langle \zeta, \int_A \lambda^\infty \, d\mathrm{P} \rangle = \delta^*(\zeta, \overline{\mathrm{co}}(\mathcal{H}_A))$.

2) Let $(\lambda^n)_n$ be a sequence in $\mathcal{H}$. Let $\mathcal{S}_1$ be the sub-σ-algebra generated by the maps $\omega \mapsto \lambda^n_\omega$, and $\mathcal{A}$ a countable algebra which generates $\mathcal{S}_1$. Using the relative compactness of $\mathcal{H}_A$ proved above, the Eberlein–Šmulian theorem and the diagonal process, we can prove the existence of a subsequence $(\lambda^{\alpha(n)})_n$ and of elements $\theta_A \in \mathbb{E}$ such that, for all $A \in \mathcal{A}$ and for all $\zeta \in \mathbb{E}^*$, $\langle \zeta, \int_A \lambda^{\alpha(n)} \rangle \, d\mathrm{P} \longrightarrow \langle \zeta, \theta_A \rangle$. As in 1) let $(\nu^n)_n$ be a sequence with $\nu^n \in \mathrm{co}\{\lambda^{\alpha(m)};\, m \geq n\}$ such that, for P-almost every ω, $(\nu^n_\omega)_n$ converges narrowly to λ^∞_ω. Then necessarily (see (6.4.3)) $\langle \zeta, \theta_A \rangle = \int_A \langle \zeta, \lambda^\infty_\omega \rangle \, d\mathrm{P}(\omega)$. By equi-continuity the convergence $\int_A \langle \zeta, \lambda^{\alpha(n)} \rangle \, d\mathrm{P} \to \int_A \langle \zeta, \lambda^\infty \rangle \, d\mathrm{P}$ remains valid for $A \in \mathcal{S}_1$. Thus (6.4.2) holds in the case when $h(\omega, x) = \mathbb{1}_A(\omega)\zeta$. By linear combinations and taking limits, we get (6.4.3) for $\psi \in \mathrm{L}^\infty(\Omega, \mathcal{S}_1, \mathrm{P})$. The extension to the case when $\psi \in \mathrm{L}^\infty(\Omega, \mathcal{S}, \mathrm{P})$ is straightforward (use the conditional expectation operator $E^{\mathcal{S}_1}$). □

The following is an application of Proposition 6.4.1.

Proposition 6.4.6 *Suppose that (λ^n) is a sequence in $\mathcal{Y}^1(\Omega, \mathcal{S}, \mathrm{P}; \mathbb{S})$ satisfying: For any subsequence (ν_n) of (λ^n) there is a sequence $\widetilde{\nu}^n$ with $\widetilde{\nu}^n \in \mathrm{co}\{\nu^m;\, m \geq n\}$ such that, for every $A \in \mathcal{S}$, the sequence $(\int_A \widetilde{\nu}^n \, d\mathrm{P})$ is narrowly convergent in $\mathcal{M}^+(\mathbb{S})$, then there are a subsequence $(\lambda^{n'})$ and $\lambda^\infty \in \mathcal{Y}^1(\Omega, \mathcal{S}, \mathrm{P}; \mathbb{S})$ such that*

$$\lim_n \langle u \otimes h, \lambda^{n'} \rangle = \langle u \otimes h, \lambda^\infty \rangle$$

for all $(u, h) \in \mathrm{L}^\infty_{\mathbb{R}}(\Omega, \mathcal{S}, \mathrm{P}) \times \mathrm{C}_b(\mathbb{S})$.

Proof. *Step 1.* Let (φ_p) be a sequence in $\mathrm{C}_b(\mathbb{S})$ which separates the points of $\mathcal{M}^{+,1}(\mathbb{S})$. It is obvious that for each p the sequence $(\langle \varphi_p, \lambda^n \rangle)$ defined by

$$\langle \varphi_p, \lambda^n \rangle(\omega) = \langle \varphi_p, \lambda^n_\omega \rangle,$$

for all $\omega \in \Omega$, is bounded in $\mathrm{L}^\infty_{\mathbb{R}}(\Omega, \mathcal{S}, \mathrm{P})$. So $(\langle \varphi_p, \lambda^n \rangle)$ is relatively $\sigma(\mathrm{L}^\infty, \mathrm{L}^1)$ compact. As the injection $i : \mathrm{L}^\infty \to \mathrm{L}^1$ is weak*-weak continuous, $(\langle \varphi_p, \lambda^n \rangle)$ is relatively weakly compact in $\mathrm{L}^1_{\mathbb{R}}(\Omega, \mathcal{S}, \mathrm{P})$. So $(\langle \varphi_p, \lambda^n \rangle)$ is relatively sequentially weakly compact in $\mathrm{L}^1_{\mathbb{R}}(\Omega, \mathcal{S}, \mathrm{P})$ in view of the Eberlein–Šmulian theorem. Using an appropriate diagonal procedure provides a sequence (r_p) of real-valued bounded measurable functions and a subsequence $(\lambda^{n'})$ of (λ^n) such that

$$\forall p \ \ \forall u \in \mathrm{L}^1_{\mathbb{R}}(\Omega, \mathcal{S}, \mathrm{P}) \quad \lim_{n\to\infty} \langle u \otimes \varphi_p, \lambda^{n'} \rangle = \langle u, r_p \rangle. \tag{6.4.4}$$

Step 2. Let $u \in \mathrm{L}^\infty_{\mathbb{R}}(\Omega, \mathcal{S}, \mathrm{P})$ and $h \in \mathrm{C}_b(\mathbb{S})$ be fixed. Choose a subsequence $(\nu^{n'})$ of $(\lambda^{n'})$ such that

$$\limsup_{n\to\infty} \langle u \otimes h, \lambda^{n'} \rangle = \lim_{n\to\infty} \langle u \otimes h, \nu^{n'} \rangle. \tag{6.4.5}$$

By our assumption, there is a sequence $\widetilde{\nu}^{n'}$ with $\widetilde{\nu}^{n'} \in \mathrm{co}\{\nu^{m'};\, m \geq n\}$ such that, for each $A \in \mathcal{S}$, $(\int_A \widetilde{\nu}^{n'} \, d\mathrm{P})$ narrowly converges. By Proposition 6.4.1, there is $\nu^{\infty'} \in \mathcal{Y}^1(\Omega, \mathcal{S}, \mathrm{P}; \mathbb{S})$ such that, for each $A \in \mathcal{S}$, $(\int_A \widetilde{\nu}^{n'} \, d\mathrm{P})$ narrowly converges to $\int_A \nu^{\infty'} \, d\mathrm{P}$. By (6.4.5) and the remark of Proposition 6.4.1, it follows that

$$\limsup_{n\to\infty} \langle u \otimes h, \lambda^{n'} \rangle = \lim_{n\to\infty} \langle u \otimes h, \nu^{n'} \rangle = \langle u \otimes h, \nu^{\infty'} \rangle. \tag{6.4.6}$$

Coming back to (6.4.4) we get

$$\lim_{n\to\infty} \langle v \otimes \varphi_p, \lambda^{n'} \rangle = \langle v, r_p \rangle = \langle v \otimes \varphi_p, \nu^{\infty'} \rangle \tag{6.4.7}$$

for all $v \in \mathrm{L}^\infty_{\mathbb{R}}(\Omega, \mathcal{S}, \mathrm{P})$ and for all p. Similarly, we find $\mu^{\infty'} \in \mathcal{Y}^1(\Omega, \mathcal{S}, \mathrm{P}; \mathbb{S})$ such that

$$\liminf_{n\to\infty} \langle u \otimes h, \lambda^{n'} \rangle = \langle u \otimes h, \mu^{\infty'} \rangle \tag{6.4.8}$$

and

$$\lim_{n\to\infty} \langle v \otimes \varphi_p, \lambda^{n'} \rangle = \langle v, r_p \rangle = \langle v \otimes \varphi_p, \mu^{\infty'} \rangle \tag{6.4.9}$$

for all $v \in L^{\infty}_{\mathbb{R}}(\Omega, \mathcal{S}, P)$ and for all p. By (6.4.7) and (6.4.9) we get

$$\langle v \otimes \varphi_p, \nu^{\infty'} \rangle = \langle v \otimes \varphi_p, \mu^{\infty'} \rangle$$

for all $v \in L^{\infty}_{\mathbb{R}}(\Omega, \mathcal{S}, P)$ and for all p. So we can conclude that $\nu^{\infty'} = \mu^{\infty'}$ a.e.

Step 3. Finally applying the results obtained in the preceding steps to any $(u', h') \in L^{\infty}_{\mathbb{R}}(\Omega, \mathcal{S}, P) \times C_b(\mathbb{S})$ provides $\sigma^{\infty'} \in \mathcal{Y}^1(\Omega, \mathcal{S}, P; \mathbb{S})$ such that

$$\lim_{n \to \infty} \langle u' \otimes h', \lambda^{n'} \rangle = \langle u' \otimes h', \mu^{\infty'} \rangle$$

and

$$\forall A \in \mathcal{S} \quad \forall p \quad \int_A \langle \varphi_p, \sigma^{\infty'}(\omega) \rangle \, d\,P(\omega) = \int_A r_p \, d\,P\,.$$

So $\sigma^{\infty'} = \nu^{\infty'}$ a.e., thus completing the proof. □

Comments Propositions 6.4.5–6.4.6 are the analogs for Young measures of the Ülger–Diestel–Ruess–Schachermayer characterization of weak compactness in $L^1_{\mathbb{X}}$ where $\mathbb{X}$ is a Banach space [Ülg91, DRS93]. Namely these authors proved that a bounded uniformly integrable subset $\mathcal{H}$ of $L^1_{\mathbb{X}}$ is relatively compact iff

(*) given any sequence $(u_n)_n$ in $\mathcal{H}$, there are $v_n \in \text{co}\{u_m;\, m \geq n\}$ such that the sequence $(v_n(\omega))_n$ is weakly convergent in $\mathbb{X}$ for almost all $\omega \in \Omega$.

In this spirit, combining Propositions 6.4.5–6.4.6 and the techniques developed above leads to new compactness results in the space $L^1_{\mathbb{E}}$ (where $\mathbb{E}$ is a separable Banach space) that we present in the next paragraph.

Before going further let us mention some significant applications of the preceding results.

Proposition 6.4.7 *Suppose that $\mathbb{E}$ is a Banach space with strongly separable dual and $\mathbb{S}$ is a closed convex bounded subset of $\mathbb{E}$. Let $T : \mathcal{M}^{+,1}(\mathbb{S}) \to \mathbb{E}$ denote the map $\mu \mapsto \text{bar}\,(\mu)$ and $\widetilde{T}$ its natural extension $\widetilde{T} : \mathcal{Y}^1(\Omega, P; \mathbb{S}) \to L^1_{\mathbb{E}}(\Omega, \mathcal{S}, P)$. Let $(\lambda^n)_n$ be a sequence in $\mathcal{Y}^1(\Omega, P; \mathbb{S})$ such that for each $A \in \mathcal{S}$, $\int_A \lambda^n_\omega \, d\,P(\omega)$ is narrowly convergent in $\mathcal{M}^+(\mathbb{S})$. Then $\widetilde{T}$ transforms $(\lambda^n)_n$ into a weakly convergent sequence in $L^1_{\mathbb{E}}(\Omega, \mathcal{S}, P)$.*

Remarks The existence of $\text{bar}\,(\mu)$ $(\mu \in \mathcal{M}^{+,1}(\mathbb{S}))$ is ensured by [Bou83, Lemma 6.2.2 page 178] because $\mathbb{S}$ is a closed convex bounded subset of $\mathbb{E}$. Obviously $\widetilde{T}$ operates as $[\widetilde{T}(\lambda_{(.)})](\omega) = T(\lambda_\omega) = \text{bar}\,(\lambda_\omega)$.

Proof. 1) By Proposition 6.4.1 there exists $\lambda^\infty \in \mathcal{Y}^1(\Omega, \mathcal{S}, P; \mathbb{S})$ such that for each $A \in \mathcal{S}$, $\int_A \lambda^n \, d\,P$ narrowly converges to $\int_A \lambda^\infty \, d\,P$.

2) Let $A \in \mathcal{S}$ with $\mathrm{P}(A) > 0$ and $x' \in \mathbb{E}^*$. Then

$$\begin{aligned}\left\langle x', \int_A [\widetilde{T}(\lambda^n_{(.)})](\omega)\, d\,\mathrm{P}(\omega)\right\rangle &= \left\langle x', \int_A \mathrm{bar}\,(\lambda^n_\omega)\; d\,\mathrm{P}(\omega)\right\rangle \\ &= \int_A \langle x', \mathrm{bar}\,(\lambda^n_\omega)\rangle\, d\,\mathrm{P}(\omega) \\ &= \int_A \left[\int_{\mathbb{S}} x'_{|\mathbb{S}} d\lambda^n_\omega\right] d\,\mathrm{P}(\omega) \\ &= \mathrm{P}(A)\left\langle x'_{|\mathbb{S}}, \frac{1}{\mathrm{P}(A)}\int_A \lambda^n\, d\,\mathrm{P}\right\rangle\end{aligned}$$

and, since $x'_{|\mathbb{S}}$ belongs to $\mathrm{C}_b(\mathbb{S})$ (even to $\mathrm{BL}(\mathbb{S}, d)$), we have got

$$\langle \psi, \widetilde{T}(\lambda^n)\rangle \longrightarrow \langle \psi, \widetilde{T}(\lambda^\infty)\rangle \tag{6.4.10}$$

in the case when $\psi \in \mathrm{L}^\infty_{\mathbb{E}^*}(\Omega, \mathcal{S}, \mathrm{P})$ has the form $\psi(\omega) = \mathbf{1}_A(\omega)\, x'$.

3) By linearity (6.4.10) extends to step functions ψ. Then if $\psi \in \mathrm{L}^\infty_{\mathbb{E}^*}(\Omega, \mathcal{S}, \mathrm{P})$ is countably valued with $\psi(\omega) = x'_k$ on disjoint sets A_k,

$$\langle \psi, \widetilde{T}(\lambda^n)\rangle = \sum_{k=0}^{\infty} \int_{A_k} \langle x'_k, \mathrm{bar}\,(\lambda^n_\omega)\rangle\, d\,\mathrm{P}(\omega)$$

and the convergence (6.4.10) still holds. Since $\mathbb{E}^*$ is separable, any element ψ of $\mathrm{L}^\infty_{\mathbb{E}^*}(\Omega, \mathcal{S}, \mathrm{P})$ can be uniformly approximated by countably valued ψ_p, and the preceding equality holds for any $\psi \in \mathrm{L}^\infty_{\mathbb{E}^*}(\Omega, \mathcal{S}, \mathrm{P})$. The proof is therefore complete.

□

Now is a variant of the preceding result. Let $T : \mathrm{BL}(\mathbb{S}, d)^* \to \mathbb{E}$ be a bounded linear operator from the Banach space $\mathrm{BL}(\mathbb{S}, d)^*$ into a Banach space $\mathbb{E}$. We denote by $\widehat{T} : \mathrm{L}^1_{\mathrm{BL}(\mathbb{S},d)^*}(\Omega, \mathcal{S}, \mathrm{P}) \to \mathrm{L}^1_{\mathbb{E}}(\Omega, \mathcal{S}, \mathrm{P})$ the natural linear extension of T to a bounded linear operator from $\mathrm{L}^1_{\mathrm{BL}(\mathbb{S},d)^*}$ to $\mathrm{L}^1_{\mathbb{E}}$.

Proposition 6.4.8 *Suppose that $\mathbb{E}$ is a Banach space with strongly separable dual, $(\lambda^n)_n$ is a sequence in $\mathcal{Y}^1(\Omega, \mathrm{P}; \mathbb{S})$ such that $\forall A \in \mathcal{S}$, $\int_A \lambda^n_\omega\, d\,\mathrm{P}(\omega)$ is narrowly convergent in $\mathcal{M}^+(\mathbb{S})$. Then the mapping $\widehat{T}$ transforms $(\lambda^n)_n$ into a weakly convergent sequence in $\mathrm{L}^1_{\mathbb{E}}(\Omega, \mathcal{S}, \mathrm{P})$.*

Proof. By Proposition 6.4.1 and Dudley's homeomorphism theorem there is $\lambda^\infty \in \mathcal{Y}^1(\Omega, \mathcal{S}, \mathrm{P}; \mathbb{S})$ such that

$$\forall A \in \mathcal{S} \quad \int_A \lambda^n\, d\,\mathrm{P} \to \int_A \lambda^\infty\, d\,\mathrm{P}$$

in the Banach space $\mathrm{BL}(\mathbb{S},d)^*$. It follows that $\int_A \widehat{T}\lambda^n \, d\mathrm{P} \to \int_A \widehat{T}\lambda^\infty \, d\mathrm{P}$ in the Banach $\mathbb{E}$ for every $A \in \mathcal{S}$. The conclusion that $(\widehat{T}\lambda^n)_n$ converges $\sigma(\mathrm{L}^1_{\mathbb{E}}, \mathrm{L}^\infty_{\mathbb{E}^*})$ to $\widehat{T}\lambda^\infty$ is obtained as in the proof of Proposition 6.4.7. □

The following is an application of the preceding result to best approximation in $\mathrm{L}^1_{\mathbb{E}}$. Let $\mathcal{B}$ be a complete sub-σ-algebra of $\mathcal{S}$ and let $\mathcal{Y}^1(\Omega,\mathcal{B},\mathrm{P};\mathbb{S})$ be the set of Young measures defined over the complete probability space $(\Omega,\mathcal{B},\mathrm{P})$.

Proposition 6.4.9 *Suppose that $\mathbb{E}$ is a Banach space with strongly separable dual, $\mathcal{H}$ is a $\sigma(\mathcal{H}, \mathrm{L}^\infty(\Omega,\mathcal{B},\mathrm{P}) \otimes \mathrm{BL}(\mathbb{S},d))$ closed subset of $\mathcal{Y}^1(\Omega,\mathcal{B},\mathrm{P};\mathbb{S})$ such that for every sequence $(\lambda^n)_n$ in $\mathcal{H}$ there exists a sequence $(\nu^n)_n$ in $\mathcal{Y}^1(\Omega,\mathcal{B},\mathrm{P};\mathbb{S})$ with $\nu^n \in \mathrm{co}\{\lambda^m;\, m \geq n\}$ such that for P-almost every ω, $(\nu^n_\omega)_n$ narrowly converges in $\mathcal{M}^{+,1}(\mathbb{S})$. Let $f \in \mathrm{L}^1_{\mathbb{E}}(\Omega,\mathcal{S},\mathrm{P})$. Then there exists $\bar\lambda \in \mathcal{H}$ such that*

$$\inf_{\lambda\in\mathcal{H}} \int_\Omega \|f - \widehat{T}\lambda\| \, d\mathrm{P} = \int_\Omega \|f - \widehat{T}\bar\lambda\| \, d\mathrm{P}.$$

Proof. Let $(\lambda^n)_n$ be a minimizing sequence in $\mathcal{H}$, that is,

$$\lim_{n\to\infty} \int_\Omega \|f - \widehat{T}\lambda^n\| \, d\mathrm{P} = \inf_{\lambda\in\mathcal{H}} \int_\Omega \|f - \widehat{T}\lambda\| \, d\mathrm{P}.$$

Using Proposition 6.4.5 and the arguments of Proposition 6.4.8 provides a subsequence still denoted by $(\lambda^n)_n$ such that $(\widehat{T}\lambda^n)$ converges $\sigma(\mathrm{L}^1, \mathrm{L}^\infty)$ to $\widehat{T}\bar\lambda$ in $\mathrm{L}^1_{\mathbb{E}}(\Omega,\mathcal{B},\mathrm{P})$ with $\bar\lambda \in \mathcal{H}$ by hypothesis. Then it is easily proved (using the operator $E^{\mathcal{B}}$) that $(\widehat{T}\lambda^n)$ converges weakly to $\widehat{T}\bar\lambda$ in $\mathrm{L}^1_{\mathbb{E}}(\Omega,\mathcal{S},\mathrm{P})$. It follows that

$$\liminf_{n\to\infty} \int_\Omega \|f - \widehat{T}\lambda^n\| \, d\mathrm{P} \geq \int_\Omega \|f - \widehat{T}\bar\lambda\| \, d\mathrm{P}$$

and the proof is complete. □

To end this section let us mention an application of Propositions 6.4.5–6.4.6 to Komlós convergence (see also Balder [Bal91]).

Proposition 6.4.10 *With the notations and hypotheses of Proposition 6.4.5 (resp. 6.4.6) there are a subsequence $(\lambda^{\alpha(n)})_n$ of $(\lambda^n)_n$ and $\lambda^\infty \in \mathcal{Y}^1(\Omega,\mathrm{P};\mathbb{S})$ such that for each further subsequence $(\lambda^{\beta(n)})_n$, the following holds:*

$$\frac{1}{n}\sum_{j=1}^{n} \lambda^{\beta(j)}_\omega \xrightarrow{\text{stably}} \lambda^\infty_\omega$$

for almost every $\omega \in \Omega$ (the negligible set depends on the subsequence).

Proof. 1) By Proposition 6.4.5 there exist a subsequence $(\lambda^{\alpha(n)})_n$ of $(\lambda^n)_n$ and $\lambda^\infty \in \mathcal{Y}^1(\Omega,\mathrm{P};\mathbb{S})$ such that, for every $h \in \mathrm{L}^\infty(\Omega,\mathcal{S},\mathrm{P}) \otimes \mathrm{BL}(\mathbb{S},d)$, we have $\lim_{n\to\infty}\langle \lambda^{\alpha(n)}, h\rangle = \langle \lambda^\infty, h\rangle$. By Lemma 6.4.2 this implies

$$\forall h \in \mathrm{L}^\infty(\Omega,\mathcal{S},\mathrm{P}) \otimes \mathrm{C}_b(\mathbb{S}) \quad \lim_{n\to\infty}\langle \lambda^{\alpha(n)}, h\rangle = \langle \lambda^\infty, h\rangle. \tag{6.4.11}$$

2) Let $(\varphi_p)_p$ be a sequence of bounded continuous functions such that for any sequence $(\theta_n)_n$ in $\mathcal{M}^{+,1}(\mathbb{S})$, $[\forall p, \int_{\mathbb{S}} \varphi_p \, d\theta_n \to \int_{\mathbb{S}} \varphi_p \, d\theta_\infty]$ is equivalent to the narrow convergence. Such a sequence $(\varphi_p)_p$ does exist: see e.g. [Par67, Theorem 6.6 page 47].

3) Now the Komlós theorem [Kom67] and an appropriate diagonal procedure provide a subsequence still denoted by $(\lambda^{\alpha(n)})_n$ and functions $\zeta_p \in \mathrm{L}^1_{\mathbb{R}}$ such that

$$\lim_{n\to\infty} \frac{1}{n} \sum_{j=1}^{n} \langle \lambda_\omega^{\beta(j)}, \varphi_p \rangle \overset{\text{a.e.}}{=} \zeta_p(\omega) \tag{6.4.12}$$

for each subsequence $(\lambda^{\beta(n)})_n$. By (6.4.11) it follows that $\frac{1}{n}\sum_{j=1}^{n} \langle \lambda^{\beta(j)}, \varphi_p \rangle$ converges $\sigma(\mathrm{L}^1, \mathrm{L}^\infty)$ to $\langle \lambda^\infty, \varphi_p \rangle$. From (6.4.12) we deduce that $\zeta_p(\omega) \overset{\text{a.e.}}{=} \langle \lambda_\omega^\infty, \varphi_p \rangle$ for all p.

Under the hypotheses of Proposition 6.4.6 there exist a subsequence $(\lambda^{\alpha(n)})_n$ of $(\lambda^n)_n$ and $\lambda^\infty \in \mathcal{Y}^1(\Omega, \mathrm{P}; \mathbb{S})$ such that $\lim_{n\to\infty} \langle \lambda^{\alpha(n)}, h \rangle = \langle \lambda^\infty, h \rangle$ for every $h \in \mathrm{L}^1(\Omega, \mathcal{S}, \mathrm{P}) \otimes \mathrm{C}_b(\mathbb{S})$, so the proof follows as in 2) and 3). □

Some more weak compactness in $\mathrm{L}^1_{\mathbb{E}}(\mathrm{P})$

Lemma 6.4.11 *Let $\mathbb{E}$ be a separable Banach space, $(u_n)_n$ a bounded sequence in $\mathrm{L}^1_{\mathbb{E}}$ which satisfies:*

(i) $\forall x' \in \mathbb{E}^*$, $\{\langle x', u_n(.)\rangle;\ n \in \mathbb{N}\}$ is uniformly integrable.

(ii) For any $A \in \mathcal{S}$, $\mathcal{H}_A := \{\int_A u_n \, d\mathrm{P};\ n \in \mathbb{N}\}$ is relatively weakly compact.

(iii) For any subsequence $(u'_n)_n$ of $(u_n)_n$, there exists $v_n \in \mathrm{co}\{u'_m;\ m \geq n\}$ such that $\forall A \in \mathcal{S}$, the sequence $(\int_A \delta_{v_n(.)} \, d\mathrm{P})_n$ is narrowly convergent.

Then there exists a subsequence $(u_{n_k})_k$ and $u_\infty \in \mathrm{L}^1_{\mathbb{E}}$ such that

$$\forall x' \in \mathbb{E}^* \quad \forall A \in \mathcal{S} \quad \lim_{k\to\infty} \int_A \langle x', u_{n_k}\rangle \, d\mathrm{P} = \int_A \langle x', u_\infty \rangle \, d\mathrm{P}.$$

Remark 6.4.12 1) Hypothesis *(ii)* appears in Diestel–Uhl [DU77, Theorem 1 page 101] and is exploited in [CC85, Theorem 4.1 page 354].

2) In the same line, [Cas96, Lemma 2.5] treats multifunctions but with a variant of *(iii)*, which for single-valued functions writes as

(iii′) for any subsequence $(u'_n)_n$ of $(u_n)_n$, there exists $v_n \in \mathrm{co}\{u'_m;\ m \geq n\}$ and a measurable function u_∞ satisfying $u_n(\omega) \to u_\infty(\omega)$ P-a.e.

3) The measure $\int_A \delta_{v_n(.)} \, d\mathrm{P}$ is also the image of $\mathrm{P}\lfloor_A$ by v_n. But the expression of *(iii)* is close to the statement of Propositions 6.4.1–6.4.6 ($\delta_{v_n(.)}$ is the Young measure associated with v_n) which is used in the proof.

Proof. 1) By Lemma 6.2.9 there exists a subsequence $(u_{n_k})_k =: (\tilde{u}_k)_k$ such that for any $x' \in \mathbb{E}^*$ and $A \in \mathcal{S}$, $\lim_{k\to\infty} \int_A \langle x', \tilde{u}_k\rangle \, d\mathrm{P} =: \ell_{x',A}$ exists in $\mathbb{R}$.

2) Let $(\tilde{v}_k)_k$ be a sequence such that $\tilde{v}_k \in \mathrm{co}\{\tilde{u}_m;\, m \geq k\}$ and which satisfies (iii). Let ν^k be the Young measure associated with $\tilde{v}_k$, that is, $\nu^k_\omega = \delta_{\tilde{v}_k(\omega)}$. By Proposition 6.4.1 there exists $\nu \in \mathcal{Y}^1(\Omega, \mathcal{S}, \mathrm{P}; \mathbb{S})$ such that $\forall A$, $\lim_k \int_A \nu^k \, d\mathrm{P} = \int_A \nu \, d\mathrm{P}$. By a well-known lower semicontinuity result (Proposition 2.1.12 or [Bal84a, Val90b, Val94])

$$\begin{aligned}\int_\Omega \Big[\int_{\mathbb{E}} \|x\| \, d\nu_\omega(x)\Big] \, d\mathrm{P}(\omega) &\leq \liminf_{k\to\infty} \int_\Omega \Big[\int_{\mathbb{E}} \|x\| \, d\nu^k_\omega(x)\Big] \, d\mathrm{P}(\omega) \\ &= \liminf_{k\to\infty} \int_\Omega \|\tilde{v}_k(\omega)\| \, d\mathrm{P}(\omega) \\ &\leq \sup_{n\in\mathbb{N}} \|u_n\|_{\mathrm{L}^1} < +\infty\,.\end{aligned}$$

Hence ν_ω has a barycenter: $\mathrm{bar}\,(\nu_\omega) =: u_\infty(\omega)$ and $|u_\infty(.)|$ is integrable. Thanks to hypothesis (i) (see e.g. [Bal95, Val94]),

$$\int_A \langle x', \tilde{v}_k\rangle \, d\mathrm{P} \longrightarrow \int_A \langle x', \mathrm{bar}\,(\nu_\omega)\rangle \, d\mathrm{P}(\omega) = \int_A \langle x', u_\infty\rangle \, d\mathrm{P}\,.$$

3) Now comes the conclusion: $\lim_{k\to\infty} \int_A \langle x', \tilde{v}_k\rangle \, d\mathrm{P}$ is the limit of convex combinations, so necessarily it equals $\ell_{x',A} = \lim_{k\to\infty} \int_A \langle x', u_{n_k}\rangle \, d\mathrm{P}$. □

Proposition 6.4.13 *Let $\mathbb{E}$ be a Banach space whose dual $\mathbb{E}^*$ is strongly separable and $\mathcal{H}$ a bounded subset of $\mathrm{L}^1_{\mathbb{E}}$. A necessary and sufficient condition for $\mathcal{H}$ to be relatively weakly compact is the following conditions:*

(i) *$\mathcal{H}$ is* UI.

(ii) *For any $A \in \mathcal{S}$, $\mathcal{H}_A := \{\int_A u \, d\mathrm{P};\, u \in \mathcal{H}\}$ is relatively weakly compact.*

(iii) *For any subsequence $(u'_n)_n$ of $(u_n)_n$, there exists $v_n \in \mathrm{co}\{u'_m;\, m \geq n\}$ such that $\forall A \in \mathcal{S}$, the sequence $\left(\int_A \delta_{v_n(.)} \, d\mathrm{P}\right)_n$ is narrowly convergent.*

Proof. Since $\mathbb{E}^*$ is separable, we have $\left(\mathrm{L}^1_{\mathbb{E}}\right)^* = \mathrm{L}^\infty_{\mathbb{E}^*}$ (cf. [DU77, Theorem 1 page 98]).

1) The necessity of (i) is well-known [DU77, Theorem 4 page 104]. That of (i) is easy. As to (iii), by the Eberlein–Šmulian Theorem there exists a weakly convergent subsequence $(u''_n)_n$ with limit u_∞. By the Mazur trick there exists $v_n \in \mathrm{co}\{u''_m;\, m \geq n\} \subset \mathrm{co}\{u'_m;\, m \geq n\}$ such that $\|v_n - u_\infty\|_{\mathrm{L}^1} \to 0$. This implies convergence in measure, hence $\int_A \delta_{v_n(.)} \, d\mathrm{P} \to \int_A \delta_{u_\infty(.)} \, d\mathrm{P}$ (see for example [Val94, Proposition 1], but this is an easy consequence of the Lebesgue–Vitali theorem).

2) Let $(u_n)_n$ be a sequence in $\mathcal{H}$. By Lemma 6.4.11 there exist a subsequence $(u_{n_k})_k$ and $u_\infty \in \mathrm{L}^1_{\mathbb{E}}$ such that $\forall x' \in \mathbb{E}^*$, $\forall A \in \mathcal{S}$, $\lim_{k\to\infty} \int_A \langle x', u_{n_k}\rangle \, d\mathrm{P} = \int_A \langle x', u_\infty\rangle \, d\mathrm{P}$. Thus for any step-function $\varphi : \Omega \to \mathbb{E}^*$, we have

$$\lim_{k\to\infty} \int_\Omega \langle \varphi, u_{n_k}\rangle \, d\mathrm{P} = \int_\Omega \langle \varphi, u_\infty\rangle \, d\mathrm{P}.$$

This extends from step-functions φ to $h \in \mathrm{L}^\infty_{\mathbb{E}^*} = \left(\mathrm{L}^1_{\mathbb{E}}\right)^*$. Indeed any $h \in \mathrm{L}^\infty_{\mathbb{E}^*}$ is limit of an almost everywhere convergent sequence $(h_p)_p$ of step functions satisfying $\forall p$, $\|h_p(\omega)\| \leq \|h\|_\infty$. Recall a general fact: On bounded subsets of $\mathrm{L}^\infty_{\mathbb{E}^*}$ convergence in measure coincide with uniform convergence on uniformly integrable subsets of $\mathrm{L}^1_{\mathbb{E}}$, cf. [Cas80, Proposition 1 page 5.3] and, for dimension 1, [Gro64, Proposition 1 chapter 5 §4 page 298]. As $\forall k$, $\forall p$,

$$|\langle h, u_{n_k} - u_\infty\rangle| \leq \sup_{u\in\mathcal{H}} |\langle h - h_p, u\rangle| + |\langle h - h_p, u_\infty\rangle| + |\langle h_p, u_{n_k} - u_\infty\rangle|,$$

u_{n_k} converges $\sigma(\mathrm{L}^1_{\mathbb{E}}, \mathrm{L}^\infty_{\mathbb{E}^*})$ to u_∞. □

6.5 Support theorem for Young measures

The following results have their applications in Fatou type lemmas in Mathematical Economics that we present in the next section.

Theorem 6.5.1 *Let $\mathbb{E}$ be a separable Banach space. Let (u_n) be a sequence of $(\mathcal{S}, \mathcal{B}_{\mathbb{E}})$-measurable mappings from Ω into $\mathbb{E}$. Assume that (u_n) is weakly flexibly tight, and there exist $L \in \mathcal{R}\mathrm{cw}\mathcal{K}(\mathbb{E})$ and such that $u_n(\omega) \in L$ for all n and $\omega \in \Omega$. Then there exists a subsequence (v_m) and a Young measure $\lambda \in \mathcal{Y}^1(\Omega; \mathbb{E}_\sigma)$ such that the sequence of Young measures $\underline{\delta}_{v_m}$ S-stably converges to λ in $\mathcal{Y}^1(\Omega; \mathbb{E}_\sigma)$ and, for a.e. $\omega \in \Omega$*

$$\lambda_\omega\Big(\bigcap_p \text{w–sequ cl}\,\{v_m(\omega);\, m \geq p\}\Big) = 1, \tag{6.5.1}$$

where, for any subset A of $\mathbb{E}$, w–sequ cl A *denotes the sequential closure of A in $\mathbb{E}_\sigma$.*

Proof. The first part of Theorem 6.5.1 is an immediate consequence of Prohorov's Criterion (Theorem 4.3.5). The second part needs a careful look. Let us set

$$\Gamma_p(\omega) = \text{w–sequ cl}\,\{v_m(\omega);\, m \geq p\}.$$

As L is ball-weakly compact, by repeating some arguments in [ACV92, page 178] one can check that Γ_p has its graph in $\mathcal{S} \otimes \mathcal{B}_{\mathbb{E}}$. Indeed, by Banach–Steinhaus' theorem, we have

$$\Gamma_p(\omega) = \bigcup_k \text{w–sequ cl}\,\{v_m(\omega);\, m \geq p\} \cap \overline{B}_{\mathbb{E}}(0, k) \cap L.$$

It follows that Γ_p is K_σ-valued and its graph belongs to $\mathcal{S} \otimes \mathcal{B}_{\mathbb{E}_\sigma}$ because for each k, the weakly compact valued multifunction

$$\Gamma_p^k(\omega) := \text{w–sequ cl } \{v_m(\omega);\ m \geq p\} \cap \overline{B}_{\mathbb{E}}(0,k) \cap L$$

from Ω into the $\sigma(\mathbb{E}^*, \mathbb{E})$ compact metrizable set $\overline{B}_{\mathbb{E}}(0,k) \cap L$ admits a Castaing representation on the set $\{\omega \in \Omega;\ \exists m \geq p,\ v_m(\omega) \in \overline{B}_{\mathbb{E}}(0,k)\}$. Thus, from [CV77, Proposition III.13], its graph belongs to $\mathcal{S} \otimes \mathcal{B}_{\mathbb{E}}$. So the graph of $\Gamma = \cap_p \cup_k \Gamma_p^k$ belongs to $\mathcal{S} \otimes \mathcal{B}_{\mathbb{E}}$, too. Let us consider the integrand:

$$\varphi_p(\omega, x) := \mathbb{1}_{\mathbb{E} \setminus \Gamma_p(\omega)}(x).$$

Then it is obvious that φ_p is $\mathcal{S} \otimes \mathcal{B}_{\mathbb{E}_\sigma}$-measurable and lower semicontinuous on $\mathbb{E}$. As $\underline{\delta}_{v_m(\omega)}$ is supported by $\Gamma_p(\omega)$ for $m \geq p$, and $\underline{\delta}_{v_m}$ S–stably converges to λ, by the Portmanteau Theorem 2.1.3, we get from the definition of stable convergence, $1 = \lambda_\omega(\Gamma_p(\omega))$ a.e., it follows that

$$\lambda_\omega(\Gamma(\omega)) = \lim_p \lambda_\omega(\Gamma_p(\omega)) = 1 \text{ a.e.}$$

□

Remark 6.5.2 1) Actually we have $\cap_p$ w–sequ cl $\{u_m(\omega);\ n \geq p\}$ = w-ls $u_n(\omega)$. See [BH96, page 42].

2) Theorem 6.5.1 can be applied to any bounded sequence (u_n) in $\mathrm{L}^1_{\mathbb{E}}(\Omega, \mathcal{S}, \mathrm{P})$ with $u_n(\omega) \in L$ for all n and for all $\omega \in \Omega$. Even in the particular case when (u_n) is a bounded sequence in $\mathrm{L}^1_{\mathbb{E}}(\Omega, \mathcal{S}, \mathrm{P})$ where $\mathbb{E}$ is a separable reflexive Banach space (here $L = \mathbb{E}$) the required support property (6.5.1) is not trivial.

3) A similar result was given in [BH95] for a bounded sequence (u_n) in $\mathrm{L}^1_{\mathbb{E}}(\mathrm{P})$ satisfying some tightness condition by using a different technique.

4) Let us mention that the proof of (6.5.1) given above shows that the weak sequential closure of a sequence (x_n) in a closed convex ball-weakly compact subset of a Banach space $\mathbb{E}$ is Borel, even a $K_{\sigma\delta}$ subset in the vector space $\mathbb{E}_\sigma$ so that the first member of (6.5.1) has a meaning. In establishing the support property (6.5.1) it turns out the measurability of the Borel-valued multifunction $\cap_p$ w–sequ cl $\{u_m(\omega);\ n \geq p\}$ is crucial. At this point, by combining the support property (6.5.1) and the measurability of the multifunction $\cap_p$ w–sequ cl $\{u_m(\omega);\ n \geq p\}$, it is easy to obtain a Fatou-type lemma in Mathematical Economics for unbounded multifunctions. We refer to [BH96, Theorem 5.3, Cor. 5.3, Cor 5.4] for details.

We give some applications of the preceding theorem.

Proposition 6.5.3 *Let $\mathbb{E}$ be a separable Banach space. Let L be closed convex ball-weakly compact subset in $\mathbb{E}$. Let (u^n) be a bounded sequence in $\mathrm{L}^1_{\mathbb{E}}(\Omega, \mathcal{S}, \mathrm{P})$, such that $u^n(\omega) \in L$ for all n and for all ω. Then the following hold:*

(a) *The multifunctions*

$$\omega \mapsto \bigcap_p \text{w–sequ cl}\,\{u^n(\omega);\, n \geq p\}$$

and

$$\omega \mapsto \overline{\text{co}}(\bigcap_p \text{w–sequ cl}\,\{u^n(\omega);\, n \geq p\})$$

are measurable.

(b) *Assume further that the sequence* (u^n) *is* scalarly uniformly integrable *(that is, the set* $\{\langle x', u^n(.)\rangle;\ \|x'\| \leq 1,\ n \in \mathbb{N}\}$ *is uniformly integrable in* $\mathrm{L}^1_{\mathbb{R}}(\Omega, \mathcal{S}, \mathrm{P})$*). Then there exist a subsequence* (v^m) *and a Young measure* $\lambda^\infty \in \mathcal{Y}^1(\Omega; \mathbb{E}_\sigma)$ *such that the sequence of Young measures* $(\underline{\delta}_{v^m})$ *S–stably converges to* λ^∞ *in* $\mathcal{Y}^1(\Omega; \mathbb{E}_\sigma)$*, and that*

$$\lambda^\infty_\omega(\bigcap_p \text{w–sequ cl}\,\{v^n(\omega);\, n \geq p\}) = 1 \quad \textit{and} \quad \int_{\mathbb{E}} \|x\|\, d\lambda^\infty_\omega(x) < +\infty,$$

for almost all $\omega \in \Omega$*. Moreover the function* $u^\infty : \omega \mapsto \text{bar}\,(\lambda^\infty_\omega)$ *belongs to* $\mathrm{L}^1_{\mathbb{E}}(\Omega, \mathcal{S}, \mathrm{P})$ *and the sequence* (v^m) $\sigma(\mathrm{L}^1_{\mathbb{E}}, \mathrm{L}^\infty \otimes \mathbb{E}^*)$*-converges to* u^∞*.*

Proof. (a) follows from [ACV92, Théorème 8 page 176].

(b) By Theorem 6.5.1, there is a subsequence (v^m) such that $(\underline{\delta}_{v^m})$ S–stably converges to a Young measure $\lambda^\infty \in \mathcal{Y}^1(\Omega, \mathcal{S}, \mathrm{P}; \mathbb{E}_\sigma)$, that is, $(\underline{\delta}_{v^m})$ converges $\sigma(\mathcal{Y}^1(\Omega; \mathbb{E}_\sigma), \mathcal{C}th^b(\Omega, \mathbb{E}_\sigma))$ where $\mathcal{C}th^b(\Omega, \mathbb{E}_\sigma)$ is the set of all bounded Carathéodory integrand on $\Omega \times \mathbb{E}_\sigma$. Let $A \in \mathcal{S}$ and $x' \in \mathbb{E}^*$ with $\|x'\| \leq 1$. Let ψ be the integrand $\psi : (\omega, x) \to \mathbb{1}_A(\omega)\langle x', x\rangle$ defined on $\Omega \times \mathbb{E}$. As (v^m) is scalarly uniformly integrable, the sequence $(\psi(., v^m(.))_m$ is uniformly integrable. By Theorem 6.3.5 we have

$$\lim_m \int_A \langle x', v^m(\omega)\rangle\, d\mathrm{P}(\omega) = \int_A [\int_{\mathbb{E}} \langle x', x\rangle\, d\lambda^\infty_\omega(x)]\, d\mathrm{P}(\omega)$$

and by the Portmanteau Theorem 2.1.3

$$\int_\Omega [\int_{\mathbb{E}} \|x\|\, d\lambda^\infty_\omega(x)]\, d\mathrm{P}(\omega) \leq \liminf_m \int_\Omega \|u^m(\omega)\|\, d\mathrm{P}(\omega) < +\infty.$$

Hence $\int_{\mathbb{E}} \|x\|\, d\lambda^\infty_\omega(x) < +\infty$ a.e. And the required property for the limit measure λ^∞ follows from Therem 6.5.1. So the barycenter $\text{bar}\,(\lambda^\infty_\omega)$ exists a.e. and the mapping $u^\infty : \omega \mapsto \text{bar}\,(\lambda^\infty_\omega)$ belongs to $\mathrm{L}^1_{\mathbb{E}}(\Omega, \mathcal{S}, \mathrm{P})$. The $\sigma(\mathrm{L}^1_{\mathbb{E}}, \mathrm{L}^\infty \otimes \mathbb{E}^*)$ convergence of v^m to u^∞ follows again from Theorem 6.3.5. □

Remark 6.5.4 Proposition 6.5.3 provides a variant of a weak compactness result in [ACV92, Theorem 8 page 176].

Let us recall a result due to Benabdellah [Ben91, Proposition 2.2 page 4.10] which has several applications to the "problem of norm convergence is implied by the weak." The notation $\partial_{\text{ext}}(K)$ or $\partial_{\text{ext}} K$ will denote the set of extreme points of a closed convex subset K of a Banach space. Later we will also use the set of denting points $\partial_{\text{dent}}(K)$ of K.

Proposition 6.5.5 *Let K be a closed convex subset of a Banach space $\mathbb{X}$. Let $\varphi : K \to \mathbb{R}^+$ be a convex lower semicontinuous function. Then the following three conditions are equivalent:*

(*i*) $(x_0, \varphi(x_0)) \in \partial_{\text{ext}}(\operatorname{Epi}\varphi)$, *where* $\operatorname{Epi}\varphi := \{(x,t) \in \mathbb{X} \times \mathbb{R};\, t \geq \varphi(x)\}$ *is the* epigraph *of* φ.

(*ii*) *For any pair* $(x_1, x_2) \in K \times K$ *with* $x_1 \neq x_2$ *and every* $t \in\,]0,1[$, *one has*

$$x_0 = tx_1 + (1-t)x_2 \Longrightarrow \varphi(x_0) < t\varphi(x_1) + (1-t)\varphi(x_2),$$

(*iii*) δ_{x_0} *is the unique probability Radon measure μ on K such that*

$$\int_K x\, d\mu(x) = x_0 \text{ and } \int_K \varphi(x)\, d\mu(x) = \varphi(x_0).$$

The following is an application of Proposition 6.5.3 and Proposition 6.5.5 to a Visintin-type convergence under extreme point condition [Bal91, Ben91, Val89, BCG99a, Vis84, Bal86b, Rze89, Rze92].

Theorem 6.5.6 *Let $\mathbb{E}$ be a separable Banach space. Let L be closed convex ball-weakly compact subset in $\mathbb{E}$. Let (u^n) be a bounded sequence in $\mathrm{L}^1_{\mathbb{E}}(\Omega, \mathcal{S}, \mathrm{P})$, such that $u^n(\omega) \in L$ for all n and for all ω. Let $\varphi : \Omega \times \mathbb{E} \to [0, +\infty]$ be an $\mathcal{S} \otimes \mathcal{B}_{\mathbb{E}}$-measurable integrand such that $\varphi(\omega, .)$ is convex lower semicontinuous on $\mathbb{E}$ for every fixed $\omega \in \Omega$. Assume further that*

(*i*) (u^n) *is scalarly uniformly integrable and converges* $\sigma(\mathrm{L}^1_{\mathbb{E}}, \mathrm{L}^\infty \otimes \mathbb{E}^*)$ *to* $u^\infty \in \mathrm{L}^1_{\mathbb{E}}(\Omega, \mathcal{S}, \mathrm{P})$,

(*ii*) $\limsup_{n\to\infty} \int_\Omega \varphi(\omega, u^n(\omega))\, d\mathrm{P}(\omega) \leq \int_\Omega \varphi(\omega, u^\infty(\omega))\, d\mathrm{P}(\omega) < +\infty.$

Then there is a subsequence $(\underline{\delta}_{v^m})$ which S–stably converges to a Young measure $\lambda^\infty \in \mathcal{Y}^1(\Omega, \mathcal{S}, \mathrm{P}; \mathbb{E}_\sigma)$ satisfying

(*a*) $u^\infty(\omega) = \operatorname{bar}(\lambda^\infty_\omega)$ *a.e.*,

(*b*) $\int_{\cap_p \text{w-sequ cl}\{v^m(\omega);\, n\geq p\}} \varphi(\omega, x)\, d\lambda^\infty_\omega(x) = \varphi(\omega, u^\infty(\omega))$ *a.e.*,

(*c*) *in addition, suppose that* $(u^\infty(\omega), \varphi(\omega, u^\infty(\omega)))$ *is an extremal point of* $\operatorname{Epi}\varphi_\omega$ *a.e., and* $\overline{\text{co}}(\cap_p \text{ w–sequ cl } \{v^m(\omega);\, n \geq p\}) \subset \operatorname{dom}\varphi_\omega$ *for all* $\omega \in \Omega$, *where* $\operatorname{dom}\varphi_\omega$ *is the domain of* φ_ω. *Then* $\lambda^\infty_\omega = \delta_{u^\infty(\omega)}$ *a.e. so that* $(\underline{\delta}_{v^m})$ *S–stably converges to* $\underline{\delta}_{u^\infty}$ *in* $\mathcal{Y}^1(\Omega, \mathcal{S}, \mathrm{P}; \mathbb{E}_\sigma)$.

Proof. (*a*) Let $A \in \mathcal{S}$ and $x' \in \mathbb{E}^*$. Let ψ the integrand $\psi : (\omega, x) \to \mathbb{1}_A(\omega)\langle x', x\rangle$ defined on $\Omega \times \mathbb{E}$. As (v^m) is scalarly uniformly integrable, the sequence $(\psi(., v^m(.)))$ is uniformly integrable. By (i) and Theorem 6.3.5 we have

$$\int_A \langle x', u^\infty(\omega)\rangle \, d\,\mathrm{P}(\omega) = \lim_m \int_A \langle x', v^m(\omega)\rangle \, d\,\mathrm{P}(\omega)\rangle = \int_A \Big[\int_{\mathbb{E}} \langle x', x\rangle \, d\lambda_\omega^\infty(x)\Big] \, d\,\mathrm{P}(\omega).$$

By the Portmanteau Theorem 2.1.3

$$\int_\Omega \Big[\int_{\mathbb{E}} \|x\| \, d\lambda_\omega^\infty(x)\Big] \, d\,\mathrm{P}(\omega) \le \liminf_m \int_\Omega \|v^m(\omega)\| \, d\,\mathrm{P}(\omega) < +\infty,$$

so the barycenter $\mathrm{bar}\,(\lambda_\omega^\infty)$ exists a.e. and the mapping $\omega \mapsto \mathrm{bar}\,(\lambda_\omega^\infty)$ belongs to $\mathrm{L}^1_{\mathbb{E}}(\Omega, \mathcal{S}, \mathrm{P})$. It follows that $u^\infty(\omega) = \mathrm{bar}\,(\lambda_\omega^\infty)$ a.e.

(*b*) Using (a), the Portmanteau Theorem 2.1.3 and Jensen's inequality we get

$$\begin{aligned}\liminf_n \int_\Omega \varphi(\omega, v^n(\omega)) \, d\,\mathrm{P}(\omega) &\ge \int_\Omega \Big[\int_{\cap_p \text{w-sequ cl}\{v^n(\omega);\, n\ge p\}} \varphi(\omega, x) \, d\lambda_\omega^\infty(x)\Big] \, d\,\mathrm{P}(\omega) \\ &\ge \int_\Omega \varphi(\omega, u^\infty(\omega)) \, d\,\mathrm{P}(\omega).\end{aligned}$$

Combining this inequality with (ii) yields

$$\int_{\cap_p \text{w-sequ cl}\{v^n(\omega);\, n\ge p\}} \varphi(\omega, x) \, d\lambda_\omega^\infty(x) = \varphi(\omega, u^\infty(\omega)) \text{ a.e.}$$

(*c*) By our assumption, for every fixed $\omega \in \Omega$, the $\mathbb{R}^+$-valued function $\varphi(\omega, .)$ is convex lower semicontinuous on the closed convex set $K(\omega) := \overline{\mathrm{co}}(\cap_p \text{ w–sequ cl } \{v^n(\omega);\, n \ge p\})$, moreover we have

$$\int_{K(\omega)} \varphi(\omega, x) \, d\lambda_\omega^\infty(x) = \varphi(\omega, u^\infty(\omega)) \text{ a.e.}$$

As $(u^\infty(\omega), \varphi(\omega, u^\infty(\omega))$ is an extremal point of $\mathrm{Epi}\,\varphi_\omega$ a.e., it follows from Proposition 6.5.5 that $\lambda_\omega^\infty = \delta_{u^\infty(\omega)}$ a.e. □

Corollary 6.5.7 *Let $\mathbb{E}$ be a separable Banach space. Let L be closed convex ball-weakly compact subset in $\mathbb{E}$. Let (u^n) be a bounded sequence in $\mathrm{L}^1_{\mathbb{E}}(\Omega, \mathcal{S}, \mathrm{P})$, such that $u^n(\omega) \in L$ for all n and for all ω. Assume further that (u^n) is scalarly uniformly integrable and converges $\sigma(\mathrm{L}^1_{\mathbb{E}}, \mathrm{L}^\infty \otimes \mathbb{E}^*)$ to $u^\infty \in \mathrm{L}^1_{\mathbb{E}}(\Omega, \mathcal{S}, \mathrm{P})$ with $u^\infty(\omega) \in \partial_{\text{ext}}\, \overline{\mathrm{co}}(\cap_p \text{ w–sequ cl } \{v^m(\omega);\, n \ge p\})$ a.e., then $(\underline{\delta}_{u^n})$ S-stably converges to $\underline{\delta}_{u^\infty}$ in $\mathcal{Y}^1(\Omega, \mathcal{S}, \mathrm{P}; \mathbb{E}_\sigma)$.*

Proof. Apply Theorem 6.5.6 to the convex normal integrand φ:

$$\varphi(\omega, x) := \delta\Big(x, \overline{\mathrm{co}}\big(\bigcap_p \text{w–sequ cl } \{v^m(\omega);\, m \ge p\}\big)\Big)$$

where $x \mapsto \delta(x, C)$ is the *indicator function (in the sense of Convex Analysis)* of the closed convex set C, that is, $\delta(x, C) = 0$ if $x \in C$ and $\delta(x, C) = +\infty$ if $x \notin C$. □

The following example is an application of Theorem 6.5.6 to an optimization problem.

Example 6.5.8 Let $\varphi : \Omega \times \mathbb{R}^d \to [0, +\infty]$ be a $\mathcal{S} \otimes \mathcal{B}_{\mathbb{R}^d}$-measurable integrand such that $\varphi(\omega, .)$ is convex lower semicontinuous on $\mathbb{R}^d$ for every fixed $\omega \in \Omega$. Let $(K_n)_{n \in \mathbb{N} \cup \{\infty\}}$ be a sequence of closed convex valued measurable and integrable multifunctions. For each $n \in \mathbb{N} \cup \{\infty\}$, let $\mathrm{S}^1_{K_n}$ the set of all integrable selections of K_n. Assume further that the following conditions are satisfied:

(*i*) w-ls $\mathrm{S}^1_{K_n} := \{u \in \mathrm{L}^1_{\mathbb{R}^d}(\Omega, \mathcal{S}, \mathrm{P}); \exists u_{n_k} \in \mathrm{S}^1_{K_n}$ with $u_{n_k} \to u$ weakly$\} \subset \mathrm{S}^1_{K_\infty}$.

(*ii*) The integral functional $I_\varphi : \mathrm{L}^1_{\mathbb{R}^d}(\Omega, \mathcal{S}, \mathrm{P}) \to [0, +\infty]$ associated with φ is proper, inf-weakly compact on $\mathrm{L}^1_{\mathbb{R}^d}(\Omega, \mathcal{S}, \mathrm{P})$ and strictly convex on $\mathrm{S}^1_{K_\infty}$.

(*iii*) $\inf\{I_\varphi(u);\, u \in \mathrm{S}^1_{K_n}\} \to \inf\{I_\varphi(u);\, u \in \mathrm{S}^1_{K_\infty}\} < +\infty$.

(*iv*) $\overline{\mathrm{co}}\,\mathrm{ls}\, K_n(\omega) \subset \mathrm{dom}\, \varphi_\omega$ for all $\omega \in \Omega$.

Then any optimal solution $u_n \in \mathrm{S}^1_{K_n}$ converges in $\mathrm{L}^1_{\mathbb{R}^d}(\Omega, \mathcal{S}, \mathrm{P})$ to the optimal solution $u_\infty \in \mathrm{S}^1_{K_\infty}$.

Proof. By (*i*), (*ii*) and (*iii*), it is straightforward to check that $u_n \to u_\infty$ weakly in $\mathrm{L}^1_{\mathbb{R}^d}(\Omega, \mathcal{S}, \mathrm{P})$ with

$$u_\infty(\omega) \in \overline{\mathrm{co}}\,\mathrm{ls}\, u_n(\omega) \subset \overline{\mathrm{co}}\,\mathrm{ls}\, K_n(\omega) \subset \mathrm{dom}\, \varphi_\omega$$

for almost all $\omega \in \Omega$, using Theorem 6.5.1 or [ACV92, Théorème 8 page 176] and Remark 6.5.2. In view of Theorem 6.5.6 $\underline{\delta}_{u_n} \to \underline{\delta}_{u_\infty}$ stably in $\mathcal{Y}^1(\Omega, \mathcal{S}, \mathrm{P}; \mathbb{R}^d)$. By Part 3 of Theorem 3.1.2 $u_n \to u_\infty$ in measure. Since weak convergence implies uniform integrability, we deduce that $u_n \to u_\infty$ in $\mathrm{L}^1_{\mathbb{R}^d}(\Omega, \mathcal{S}, \mathrm{P})$. □

Some weak compactness and convergences results in $\mathrm{L}^1_{\mathbb{E}^*}[\mathbb{E}]$ We present some weak compactness result in the space $\mathrm{L}^1_{\mathbb{E}^*}[\mathbb{E}](\Omega, \mathcal{S}, \mathrm{P})$ of scalarly integrable mappings $f : \Omega \to \mathbb{E}^*$ such that $|f| : \omega \mapsto \|f(\omega)\|$ is integrable. The following is a support theorem for the Young measure limit in $\mathcal{Y}^1(\Omega, \mathcal{S}, \mathrm{P}; \mathbb{E}^*_\sigma)$ generated by a bounded sequence in $\mathrm{L}^1_{\mathbb{E}^*}[\mathbb{E}](\Omega, \mathcal{S}, \mathrm{P})$.

Theorem 6.5.9 *Suppose that* $\mathbb{E}$ *is a separable Banach space and* (u_n) *is a bounded sequence in* $\mathrm{L}^1_{\mathbb{E}^*}[\mathbb{E}](\Omega, \mathcal{S}, \mathrm{P})$ *and* h *is a Carathéodory integrand defined on* $\Omega \times \mathbb{E}^*_\sigma$

such that the sequence $(h(u_n))_n = (h(.,u_n(.))_n$ *is uniformly integrable, then there are a subsequence* (v_n) *and a Young measure* $\lambda \in \mathcal{Y}^1(\Omega, \mathcal{S}, \mathrm{P}; \mathbb{E}^*_\sigma)$ *such that*

$$\lim_{n\to\infty} \int_\Omega h(\omega, v_n(\omega))\, d\,\mathrm{P}(\omega) = \int_\Omega \Big[\int_{\mathbb{E}} h(\omega, x)\, d\lambda^\infty_\omega(x)\Big]\, d\,\mathrm{P}(\omega), \tag{6.5.2}$$

and

$$\lambda_\omega\Big(\bigcap_p w^*\text{-}\mathrm{cl}[\{v_m(\omega);\ m \geq p\}]\Big) = 1 \text{ a.e.} \tag{6.5.3}$$

Assume further that (u_n) *is uniformly integrable, then* (v_n) $\sigma(\mathrm{L}^1_{\mathbb{E}^*}[\mathbb{E}], \mathrm{L}^\infty_{\mathbb{E}})$ *converges to* $u \in \mathrm{L}^1_{\mathbb{E}^*}[\mathbb{E}]$ *with* $u(\omega) = \mathrm{bar}\,(\lambda_\omega)$ *a.e.*

Proof. By Markov's inequality the sequence $(\underline{\delta}_{u_n})$ is strictly tight in $\mathcal{Y}^1(\Omega, \mathcal{S}, \mathrm{P}; \mathbb{E}^*_\sigma)$. In view of Theorem 4.3.5, there is a subsequence $(\underline{\delta}_{v_n})$ that S–stably converges to $\lambda \in \mathcal{Y}^1(\Omega, \mathcal{S}, \mathrm{P}; \mathbb{E}^*_\sigma)$. Since $\mathbb{E}^*_\sigma$ is a Lusin space, in view of Theorem 6.3.5, we get (6.5.2). At this point we may also remark that this convergence holds for the topology $\sigma(\mathcal{Y}^1(\Omega; \mathbb{E}^*_\sigma), \mathcal{C}th^1(\Omega, \mathbb{E}^*_\sigma))$ where $\mathcal{C}th^1(\Omega, \mathbb{E}^*_\sigma)$ denotes the set of all Carathéodory integrands h of first order defined on $\Omega \times \mathbb{E}^*_\sigma$, that is, $|h(\omega, x')| \leq c(1 + \|x'\|), \forall(\omega, x') \in \Omega \times \mathbb{E}^*_\sigma$, by using the techniques developed in Lemma 6.2.1. Repeating the arguments in Theorem 6.5.1, it is not difficult to see that

$$\lambda_\omega\Big(\bigcap_p w^*\text{-}\mathrm{cl}[\{v_m(\omega);\ m \geq p\}]\Big) = 1 \text{ a.e.}$$

using the fact that $\mathbb{E}^* = \bigcup_k k\,\overline{B}_{\mathbb{E}^*}$, and $B_{\mathbb{E}^*}$ is $\sigma(\mathbb{E}^*, \mathbb{E})$ compact metrizable, namely

$$\bigcap_p w^*\text{-}\mathrm{cl}[\{v_m(\omega);\ m \geq p\} = \bigcap_p \Big[\bigcup_k w^*\text{-}\mathrm{cl}[\{v_m(\omega);\ m \geq p\} \cap k\overline{B}_{\mathbb{E}^*}\Big].$$

Now suppose that (u_n) is uniformly integrable. Applying (6.5.2) by taking $h(\omega, y) = \mathbb{1}_A(\omega)|\langle x, y\rangle|$ with $A \in \mathcal{S}$ and $x \in \overline{B}_{\mathbb{E}}$ gives

$$\int_A \Big[\int_{\mathbb{E}} |\langle x, y\rangle|\, d\lambda^\infty_\omega(y)\Big]\, d\,\mathrm{P}(\omega) = \lim_{n\to\infty} \int_A |\langle x, v_n(\omega)\rangle|\, d\,\mathrm{P}(\omega) \leq \sup_n \int_\Omega \|v_n(\omega)\|\, d\,\mathrm{P}(\omega).$$

It follows that the barycenter $\mathrm{bar}\,(\lambda_\omega)$ exists and satisfies

$$\langle x, \mathrm{bar}\,(\lambda_\omega)\rangle = \int_{\Gamma(\omega)} \langle x, y\rangle\, d\lambda_\omega(y)$$

where $\Gamma(\omega) := \cap_p\, w^*\text{-}\mathrm{cl}[\{v_m(\omega);\ m \geq p\}]$. By the Portmanteau Theorem 2.1.3 we have

$$\int_\Omega \Big[\int_{\Gamma(\omega)} \|y\|\, d\lambda_\omega(y)\Big]\, d\,\mathrm{P}(\omega) \leq \sup_n \int_\Omega \|v_n(\omega)\|\, d\,\mathrm{P}(\omega) < +\infty.$$

Hence the mapping $\omega \mapsto \text{bar}(\lambda_\omega)$ belongs to $\mathrm{L}^1_{\mathbb{E}^*}[\mathbb{E}](\Omega, \mathcal{S}, \mathrm{P})$ with $\text{bar}(\lambda_\omega) \in \overline{\text{co}}\,\Gamma(\omega)$ a.e. Now let $g \in \mathrm{L}^\infty_{\mathbb{E}}(\Omega, \mathcal{S}, \mathrm{P})$. Then the integrand $j : (\omega, y) \mapsto \langle g(\omega), y \rangle$ defined on $\Omega \times \mathbb{E}^*_\sigma$ belongs to $\mathcal{C}th^1(\Omega, \mathbb{E}^*_\sigma)$. Applying again (6.5.2) by taking $h = j$, gives

$$\lim_{m\to\infty} \int_\Omega \langle g(\omega), u_m(\omega)\rangle \, d\,\mathrm{P}(\omega) = \int_\Omega \Big[\int_{\Gamma(\omega)} \langle g(\omega), y\rangle \, d\lambda_\omega(y)\Big] \, d\,\mathrm{P}(\omega)$$

$$= \int_\Omega \langle g(\omega), \text{bar}(\lambda_\omega)\rangle \, d\,\mathrm{P}(\omega).$$

Taking $u : \omega \mapsto \text{bar}(\lambda_\omega)$ completes the proof. □

Corollary 6.5.10 *Let $\Phi : \Omega \to \text{cw}\mathcal{K}(\mathbb{E}^*_\sigma)$ be a convex $\sigma(\mathbb{E}^*, \mathbb{E})$-compact valued measurable and integrably bounded multifunction, that is, there exists $\beta \in \mathrm{L}^1_{\mathbb{R}^+}$ such that $\Phi(\omega) \subset \beta(\omega)\, \overline{B}_{\mathbb{E}^*}$ for all $\omega \in \Omega$. Then the set S^1_Φ of all scalarly integrable selections of Φ is sequentially $\sigma(\mathrm{L}^1_{\mathbb{E}^*}[\mathbb{E}], \mathrm{L}^\infty_{\mathbb{E}})$ compact.*

The following result is a combined effort of Theorem 6.5.9 and the Biting Lemma.

Proposition 6.5.11 *Suppose that (u_n) is a bounded sequence in $\mathrm{L}^1_{\mathbb{E}^*}[\mathbb{E}]$, then there exist a subsequence (v_n) of (u_n) and $u_\infty \in \mathrm{L}^1_{\mathbb{E}^*}[\mathbb{E}]$ such that (v_n) biting weakly converges to u_∞, that is, there exists an increasing sequence (A_p) in $\mathcal{S}$ such that $\lim_{p\to\infty} \mathrm{P}(A_p) = 1$, and such that, for each p and for each $h \in \mathrm{L}^\infty_{\mathbb{E}}(A_p, A_p \cap \mathcal{S}, \mathrm{P}\,|_{A_p})$, the following holds:*

$$\lim_{n\to\infty} \int_{A_p} \langle v_n, h\rangle \, d\,\mathrm{P} = \int_{A_p} \langle u_\infty, h\rangle \, d\,\mathrm{P}$$

and

$$u_\infty(\omega) \in \overline{\text{co}}\Big(\bigcap_p w^*\text{-}\mathrm{cl}\{u_m(\omega);\ m \geq p\}\Big) \text{ a.e.}$$

Proof. By the Biting Lemma there are an increasing sequence (A_p) in $\mathcal{S}$ with $\lim_{p\to\infty} \mathrm{P}(A_p) = 1$ and a subsequence (u'_n) of (u_n) such that $(u'_n|_{A_p})_n$ is uniformly integrable. Using Theorem 6.5.9 and a diagonal procedure, it is not difficult to produce a subsequence (v_n) of (u'_n) and a sequence (γ_p) with $\gamma_p \in \mathrm{L}^1_{\mathbb{E}^*}[\mathbb{E}](A_p, A_p \cap \mathcal{S}, \mathrm{P}\,|_{A_p})$, such that

$$\lim_n \int_A \langle v_n(\omega), h(\omega)\rangle \, d\,\mathrm{P}(\omega) = \int_A \langle \gamma_p(\omega), h(\omega)\rangle \, d\,\mathrm{P}(\omega),$$

for all $A \in A_p \cap \mathcal{S}$ and for all $h \in \mathrm{L}^\infty_{\mathbb{E}}(\Omega, \mathcal{S}, \mathrm{P})$. As (A_p) is increasing, it is obvious that $\gamma_{p+1} = \gamma_p$ for a.e. in A_p. It is obvious that the function u_∞ defined by $u_\infty(\omega) = \gamma_p(\omega)$ if $\omega \in A_p$ and $u_\infty(\omega) = 0$ if $\omega \notin \cup_p A_p$ belongs to $\mathrm{L}^1_{\mathbb{E}^*}[\mathbb{E}](\Omega, \mathcal{S}, \mathrm{P})$ and that it is the biting weak limit of (v_n), whereas the required inclusion follows easily from (6.5.3). □

As a corollary of Proposition 6.5.11 we provide a Fatou-type lemma for bounded sequence in $\mathrm{L}^1_{\mathbb{E}^*}[\mathbb{E}]$.

Proposition 6.5.12 *Suppose that (h_n) is a bounded sequence in $\mathrm{L}^\infty_{\mathbb{E}}$ such that (h_n) converges in measure to $h_\infty \in \mathrm{L}^\infty_{\mathbb{E}}$ and (u_n) is a bounded sequence in $\mathrm{L}^1_{\mathbb{E}^*}[\mathbb{E}]$ such that the sequence $(\langle h_n, u_n\rangle^-)$ is uniformly integrable, then there exists $u_\infty \in \mathrm{L}^1_{\mathbb{E}^*}[\mathbb{E}]$ such that*

$$\liminf_n \int_\Omega \langle h_n, u_n\rangle \, d\mathrm{P} \geq \int_\Omega \langle h_\infty, u_\infty\rangle \, d\mathrm{P}$$

with

$$u_\infty(\omega) \in \overline{\mathrm{co}}(\bigcap_p w^*\text{-}\mathrm{cl}\{u_m(\omega);\, m \geq p\}) \text{ a.e.}$$

Proof. We may suppose that

$$a := \lim_{n\to\infty} \int \langle h_n, u_n\rangle \, d\mathrm{P} \in \mathbb{R}.$$

Furthermore, by Proposition 6.5.11 we may suppose that (u_n) weakly biting converges to $u_\infty \in \mathrm{L}^1_{\mathbb{E}^*}[\mathbb{E}]$, that is, there exist $u_\infty \in \mathrm{L}^1_{\mathbb{E}^*}[\mathbb{E}]$ and a subsequence (v_n) of (u_n) and an increasing sequence (A_p) in $\mathcal{S}$ such that $\lim_{p\to\infty} \mathrm{P}(A_p) = 1$, and such that, for each p and for each $h \in \mathrm{L}^\infty_{\mathbb{E}}(A_p, A_p \cap \mathcal{S}, \mathrm{P}\,|_{A_p})$, the following holds:

$$\lim_{n\to\infty} \int_{A_p} \langle v_n, h\rangle \, d\mathrm{P} = \int_{A_p} \langle u_\infty, h\rangle \, d\mathrm{P}$$

with $u_\infty \in \overline{\mathrm{co}}(\cap_p w^*\text{-}\mathrm{cl}[\{u_m(\omega);\, m \geq p\}])$ a.e. Let $\varepsilon > 0$ be given. Pick $N \in \mathbb{N}$ such that

$$\int_{A_N} \langle h_\infty, u_\infty\rangle \, d\mathrm{P} \geq \int_\Omega \langle h_\infty, u_\infty\rangle \, d\mathrm{P} - \varepsilon,$$

and that

$$\limsup_{n\to\infty} \int_{\Omega\setminus A_N} \langle h_n, u_n\rangle^- \, d\mathrm{P} \leq \varepsilon,$$

because $(\langle h_n, u_n\rangle^-)_n$ is uniformly integrable by hypothesis. As $\|h_n(.) - h_\infty(.)\| \to 0$ in measure, $\|h_n(.) - h_\infty(.)\| \to 0$ uniformly on uniformly integrable subsets of $\mathrm{L}^1_{\mathbb{R}}(\Omega, \mathcal{S}, \mathrm{P})$, cf. [Cas80, Proposition 1 page 5.3] and [Gro64, Proposition 1 chapter 5 §4 page 298]. It follows that

$$\lim_{n\to\infty} \int_{A_N} \|h_n(\omega) - h_\infty(\omega)\| \, \|u_n(\omega)\| \, d\mathrm{P}(\omega) = 0.$$

Hence

$$\lim_{n\to\infty} \Big[\int_{A_N} \langle h_n, u_n\rangle \, d\mathrm{P} - \int_{A_N} \langle h_\infty, u_n\rangle \, d\mathrm{P}\Big] = 0.$$

An easy computation gives

$$a \geq \lim_{n\to\infty} \int_{A_N} \langle h_n, v_n\rangle - \limsup_{n\to\infty} \int_{\Omega\setminus A_N} \langle h_n, v_n\rangle^- \, d\mathrm{P} \geq \lim_{n\to\infty} \int_{A_N} \langle h_n, v_n\rangle \, d\mathrm{P} - \varepsilon.$$

Finally we get

$$a \geq \lim_{n\to\infty} \int_{A_N} \langle h_n, v_n \rangle \, d\,\mathrm{P} - \varepsilon = \lim_{n\to\infty} \int_{A_N} \langle h_\infty, v_n \rangle \, d\,\mathrm{P} - \varepsilon$$
$$= \int_{A_N} \langle h_\infty, u_\infty \rangle \, d\,\mathrm{P} - \varepsilon \geq \int_\Omega \langle h_\infty, u_\infty \rangle) \, d\,\mathrm{P} - 2\varepsilon.$$

□

Now let us focus our attention to the particular case when $\mathbb{E} = \mathrm{C}_0(\mathbb{R}^d)$ where $\mathrm{C}_0(\mathbb{R}^d)$ is the separable Banach space of all continuous mappings $f : \mathbb{R}^d \to \mathbb{R}$ tending to 0 when $\|x\| \to +\infty$ equipped with the sup norm. Then the dual $\mathbb{E}^* = \mathrm{C}_0(\mathbb{R}^d)^*$ is identified with the Banach space $\mathcal{M}(\mathbb{R}^d) = \mathrm{ca}(\mathbb{R}^d)$ of bounded measures on $\mathbb{R}^d$ equipped with the norm $\|\nu\| = \int_{\mathbb{R}^d} d|\nu|$. We present some relationships betwen convergence results stated above in the context of Young measures on $\mathbb{R}^d$ with those using the duality $(\mathrm{L}^\infty_{\mathrm{ca}(\mathbb{R}^d)}(\mathrm{P}), \mathrm{L}^1_{\mathrm{C}_0(\mathbb{R}^d)}(\mathrm{P}))$. Let us mention first the following:

Proposition 6.5.13 *Let (λ^n) be a bounded sequence in $\mathrm{L}^1_{\mathrm{ca}(\mathbb{R}^d)}[\mathrm{C}_0(\mathbb{R}^d)]$ such that $\lambda^n_\omega \in \mathcal{M}^+(\mathbb{R}^d)$ for all n and for all $\omega \in \Omega$. If (λ^n) converges $\sigma(\mathrm{L}^1_{\mathrm{ca}(\mathbb{R}^d)}[\mathrm{C}_0(\mathbb{R}^d)], \mathrm{L}^\infty \otimes \mathrm{C}_0(\mathbb{R}^d))$ to $\lambda^\infty \in \mathrm{L}^1_{\mathrm{ca}(\mathbb{R}^d)}[\mathrm{C}_0(\mathbb{R}^d)]$, then*

$$\lambda^\infty_\omega \in \mathcal{M}^+(\mathbb{R}^d) \text{ a.e.} \tag{6.5.4}$$

Proof. Let us observe that $\mathcal{M}^+(\mathbb{R}^d)$ is $\sigma(\mathcal{M}(\mathbb{R}^d), \mathrm{C}_0(\mathbb{R}^d))$ closed convex, locally compact and contains no lines. Suppose by contradiction that (6.5.4) does not holds. By [CV77, Lemma III.34] there is an element $f \in \mathrm{C}_0(\mathbb{R}^d)$ and a measurable set $A \in \mathcal{S}$ with $\mathrm{P}(A) > 0$ such that

$$\langle f, \lambda^\infty_\omega \rangle > \delta^*(f, \mathcal{M}^+(\mathbb{R}^d))$$

for all $\omega \in A$. By hypothesis, we deduce that

$$\int_A \langle f, \lambda^\infty_\omega \rangle \, d\,\mathrm{P}(\omega) = \lim_{n\to\infty} \int_A \langle f, \lambda^n_\omega \rangle \, d\,\mathrm{P}(\omega)$$
$$\leq \int_A \delta^*(f, \mathcal{M}^+(\mathbb{R}^d)) \, d\,\mathrm{P}(\omega)$$
$$= \mathrm{P}(A) \, \delta^*(f, \mathcal{M}^+(\mathbb{R}^d))$$

which contradicts the inequality

$$\int_A \langle f, \lambda^\infty_\omega \rangle \, d\,\mathrm{P}(\omega) > \mathrm{P}(A) \, \delta^*(f, \mathcal{M}^+(\mathbb{R}^d)).$$

□

Remark 6.5.14 1) If $\|\lambda^n_\omega\| \leq \alpha$ (α being a finite number) for all n and for almost all $\omega \in \Omega$, then it is easy to check that $\|\lambda^\infty_\omega\| \leq \alpha$ for almost all $\omega \in \Omega$.

2) If (u_n) is a tight sequence of $\mathbb{R}^d$-valued $\mathcal{S}$-measurable functions defined on Ω, then it is well-known that the sequence $(\lambda^n) = (\underline{\delta}_{u_n})$ (up to an extracted subsequence) converges stably to a Young measure $\nu^\infty \in \mathcal{Y}^1(\Omega, \mathcal{S}, \mathrm{P}; \mathbb{R}^d)$. Since the sequence (λ^n) (up to an extracted subsequence) converges to $\lambda^\infty \in \mathrm{L}^1_{\mathrm{ca}(\mathbb{R}^d)}(\Omega, \mathcal{S}, \mathrm{P})$, for the $\sigma(\mathrm{L}^1_{\mathrm{ca}(\mathbb{R}^d)}, \mathrm{L}^\infty \otimes \mathrm{C}_0(\mathbb{R}^d))$ topology, we have $\lambda^\infty = \nu^\infty$ a.e. because $\mathrm{L}^\infty \otimes \mathrm{C}_0(\mathbb{R}^d) \subset \mathrm{L}^\infty \otimes \mathrm{C}_b(\mathbb{R}^d)$.

Best approximants in $\mathrm{L}^1_{\mathbb{E}^*}[\mathbb{E}]$ We now consider a problem of best approximation in $\mathrm{L}^1_{\mathbb{E}^*}[\mathbb{E}](\Omega, \mathcal{S}, \mathrm{P})$. This problem has been studied essentially for the space $\mathrm{L}^1_{\mathbb{E}}(\Omega, \mathcal{S}, \mathrm{P})$ when $\mathbb{E}$ is a reflexive separable Banach space, and even for the space $\mathcal{L}_{\mathrm{cw}\mathcal{K}(\mathbb{E})}(\Omega, \mathcal{S}, \mathrm{P})$ of convex weakly compact valued measurable and integrably bounded multifunctions, see [CC85, Proposition 5.4] and the references therein. In a recent paper [BC01, Theorem 5.10 page 36] the authors gave a result of best approximation in $\mathrm{L}^1_{\mathbb{E}^*}[\mathbb{E}]$ using some new structure results of this space, mainly the characterization of weak compactness in this Banach space. This study is quite delicate because of the lack of the characterization of the dual of $\mathrm{L}^1_{\mathbb{E}^*}[\mathbb{E}]$ and is independent of the theory of Young measures. We aim to present a new variant of this result by exploiting the sequential $\sigma(\mathrm{L}^1_{\mathbb{E}^*}[\mathbb{E}], \mathrm{L}^\infty_{\mathbb{E}})$ compactness result in Theorem 6.5.9 and a new characterization of the norm N_1 of $\mathrm{L}^1_{\mathbb{E}^*}[\mathbb{E}](\Omega, \mathcal{S}, \mathrm{P})$ given in [BC01, Theorem 4.1]. We consider the spaces $\mathrm{L}^1_{\mathbb{E}^*}[\mathbb{E}](\Omega, \mathcal{S}, \mathrm{P})$ (shortly $\mathrm{L}^1_{\mathbb{E}^*}[\mathbb{E}](\mathcal{S})$) and $\mathrm{L}^1_{\mathbb{E}^*}[\mathbb{E}](\Omega, \mathcal{B}, \mathrm{P})$ (shortly $\mathrm{L}^1_{\mathbb{E}^*}[\mathbb{E}](\mathcal{B})$) where $\mathcal{B}$ is a complete sub-σ algebra of $\mathcal{S}$. We need first a crucial lemma.

Lemma 6.5.15 *Suppose that $\mathcal{B}$ is a complete sub-σ-algebra of $\mathcal{S}$ and $f \in \mathrm{L}^1_{\mathbb{E}^*}[\mathbb{E}](\mathcal{S})$, then any minimizing sequence $(g_n)_{n\geq 1}$ in $\mathrm{L}^1_{\mathbb{E}^*}[\mathbb{E}](\mathcal{B})$, that is,*

$$\lim_{n\to\infty} \int_\Omega \|f - g_n\| \, d\mathrm{P} = \inf\{\int_\Omega \|f - g\| \, d\mathrm{P};\ g \in \mathrm{L}^1_{\mathbb{E}^*}[\mathbb{E}](\mathcal{B})\},$$

is relatively sequentially $\sigma(\mathrm{L}^1_{\mathbb{E}^*}[\mathbb{E}], \mathrm{L}^\infty_{\mathbb{E}})$ compact in $\mathrm{L}^1_{\mathbb{E}^*}[\mathbb{E}](\mathcal{B})$.

Proof. *Step* 1. For any sequence (B_n) in $\mathcal{B}$ with $\lim_{n\to\infty} \mathrm{P}(B_n) = 0$, we have

$$\lim_{n\to\infty} \int_{B_n} \|f(\omega) - g_n(\omega)\| \, d\mathrm{P}(\omega) = 0.$$

Suppose by contradiction that there exists a sequence (B_n) in $\mathcal{B}$ such that

$$\int_{B_n} \|f(\omega) - g_n(\omega)\| \, d\mathrm{P}(\omega) \not\to 0.$$

Then there exist $\varepsilon > 0$ and a subsequence (g_{n_k}) of (g_n) and a subsequence (B_{n_k}) of (B_n) such that

$$\int_{B_{n_k}} \|f(\omega) - g_{n_k}(\omega)\| \, d\mathrm{P}(\omega) \geq \varepsilon$$

for all k. Let us consider the sequence (h_{n_k}) in $L^1_{\mathbb{E}^*}[\mathbb{E}](\mathcal{B})$ defined by

$$h_{n_k} = \mathbb{1}_{\Omega\setminus B_{n_k}} g_{n_k} + \mathbb{1}_{B_{n_k}} g_1.$$

Then we have $h_{n_k} \in L^1_{\mathbb{E}^*}[\mathbb{E}](\mathcal{B})$ for all k so that

$$\inf\{\int_\Omega \|f-g\|\,dP;\ g \in L^1_{\mathbb{E}^*}[\mathbb{E}](\mathcal{B})\} \le \int_\Omega \|f-h_{n_k}\|\,dP$$

for all k. Hence

$$\inf\{N_1(f-g);\ g\in\mathcal{H}\} \le \liminf_{k\to\infty}[N_1(\mathbb{1}_{\Omega\setminus B_{n_k}}(f-g_{n_k})) + N_1(\mathbb{1}_{B_{n_k}}(f-g_1))].$$

Since $\lim_{k\to\infty} P(B_{n_k}) = 0$ we have $\lim_{k\to\infty} N_1(\mathbb{1}_{B_{n_k}}(f-g_1)) = 0$. Therefore we get

$$\begin{aligned}\inf\{N_1(f-g);\ g\in L^1_{\mathbb{E}^*}[\mathbb{E}](\mathcal{B})\} &\le \liminf_{k\to\infty} N_1(\mathbb{1}_{\Omega\setminus B_{n_k}}(f-g_{n_k}))\\ &\le \liminf_{k\to\infty} N_1(f-g_{n_k}) - \varepsilon.\end{aligned}$$

That is a contradiction.

Step 2. Any minimizing sequence is relatively sequentially $\sigma(L^1_{\mathbb{E}^*}[\mathbb{E}], L^\infty_{\mathbb{E}})$ compact in $L^1_{\mathbb{E}^*}[\mathbb{E}](\mathcal{B})$. Using the notations of Step 1 and the triangular inequality:

$$\int_{B_n} \|g_n\|\,dP \le \int_{B_n} \|f-g_n\|\,dP + \int_{B_n} \|f\|\,dP$$

we see that

$$\lim_{n\to\infty} \int_{B_n} \|g_n\|\,dP = 0$$

for any sequence (B_n) in $\mathcal{B}$ with $\lim_{n\to\infty} P(B_n) = 0$. Therefore (g_n) is uniformly integrable in $L^1_{\mathbb{E}^*}[\mathbb{E}](\mathcal{B})$. In view of Theorem 6.5.9 we conclude that (g_n) is relatively sequentially $\sigma(L^1_{\mathbb{E}^*}[\mathbb{E}], L^\infty_{\mathbb{E}})$ compact in $L^1_{\mathbb{E}^*}[\mathbb{E}](\mathcal{B})$. The proof is therefore complete. □

Now we are able to state the following best approximation result.

Theorem 6.5.16 *Suppose that $\mathcal{B}$ is a complete sub-σ-algebra of $\mathcal{S}$, then for any $f \in L^1_{\mathbb{E}^*}[\mathbb{E}](\mathcal{S})$, there exists $g \in L^1_{\mathbb{E}^*}[\mathbb{E}](\mathcal{B})$ such that*

$$\int_\Omega \|f-g\|\,dP = \inf\{\int_\Omega \|f-h\|\,dP;\ h\in L^1_{\mathbb{E}^*}[\mathbb{E}](\mathcal{B})\}.$$

Proof. Let (g_n) be a minimizing sequence in $L^1_{\mathbb{E}^*}[\mathbb{E}](\mathcal{B})$. By Lemma 6.5.15, (g_n) is relatively sequentially $\sigma(L^1_{\mathbb{E}^*}[\mathbb{E}], L^\infty_{\mathbb{E}})$ compact in $L^1_{\mathbb{E}^*}[\mathbb{E}](\mathcal{B})$. Hence we may suppose that (g_n) converges $\sigma(L^1_{\mathbb{E}^*}[\mathbb{E}](\mathcal{B}), L^\infty_{\mathbb{E}}(\mathcal{B}))$ in $L^1_{\mathbb{E}^*}[\mathbb{E}](\mathcal{B})$ to a function $g \in L^1_{\mathbb{E}^*}[\mathbb{E}](\mathcal{B})$.

Step 1. Claim: (g_n) converges $\sigma(\mathrm{L}^1_{\mathbb{E}^*}[\mathbb{E}](\Omega,\mathcal{S},\mathrm{P}),\mathrm{L}^\infty_{\mathbb{E}}(\Omega,\mathcal{S},\mathrm{P}))$ to g. By the $\sigma(\mathrm{L}^1_{\mathbb{E}^*}[\mathbb{E}](\mathcal{B}),\mathrm{L}^\infty_{\mathbb{E}}(\mathcal{B}))$ convergence of (g_n) in $\mathrm{L}^1_{\mathbb{E}^*}[\mathbb{E}](\mathcal{B})$ we have

$$\forall v\in \mathrm{L}^\infty_{\mathbb{E}}(\mathcal{B}) \quad \lim_{n\to\infty}\int_\Omega \langle g_n(\omega),v(\omega)\rangle\,d\mathrm{P}(\omega)=\int_\Omega \langle g(\omega),v(\omega)\rangle\,d\mathrm{P}(\omega).$$

Now let $v\in \mathrm{L}^\infty_{\mathbb{E}}(\mathcal{S})$ and let $E^{\mathcal{B}}(v)$ be the conditional expectation of v w.r.t. $\mathcal{B}$. Then $E^{\mathcal{B}}(v)\in \mathrm{L}^\infty_{\mathbb{E}}(\mathcal{B})$ and

$$\forall u\in \mathrm{L}^1_{\mathbb{E}^*}[\mathbb{E}](\mathcal{B}) \quad \int_\Omega \langle u,E^{\mathcal{B}}(v)\rangle\,d\mathrm{P}=\int_\Omega \langle u,v\rangle\,d\mathrm{P}.$$

It follows that

$$\begin{aligned}\lim_{n\to\infty}\int_\Omega \langle g_n(\omega),v(\omega)\rangle\,d\mathrm{P}(\omega)&=\lim_{n\to\infty}\int_\Omega \langle g_n(\omega),E^{\mathcal{B}}(v)(\omega)\rangle\,d\mathrm{P}(\omega)\\&=\int_\Omega \langle g(\omega),E^{\mathcal{B}}(v)(\omega)\rangle\,d\mathrm{P}(\omega)\\&=\int_\Omega \langle g(\omega),v(\omega)\rangle\,d\mathrm{P}(\omega),\end{aligned}$$

for all $v\in \mathrm{L}^\infty_{\mathbb{E}}(\mathcal{S})$.

Step 2. Claim: $\int_\Omega \|f-g\|\,d\mathrm{P}=\inf\{\int_\Omega \|f-h\|\,d\mathrm{P};\ h\in \mathrm{L}^1_{\mathbb{E}^*}[\mathbb{E}](\mathcal{B})\}$. Now let us consider the subset $\mathrm{Step}(\mathcal{S})$ of the dual $\mathrm{L}^1_{\mathbb{E}^*}[\mathbb{E}](\mathcal{S})^*$ defined by

$$\mathrm{Step}(\mathcal{S})=\{\hat{s}\in \mathrm{L}^1_{\mathbb{E}^*}[\mathbb{E}](\mathcal{S})^*;\ s:\Omega\to\overline{B}_{\mathbb{E}} \text{ is a step } \mathcal{S}\text{-measurable function}\}$$

where

$$\hat{s}(f):=\int_\Omega \langle f,s\rangle\,d\mathrm{P},\qquad \forall f\in \mathrm{L}^1_{\mathbb{E}^*}[\mathbb{E}](\mathcal{S}).$$

Then $\mathrm{Step}(\mathcal{S})$ is an absolutely convex subset of $\mathrm{L}^1_{\mathbb{E}^*}[\mathbb{E}](\mathcal{S})^*$ and is included in the closed unit ball U_1 of $(\mathrm{L}^1_{\mathbb{E}^*}[\mathbb{E}](\mathcal{S})^*$ because $\|s(\omega)\|\le 1$ for all $s\in\mathbb{S}$ and for all $\omega\in\Omega$. In view of the proof of [BC01, Theorem 4.1], we have

$$\overline{N}_1(f)=\sup_{l\in\mathrm{Step}(\mathcal{S})} l(f),\qquad \forall f\in \mathrm{L}^1_{\mathbb{E}^*}[\mathbb{E}](\mathcal{S}).$$

Hence we have

$$\int_\Omega \|f-g_n\|\,d\mathrm{P}\ge\langle\hat{s},f-g_n\rangle,$$

for all n and for all $\hat{s}\in\mathrm{Step}(\mathcal{S})$. It follows that

$$\liminf_{n\to\infty}\int_\Omega \|f-g_n\|\,d\mathrm{P}\ge\langle\hat{s},f-g\rangle=\int_\Omega\langle f-g,s\rangle\,d\mathrm{P}$$

for all $\hat{s} \in \text{Step}(\mathcal{S})$. Hence by taking the supremum over $\text{Step}(\mathcal{S})$ in the preceding inequality we get

$$\liminf_{n\to\infty} \int_\Omega \|f - g_n\| \, d\mathrm{P} \geq \int_\Omega \|f - g\| \, d\mathrm{P} \geq \inf\{\int_\Omega \|f - h\| \, d\mathrm{P};\ h \in \mathrm{L}^1_{\mathbb{E}^*}[\mathbb{E}](\mathcal{B})\}.$$

□

References See Shintani–Ando [SA76], Valadier [Val84] for the case of $\mathrm{L}^1_{\mathbb{E}}$ spaces and Herrndorf [Her81] for the case of Orlicz spaces of $\mathbb{E}$-valued functions where $\mathbb{E}$ is a separable reflexive Banach space. Note that in Castaing–Clauzure [CC85] the authors deal with both Orlicz spaces and the space $\mathrm{L}^1(\Omega, \mathcal{S}, \mathrm{P}; \mathrm{cw}\mathcal{K}(\mathbb{E}))$ of measurable and integrably bounded convex weakly compact valued multifunctions in a separable reflexive Banach space $\mathbb{E}$.

Now we focus our attention to the case when $\mathbb{E}$ is a separable reflexive Banach space. The following result shows the relationship between the modes of convergence encountered along this section.

Proposition 6.5.17 *Suppose that $\mathbb{E}$ is a separable reflexive Banach space, (u_n) is a bounded sequence in $\mathrm{L}^1_{\mathbb{E}}(\Omega, \mathcal{S}, \mathrm{P})$. Then the following holds:*

(a) *$(\underline{\delta}_{u_n})$ (up to an extracted subsequence) $\sigma(\mathcal{Y}^1(\Omega; \mathbb{E}_\sigma), \mathcal{C}th^b(\Omega, \mathbb{E}_\sigma))$ converges to a Young measure λ^∞ in $\mathcal{Y}^1(\Omega; \mathbb{E}_\sigma)$.*

(b) *(u_n) (up to an extracted subsequence) Mazur strongly converges a.e. to a function $u_\infty \in \mathrm{L}^1_{\mathbb{E}}(\Omega, \mathcal{S}, \mathrm{P})$.*

(c) *(u_n) (up to an extracted subsequence) biting weakly converges to a function $v_\infty \in \mathrm{L}^1_{\mathbb{E}}(\Omega, \mathcal{S}, \mathrm{P})$.*

(d) *If, in adddition, $\mathbb{E}$ is B–convex reflexive, (u_n) (up to an extracted subsequence) Komlós strongly converges to a function $w_\infty \in \mathrm{L}^1_{\mathbb{E}}(\Omega, \mathcal{S}, \mathrm{P})$.*

If we have considered the same extracted subsequence in (a), (b), (c) and (d), then

$$\mathrm{bar}\,(\lambda^\infty)\,(\omega) = u_\infty(\omega) = v_\infty(\omega) = w_\infty(\omega) \ \text{ a.e.}$$

The proof is straightforward since it follows directly from the above mentioned results.

In the following two sections we treat some problems of convergence in the spaces of scalarly integrable $\mathbb{E}$-valued functions $\mathrm{P}^1_{\mathbb{E}}(\Omega, \mathcal{S}, \mathrm{P})$ and $\mathrm{L}^1_{\mathbb{E}^*}[\mathbb{E}](\Omega, \mathcal{S}, \mathrm{P})$ under extreme point condition.

6.6 Visintin-type theorem in $\mathrm{P}^1_{\mathbb{E}}(\Omega, \mathcal{S}, \mathrm{P})$

We begin with some preliminary results. Some parts of the results presented here are borrowed from [ACV92, ACV98, BCG99b, BC01], see also [BS03]. If $f : \Omega \to \mathbb{E}$ is a scalarly integrable function, the Pettis norm $\|f\|_{\mathrm{Pe}}$ of f [Gei81, Huf86, Mus91] is defined by $\|f\|_{\mathrm{Pe}} = \sup_{x' \in \overline{B}_{\mathbb{E}^*}} \int_\Omega |\langle x', f\rangle| \, d\mathrm{P}$. The space $\mathrm{P}^1_{\mathbb{E}}(\Omega, \mathcal{S}, \mathrm{P})$ of $\mathbb{E}$-valued Pettis integrable functions is endowed with the Pettis norm $\|.\|_{\mathrm{Pe}}$. A subset $\mathcal{H} \subset \mathrm{P}^1_{\mathbb{E}}(\Omega, \mathcal{S}, \mathrm{P})$ is *Pettis uniformly integrable* (PUI for short) if, for every $\varepsilon > 0$, there exists $\delta > 0$ such that

$$\mathrm{P}(A) \leq \delta \Longrightarrow \sup_{u \in \mathcal{H}} \| \mathbb{1}_A u\|_{\mathrm{Pe}} \leq \varepsilon.$$

If $f \in \mathrm{P}^1_{\mathbb{E}}$, the singleton $\{f\}$ is PUI since the set $\{\langle x', f\rangle;\ \|x'\| \leq 1\}$ is uniformly integrable [Gei81]. More generally, a subset $\mathcal{H} \subset \mathrm{P}^1_{\mathbb{E}}(\Omega, \mathcal{S}, \mathrm{P})$ is *scalarly Pettis uniformly integrable* if the set $\{\langle x', f\rangle;\ f \in \mathcal{H}, \|x'\| \leq 1\}$ is uniformly integrable in the space $\mathrm{L}^1_{\mathbb{R}}(\Omega, \mathcal{S}, \mathrm{P})$ (this property, in the context of locally convex spaces, is called *$\mathfrak{S}$-uniform scalar integrability* in [CRdF00]). If $\mathcal{H}$ is scalarly Pettis uniformly integrable, then it is Pettis uniformly integrable.

Proof. Indeed, we have

$$\lim_{a \to +\infty} \sup_{f \in \mathcal{H}} \sup_{x' \in \overline{B}_{\mathbb{E}^*}} \int_{\{|\langle x', f\rangle| > a\}} |\langle x', f\rangle| \, d\mathrm{P} = 0.$$

For any $x' \in \overline{B}_{\mathbb{E}^*}$, one has

$$(^*) \quad \int_A |\langle x', u\rangle| \, d\mathrm{P} = \int_{A \cap \{|\langle x', u\rangle| \leq a\}} |\langle x', u\rangle| \, d\mathrm{P} + \int_{A \cap \{|\langle x', u\rangle| > a\}} |\langle x', u\rangle| \, d\mathrm{P}.$$

Let a be large enough in order to ensure

$$\forall x' \in \overline{B}_{\mathbb{E}^*} \quad \forall u \in \mathcal{H} \quad \int_{A \cap \{|\langle x', u\rangle| > a\}} |\langle x', u\rangle| \, d\mathrm{P} \leq \varepsilon/2.$$

Thus, the last term of $(^*)$ is $\leq \varepsilon/2$. Now, if δ is small enough in order to ensure $a\delta \leq \varepsilon/2$, we obtain

$$\int_{A \cap \{|\langle x', u\rangle| \leq a\}} |\langle x', u\rangle| \, d\mathrm{P} \leq a\,\mathrm{P}(A) \leq \varepsilon/2$$

as soon as $\mathrm{P}(A) \leq \delta$. □

Let K be a convex subset of $\mathbb{E}$. Recall that a point $x \in K$ is a *denting point of* K if for any $\varepsilon > 0$, $x \notin \overline{\mathrm{co}}(K \setminus B_{\mathbb{E}}(0, \varepsilon))$. The set of all denting points of K is denoted by $\partial_{\mathrm{dent}}(K)$. Let $\Gamma : \Omega \to \mathrm{cw}\mathcal{K}(\mathbb{E})$ be a scalarly integrable multifunction.

We denote by $\mathrm{S}^{\mathrm{Pe}}_\Gamma$ the set of all Pettis integrable selections of Γ. If $\mathrm{S}^{\mathrm{Pe}}_\Gamma$ is nonempty, the *integral of* Γ *over a* $\mathcal{S}$*-measurable set* A is defined by

$$\int_A \Gamma \, d\mathrm{P} := \{\int_A f \, d\mathrm{P};\ f \in \mathrm{S}^{\mathrm{Pe}}_\Gamma\}$$

where $\int_A f \, d\mathrm{P}$ is the Pettis integral of f over A.

A scarlarly integrable multifunction $\Gamma : \Omega \to \mathrm{cw}\mathcal{K}(\mathbb{E})$ is *Pettis-integrable* if the set

$$\{\delta^*(x', \Gamma);\ \|x'\| \leq 1\}$$

is uniformly integrable in $\mathrm{L}^1_{\mathbb{R}}(\mathrm{P})$.

A sequence (u_n) in $\mathrm{P}^1_{\mathbb{E}}$ *weakly* converges to $u \in \mathrm{P}^1_{\mathbb{E}}$ if $u_n \to u$ in the topology of pointwise convergence on $\mathrm{L}^\infty_{\mathbb{R}}(\mathrm{P}) \otimes \mathbb{E}^*$.

A subset $\mathcal{H}$ in $\mathrm{P}^1_{\mathbb{E}}$ is $\mathrm{cw}\mathcal{K}(\mathbb{E})$*-tight* (resp. $\mathrm{c}\mathcal{K}(\mathbb{E})$*-tight*) if, for every $\varepsilon > 0$, there exists a $\mathrm{cw}\mathcal{K}(\mathbb{E})$-valued (resp. $\mathrm{c}\mathcal{K}(\mathbb{E})$-valued) Pettis-integrable multifunction Γ_ε satisfying:

$$\sup_{u \in \mathcal{H}} \mathrm{P}(\{\omega \in \Omega;\ u(\omega) \notin \Gamma_\varepsilon(\omega)\}) \leq \varepsilon.$$

The following is a version of the Lebesgue–Vitali theorem for scalarly Pettis uniformly integrable functions.

Proposition 6.6.1 *Suppose* $(u_n)_{n \in \mathbb{N}}$ *is a* PUI *sequence of scalarly integrable* $\mathbb{E}$*-valued functions converging in measure to a scalarly integrable* $\mathbb{E}$*-valued function* u_∞*, then* $\|u_n - u_\infty\|_{\mathrm{Pe}} \to 0$.

Proof. Suppose the conclusion does not hold. There exists $\varepsilon > 0$ such that, for a subsequence still denoted by $(u_n)_n$, $\forall n$, $\|u_n - u_\infty\|_{\mathrm{Pe}} > \varepsilon$. Since $(u_n)_n$ is PUI, there exists $\delta > 0$ such that $\mathrm{P}(A) < \delta$ implies $\forall n \in \mathbb{N} \cup \{\infty\}$, $\|\mathbb{1}_A u_n\|_{\mathrm{Pe}} \leq \varepsilon/3$. There exists a subsequence still not relabeled such that $\|u_n - u_\infty\| \to 0$ a.e. By virtue of Egorov's theorem, there exists $B \in \mathcal{S}$ such that $\mathrm{P}(\Omega \setminus B) < \delta$ and $\|u_n - u_\infty\| \to 0$ uniformly on B. Let n_0 be such that $\forall n \geq n_0$, $\|u_n(\omega) - u_\infty(\omega)\| \leq \varepsilon/3$ on B. Then $\forall n \geq n_0$, we have

$$\begin{aligned}\|u_n - u_\infty\|_{\mathrm{Pe}} &\leq \|\mathbb{1}_B(u_n - u_\infty)\|_{\mathrm{Pe}} + \|\mathbb{1}_{\Omega \setminus B}(u_n - u_\infty)\|_{\mathrm{Pe}} \\ &\leq \varepsilon/3 + \|\mathbb{1}_{\Omega \setminus B} u_n\|_{\mathrm{Pe}} + \|\mathbb{1}_{\Omega \setminus B} u_\infty\|_{\mathrm{Pe}} \leq \varepsilon.\end{aligned}$$

This contradicts the initial assumption. □

Proposition 6.6.2 *Let* $\mathbb{E}$ *be a separable Banach space,* $(u^n)_{n \in \mathbb{N}}$ *be a scalarly Pettis uniformly integrable sequence of* $\mathbb{E}$*-valued Pettis integrable functions, that is, the set*

$$\{\langle x', u^n(.)\rangle;\ \|x'\| \leq 1,\ n \in \mathbb{N}\}$$

is uniformly integrable in $L^1_{\mathbb{R}}(\Omega, \mathcal{S}, P)$. *Suppose further that the sequence* $(u^n)_{n\in\mathbb{N}}$ *is norm tight, then there exists a subsequence* $(\lambda^{\alpha(n)})_{n\in\mathbb{N}}$ *of the sequence* $(\lambda^n)_{n\in\mathbb{N}} = (\underline{\delta}_{u^n})_{n\in\mathbb{N}}$ *and* $\lambda^\infty \in \mathcal{Y}^1(\Omega, \mathcal{S}, P; \mathbb{E})$ *such that the following hold:*

(a) $\lim_n \langle \psi, \lambda^{\alpha(n)} \rangle = \langle \psi, \lambda^\infty \rangle$ *for every bounded Carathéodory integrand* ψ *on* $\Omega \times \mathbb{E}$.

(b) *There is a scalarly* $\mathbb{E}^*$*-integrable* $\mathbb{E}^{**}$*-valued mapping* $b_{\lambda^\infty} : \omega \mapsto b_{\lambda^\infty_\omega}$ *defined by*

$$\langle b_{\lambda^\infty_\omega}, x' \rangle = \int_{\mathbb{E}} \langle x', x \rangle \, d\lambda^\infty_\omega(x),$$

for all $x' \in \mathbb{E}^*$ *and for all* $\omega \in \Omega$ *which satisfies*

$$\lim_{n\to\infty} \int_\Omega g(\omega) \langle x', u^{\alpha(n)}(\omega) \rangle \, d P(\omega) = \int_\Omega g(\omega) \langle x', b_{\lambda^\infty_\omega} \rangle \, d P(\omega)$$

for all $g \in L^\infty_{\mathbb{R}}(\Omega, \mathcal{S}, P)$ *and for all* $x' \in \overline{B}_{\mathbb{E}^*}$.

Proof. (a) follows from Theorem 4.3.5. We claim that

$$\begin{aligned} \int_A \Big[\int_{\mathbb{E}} |\langle x', x \rangle| \, d\lambda^\infty_\omega(x)\Big] \, d P(\omega) &= \lim_n \int_A |\langle x', u^{\alpha(n)}(\omega) \rangle| \, d P(\omega) \\ &\leq \sup_{\|x'\| \leq 1,\, n\in\mathbb{N}} \int_\Omega |\langle x', u^n(\omega) \rangle| \, d P(\omega) < +\infty \end{aligned} \tag{6.6.1}$$

for all $A \in \mathcal{S}$ and for all $x' \in \overline{B}_{\mathbb{E}^*}$. Indeed since $\{\langle x', u^n(.) \rangle;\ \|x'\| \leq 1,\ n \in \mathbb{N}\}$ is uniformly integrable in $L^1_{\mathbb{R}}$ the second equality is obvious whereas the first equality need a careful look. Let $A \in \mathcal{S}$ and $\|x'\| \leq 1$ and let h be the positive Carathéodory integrand $h(\omega, x) := \mathbb{1}_A(\omega) |\langle x', x \rangle|$ defined on $\Omega \times \mathbb{E}$. Then the sequence $(h(u_n)) = (h(., u_n(.)))$ is uniformly integrable. Now using (a) and Theorem 6.3.5 yields

$$\lim_{n\to\infty} \int_\Omega h(\omega, u_n(\omega)) \, d P(\omega) = \int_\Omega \Big[\int_{\mathbb{E}} h(\omega, s) \, d\lambda^\infty_\omega(s)\Big] \, d P(\omega).$$

Hence (6.6.1) shows that the mapping: $b_{\lambda^\infty} : \omega \mapsto b_{\lambda^\infty_\omega}$ from Ω into $\mathbb{E}^{**}$ defined by

$$\langle b_{\lambda^\infty_\omega}, x' \rangle = \int_{\mathbb{E}} \langle x', x \rangle \, d\lambda^\infty_\omega(x)$$

for all $x' \in \mathbb{E}^*$ and for all $\omega \in \Omega$, is scalarly integrable, i.e. for every $x' \in \mathbb{E}^*$ the function $\omega \mapsto \langle b_{\lambda^\infty_\omega}, x' \rangle$ is integrable. Similarly we have

$$\begin{aligned} \lim_n \int g(\omega) \langle x', u^{\alpha(n)}(\omega) \rangle \, d P(\omega) &= \int_\Omega \Big[\int_{\mathbb{E}} g(\omega) \langle x', x \rangle \, d\lambda^\infty_\omega(x)\Big] \, d P(\omega) \\ &= \int_\Omega g(\omega) \langle x', b_{\lambda^\infty_\omega} \rangle \, d P(\omega) \end{aligned} \tag{6.6.2}$$

for all $g \in L^\infty_{\mathbb{R}}(\Omega, \mathcal{S}, P)$ and for all $x' \in \overline{B}_{\mathbb{E}^*}$, thus proving (b). □

Proposition 6.6.3 *Assume that* $\mathbb{E}^*$ *is strongly separable and* (u^n) *is a scalarly Pettis uniformly integrable and norm tight sequence in* $\mathrm{P}^1_{\mathbb{E}}$ *which* $\sigma(\mathrm{P}^1_{\mathbb{E}}, \mathrm{L}^\infty \otimes \mathbb{E}^*)$ *converges to a Pettis integrable function* $u^\infty \in \mathrm{P}^1_{\mathbb{E}}(\Omega, \mathcal{S}, \mathrm{P})$*, then there exists a subsequence* $(\lambda^{\alpha(n)})_{n\in\mathbb{N}}$ *of* $(\lambda^n)_{n\in\mathbb{N}} = (\underline{\delta}_{u^n})_{n\in\mathbb{N}}$ *and* $\lambda^\infty \in \mathcal{Y}^1(\Omega, \mathcal{S}, \mathrm{P}; \mathbb{E})$ *such that* $(\lambda^{\alpha(n)})_{n\in\mathbb{N}}$ *S-stably converges to* λ^∞ *which satisfies* $\lambda^\infty_\omega(\bigcap_p \mathrm{cl}\{u_m(\omega);\, n \geq p\}) = 1$ *and* $u^\infty(\omega) = \int_{\mathbb{E}} x\, d\lambda^\infty_\omega(x)$ *a.e.*

Proof. The existence of λ^∞ satisfying the required support property follows easily from the norm tighness of (u^n). According to the preceding proposition, the barycenter $\mathrm{bar}\,(\lambda^\infty_\omega) = \int_{\mathbb{E}} x\, d\lambda^\infty_\omega(x)$ belongs to $\mathbb{E}^{**}$ and the mapping $\mathrm{bar}\,(\lambda^\infty) : \omega \mapsto \mathrm{bar}\,(\lambda^\infty_\omega)$ is scalarly $\mathbb{E}^*$-integrable. As $\mathbb{E}^*$ is strongly separable there is a sequence $(x'_j)_j$ in $\overline{B}_{\mathbb{E}^*}$ which separates the points of $\mathbb{E}^{**}$. In view of the preceding proposition and the weak convergence of (u^n) to u^∞ we get

$$\begin{aligned}\lim_{n\to\infty} \int_A \langle x'_j, u^{\alpha(n)}(\omega)\rangle\, d\mathrm{P}(\omega) &= \int_A \langle x'_j, u_\infty(\omega)\rangle\, d\mathrm{P}(\omega)\\ &= \int_A \langle x'_j, \mathrm{bar}\,(\lambda^\infty_\omega)\rangle\, d\mathrm{P}(\omega)\end{aligned}$$

for all j and for all $A \in \mathcal{S}$. Hence $u^\infty(\omega) = \mathrm{bar}\,(\lambda^\infty_\omega)$ a.e. □

The following is a Pettis variant of a result due to Benabdellah [Ben91, Theorem 3.2] who treated a similar result for the space $\mathrm{L}^1_{\mathbb{E}}(\Omega, \mathcal{S}, \mathrm{P})$. See also [Val89, Theorem 9].

Theorem 6.6.4 *Assume that* $\mathbb{E}^*$ *is strongly separable,* (u^n) *is a scalarly Pettis uniformly integrable and norm tight sequence in* $\mathrm{P}^1_{\mathbb{E}}$ *which* $\sigma(\mathrm{P}^1_{\mathbb{E}}, \mathrm{L}^\infty \otimes \mathbb{E}^*)$ *converges to a scalarly integrable* $\mathbb{E}$*-valued function* u^∞ *and* $\varphi : \Omega \times \mathbb{E} \to [0, +\infty]$ *is a convex normal integrand satisfying*

(*i*) $K(\omega) := \overline{\mathrm{co}}(\bigcap_p \mathrm{cl}\{u^m(\omega);\, n \geq p\}) \subset \mathrm{dom}\,\varphi_\omega$,

(*ii*) $\limsup_n \int_\Omega \varphi(\omega, u^n(\omega))\, d\mathrm{P}(\omega) \leq \int_\Omega \varphi(\omega, u^\infty(\omega))\, d\mathrm{P}(\omega) < +\infty$,

(*iii*) $(u^\infty(\omega), \varphi(\omega, u^\infty(\omega)))$ *is an extremal point of* $\mathrm{Epi}\,\varphi_\omega$ *a.e.*,

then $u^\infty(\omega) \in K(\omega)$ *almost everywhere, and* (u^n) *converges to* u^∞ *for the Pettis norm.*

Proof. From (*ii*) and Jensen's inequality we deduce that

$$\int_{K(\omega)} \varphi(\omega, x)\, d\lambda^\infty_\omega(x) = \varphi(\omega, u^\infty(\omega)) \text{ a.e.}$$

because $\lambda^\infty_\omega(K(\omega)) = 1$ and $u^\infty(\omega) = \mathrm{bar}\,(\lambda^\infty_\omega)$ a.e., where λ^∞ is the Young measure limit given in Proposition 6.6.3. If $(u^\infty(\omega), \varphi(\omega, u^\infty(\omega)))$ is an extremal point of $\mathrm{Epi}\,\varphi_\omega$ a.e., $\delta_{u^\infty(\omega)} = \lambda^\infty_\omega$ a.e. by applying Proposition 6.5.5. By Part 3 of

Theorem 3.1.2 $u^{\alpha(n)}$ converges to u^∞ in measure. By virtue of Proposition 6.6.1, $u^{\alpha(n)}$ converges to u^∞ for the Pettis norm since (u^n) is scalarly Pettis uniformly integrable. The conclusion follows. □

Corollary 6.6.5 *Assume that* $\mathbb{E}^*$ *is strongly separable,* (u^n) *is a scalarly Pettis uniformly integrable and norm tight sequence in* $\mathrm{P}^1_{\mathbb{E}}$ *which* $\sigma(\mathrm{P}^1_{\mathbb{E}}, \mathrm{L}^\infty \otimes \mathbb{E}^*)$ *converges to a scalarly integrable* $\mathbb{E}$*-valued function* u^∞ *such that* $u^\infty(\omega)$ *is an extremal point of* $\overline{\mathrm{co}}(\bigcap_p \mathrm{cl}\{u_m(\omega);\, n \geq p\})$ *a.e., then* (u^n) *converges to* u_∞ *for the Pettis norm.*

Proof. Apply the preceding theorem by taking

$$\varphi(\omega, x) = \delta(x, \overline{\mathrm{co}}(\bigcap_p \mathrm{cl}\{u_m(\omega);\, n \geq p\}))$$

where $\delta(.,C)$ is the indicator function (in the sense of Convex Analysis) of C. □

Now we proceed to a variant of Theorem 6.6.4. A multifunction $\Gamma : \Omega \to \mathrm{cw}\mathcal{K}(\mathbb{E})$ is *Pettis integrable* if it is scalarly integrable, that is, if the function $\delta^*(x', \Gamma(.))$ is integrable for every $x' \in \mathbb{E}^*$ and the set $\{\delta^*(x', \Gamma(.));\, \|x'\| \leq 1\}$ is uniformly integrable in $\mathrm{L}^1_{\mathbb{R}}(\Omega, \mathcal{S}, \mathrm{P})$. We begin with a lemma formulated for simplicity in a special case. See [ACV98, Lemma 2.2 page 326].

Lemma 6.6.6 *Let* $\mathbb{E}$ *be a separable Banach space. Let* $\Gamma : \Omega \to \mathrm{c}\mathcal{K}(\mathbb{E})$ *be a Pettis integrable multifunction, and* (u_n) *be a sequence of Pettis integrable selection of* Γ *which converges* $\sigma(\mathrm{P}^1_{\mathbb{E}}, \mathrm{L}^\infty \otimes \mathbb{E}^*)$ *to a Pettis integrable selection* u *of* Γ*. Suppose* $u(\omega)$ *is an extremal point of* $\Gamma(\omega)$ *for a.e.* $\omega \in \Omega$*, then* $\|u_n - u\|_{\mathrm{Pe}} \to 0$.

Proof. Without lost we may suppose that $u = 0$ so that $0 \in \partial_{\mathrm{ext}}\,(\Gamma(\omega))$ a.e. Since the sequence (u_n) is scalarly Pettis uniformly integrable, we need only to prove that $\|u_n(.)\| \to 0$ in measure. Suppose not. Then there exist $\varepsilon > 0$ and $\eta > 0$ such that

$$\mathrm{P}(\{\omega \in \Omega;\, \|u_n(\omega)\| \geq \eta\}) \geq \eta$$

for infinitely many n, namely there exists an infinite subset $S \subset \mathrm{N}$ such that the preceding inequality holds for all $n \in S$. Let us consider the following Pettis integrable functions

$$v_n := \mathbb{1}_{\{\omega \in \Omega;\, u_n(\omega) \notin B_{\mathbb{E}}(0,\varepsilon)\}} u_n$$

and $w_n := u_n - v_n$. By virtue of Theorem 6.3.4 the sequence (v_n) is relatively sequentially $\sigma(\mathrm{P}^1_{\mathbb{E}}, \mathrm{L}^\infty \otimes \mathbb{E}^*)$ compact. By extracting an appropriate subsequence, we may suppose that (v_n) converges to $v \in \mathrm{P}^1_{\mathbb{E}}$ for this topology. It follows that (w_n) converges $\sigma(\mathrm{P}^1_{\mathbb{E}}, \mathrm{L}^\infty \otimes \mathbb{E}^*)$ to $-v$. As $0 \in \partial_{\mathrm{ext}}\,(\Gamma(\omega))$ a.e. implies $0 \in \partial_{\mathrm{ext}}\,(\Gamma(\omega))$ a.e. and u_n converges $\sigma(\mathrm{P}^1_{\mathbb{E}}, \mathrm{L}^\infty \otimes \mathbb{E}^*)$ to 0, we get $v = 0$. Since $\partial_{\mathrm{ext}}\,(\Gamma(\omega)) = \partial_{\mathrm{dent}}(\Gamma(\omega))$ because $\Gamma(\omega)$ is convex norm compact, $0 \in \partial_{\mathrm{dent}}(\Gamma(\omega))$ a.e. Hence, $0 \notin \overline{\mathrm{co}}(\Gamma(\omega) \setminus B_{\mathbb{E}}\,(0,\varepsilon))$ a.e. As the multifunction Γ is scalarly $\mathcal{S}$-measurable, its graph belongs $\mathcal{S} \otimes \mathcal{B}_{\mathbb{E}}$. Hence the graph of the multifunction $\Delta := \Gamma(.) \setminus B_{\mathbb{E}}(0, \varepsilon)$ from Ω to norm compact sets of $\mathbb{E}$ belongs to $\mathcal{S} \otimes \mathcal{B}_{\mathbb{E}}$, too. Consequently the set $A = \{\omega \in \Omega;\, \Gamma(\omega) \setminus B_{\mathbb{E}}(0, \varepsilon) \neq \emptyset\}$ is $\mathcal{S}$-measurable by a classical mesurable projection theorem [CV77, Theorem III.23] (cf. Remark 1.2.1) and it is obvious that the multifunction $\Sigma : \omega \to \overline{\mathrm{co}}(\Gamma(\omega) \setminus B_{\mathbb{E}}(0, \varepsilon))$ defined on A is convex norm compact-valued and has its graph in $(A \cap \mathcal{S}) \otimes \mathcal{B}_{\mathbb{E}}$. Hence the multifunction Ψ defined on A with nonempty values in the closed unit ball $\overline{B}_{\mathbb{E}^*}$ of $\mathbb{E}^*$ (thanks to Hahn–Banach's theorem):

$$\Psi(\omega) = \{x' \in \overline{B}_{\mathbb{E}^*};\, \delta^*(x', \Sigma(\omega)) < 0\}$$

has its graph in $(A \cap \mathcal{S}) \otimes \mathcal{B}_{\overline{B}_{\mathbb{E}^*}}$, where $\overline{B}_{\mathbb{E}^*}$ is endowed with the topology of compact convergence on $\mathbb{E}^*$. Since $\mathbb{E}$ is separable, $\overline{B}_{\mathbb{E}^*}$ is compact metrisable for this topology. Hence Ψ admits a $\mathcal{S}$ measurable selection $\sigma : A \to \overline{B}_{\mathbb{E}^*}$ and there is a sequence (σ_k) of simple $\mathcal{S}$-measurable mappings from A to $\overline{B}_{\mathbb{E}^*}$ such that σ_k pointwise converges to σ for the compact convergence. It follows that

$$\lim_{k\to\infty} \delta^*(\sigma_k(\omega), \Sigma(\omega)) = \delta^*(\sigma(\omega), \Sigma(\omega)) < 0$$

for every $\omega \in A$. In view of [ACV92, Lemma 3], there are $a < 0$ and $k_1 \in \mathbb{N}$ such that

$$\forall k \geq k_1 \quad \mathrm{P}(\{\omega \in A;\, \delta^*(\sigma_k(\omega), \Sigma(\omega)) > a\}) < \frac{\eta}{2}.$$

Let $k \geq k_1$ be fixed and set

$$A_k = \{\omega \in A;\, \delta^*(\sigma_k(\omega), \Sigma(\omega)) \leq a\} \text{ and } B_k = A \setminus A_k.$$

Then we have

$$\limsup_{n\to\infty} \langle \sigma_k(\omega), v_n(\omega)\rangle \leq 0$$

for all $\omega \in A_k$. Since σ_k is a simple function with values in $\overline{B}_{\mathbb{E}^*}$ and v_n converges $\sigma(\mathrm{P}^1_{\mathbb{E}}, \mathrm{L}^\infty \otimes \mathbb{E}^*)$ to 0, $(\langle \sigma_k, v_n\rangle)_n$ converges $\sigma(\mathrm{L}^1_{\mathbb{R}}(A_k), \mathrm{L}^\infty_{\mathbb{R}}(A_k))$ to 0. It follows that $(\langle \sigma_k, v_n\rangle)_n$ converges to 0 in measure on A_k. Consequently there exists N_1 such that

$$\forall n \geq N_1 \quad \mathrm{P}(\{\omega \in A_k;\, \langle \sigma_k(\omega), v_n(\omega)\rangle \leq a\}) < \frac{\eta}{2}.$$

We have

$$\begin{aligned}\{\omega \in A;\, v_n(\omega) \neq 0\} &= \{\omega \in A_k;\, v_n(\omega) \neq 0\} \cup \{\omega \in B_k;\, v_n(\omega) \neq 0\} \\ &\subset \{\omega \in A_k;\, v_n(\omega) \neq 0\} \cup B_k \\ &\subset \{\omega \in A_k;\, \langle \sigma_k(\omega), v_n(\omega)\rangle \leq a\} \cup B_k.\end{aligned}$$

For $n \geq N_1$, we have

$$\begin{aligned}\mathrm{P}(\{\omega \in A;\, v_n(\omega) \neq 0\}) &\leq \mathrm{P}(\{\omega \in A_k;\, \langle \sigma_k(\omega), v_n(\omega)\rangle \leq a\}) + \mathrm{P}(B_k) \\ &< \frac{\eta}{2} + \frac{\eta}{2} = \eta.\end{aligned}$$

Hence for $n \geq N_1$, $n \in S$, we get the contradiction

$$\mathrm{P}(\{\omega \in \Omega;\ u_n(\omega) \notin B_{\mathbb{E}^*}(0,\varepsilon)\}) = \mathrm{P}(\{\omega \in A;\ v_n(\omega) \neq 0\}) < \eta.$$

□

The above results lead to other variants. We mention first the following which extends Lemma 6.6.6.

Theorem 6.6.7 *Let $\mathbb{E}$ be a separable Banach space and Γ be a measurable multifunction on Ω with bounded closed convex values in $\mathbb{E}$. Let (u^n) be a scalarly Pettis uniformly integrable and norm tight sequence in $\mathrm{P}^1_{\mathbb{E}}$ such that $u^n(\omega) \in \Gamma(\omega)$ for all n and for all $\omega \in \Omega$. Assume further that (u^n) converges $\sigma(\mathrm{P}^1_{\mathbb{E}}, \mathrm{L}^\infty \otimes \mathbb{E}^*)$ to a scalarly integrable $\mathbb{E}$-valued function u^∞. Let $\varphi : \Omega \times \mathbb{E} \to [0, +\infty]$ be a convex normal integrand satisfying*

(i) $K(\omega) := \overline{\mathrm{co}}(\bigcap_p \mathrm{cl}[\{u^m(\omega);\ n \geq p\}] \subset \mathrm{dom}\,\varphi_\omega,$

(ii) $\limsup_n \int_\Omega \varphi(\omega, u^n(\omega))\, d\,\mathrm{P}(\omega) \leq \int_\Omega \varphi(\omega, u^\infty(\omega))\, d\,\mathrm{P}(\omega) < +\infty,$

(iii) $(u^\infty(\omega), \varphi(\omega, u^\infty(\omega))$ *is an extremal point of* $\mathrm{Epi}\,\varphi_\omega$ *a.e.*

Then $u^\infty(\omega) \in K(\omega)$ almost everywhere, and (u^n) converges to u^∞ for the Pettis norm.

Proof. We use some arguments given in the proof of Propositions 6.6.2, 6.6.3, 6.6.4 with appropriated modifications. Since (u^n) is norm tight there exist a subsequence $(\lambda^{\alpha(n)})_{n\in\mathbb{N}}$ of the sequence $(\lambda^n)_{n\in\mathbb{N}} = (\delta_{u^n})$ and $\lambda^\infty \in \mathcal{Y}^1(\Omega, \mathcal{S}, \mathrm{P}; \mathbb{E})$ such that the following hold:

(*) $\lim_n \langle h, \lambda^{\alpha(n)}\rangle = \langle h, \lambda^\infty\rangle$ for every bounded Carathéodory integrand h on $\Omega\times\mathbb{E}$.

(**) there is a scalarly $\mathbb{E}^*$-integrable $\mathbb{E}^{**}$-valued mapping $b_{\lambda^\infty} : \omega \mapsto b_{\lambda^\infty_\omega}$ defined by

$$\langle b_{\lambda^\infty_\omega}, x'\rangle = \int_{\mathbb{E}} \langle x', x\rangle\, d\lambda^\infty_\omega(x)$$

for all $x' \in \mathbb{E}^*$ and for all $\omega \in \Omega$ which satisfies

$$(***) \qquad \lim_{n\to\infty} \int_\Omega g(\omega)\langle x', u^{\alpha(n)}(\omega)\rangle\, d\,\mathrm{P}(\omega) = \int_\Omega g(\omega)\langle x', b_{\lambda^\infty_\omega}\rangle\, d\,\mathrm{P}(\omega)$$

for all $g \in \mathrm{L}^\infty_{\mathbb{R}}(\Omega, \mathcal{S}, \mathrm{P})$ and for all $x' \in \overline{B}_{\mathbb{E}^*}$.

Since $u^n(\omega) \in \Gamma(\omega)$ for all n and for all $\omega \in \Omega$ and by (*) $\lambda^{\alpha(n)}$ stably converges to λ^∞, $\lambda^\infty_\omega(\Gamma(\omega)) = 1$ a.e. As $\Gamma(\omega)$ is bounded closed convex subset of $\mathbb{E}$, the barycenter of λ^∞_ω exists [Bou83, Lemma 6.2.2 page 178] and belongs to $\Gamma(\omega)$ a.e. Now (**) and (***) follow by repeating the same arguments as in the proof of Proposition 6.6.3. Using the separability of $\mathbb{E}$ and (***) we get $u^\infty(\omega) = \int_{\mathbb{E}} x\, d\lambda^\infty_\omega(x)$ a.e. The proof ends as that of Theorem 6.6.4. □

Remark 6.6.8 Theorem 6.6.7 provides an alternative proof of Lemma 6.6.6 via Young measures, because Γ is convex norm compact valued and the set $\mathrm{S}_\Gamma^{\mathrm{Pe}}$ of all Pettis integrable selections of Γ is sequentially compact for the topology $\sigma(\mathrm{P}^1_{\mathbb{E}}, \mathrm{L}^\infty \otimes \mathbb{E}^*)$. But the techniques of Lemma 6.6.6 can be applied to other places, mainly when the use of Young measures is no longer available. At this point it is worth to mention that in the results given above, the "scalarly Pettis uniformly integrable " assumption combined with the norm tightness of the sequence (u_n) are crucial in the use of Young measures. In the further results, dealing with "Pettis uniformly integrable sequences", we need another method using an appropriate truncation technique. It is worthy to address the question: What happens if one assumes in the framework of Lemma 6.6.6 that the multifunction Γ is convex weakly compact valued? This leads to the following variant.

Theorem 6.6.9 *Let $\mathbb{E}$ be a separable Banach space and Γ be a measurable multifunction on Ω with bounded closed convex values in $\mathbb{E}$. Let (u^n) be a scalarly Pettis uniformly integrable and weakly strictly tight sequence in $\mathrm{P}^1_{\mathbb{E}}$ such that $u^n(\omega) \in \Gamma(\omega)$ for all n and for all $\omega \in \Omega$. Assume further that (u^n) converges $\sigma(\mathrm{P}^1_{\mathbb{E}}, \mathrm{L}^\infty \otimes \mathbb{E}^*)$ to a scalarly integrable $\mathbb{E}$-valued function u^∞. Let $\varphi : \Omega \times \mathbb{E} \to [0, +\infty]$ be a convex normal integrand satisfying*

(*i*) $\Gamma(\omega) \subset \operatorname{dom} \varphi$,

(*ii*) $\limsup_n \int_\Omega \varphi(\omega, u^n(\omega))\, d\mathrm{P}(\omega) \leq \int_\Omega \varphi(\omega, u^\infty(\omega))\, d\mathrm{P}(\omega) < +\infty$,

(*iii*) $(u^\infty(\omega), \varphi(\omega, u^\infty(\omega)))$ *is an extremal point of* $\operatorname{Epi} \varphi_\omega$ *a.e.*

Then $u^\infty(\omega) \in \Gamma(\omega)$ almost everywhere, and $(\underline{\delta}_{u^n})$ M–stably converges to $\underline{\delta}_{u^\infty}$ in the space $\mathcal{Y}^1(\Omega, \mathcal{S}, \mathrm{P}; \mathbb{E}_\sigma)$.

Proof. We use some arguments given in the proof of Propositions 6.6.2 and 6.6.3 and Theorem 6.6.4 with appropriated modification. Since (u^n) is weakly strictly tight in view of Theorem 4.3.5 there exist a subsequence $\lambda^{\alpha(n)} = (\underline{\delta}_{u^{\alpha(n)}})$ of the sequence $(\lambda^n) = (\underline{\delta}_{u^n})$ and $\lambda^\infty \in \mathcal{Y}^1(\Omega, \mathcal{S}, \mathrm{P}; \mathbb{E})$ such that the following hold:

(*)

$$\lim_n \langle h, \lambda^{\alpha(n)} \rangle = \langle h, \lambda^\infty \rangle,$$

for every bounded Carathéodory integrand h on $\Omega \times \mathbb{E}_\sigma$.

(**) There is a scalarly integrable $\mathbb{E}$-valued mapping $b_{\lambda^\infty} : \omega \mapsto b_{\lambda^\infty_\omega}$ defined by

$$\langle b_{\lambda^\infty_\omega}, x' \rangle = \int_{\mathbb{E}} \langle x', x \rangle \, d\lambda^\infty_\omega(x),$$

for all $x' \in \mathbb{E}^*$ and for all $\omega \in \Omega$ which satisfies

(***) $$\lim_{n \to \infty} \int_\Omega g(\omega) \langle x', u_{\alpha(n)}(\omega) \rangle \, d\mathrm{P}(\omega) = \int_\Omega g(\omega) \langle x', b_{\lambda^\infty_\omega} \rangle \, d\mathrm{P}(\omega),$$

for all $g \in L^\infty_{\mathbb{R}}(\Omega, \mathcal{S}, P)$ and for all $x' \in \overline{B}_{\mathbb{E}^*}$.

Since $u^n(\omega) \in \Gamma(\omega)$ for all n and for all $\omega \in \Omega$ and by (*) $\lambda^{\alpha(n)}$ M–stably converges to λ^∞, we have that $\lambda^\infty_\omega(\Gamma(\omega)) = 1$ a.e. As $\Gamma(\omega)$ is a bounded closed convex subset of $\mathbb{E}$, the barycenter of λ^∞_ω exists [Bou83, Lemma 6.2.2 page 178] and belongs to $\Gamma(\omega)$ a.e. Now (**) and (***) follow by using the same arguments as in the proof of Proposition 6.6.3. Repeating the arguments given in the end of the proof of Theorem 6.6.4 shows that $\underline{\delta}_{u^\infty(\omega)} = \lambda^\infty(\omega)$ a.e. so that by (*) $\underline{\delta}_{u^{\alpha(n)}}$ M-stably converges to $\underline{\delta}_{u^\infty}$ in $\mathcal{Y}^1(\Omega, \mathcal{S}, P; \mathbb{E}_\sigma)$. □

Now we proceed to a generalization of Lemma 6.6.6 and its applications. Let us recall (page 181) that a subset $\mathcal{H}$ in $P^1_{\mathbb{E}}$ is c$\mathcal{K}(\mathbb{E})$-tight if, for every $\varepsilon > 0$, there exists a c$\mathcal{K}(\mathbb{E})$-valued Pettis-integrable multifunction Γ_ε satisfying:

$$\sup_{u \in \mathcal{H}} P(\{\omega \in \Omega;\ u(\omega) \notin \Gamma_\varepsilon(\omega)\}) \le \varepsilon.$$

Theorem 6.6.10 *Suppose that $\mathbb{E}$ is a separable Banach space, $\Phi : \Omega \to \mathcal{L}\mathrm{cw}\mathcal{K}(\mathbb{E})$ is an $\mathcal{L}\mathrm{cw}\mathcal{K}(\mathbb{E})$-valued measurable multifunction, $(u_n)_{n\in\mathbb{N}}$ is a Pettis uniformly integrable and c$\mathcal{K}(\mathbb{E})$-tight sequence in $P^1_{\mathbb{E}}$ which converges $\sigma(P^1_{\mathbb{E}}, L^\infty \otimes \mathbb{E}^*)$ to $u \in P^1_{\mathbb{E}}$ with $u_n(\omega) \in \Phi(\omega)$ $(n \in \mathbb{N})$ and $u(\omega) \in \partial_{\mathrm{ext}}\,(\Phi(\omega))$ a.e., then $\|u_n - u\|_{\mathrm{Pe}} \to 0$.*

Proof. We may suppose that $u \equiv 0$ and $0 \in \Phi(\omega)$ for all $\omega \in \Omega$ because $\overline{\mathrm{co}}(\Phi(\omega) \cup \{0\})$ still belongs to $\mathcal{L}\mathrm{cw}\mathcal{K}(\mathbb{E})$. So we have $0 \in \partial_{\mathrm{ext}}\,(\Phi(\omega))$ for a.e. $\omega \in \Omega$. Let $\varepsilon > 0$. Since $(u_n)_{n\in\mathbb{N}}$ is Pettis uniformly integrable, there exists $\delta > 0$ such that

$$P(A) < \delta \Longrightarrow \sup_{n\in\mathbb{N}} \|\,\mathbb{1}_A u_n\|_{\mathrm{Pe}} \le \varepsilon.$$

By tightness hypothesis, we find a c$\mathcal{K}(\mathbb{E})$-valued Pettis integrable multifunction Γ_δ such that

$$\sup_{n\in\mathbb{N}} P(\{\omega \in \Omega;\ u_n(\omega) \notin \Gamma_\delta(\omega)\}) \le \delta.$$

For every $n \in \mathbb{N}$, we set

$$v_n := \mathbb{1}_{\{\omega\in\Omega;\ u_n(\omega)\in\Gamma_\delta(\omega)\}} u_n$$
$$w_n := \mathbb{1}_{\{\omega\in\Omega;\ u_n(\omega)\notin\Gamma_\delta(\omega)\}} u_n$$

and

$$\Delta(\omega) := \Phi(\omega) \cap \overline{\mathrm{co}}(\Gamma_\delta(\omega) \cup \{0\})$$

for all $\omega \in \Omega$. Then it is clear that the multifunction Δ has nonempty convex compact values and is Pettis integrable because

$$\forall x' \in \mathbb{E}^* \quad 0 \le \delta^*(x', \Delta) \le \delta^*(x', \Gamma_\delta \cup \{0\}),$$

so that $\{\delta^*(x', \Delta);\ \|x'\| \leq 1\}$ is uniformly integrable. Thus $v_n \in \mathrm{S}^{\mathrm{Pe}}_{\Delta}$. By Corollary 6.3.3, we may suppose that v_n converges $\sigma(\mathrm{P}^1_{\mathbb{E}}, \mathrm{L}^\infty \otimes \mathbb{E}^*)$ to $v \in \mathrm{S}^{\mathrm{Pe}}_{\Delta}$, by extracting a subsequence if necessary. Hence we have

$$0 = \lim_{n\to\infty} u_n = \lim_{n\to\infty} [v_n + w_n] = v + w$$

where the limits are taken for $\sigma(\mathrm{P}^1_{\mathbb{E}}, \mathrm{L}^\infty \otimes \mathbb{E}^*)$, and $w \in \mathrm{S}^{\mathrm{Pe}}_{\Phi}$ using Lemma 6.3.8. Since $0 \in \partial_{\mathrm{ext}}\ (\Phi(\omega))$ almost everywhere, it follows that $v = w = 0$ a.e. and $0 \in \partial_{\mathrm{ext}}\ (\Delta(\omega))$ a.e. This allows to apply Lemma 6.6.6 to the sequence $(v_n)_n$ showing that $\|v_n\|_{\mathrm{Pe}} \to 0$. Since

$$\|u_n\|_{\mathrm{Pe}} \leq \|v_n\|_{\mathrm{Pe}} + \|w_n\|_{\mathrm{Pe}} \leq \|v_n\|_{\mathrm{Pe}} + \varepsilon$$

for all $n \in \mathbb{N}$, and ε is arbitrary > 0, $\|u_n\|_{\mathrm{Pe}} \to 0$. □

Remark 6.6.11 It is worthy to mention two useful properties of Pettis uniformly integrable sequences in $\mathrm{P}^1_{\mathbb{E}}$. Namely we have the following: If $(u_n)_{n\in\mathbb{N}}$ is Pettis uniformly integrable and $\mathrm{cw}\mathcal{K}(\mathbb{E})$-tight (resp. $\mathrm{c}\mathcal{K}(\mathbb{E})$-tight), then, for every $A \in \mathcal{S}$, the sequence $(\int_A u_n\, d\mathrm{P})_{n\in\mathbb{N}}$ is relatively weakly compact (resp. norm compact). It is enough to prove this fact when $A = \Omega$. Let $\varepsilon > 0$. An easy inspection of the proof of the preceding theorem shows that $u_n = v_n + w_n$, where $(\int_\Omega v_n\, d\mathrm{P})_{n\in\mathbb{N}}$ is relatively weakly compact (resp. norm compact) and $\|w_n\|_{\mathrm{Pe}} \leq \varepsilon$ for all $n \in \mathbb{N}$. It follows that the sequence $(\int_\Omega u_n\, d\mathrm{P})_{n\in\mathbb{N}}$ is relatively weakly compact (resp. norm compact) too. At this point, let us mention that the preceding norm compactness results can be also deduced from Proposition 6.3.6 when one deals with the scalar Pettis uniform integrability. Even in Bochner integration the norm compactness of $(\int_A u_n\, d\mathrm{P})_{n\in\mathbb{N}}$ is useful in several places. See for example [BGJ94] in which the authors state the Pettis-norm convergence via Bocce criterion and the preceding compactness result.

The following result shows that Pettis-norm convergence is implied by strict convexity.

Theorem 6.6.12 *Let $\varphi : \Omega \times \mathbb{E} \to\ ]-\infty, +\infty]$ be a $\mathcal{S} \otimes \mathcal{B}_{\mathbb{E}}$-measurable integrand such that $\varphi(\omega, .)$ is convex lower semicontinuous on $\mathbb{E}$ for every fixed $\omega \in \Omega$. Let $(u_n)_{n\in\mathbb{N}}$ be a Pettis uniformly integrable and $\mathrm{c}\mathcal{K}(\mathbb{E})$-tight sequence in $\mathrm{P}^1_{\mathbb{E}}$. Suppose that* $\mathrm{Epi}\,\varphi_\omega \in \mathcal{L}\mathrm{cw}\mathcal{K}(\mathbb{E})$ *for every $\omega \in \Omega$, u_n weakly converges to $u \in \mathrm{P}^1_{\mathbb{E}}$, the functions $\varphi(., u_n(.))$ and $\varphi(., u(.))$ are integrable, and $\varphi(., u_n(.))$ converges $\sigma(\mathrm{L}^1, \mathrm{L}^\infty)$ to $\varphi(., u(.))$ with*

$$(u(\omega), \varphi(\omega, u(\omega))) \in \partial_{\mathrm{ext}}\ (\mathrm{Epi}\,\varphi(\omega, .))\ \mathrm{P}\text{-a.e.},$$

then $\|u_n - u\|_{\mathrm{Pe}} \to 0$.

Remark 6.6.13 If $\varphi(\omega, .)$ is strictly convex, $(x, \varphi(\omega, x))$ is always an extremal point of $\mathrm{Epi}\,\varphi(\omega, .)$ for all $x \in \mathbb{E}$.

Proof. Since $\varphi(.,u_n(.))$ converges $\sigma(\mathrm{L}^1,\mathrm{L}^\infty)$ to $\varphi(.,u(.))$, applying Theorem 6.6.10 to the Pettis uniformly integrable $\mathbb{E}\times\mathbb{R}$-valued functions

$$(u_n,\varphi(.,u_n(.))) \text{ and } (u,\varphi(.,u(.)))$$

and to the multifunction $\Phi(\omega,.) = \operatorname{Epi}\varphi(\omega,.)$ yields the desired result. The details are left to the reader. □

6.7 Visintin-type theorem in $\mathrm{L}^1_{\mathbb{E}^*}[\mathbb{E}](\Omega,\mathcal{S},\mathrm{P})$

Let $\mathbb{E}$ be a Banach space. For the sake of completeness we will recall the following notations and notions and summarize some useful results [BC01] in the space $\mathrm{L}^1_{\mathbb{E}^*}[\mathbb{E}](\mathrm{P})$ (shortly $\mathrm{L}^1_{\mathbb{E}^*}[\mathbb{E}]$) before we state the main result in this section. We denote by $\mathcal{L}^1_{\mathbb{E}^*}[\mathbb{E}]$ the vector space of scalarly measurable functions $f:\Omega\to\mathbb{E}^*$ such that there exists a positive integrable function h (depending on f) satisfying $\forall\omega\in\Omega$, $\|f(\omega)\|\le h(\omega)$. A seminorm on $\mathcal{L}^1_{\mathbb{E}^*}[\mathbb{E}]$ is defined by

$$N_1(f) = \int_\Omega^* \|f(\omega)\|\,d\mathrm{P}(\omega) = \inf\{\int_\Omega h\,d\mathrm{P};\ h \text{ is integrable, } h\ge\|f(.)\|\}.$$

Two functions $f,g\in\mathcal{L}^1_{\mathbb{E}^*}[\mathbb{E}]$ are *equivalent* (shortly $f\equiv g\ (w^*)$) if $\langle f(.),x\rangle = \langle g(.),x\rangle$ a.e. for every $x\in\mathbb{E}$. The equivalence class of f is denoted by $\bar f$. The quotient space $\mathrm{L}^1_{\mathbb{E}^*}[\mathbb{E}]$ is equipped with the norm $\overline{N}_1$ given by

$$\overline{N}_1(\bar f) = \inf\{\overline{N}_1(g);\ g\in\bar f\}.$$

Let $\dot\rho$ be the lifting in $\mathcal{L}^\infty_{\mathbb{E}^*}[\mathbb{E}]$ associated with a lifting ρ in $\mathcal{L}^\infty_{\mathbb{R}}(\mathrm{P})$ [ITIT69, VI.4]. We denote by $\mathcal{L}^{1,\rho}_{\mathbb{E}^*}[\mathbb{E}]$ the vector space of all mappings $f\in\mathcal{L}^1_{\mathbb{E}^*}[\mathbb{E}]$ such that there exists a sequence $(A_n)_{n\ge1}$ in $\mathcal{S}$ satisfying:

$$\bigcup_{n\ge1} A_n = \Omega,\quad \forall n\ge1\quad \mathbb{1}_{A_n}f\in\mathcal{L}^\infty_{\mathbb{E}^*}[\mathbb{E}],\ \text{and } \dot\rho(\mathbb{1}_{A_n}f) = \mathbb{1}_{\rho(A_n)}f.$$

If $f\in\mathcal{L}^{1,\rho}_{\mathbb{E}^*}[\mathbb{E}]$, $\|f(.)\|$ is measurable [BC01, Proposition 3.1(c)] and the quotient space $\mathrm{L}^{1,\rho}_{\mathbb{E}^*}[\mathbb{E}]$ is equipped with the norm

$$N_{1,\rho}(\bar f) = N_1(\|f\|) = \int_\Omega \|f\|\,d\mathrm{P}.$$

By [BC01, Theorem 3.2(b)] there is a linear isometric isomorphism

$$\widetilde\rho : (\mathrm{L}^1_{\mathbb{E}^*}[\mathbb{E}],\overline{N}_1)\to(\mathrm{L}^{1,\rho}_{\mathbb{E}^*}[\mathbb{E}],N_{1,\rho})$$

so that $\mathrm{L}^1_{\mathbb{E}^*}[\mathbb{E}]$ and $\mathrm{L}^{1,\rho}_{\mathbb{E}^*}[\mathbb{E}]$ can be identified. In this identification $\bar f\in\mathrm{L}^1_{\mathbb{E}^*}[\mathbb{E}]$ is identified with $\widetilde\rho(\bar f)$ and for notational convenience, $\bar f$ is identified with a function

$f \in \mathcal{L}^{1,\rho}_{\mathbb{E}^*}[\mathbb{E}]$. Let $c\mathcal{K}(\mathbb{E}^*)$ (resp. $cw\mathcal{K}(\mathbb{E}^*)$) be the set of all nonempty convex norm compact (resp. $\sigma(\mathbb{E}^*, \mathbb{E}^{**})$ compact) subsets of the Banach space $\mathbb{E}^*$. A $cw\mathcal{K}(\mathbb{E}^*)$-valued multifunction Γ on Ω is scalarly measurable (resp. integrable) if, for every $x \in \mathbb{E}$, the function $\delta^*(x, \Gamma(.))$ is measurable (resp. integrable), where $\delta^*(x, K)$ denotes the support function of $K \in cw\mathcal{K}(\mathbb{E}^*)$.

Proposition 6.7.1 *Suppose that Γ is a $cw\mathcal{K}(\mathbb{E}^*)$-valued multifunction on Ω and (f_n) is a uniformly integrable sequence in $L^1_{\mathbb{E}^*}[\mathbb{E}]$ such that $f_n(\omega) \in \Gamma(\omega)$ for a.e. $\omega \in \Omega$ and for all n, then (f_n) is relatively $\sigma(L^1_{\mathbb{E}^*}[\mathbb{E}](P), (L^1_{\mathbb{E}^*}[\mathbb{E}](P))^*)$ (weakly) compact in $L^1_{\mathbb{E}^*}[\mathbb{E}]$.*

We only sketch the proof. The details are in [BC01, Proposition 5.1]. By [BC01, Theorem 4.9] there are a sequence (g_n) with $g_n \in \mathrm{co}\{f_m;\, m \geq n\}$ and two measurable sets A and B in Ω with $P(A \cup B) = 1$ such that

(a) $\forall \omega \in A$, $(g_n(\omega))$ is $\sigma(\mathbb{E}^*, \mathbb{E}^{**})$ Cauchy in $\mathbb{E}^*$,

(b) $\forall \omega \in B$, there exists $k \in \mathbb{N}$ such that the sequence $(g_n(\omega))_{n \geq k}$ is equivalent to the vector unit basis of l^1.

As $\Gamma(\omega)$ is $\sigma(\mathbb{E}^*, \mathbb{E}^{**})$-compact for all $\omega \in \Omega$, using (b) one has $P(B) = 0$. Hence there is a sequence (g_n) in $L^1_{\mathbb{E}^*}[\mathbb{E}]$ with $g_n \in \mathrm{co}\{f_m;\, m \geq n\}$ such that $(g_n(\omega))$ is $\sigma(\mathbb{E}^*, \mathbb{E}^{**})$-convergent a.e. By [BC01, Theorem 4.5] (g_n) converges for $\sigma(L^1_{\mathbb{E}^*}[\mathbb{E}], (L^1_{\mathbb{E}^*}[\mathbb{E}])^*)$. Hence (f_n) is relatively $\sigma(L^1_{\mathbb{E}^*}[\mathbb{E}], (L^1_{\mathbb{E}^*}[\mathbb{E}])^*)$–compact by a general criterion for weak compactness in Banach spaces.

In the remainder of this section we shall suppose that $\mathbb{E}$ is a *separable* Banach space. By *weak convergence*, we mean convergence for $\sigma(L^1_{\mathbb{E}^*}[\mathbb{E}], (L^1_{\mathbb{E}^*}[\mathbb{E}])^*)$. Using Proposition 6.7.1 and the separability of $\mathbb{E}$ we have

Corollary 6.7.2 *Suppose that Γ is a scalarly measurable $cw\mathcal{K}(\mathbb{E}^*)$-valued multifunction on Ω and there is $g \in L^1_{\mathbb{R}+}$ such that $\Gamma(\omega) \subset g(\omega)\, \overline{B}_{\mathbb{E}^*}$ for all ω in Ω, then the set S_Γ of all scalarly integrable selections of Γ is convex weakly compact in $L^1_{\mathbb{E}^*}[\mathbb{E}]$.*

Proof. See [BC01, Proposition 5.1]. □

For notational convenience such a multifunction Γ is said to be *integrably bounded.* Unlike the space $L^1_{\mathbb{E}}(P)$, the preceding results are not standard and rely on a deep result involving the Talagrand decomposition in $L^{1,\rho}_{\mathbb{E}^*}[\mathbb{E}](P)$ [BC01, Theorem 4.9].

A uniformly integrable sequence (u_n) in $L^1_{\mathbb{E}^*}[\mathbb{E}](P)$ is *norm-tight* if for every $\varepsilon > 0$ there is a scalarly $c\mathcal{K}(\mathbb{E}^*)$-valued measurable and integrably bounded multifunction Φ_ε on Ω with $0 \in \Phi_\varepsilon(\omega)$ for all $\omega \in \Omega$ such that

$$\sup_n P(\{\omega \in \Omega;\, u_n(\omega) \notin \Phi_\varepsilon(\omega)\}) \leq \varepsilon.$$

It is easily seen that u_n can be written as $u_n = \mathbb{1}_{A_n} u_n + \mathbb{1}_{\Omega\setminus A_n} u_n$ where $A_n \in \mathcal{S}$ and $\mathbb{1}_{A_n} u_n \in \mathrm{S}_{\Phi_\varepsilon}$ and $\| \mathbb{1}_{\Omega\setminus A_n} u_n\|_{\mathrm{L}^1_{\mathbb{E}^*}[\mathbb{E}]} \leq \varepsilon$, so that a uniformly integrable norm-tight sequence (u_n) in $\mathrm{L}^1_{\mathbb{E}^*}[\mathbb{E}](\mathrm{P})$ is relatively weakly compact in view of Proposition 6.7.1 and Grothendiek lemma, see [ACV92, page 183] for details.

Now we are able to present a version of Visintin's theorem in $\mathrm{L}^1_{\mathbb{E}^*}[\mathbb{E}](\mathrm{P})$ in same style as in [ACV92, Lemme 10 and Théorème 11] and [Bal91]. Since the proof follows the same lines, we don't want to give details so much. Yet this needs a careful look.

Theorem 6.7.3 *Suppose that (u_n) is a uniformly integrable norm-tight sequence in $\mathrm{L}^1_{\mathbb{E}^*}[\mathbb{E}](\mathrm{P})$ which weakly converges to $u \in \mathrm{L}^1_{\mathbb{E}^*}[\mathbb{E}](\mathrm{P})$ and such that $u(\omega) \in \partial_{\mathrm{ext}}$ $(\bigcap_{n\in\mathbb{N}} \overline{\mathrm{co}}\{u_k(\omega);\, k \geq n\})$ a.e. Then $\int_\Omega \|u_n(\omega) - u(\omega)\|\, d\mathrm{P}(\omega) \to 0$.*

Proof. We will divide the proof in two steps.

Step 1. We will prove the theorem in the particular case when (u_n) is a uniformly integrable sequence in $\mathrm{L}^1_{\mathbb{E}^*}[\mathbb{E}](\mathrm{P})$ weakly converging to $u \in \mathrm{L}^1_{\mathbb{E}^*}[\mathbb{E}](\mathrm{P})$ and satisfying:

(a) There is a convex norm compact valued multifunction Γ such that $u_n(\omega) \in \Gamma(\omega)$ for all n and all $\omega \in \Omega$.

(b) $u(\omega) \in \partial_{\mathrm{ext}} (\bigcap_{n\in\mathbb{N}} \overline{\mathrm{co}}\{u_k(\omega);\, k \geq n\})$.

Since $(\|u_n\|)_n$ is uniformly integrable it suffices to prove that $\|u_n - u\| \to 0$ in measure. We may suppose $u \equiv 0$. Suppose not. Then there exist $\varepsilon > 0$ and $\eta > 0$ such that

$$\mathrm{P}(\{\omega \in \Omega;\, u_n(\omega) \notin B_{\mathbb{E}^*}(0,\varepsilon)\}) \geq \eta \tag{6.7.1}$$

for infinitely many n (the measurability of $\{\omega \in \Omega;\, u_n(\omega) \notin B_{\mathbb{E}^*}(0,\varepsilon)\}$ will be proved later); namely there exists an infinite subset $S_1 \subset \mathbb{N}$ such that the preceding inequality holds for all $n \in S_1$. For every $\omega \in \Omega$, let

$$\Sigma_n(\omega) := \overline{\mathrm{co}}\{u_k(\omega);\, k \geq n\} \quad \text{and} \quad \Sigma(\omega) := \bigcap_{n\in\mathbb{N}} \Sigma_n(\omega).$$

Since the function $x' \mapsto \|x'\|$ is lower semicontinuous on $\mathbb{E}^*_{w^*}$, $B_{\mathbb{E}^*}(0,r)$ is a Borel subset of $\mathbb{E}^*_{w^*}$. As $\mathbb{E}^*_{w^*}$ is a Lusin space and any scalarly integrable multifunction from Ω to the set $c\mathcal{K}(\mathbb{E}^*_{w^*})$ of nonempty convex compact subsets of $\mathbb{E}^*_{w^*}$ has its graph in $\mathcal{S} \otimes \mathcal{B}_{\mathbb{E}^*_{w^*}}$. Now set

$$\begin{aligned} A &:= \{\omega \in \Omega;\, \Sigma(\omega) \setminus B_{\mathbb{E}^*}(0,\varepsilon) \neq \emptyset\},\\ B &:= \{\omega \in \Omega;\, \Sigma(\omega) \subset B_{\mathbb{E}^*}(0,\varepsilon)\},\\ B_n &:= \{\omega \in \Omega;\, \Sigma_n(\omega) \subset B_{\mathbb{E}^*}(0,\varepsilon)\}. \end{aligned}$$

As the graph of the multifunctions Σ and Σ_n belong to $\mathcal{S}\otimes\mathcal{B}_{\mathbb{E}^*_{w^*}}$ and $B_{\mathbb{E}^*}(0,r)$ is a Borel subset of $\mathbb{E}^*_{w^*}$, by a classical mesurable projection theorem [CV77, Theorem III.23], (cf. Remark 1.2.1) we see that $\{w\in\Omega;\ u_n(\omega)\notin B_{\mathbb{E}^*}(0,\varepsilon)\}$, A, B_n, B are $\mathcal{S}$-measurable. Furthermore we have $B_n\uparrow B$ because if $\omega\in B$ we have that $\bigcap_n\Sigma_n(\omega)\setminus B_{\mathbb{E}^*}(0,\varepsilon)=\emptyset$ so that by finite intersection property of compact spaces there is an integer m such that $\bigcap_{n\geq m}\Sigma_n(\omega)\setminus B_{\mathbb{E}^*}(0,\varepsilon)=\emptyset$. Pick N_1 such that $n\geq N_1$ implies $\mathrm{P}(B\setminus B_n)<\dfrac{\eta}{2}$. Since

$$\{\omega\in B;\ u_n(\omega)\notin B_{\mathbb{E}^*}(0,\varepsilon)\}\subset B\setminus B_n$$

we get

$$n\geq N_1 \Longrightarrow \mathrm{P}(\{\omega\in B;\ u_n(\omega)\notin B_{\mathbb{E}^*}(0,\varepsilon)\}<\frac{\eta}{4}. \tag{6.7.2}$$

Let us write $u_n=v_n+w_n$ where

$$v_n:=\mathbb{1}_{\{\omega\in\Omega;\ u_n(\omega)\notin B_{\mathbb{E}^*}(0,\varepsilon)\}}u_n \quad\text{and}\quad w_n:=\mathbb{1}_{\{\omega\in\Omega;\ u_n(\omega)\in B_{\mathbb{E}^*}(0,\varepsilon)\}}u_n.$$

Then the sequence (v_n) is relatively sequentially $\sigma(\mathrm{L}^1_{\mathbb{E}^*}[\mathbb{E}],(\mathrm{L}^1_{\mathbb{E}^*}[\mathbb{E}])^*)$ compact in view of Proposition 6.7.1 and Eberlein–Šmulian's theorem. There is a subsequence $(v_n)_{n\in S_2}$ where S_2 is an infinite subset of S_1 such that $(v_n)_{n\in S_2}$ converges to $v\in\mathrm{L}^1_{\mathbb{E}^*}[\mathbb{E}](A\cap\mathcal{S},\mathrm{P}|_{A\cap\mathcal{S}})$ for this topology. It follows that $(w_n)_{n\in S_2}$ weakly converges to $u-v$. Using a version of Mazur's theorem in $\mathrm{L}^1_{\mathbb{E}^*}[\mathbb{E}]$ ([BC01, Lemma 4.12] and [ACV92, Lemme 4]), we get

$$v(\omega)\in\bigcap_{n\in S_2}\overline{\mathrm{co}}\{v_k(\omega);\ k\geq n, k\in S_2\}\subset\Sigma(\omega)\ \text{ a.e.}$$

Similarly we have $w(\omega)\in\Sigma(\omega)$ a.e. Since $0\in\partial_{\mathrm{ext}}\,(\Sigma(\omega))$ a.e. and u_n weakly converges to 0, we get $v=w=0$. As $\Sigma(\omega)$ is norm compact and convex in $\mathbb{E}^*$ we have that $\partial_{\mathrm{ext}}\,(\Sigma(\omega))=\ \partial_{\mathrm{dent}}(\Sigma(\omega))$ in view of [ACV92, Lemme 1 page 171]. So we have

$$0\notin\overline{\mathrm{co}}\,[\Sigma(\omega)\setminus B_{\mathbb{E}^*}(0,\varepsilon)]\ \text{ a.e.}$$

It is obvious that $\overline{\mathrm{co}}\,[\Sigma(\omega)\setminus B_{\mathbb{E}^*}(0,\varepsilon)]$ is nonempty convex norm compact (a fortiori $\sigma(\mathbb{E}^*,\mathbb{E})$ compact in $\mathbb{E}^*$) whenever $\omega\in A$. Hence the multifunction Ψ defined from A with nonempty values in the closed unit ball $\overline{B}_{\mathbb{E}}$ of $\mathbb{E}$ (thanks to the Hahn–Banach Theorem):

$$\Psi(\omega):=\{x\in\overline{B}_{\mathbb{E}};\ \delta^*(x,\overline{\mathrm{co}}[\Sigma(\omega)\setminus B_{\mathbb{E}^*}(0,\varepsilon)]<0\}$$

has its graph in $(A\cap\mathcal{S})\otimes\mathcal{B}_{\overline{B}_{\mathbb{E}}}$ [CV77, Lemma III.14]. By [CV77, Theorem III.22], Ψ admits an $\mathcal{S}$-measurable selection $\sigma: A\to\overline{B}_{\mathbb{E}}$. Since $\Sigma_n(\omega)\setminus B_{\mathbb{E}^*}(0,\varepsilon)\downarrow\Sigma(\omega)\setminus B_{\mathbb{E}^*}(0,\varepsilon)$ we get

$$\delta^*(\sigma(\omega),\overline{\mathrm{co}}[\Sigma_n(\omega)\setminus B_{\mathbb{E}^*}(0,\varepsilon)])\to\delta^*(\sigma(\omega),\overline{\mathrm{co}}\,[\Sigma(\omega)\setminus B_{\mathbb{E}^*}(0,\varepsilon)]). \tag{6.7.3}$$

By [ACV92, Lemme 3], there are $a < 0$ and N_2 such that

$$n \geq N_2 \implies \mathrm{P}(\{\omega \in A;\ \delta^*(\sigma(\omega), \overline{\mathrm{co}}[\Sigma_n(\omega) \setminus B_{\mathbb{E}^*}(0,\varepsilon)]) > a\}) < \frac{\eta}{4}. \tag{6.7.4}$$

As v_n is either $= 0$ or belongs to $[\Sigma_n(\omega) \setminus B_{\mathbb{E}^*}(0,\varepsilon)]$, we have $\limsup_n \langle \sigma(\omega), v_n(\omega)\rangle \leq 0$. Now since $v_n \to 0$ for $\sigma(\mathrm{L}^1_{\mathbb{E}^*}[\mathbb{E}], (\mathrm{L}^1_{\mathbb{E}^*}[\mathbb{E}])^*)$ and $\mathrm{L}^\infty_{\mathbb{E}}(\mathrm{P}) \subset (\mathrm{L}^1_{\mathbb{E}^*}[\mathbb{E}](\mathrm{P}))^*$ in view of [BC00], for every $h \in \mathrm{L}^\infty_{\mathbb{R}}(A \cap \mathcal{S}, \mathrm{P}\lfloor_{A \cap \mathcal{S}})$ we get

$$\int_A h(\omega)\langle \sigma(\omega), v_n(\omega)\rangle \, d\mathrm{P}(\omega) \to 0. \tag{6.7.5}$$

Therefore by [ACV92, Corollaire D], $\langle \sigma(.), v_n(.)\rangle \to 0$ in measure. Consequently there exist N_3 such that

$$n \geq N_3\ (n \in S_2) \implies \mathrm{P}(\{\omega \in A;\ \langle \sigma(\omega), v_n(\omega)\rangle \leq a\}) < \frac{\eta}{4}. \tag{6.7.6}$$

Now observe that

$$\begin{aligned}
&\{\omega \in A;\ v_n(\omega) \neq 0\}\\
&\subset \{\omega \in A;\ v_n(\omega) \neq 0 \text{ and } \langle \sigma(\omega), v_n(\omega)\rangle > a\} \cup \{\omega \in A;\ \langle \sigma(\omega), v_n(\omega)\rangle \leq a\}\\
&\subset \{\omega \in A;\ \delta^*(\sigma(\omega), \overline{\mathrm{co}}[\Sigma_n(\omega) \setminus B_{\mathbb{E}^*}(0,\varepsilon)] > a\} \cup \{\omega \in A;\ \langle \sigma(\omega), v_n(\omega)\rangle \leq a\}.
\end{aligned}$$

From (6.7.6) we deduce that

$$n \geq \max(N_2, N_3) \implies \mathrm{P}(\{\omega \in A;\ v_n(\omega) \neq 0\}) < \frac{\eta}{2}. \tag{6.7.7}$$

But

$$\begin{aligned}
&\{\omega \in \Omega;\ u_n(\omega) \notin B_{\mathbb{E}^*}(0,\varepsilon)\}\\
&\qquad = \{\omega \in A;\ u_n(\omega) \notin B_{\mathbb{E}^*}(0,\varepsilon)\} \cup \{\omega \in B;\ u_n(\omega) \notin B_{\mathbb{E}^*}(0,\varepsilon)\}\\
&\qquad = \{\omega \in A;\ v_n(\omega) \neq 0\} \cup \{\omega \in B;\ u_n(\omega) \notin B_{\mathbb{E}^*}(0,\varepsilon)\}.
\end{aligned}$$

So by (6.7.2) and (6.7.7) and for $n \geq \max(N_1, N_2, N_3)$, we get

$$n \geq \max(N_1, N_2, N_3) \implies \mathrm{P}(\{\omega \in A;\ u_n(\omega) \notin B_{\mathbb{E}^*}(0,\varepsilon)\}) < \frac{\eta}{2} + \frac{\eta}{2} = \eta.$$

That contradicts (6.7.4).

Step 2. Now we pass to the general case. We suppose that (u_n) is *norm-tight.* Let $\varepsilon > 0$. There is a scalarly $c\mathcal{K}(\mathbb{E}^*)$-valued measurable and an integrably bounded multifunction Φ_ε with $0 \in \Phi_\varepsilon(\omega)$ for all $\omega \in \Omega$ such that u_n can be written as

$$u_n = \mathbb{1}_{A_n} u_n + \mathbb{1}_{\Omega \setminus A_n} u_n$$

where $A_n \in \mathcal{S}$ and $\mathbb{1}_{A_n} u_n \in \mathrm{S}_{\Phi_\varepsilon}$ and $\| \mathbb{1}_{\Omega \setminus A_n} u_n \|_{L^1_{\mathbb{E}^*}[\mathbb{E}]} \leq \varepsilon$. By Corollary 6.7.2, we may suppose that $(v_n) = (\mathbb{1}_{A_n} u_n)$ weakly converges to $v \in \mathrm{S}^{\mathrm{Pe}}_{\Phi_\varepsilon}$, by extracting a subsequence if necessary. Hence we have

$$0 = \text{weak-}\lim_{n \to \infty} u_n = \text{weak-}\lim_{n \to \infty} [v_n + w_n] = v + w$$

with $w_n = \mathbb{1}_{\Omega \setminus A_n} u_n$ and $w \in \mathrm{S}^{\mathrm{Pe}}_{\Phi_\varepsilon}$ similarly. By Mazur's theorem in $L^1_{\mathbb{E}^*}[\mathbb{E}]$ ([BC01, Lemma 4.12] and [ACV92, Lemme 4 page 173]), we have

$$v(\omega) \in \bigcap_{n \in \mathbb{N}} \overline{\mathrm{co}}\{v_k(\omega);\, k \geq n\} \subset \bigcap_{n \in \mathbb{N}} \overline{\mathrm{co}}\{u_k(\omega);\, k \geq n\} \text{ a.e.} \tag{6.7.8}$$

Similarly

$$w(\omega) \in \bigcap_{n \in \mathbb{N}} \overline{\mathrm{co}}\{u_k(\omega);\, k \geq n\}. \tag{6.7.9}$$

As $0 \in \partial_{\mathrm{ext}} \left(\cap_{n \in \mathbb{N}} \overline{\mathrm{co}}\{u_k(\omega);\, k \geq n\}\right)$ by hypothesis, applying the arguments of Step 1 to v_n and w_n gives $v = w = 0$. Again by [ACV92, Lemme 4 page 173] we get

$$0 = v(\omega) \in \partial_{\mathrm{ext}} \Big(\bigcap_{n \in \mathbb{N}} \overline{\mathrm{co}}\{v_k(\omega);\, k \geq n\}\Big) \text{ a.e.} \tag{6.7.10}$$

By (6.7.10) we can apply the results stated in Step 1 to the sequence $(v_n)_n$ showing that $\|v_n\|_{L^1_{\mathbb{E}^*}[\mathbb{E}]} \to 0$. Since

$$\|u_n\|_{L^1_{\mathbb{E}^*}[\mathbb{E}]} \leq \|v_n\|_{L^1_{\mathbb{E}^*}[\mathbb{E}]} + \|w_n\|_{L^1_{\mathbb{E}^*}[\mathbb{E}]} \leq \|v_n\|_{L^1_{\mathbb{E}^*}[\mathbb{E}]} + \varepsilon$$

for all $n \in \mathbb{N}$ and ε is arbitrary > 0, $\|u_n\|_{L^1_{\mathbb{E}^*}[\mathbb{E}]} \to 0$. □

Chapter 7

Applications in Control Theory

7.1 Measurable selection results

We aim to present in this section some problems in Control Theory governed by an evolution equation.

Let us recall a known fact (see e.g. [Val90b]). Given an abstract measurable space $(\Omega, \mathcal{S})$, a Polish space $\mathbb{S}$, and the set $\mathcal{M}^{+,1}(\mathbb{S})$ of (Radon) probability measures on $\mathbb{S}$ equipped with the narrow topology. Let $\nu : (\Omega, \mathcal{S}) \to \mathcal{M}^{+,1}(\mathbb{S})$. Then the following are equivalent:

(*a*) For every open set O in $\mathbb{S}$, $\omega \mapsto \nu_\omega(O)$ is $\mathcal{S}$-measurable.

(*b*) For every Borel set B in $\mathbb{S}$, $\omega \mapsto \nu_\omega(B)$ is $\mathcal{S}$-measurable.

(*c*) For every positive $\mathcal{B}_\mathbb{S}$-measurable function f defined on $\mathbb{S}$, the function $\omega \mapsto \int_\mathbb{S} f(s)\, d\nu_\omega(s)$ is $\mathcal{S}$-measurable.

Let $\mathbb{U}$ be a compact metric space, let $\mathrm{C}(\mathbb{U}) = \mathrm{C}(\mathbb{U}, \mathbb{R})$ denote the space of real valued functions on U equipped with the topology of uniform convergence, and let $\mathcal{M}^{+,1}(\mathbb{U})$ be the set of all probability Radon measures on $\mathbb{U}$. It is well-known that $\mathcal{M}^{+,1}(\mathbb{U})$ is compact metrizable for the $\sigma(\mathrm{C}(\mathbb{U})^*, \mathrm{C}(\mathbb{U}))$-topology. We need first a denseness lemma which differs from Theorem 2.2.3 by some hypotheses.

Lemma 7.1.1 (Castaing–Valadier 1971, unpublished) *Let $(T, \mathcal{T}, \mu)$ be a complete probability space and let $\mathbb{U}$ be a compact metrizable space. Let $\Gamma : T \to \mathcal{K}(\mathbb{U})$ be a compact-valued measurable multifunction and let*

$$\Sigma(t) := \{\nu \in \mathcal{M}^{+,1}(\mathbb{U});\ \nu(\Gamma(t)) = 1\}$$

for all $t \in T$. Let $\lambda \in \mathrm{S}_\Sigma$ (where S_Σ is the set of all $\mathcal{T}$-measurable selections of Σ). Recall that $\lambda = \int_T \delta_t \otimes \lambda_t \, d\mu(t)$. Let $\varepsilon > 0$, $(g_i)_{i\in I}$ and $(h_i)_{i\in I}$ two finite sequences in $\mathrm{L}^1(T, \mathcal{T}, \mu)$ and $\mathrm{C}(\mathbb{U})$ respectively. If μ is nonatomic, then there exists a $\mathcal{T}$-measurable selection of Γ, ρ, such that

$$\forall i \in I \quad |\langle \lambda, g_i \otimes h_i\rangle - \langle \underline{\delta}_\rho, g_i \otimes h_i\rangle| \leq \varepsilon$$

where $\underline{\delta}_\rho = \int_T \delta_t \otimes \delta_{\rho(t)} \, d\mu(t)$.

Remark 7.1.2 If $(T, \mathcal{T}, \mu)$ is a compact space equipped with a positive Radon measure μ, the preceding lemma provides a denseness result because $\mathrm{C}(T, \mathrm{C}(\mathbb{U}))$ is identified with $\mathrm{C}(T \times \mathbb{U})$ and $\mathrm{L}^1(T, \mu) \otimes \mathrm{C}(\mathbb{U})$ is dense in the Banach space $\mathrm{L}^1_{\mathrm{C}(\mathbb{U})}(T, \mu)$ so that the following topologies: $\sigma(\mathrm{L}^\infty_{\mathrm{ca}(\mathbb{U})}, \mathrm{L}^1_{\mathrm{C}(\mathbb{U})})$, $\sigma(\mathrm{L}^\infty_{\mathrm{ca}(\mathbb{U})}, \mathrm{L}^1 \otimes \mathrm{C}(\mathbb{U}))$, $\sigma(\mathrm{L}^\infty_{\mathrm{ca}(\mathbb{U})}, \mathrm{C}(T) \otimes \mathrm{C}(\mathbb{U}))$, $\sigma(\mathrm{L}^\infty_{\mathrm{ca}(\mathbb{U})}, \mathrm{C}(T \times \mathbb{U}))$ coincide on $\mathrm{S}_\Sigma \subset \mathrm{L}^\infty_{\mathrm{ca}(\mathbb{U})}(T, \mu)$.

Proof of Lemma 7.1.1. Let us set $M := \sup_{i\in I} \int_T |g_i(t)| \, d\mu(t)$. As $\mathbb{U}$ is compact, there exists a finite Borel partition $(\Omega_j, j \in J)$ of $\mathbb{U}$ such that for any pair u, u' in Ω_j, the following holds:

$$|h_i(u) - h_i(u')| \leq \frac{\varepsilon}{M}, \quad \forall i \in I. \tag{7.1.1}$$

Let $\mathbb{1}_{\Omega_j}$ be the indicator function of Ω_j, $(j \in J)$. Let $u_j : \Omega \to \mathbb{U}$ be $\mathcal{T}_\mu$-mesurable mappings satisfying: $u_j(t) \in \Gamma(t) \cap \Omega_j$ if $\Gamma(t) \cap \Omega_j \neq \emptyset$, and $u_j(t) \in \Gamma(t)$ otherwise. The existence of u_j is ensured by the Sainte Beuve–von Neumann–Aumann Selection Theorem (see e.g. [SB74], [CV77, Theorem III.22]). Let us set $\forall t \in T$, $\nu_t := \sum_{j\in J} \lambda_t(\Omega_j)\delta_{u_j(t)}$. Then it is obvious that $t \mapsto \nu_t$ is scalarly measurable because $t \mapsto \lambda_t(\Omega_j)$ is μ-measurable. Then

$$\langle \nu, g_i \otimes h_i\rangle = \int_T g_i(t)\langle \nu_t, h_i\rangle \, d\mu(t) = \sum_{j\in J} \int_T \lambda_t(\Omega_j) g_i(t) h(u_j(t)) \, d\mu(t). \tag{7.1.2}$$

Further we have

$$\langle \lambda - \nu, g_i \otimes h_i\rangle = \int_T g_i(t)\langle \lambda_t - \nu_t, h_i\rangle \, d\mu(t), \tag{7.1.3}$$

with

$$\langle \lambda_t - \nu_t, h_i\rangle = \int_{\Gamma(t)} \Big[h_i(u) - \sum_{j\in J} \mathbb{1}_{\Omega_j}(u) h_i(u_j(t))\Big] \, d\lambda_t(u) \tag{7.1.4}$$

because λ_t is supported by $\Gamma(t)$. But the integrand in (7.1.4) is estimated by $\dfrac{\varepsilon}{M}$. Indeed if $u \in \Gamma(t)$, there exists j_0 such that $u \in \Omega_{j_0}$, so that

$$|h_i(u) - \sum_{j\in J} \mathbb{1}_{\Omega_j}(u) h_i(u_j(t))| = |h_i(u) - h_i(u_{j_0}(t))| \leq \frac{\varepsilon}{M}$$

because $\Gamma(t) \cap \Omega_{j_0} \neq \emptyset$ implies $u_{j_0}(t) \in \Omega_{j_0}$ by construction. Taking (7.1.1) and (7.1.3) into account we have

$$|\langle \lambda - \nu, g_i \otimes h_i \rangle| \leq \varepsilon \tag{7.1.5}$$

for all $i \in I$. As μ is nonatomic, by Liapunov's theorem, there exists a $\mathcal{T}$-measurable partition $(T_j,\ j \in J)$ of T such that

$$\int_{\mathbb{T}} \lambda_t(\Omega_j) g_i(t) h(u_j(t))\, d\mu(t) = \int_{T_j} g_i(t) h_i(u_j(t))\, d\mu(t) \tag{7.1.6}$$

for all $i \in I$ and $j \in J$. Let us set $\rho(t) = u_j(t)$ if $t \in T_j$ $(j \in J)$ and recall

$$\underline{\delta}_\rho = \int_{\mathbb{T}} \delta_t \otimes \delta_{\rho(t)}\, d\mu(t).$$

Then by (7.1.2) and(7.1.6) we have

$$\langle \nu, g_i \otimes h_i \rangle = \langle \underline{\delta}_\rho, g_i \otimes h_i \rangle \tag{7.1.7}$$

for all $i \in I$. From (7.1.5) and (7.1.7) it follows that

$$|\langle \lambda, g_i \otimes h_i \rangle - \langle \underline{\delta}_\rho, g_i \otimes h_i \rangle| \leq \varepsilon$$

for all $i \in I$. □

Remark 7.1.3 1) The preceding result shows that the set S_Γ of all Lebesgue-measurable selections (alias original controls) of the Lebesgue-measurable multifunction $\Gamma : [0, T] \to \mathcal{K}(\mathbb{U})$ is dense for the above mentioned topologies in the set S_Σ of Lebesgue-measurable selections of the Lebesgue-measurable multifunction $\Sigma : [0, T] \to \mathcal{M}^{+,1}(\mathbb{U})$ defined by

$$\Sigma(t) := \{\nu \in \mathcal{M}^{+,1}(\mathbb{U});\ \nu(\Gamma(t)) = 1\}$$

for all $t \in [0, T]$.

2) For further references on the above denseness lemma, see [GH67] and [War67].

Let us mention an easy corollary of Lemma 7.1.1.

Corollary 7.1.4 *Let $(T, \mathcal{T}, \mu)$ be a complete probability space, (L, d) be a Lusin metrizable space and $\mathrm{BL}(L, d)$ the set of all bounded real-valued Lipschitz function defined on L. Let $\Gamma : T \to \mathcal{B}(L)$ be a measurable multifunction and let*

$$\Sigma(t) := \{\nu \in \mathcal{M}^{+,1}(L);\ \nu(\Gamma(t)) = 1\}$$

for all $t \in T$. Let $\lambda \in \mathrm{S}_\Sigma$ and $\lambda = \int_T \delta_t \otimes \lambda_t\, d\mu(t)$. Let $\varepsilon > 0$, $(g_i)_{i \in I}$ and $(h_i)_{i \in I}$ two finite sequences in $\mathrm{L}^1(T, \mathcal{T}, \mu)$ and $\mathrm{BL}(L, d)$ respectively. If μ is nonatomic, then there exists a $\mathcal{B}_T$-measurable selection of Γ, ρ, such that

$$\left|\langle \lambda, g_i \otimes h_i \rangle - \langle \underline{\delta}_\rho, g_i \otimes h_i \rangle\right| \leq \varepsilon, \quad \forall i \in I.$$

Proof. We follow some arguments in [Cas85, Théorème 2.2]. The space L can be viewed as a Borel subset of a compact metrizable space $(\mathbb{X}, d)$. Let j be the injection $j : L \to \mathbb{X}$. Set $j(\lambda)_t = j(\lambda_t)$. Then $j(\lambda) \in \mathcal{Y}^1(T, \mathcal{T}, \mu; \mathbb{X})$. Let $\varepsilon > 0$, $(g_i)_{i\in I}$ and $(h_i)_{i\in I}$ two finite sequences in $\mathrm{L}^1(T, \mathcal{T}, \mu)$ and $\mathrm{BL}(L)$ respectively. For each $i \in I$, let $\hat{h}_i$ be the Lipschitz extension of h_i by setting

$$\hat{h}_i(x) = \inf_{s\in L}[l_{h_i} d(s, x) + h_i(s)]$$

for all $x \in \mathbb{X}$, where

$$|h_i(s) - h_i(s')| \le l_{h_i}\, d(s, s')$$

for all $s, s' \in L$, and l_{h_i} is the Lipschitz modulus of the function h_i. Now apply the preceding denseness lemma to $j(\lambda)$ gives a $\mathcal{T}$-measurable selection of Γ, ρ, such that

$$\left|\langle j(\lambda), g_i \otimes \hat{h}_i\rangle - \langle \delta_\rho, g_i \otimes \hat{h}_i\rangle\right| \le \varepsilon, \quad \forall i \in I.$$

Note that

$$\begin{aligned}\langle j(\lambda), g_i \otimes \hat{h}_i\rangle &= \int_T g_i(t)\Big[\int_{\mathbb{X}} \hat{h}_i(x)\, dj(\lambda)_t(x)\Big]\, d\mu(t)\\ &= \int_T g_i(t)\Big[\int_L \hat{h}_i(x)\, d\lambda_t(x)\Big]\, d\mu(t)\\ &= \langle \lambda, g_i \otimes h_i\rangle,\end{aligned}$$

because $\lambda_t(L) = 1$ for almost every $t \in T$ and $\hat{h}_i = h_i (i \in I)$ on L. This ends the proof. □

7.2 Relaxed trajectories of an evolution equation governed by a maximal monotone operator

Now let us recall and summarize the following result [CFS00, Theorem 2.5].

Theorem 7.2.1 *Let $\mathbb{H}$ be a Hilbert space and let $\varphi : [0, T] \times \mathbb{H} \to [0, +\infty]$ be such that, for each $t \in [0, T]$, $\varphi(t, .)$ is convex proper l.s.c. and inf-ball-compact, (that is, for every $t \in [0, T]$ and for every $r > 0$, the set $\{x \in \mathbb{H};\ \varphi_t(x) \le r\}$ is ball-compact i.e. its intersection with any closed ball in $\mathbb{H}$ is compact). Assume that there are a Lipschitz function $k : \mathbb{H} \to \mathbb{R}^+$ and an absolutely continuous function $\alpha : [0, T] \longrightarrow \mathbb{R}^+$ with $\frac{d\alpha}{dt} \in \mathrm{L}^2_{\mathbb{R}^+}([0, T])$, such that for all $(s, t, x) \in [0, T] \times [0, T] \times \mathbb{H}$*

$$\varphi^*(t, x) - \varphi^*(s, x) \le k(x)|\alpha(t) - \alpha(s)|$$

where $\varphi^*(t, .) := \sup_{y\in\mathbb{H}} (\langle y, .\rangle - \varphi(t, y))$ denotes the Fenchel conjugate *of $\varphi(t, .)$. Let $F : [0, T] \times \mathbb{H} \to \mathrm{cw}\mathcal{K}(\mathbb{H})$ be a $\mathrm{cw}\mathcal{K}(\mathbb{H})$-valued multifunction such that* (*i*) $\forall x \in \mathbb{H}$, $F(.,x)$ *is Lebesgue-measurable on* $[0,T]$,

(*ii*) $\forall t \in [0,T]$, $F(t,.)$ *is upper semicontinuous on* $\mathbb{H}$,

(*iii*) *there exists a function* $c \in \mathrm{L}^2_{\mathbb{R}^+}([0,T])$ *such that*

$$F(t,x) \subset c(t)(1+\|x\|)\overline{B}_{\mathbb{H}}, \quad \forall t \in [0,T], \quad \forall x \in \mathbb{H}.$$

Then, for each $u_0 \in \operatorname{dom}\varphi(0,.)$, *there is an absolutely continuous solution* $u : [0,T] \to \mathbb{H}$ *to the problem* $\frac{du}{dt} \in -\partial\varphi_t(u(t)) + F(t,u(t))$ *a.e. with the initial condition* $u(0) = u_0 \in \operatorname{dom}\varphi(0,.)$. *Moreover the solutions set of this equation is compact for the topology of uniform convergence.*

Example 7.2.2 In the following $\mathbb{U}$ is a compact metric space, $\Gamma : [0,T] \to \mathcal{K}(\mathbb{U})$ is a Lebesgue-measurable compact valued multifunction, $\mathbb{H}$ is a separable Hilbert space, and $\varphi : [0,T] \times \mathbb{H} \to [0,+\infty]$ is such that, for each $t \in [0,T]$, $\varphi(t,.)$ is convex proper l.s.c. and inf-ball-compact.

Let us consider a mapping $g : [0,T] \times \mathbb{H} \times \mathbb{U} \to \mathbb{H}$ satisfying:

(*i*) For every fixed $t \in [0,T]$, $g(t,.,.)$ is continuous on $\mathbb{H} \times \mathbb{U}$.

(*ii*) For every $(x,u) \in \mathbb{H} \times \mathbb{U}$, $g(.,x,u)$ is Lebesgue-measurable on $[0,T]$.

(*iii*) For every $\eta > 0$, there exists $l(\eta) > 0$ such that $\|g(t,x,u) - g(t,y,u)\| \leq l(\eta)\|x-y\|$ for all $t \in [0,T]$ and for all $(x,y) \in \overline{B}_{\mathbb{H}}(0,\eta) \times \overline{B}_{\mathbb{H}}(0,\eta)$.

(*iv*) There exists $c > 0$ such that $g(t,x,u) \subset c(1+\|x\|)\overline{B}_{\mathbb{H}}$ for all $(t,x,u) \in [0,T] \times \mathbb{H} \times \mathbb{U}$.

We aim to compare the solutions set of the two following evolution equations:

$$(P_{\mathcal{O}}) \begin{cases} \dot{u}_\rho(t) \in -\partial\varphi_t(u_\rho(t))) + g(t,u_\rho(t),\rho(t)) \ \text{ a.e. } t \in [0,T], \\ u_\rho(0) = x_0 \in \operatorname{dom}\varphi(0,.), \end{cases}$$

where ρ belongs to the set $\mathcal{U}$ of all original controls, i.e. ρ is a Lebesgue-measurable mapping from $[0,T]$ into $\mathbb{U}$ with $\rho(t) \in \Gamma(t)$ for a.e. $t \in [0,T]$, and

$$(P_{\mathcal{R}}) \begin{cases} \dot{u}_\lambda(t) \in -\partial\varphi_t(u_\lambda(t))) + \int_{\Gamma(t)} g(t,u_\lambda(t),u)\, d\lambda_t(u) \ \text{ a.e. } t \in [0,T], \\ u_\lambda(0) = x_0 \in \operatorname{dom}\varphi(0,.), \end{cases}$$

where λ belongs to the set S_Σ of all relaxed controls, i.e. λ is a Lebesgue-measurable selection of the multifunction Σ defined by

$$\Sigma(t) := \{\nu \in \mathcal{M}^{+,1}(\mathbb{U});\ \nu(\Gamma(t)) = 1\}$$

for all $t \in [0,T]$. Note that when $A(t)$ is a bounded linear operator and $\mathbb{H}$ is finite dimensional, the problem under consideration is quite classical in the theory of

Optimal Control (see, for instance, [GH67, War67, War72]). So we don't want to go into details, but only show the main facts. At this point, we use similar arguments as in [Jaw84] exploiting the compactness of solutions set obtained in Theorem 7.2.1. For each $\lambda \in \mathrm{S}_\Sigma$, let us set

$$h_\lambda(t,x) := \int_{\Gamma(t)} g(t,x,u)\, d\lambda_t(u)$$

for all $(t,x) \in [0,T] \times \mathbb{H}$. Then h_λ is measurable on $[0,T]$ and satisfies:

(*) For every $\eta > 0$, there exists $l(\eta) > 0$ such that $\|h_\lambda(t,x) - h_\lambda(t,y)\| \le l(\eta)\|x - y\|$ for all $t \in [0,T]$ and for all $(x,y) \in \overline{B}_{\mathbb{H}}(0,\eta) \times \overline{B}_{\mathbb{H}}(0,\eta)$ thanks to (iii).

Moreover since $\lambda_t(\Gamma(t)) = 1$, we have

$$(^{**}) \qquad h_\lambda(t,x) \in \overline{\mathrm{co}}\, g(t,x,\Gamma(t)) \subset c(1+\|x\|)\, \overline{B}_{\mathbb{H}}, \quad \forall (t,x) \in [0,T] \times \mathbb{H}.$$

So, in view of Theorem 7.2.1 and the mononicity of the subdifferential operator $\partial\varphi_t$, for each $\lambda \in \mathrm{S}_\Sigma$, there is a unique solution u_λ for the problem

$$(P_{\mathcal{R}}) \begin{cases} \dot{u}_\lambda(t) \in -\partial\varphi_t(u_\lambda(t))) + \int_{\Gamma(t)} g(t,u_\lambda(t),u)\, d\lambda_t(u) & \text{a.e. } t \in [0,T], \\ u_\lambda(0) = x_0 \in \ \mathrm{dom}\, \varphi(0,.). \end{cases}$$

By (**) and Theorem 7.2.1, we see that the solutions set $\{u_\lambda;\, \lambda \in \mathrm{S}_\Sigma\}$ of $(P_{\mathcal{R}})$ is relatively compact in the space $\mathrm{C}([0,T],\mathbb{H})$ of continuous mappings from $[0,T]$ to $\mathbb{H}$, endowed with the topology of uniform convergence. It is obvious that the solutions set $\{u_\rho;\, \rho \in \mathcal{U}\}$ is a nonempty subset of $\{u_\lambda;\, \lambda \in \mathrm{S}_\Sigma\}$.

Proposition 7.2.3 *The set $\{u_\rho;\, \rho \in \mathcal{U}\}$ is dense for the uniform norm in the compact solutions set $\{u_\lambda;\, \lambda \in \mathrm{S}_\Sigma\}$.*

It is obvious that Proposition 7.2.3 follows from

Lemma 7.2.4 *The mapping $\lambda \mapsto u_\lambda$ defined on the $\sigma(\mathrm{L}^\infty_{\mathrm{C}(\mathbb{U})^*}([0,T]), \mathrm{L}^1_{\mathrm{C}(\mathbb{U})}([0,T]))$ compact set S_Σ has a closed graph.*

Proof. Let $\lambda^n \to \lambda^\infty$ for the $\sigma(\mathrm{L}^\infty_{\mathrm{C}(\mathbb{U})^*}([0,T]), \mathrm{L}^1_{\mathrm{C}(\mathbb{U})}([0,T]))$ topology and $u_{\lambda^n} \to u_\infty$ in $\mathrm{C}([0,T],\mathbb{H})$ where u_{λ^n} $(n \in \mathbb{N} \cup \{\infty\})$ is the unique absolutely continuous solution of the equation

$$\begin{cases} \dot{u}_{\lambda^n}(t) \in -\partial\varphi_t(u_{\lambda^n}(t)) + \int_{\Gamma(t)} g(t,u_{\lambda^n}(t),u)\, d\lambda^n_t(u) & \text{a.e. } t \in [0,T], \\ u_{\lambda^n}(0) = x_0 \in \ \mathrm{dom}\, \varphi(0,.), \end{cases}$$

then $u_\infty = u_{\lambda^\infty}$. We have

$$h_{\lambda^n}(t,u_{\lambda^n}(t)) - \dot{u}_{\lambda^n}(t) \in \partial\varphi_t(u_{\lambda^n}(t))$$

and

$$h_{\lambda^\infty}(t, u_{\lambda^\infty}(t)) - \dot{u}_{\lambda^\infty}(t) \in \partial\varphi_t(u_{\lambda^\infty}(t))$$

for a.e. $t \in [0, T]$. Since $\partial\varphi_t$ is monotone,

$$\langle u_{\lambda^n}(t) - u_{\lambda^\infty}(t), h_{\lambda^n}(t, u_{\lambda^n}(t)) - \dot{u}_{\lambda^n}(t) - h_{\lambda^\infty}(t, u_{\lambda^\infty}(t)) + \dot{u}_{\lambda^\infty}(t)\rangle \geq 0$$

a.e. in $[0, T]$. So

$$\frac{1}{2}\frac{d}{dt}\|u_{\lambda^n}(t) - u_{\lambda^\infty}(t)\|^2 \leq \langle u_{\lambda^n}(t) - u_{\lambda^\infty}(t), h_{\lambda^n}(t, u_{\lambda^n}(t)) - h_{\lambda^\infty}(t, u_{\lambda^\infty}(t))\rangle$$

a.e. in $[0, T]$. Integrating on $[0, t]$ gives

$$\frac{1}{2}\|u_{\lambda^n}(t) - u_{\lambda^\infty}(t)\|^2 \leq \int_0^t \langle u_{\lambda^n}(s) - u_{\lambda^\infty}(s), h_{\lambda^n}(s, u_{\lambda^n}(s)) - h_{\lambda^\infty}(s, u_{\lambda^\infty}(s))\rangle\, ds.$$

Let us set

$$L_n(t) = \int_0^t \langle u_{\lambda^n}(s) - u_{\lambda^\infty}(s), h_{\lambda^n}(s, u_{\lambda^n}(s)) - h_{\lambda^\infty}(s, u_{\lambda^\infty}(s))\rangle\, ds.$$

Then $L_n(t) = L_n^1(t) + L_n^2(t) + L_n^3(t)$ where

$$L_n^1(t) = \int_0^t \langle u_{\lambda^n}(s) - u_{\lambda^\infty}(s), h_{\lambda^n}(s, u_{\lambda^n}(s)) - h_{\lambda^n}(s, u_{\lambda^\infty}(s))\rangle\, ds,$$

$$L_n^2(t) = \int_0^t \langle u_{\lambda^n}(s) - u_\infty(t), h_{\lambda^n}(s, u_{\lambda^\infty}(s)) - h_{\lambda^\infty}(s, u_{\lambda^\infty}(s))\rangle\, ds,$$

$$L_n^3(t) = \int_0^t \langle u_\infty(s) - u_{\lambda^\infty}(s), h_{\lambda^n}(s, u_{\lambda^\infty}(s)) - h_{\lambda^\infty}(s, u_{\lambda^\infty}(s))\rangle\, ds.$$

As $\|h_\lambda(t, x)\| \leq c(1 + \|x\|)$ for all $\lambda \in \mathrm{S}_\Sigma$ and for all $(t, x) \in [0, 1] \times \mathbb{H}$, using (**) gives the estimate

$$|L_n^2(t)| \leq 2c(1 + \eta)\|u_{\lambda^n} - u_\infty\|_{\mathrm{C}([0,1],\mathbb{H})},$$

with $\sup\{\|u_\lambda\|_{\mathrm{C}([0,1],\mathbb{H})};\ \lambda \in \mathrm{S}_\Sigma\} < \eta < +\infty$. Thus $L_n^2(t) \to 0$ when $n \to \infty$ uniformly in $[0, 1]$. Similarly by (iv) the integrand

$$f(s, v) := \langle u_\infty(s) - u_{\lambda^\infty}(s), g(s, u_{\lambda^\infty}(s), v)\rangle$$

is estimated by

$$|f(s, v)| \leq c(1 + \eta)\|u_\infty - u_{\lambda^\infty}\|_{\mathrm{C}([0,1],\mathbb{H})}$$

for all $(s, v) \in [0, T] \times \mathbb{U}$. Hence $f \in \mathrm{L}^1_{\mathrm{C}(\mathbb{U})}([0, T])$. As (λ^n) converges in the topology $\sigma((\mathrm{L}^\infty_{\mathrm{C}(\mathbb{U})^*}, \mathrm{L}^1_{\mathrm{C}(\mathbb{U})})$ to λ^∞, it is immediate that for every $t \in [0, T]$,

$$\int_0^t \Big[\int_{\mathbb{U}} f(s, v)\, d\lambda_s^n(v)\Big]\, ds \to \int_0^t \Big[\int_{\mathbb{U}} f(s, v)\, d\lambda_s^\infty(v)\Big]\, ds$$

when $n \to \infty$. So $\lim_{n\to\infty} L_n^3(t) \to 0$. Using (*), there is $l(\eta) > 0$ such that

$$|L_n^1(t)| \leq \int_0^t l(\eta)\|u_{\lambda^n}(s) - u_{\lambda^\infty}(s)\|^2\, ds.$$

Finally we get

$$\frac{1}{2}\|u_{\lambda^n}(t) - u_{\lambda^\infty}(t)\|^2 \leq L_n^2(t) + L_n^3(t) + \int_0^t l(\eta)\|u_{\lambda^n}(s) - u_{\lambda^\infty}(s)\|^2\, ds.$$

As the functions $L_n^2(.)$ and $L_n^3(.)$ are continuous with $L_n^2(t) \to 0$ and $L_n^3(t) \to 0$, by Gronwall's lemma we conclude that

$$u_{\lambda^n}(t) \to u_{\lambda^\infty}(t) = u_\infty(t), \quad \forall t \in [0,1],$$

and so $u_{\lambda^\infty}(.) = u_\infty(.)$ and the set $\{u_\lambda;\ \lambda \in \ \mathrm{S}_\Sigma\}$ is compact in $\mathrm{C}([0,T], \mathbb{H})$.
□

Example 7.2.5 Let V be a convex weakly compact subset of $\mathbb{H}$ and let $\Delta : [0,T] \to \mathrm{cw}\mathcal{K}(V)$ be a convex weakly compact valued upper semicontinuous multifunction. In view of [CV77, Theorem IV.16], the graph of the multifunction $\partial_{\text{ext}}(\Delta) : t \mapsto \partial_{\text{ext}}(\Delta(t))$, is a Borel subset in $[0,T] \times V$. Then it is easy to see that the denseness property given in Remark 7.1.3 of Lemma 7.1.1 is still valid if we replace Γ by $\partial_{\text{ext}}(\Delta)$ and Σ by Ψ where

$$\Psi(t) := \{\nu \in \mathcal{M}^{+,1}(V);\ \nu(\partial_{\text{ext}}(\Delta(t))) = 1\}$$

for all $t \in [0,T]$. Using the continuity property stated in Lemma 7.2.4 we conclude that the solutions set of the equation:

$$(P_{\partial_{\text{ext}}(\Delta)}) \begin{cases} \dot{u}_\rho(t) \in -\partial\varphi_t(u_\rho(t)) + g(t, u_\rho(t), \rho(t)) & \text{a.e. } t \in [0,T], \\ u_\rho(0) = x_0 \in \ \operatorname{dom}\varphi(0,.), \end{cases}$$

where ρ belongs to the set of all Lebesgue-measurable mappings from $[0,T]$ into $\mathbb{U}$ with $\rho(t) \in \partial_{\text{ext}}(\Delta(t))$ for a.e. $t \in [0,T]$, is dense in the solutions set of

$$(P_\Psi) \begin{cases} \dot{u}_\lambda(t) \in -\partial\varphi_t(u_\lambda(t)) + \int_{\partial_{\text{ext}}(\Delta(t))} g(t, u_\lambda(t), u)\, d\lambda_t(u) & \text{a.e. } t \in [0,T], \\ u_\lambda(0) = x_0 \in \ \operatorname{dom}\varphi(0,.), \end{cases}$$

where λ is a measurable selection of the multifunction Ψ defined above.

Comments The variations of our techniques can be applied to other problems in Optimal Control involving Young measures. For instance, one can treat the same problem, when $\partial\varphi_t$ is replaced by a maximal monotone operator $A(t)$. More important is the study of differential inclusions governed by nonconvex sweeping process that we develop below.

7.3 Relaxed trajectories of a differential inclusion in a Banach space

Let us consider a separable Banach space $\mathbb{E}$ and a compact metric space $\mathbb{U}$. Recall (Lemma 7.1.1) that the set S_Γ of all Lebesgue-measurable selections (alias original controls) of the Lebesgue-measurable multifunction $\Gamma : [0,T] \to \mathcal{K}(\mathbb{U})$ is dense for the above mentioned topologies in the set S_Σ of Lebesgue-measurable selections of Lebesgue-measurable multifunction $\Sigma : [0,T] \to \mathcal{M}^{+,1}(\mathbb{U})$ defined by

$$\Sigma(t) := \{\nu \in \mathcal{M}^{+,1}(\mathbb{U});\ \nu(\Gamma(t)) = 1\}$$

for all $t \in [0,T]$.

Example 7.3.1 Let us consider a mapping $g : [0,T] \times \mathbb{E} \times \mathbb{U} \to \mathbb{E}$ satisfying:

(i) For every fixed $t \in [0,T]$, $g(t,.,.)$ is continuous on $\mathbb{E} \times \mathbb{U}$.

(ii) For every $(x,u) \in \mathbb{E} \times \mathbb{U}$, $g(.,x,u)$ is Lebesgue-measurable on $[0,T]$.

(iii) For every $\eta > 0$, there exists $l_\eta > 0$ such that $\|g(t,x,u) - g(t,y,u)\| \leq l_\eta \|x-y\|$ for all $t \in [0,T]$ and for all $(x,y) \in \overline{B}_{\mathbb{E}}(0,\eta) \times \overline{B}_{\mathbb{E}}(0,\eta)$.

(iv) There exists a convex compact valued integrable bounded multifunction $\Phi : [0,T] \to c\mathcal{K}(\mathbb{E})$ such that $g(t,x,\mathbb{U}) \subset (1+\|x\|)\Phi(t)$ for all $(t,x) \in [0,T] \times \mathbb{E}$.

We aim to compare the solutions set of the two following differential inclusions:

$$(P_{\mathcal{O}}) \begin{cases} \dot{u}_\rho(t) = g(t, u_\rho(t), \rho(t)) & \text{a.e. } t \in [0,T], \\ u_\rho(0) = x_0 \in \mathbb{E}, \end{cases}$$

where ρ belongs to the set $\mathbb{U}$ of all original controls, i.e. ρ is a Lebesgue-measurable mapping from $[0,T]$ into $\mathbb{U}$ with $\rho(t) \in \Gamma(t)$ for a.e. $t \in [0,T]$, and

$$(P_{\mathcal{R}}) \begin{cases} \dot{u}_\lambda(t) = \int_{\Gamma(t)} g(t, u_\lambda(t), z)\, d\lambda_t(z) & \text{a.e. } t \in [0,T], \\ u_\lambda(0) = x_0 \in \mathbb{E}, \end{cases}$$

where λ belongs to the set S_Σ of all relaxed controls, i.e. λ is the Lebesgue-measurable selection of the multifunction Σ defined by

$$\Sigma(t) := \{\nu \in \mathcal{M}^{+,1}(\mathbb{U});\ \nu(\Gamma(t)) = 1\}$$

for all $t \in [0,T]$. Note that the existence of absolutely continuous solutions for the preceding equation is well-known. Namely we have

$$u_\rho(t) = x_0 + \int_0^t g(s, u_\rho(s), \rho(s))\, ds,\ \forall t \in [0,T],$$

and

$$u_\lambda(t) = x_0 + \int_0^t [\int_{\Gamma(t)} g(s, u_\lambda(s), z)\, d\lambda_s(z)]\, ds, \ \forall t \in [0, T].$$

Moreover we will see that the solutions set $\{u_\lambda;\ \lambda \in\ \mathrm{S}_\Sigma\}$ of $(P_\mathcal{R})$ is compact in $\mathrm{C}([0,T],\mathbb{E})$ It is obvious that the solutions set $\{u_\rho;\ \rho \in \mathcal{U}\}$ is a nonempty subset of $\{u_\lambda;\ \lambda \in\ \mathrm{S}_\Sigma\}$.

Proposition 7.3.2 *The solutions set $\mathcal{S}_\mathcal{R}$ of $(P_\mathcal{R})$ is relatively compact in* $\mathrm{C}([0,T],\mathbb{E})$

Proof. Set $|\Phi(t)| := \sup\{\|x\|;\ x \in \Phi(t)\}$ for every $t \in [0,T]$. For simplicity we may suppose $x_0 = 0$. Then we have

$$\|u_\lambda(t)\| \leq \int_0^t |\Phi(s)|(1 + \|u_\lambda(s)\|)\, ds, \quad \forall t \in [0,T].$$

By Gronwall's lemma we get

$$\|u_\lambda(t)\| + 1 \leq z(t) := \exp(\int_0^t |\Phi(s)|\, ds), \quad \forall t \in [0,T],$$

this entails

$$h(s, u_\lambda(s), \lambda_s) \in (\|u_\lambda(s)\| + 1)\Phi(s) \subset z(s)\Phi(s), \quad \forall s \in [0,T].$$

Thus $\mathcal{S}_\mathcal{R}$ is included in the primitive $\mathcal{P}(\Psi)$ of the multifunction $\Psi(.) := z(.)\Phi(.)$, namely

$$\mathcal{P}(\Psi) := \{u : [0,T] \to \mathbb{E};\ u(0) = 0,\ u(t) = \int_0^t f(s)\, ds,\ f(s) \in \Psi(s) \text{ a.e.}\}.$$

Making use of Ascoli's theorem and Banach–Dieudonné's theorem (cf. Proposition 6.2.3 or Proposition 6.3.6) we see that $\mathrm{P}(\Psi)$ is compact in $\mathrm{C}([0,T],\mathbb{E})$.
□

Proposition 7.3.3 *The set $\mathcal{S}_\mathcal{O} := \{u_\rho;\ \rho \in \mathcal{U}\}$ is dense for the uniform norm in the compact solutions set $\mathcal{S}_\mathcal{R} := \{u_\lambda;\ \lambda \in\ \mathrm{S}_\Sigma\}$.*

We shall need the following Gronwall–type lemma.

Lemma 7.3.4 *Let l be a positive Lebesgue-integrable function defined on $[0,T]$, and a and h two nonnegative continuous functions defined on $[0,T]$ such that*

$$a(t) \leq h(t) + \int_0^t l(s)a(s)\, ds$$

for all $t \in [0,T]$. Then we have

$$\exp(-\int_0^t l(s)\, ds) \int_0^t l(s)a(s)\, ds \leq \int_0^t \exp(-\int_0^s l(\tau)\, d\tau)\, l(s)h(s)\, ds$$

for all $t \in [0,T]$.

Proof. Let us set $b(t) = \int_0^t l(s)a(s)\,ds$ for all $t \in [0,T]$. Then we have

$$\frac{d}{dt}[b(t)\exp(-\int_0^t l(s)\,ds)] = [l(t)a(t) - b(t)l(t)]\exp(-\int_0^t l(s)\,ds)$$

$$\leq l(t)h(t) + b(t)l(t) - b(t)l(t)]\exp(-\int_0^t l(s)\,ds) = l(t)h(t)\exp(-\int_0^t l(s)\,ds).$$

Hence the result follows by integrating on $[0,t]$

$$\exp(-\int_0^t l(s)\,ds)\int_0^t l(s)a(s)\,ds \leq \int_0^t \exp(-\int_0^s l(\tau)\,d\tau)\,l(s)h(s)\,ds.$$

□

First proof of Proposition 7.3.3. Since $\mathrm{L}^1_{C(\mathbb{U})}([0,T])$ is separable, S_Σ is compact [CV77, Theorem V.2] metrizable for the σ^*-topology on $\mathrm{L}^\infty_{C(\mathbb{U})^*}([0,T])$. Hence it is enough to show that $\lambda \mapsto u_\lambda$ is sequentially continuous on S_Σ for this topology. Let $\lambda^n \to \lambda^\infty$ in S_Σ. For every $t \in [0,T]$, we write

$$\begin{aligned} u_{\lambda^\infty}(t) - u_{\lambda^n}(t) = &\int_0^t [\int_{\mathbb{U}} g(s, u_{\lambda^\infty}(s), z)\,d\lambda^\infty_s(z)]\,ds \\ &- \int_0^t [\int_{\mathbb{U}} g(s, u_{\lambda^\infty}(s), z)\,d\lambda^n_s(z)]\,ds \\ &+ \int_0^t [\int_{\mathbb{U}} [g(s, u_{\lambda^\infty}(s), z) - g(s, u_{\lambda^n}(s), z)]\,d\lambda^n_s(z)]\,ds. \end{aligned} \tag{7.3.1}$$

Let us set

$$a_n(t) := \|u_{\lambda^\infty}(t) - u_{\lambda^n}(t)\|,$$

$$b_n(t) := \int_0^t [\int_{\mathbb{U}} g(s, u_{\lambda^\infty}(s), z)\,d\lambda^\infty_s(z)]\,ds - \int_0^t [\int_{\mathbb{U}} g(s, u_{\lambda^\infty}(s), z)\,d\lambda^n_s(z)]\,ds,$$

$$c_n(t) := \int_0^t [\int_{\mathbb{U}} g(s, u_{\lambda^\infty}(s), z)\,d\lambda^n_s(z)]\,ds - \int_0^t [\int_{\mathbb{U}} g(s, u_{\lambda^n}(s), z)\,d\lambda^n_s(z)]\,ds,$$

$$d_n(t) := \|b_n(t)\|.$$

By Proposition 7.3.2, the solutions set of $(P_{\mathcal{R}})$ is relatively compact, hence there is $\eta > 0$ such that $\sup_{\lambda \in \mathrm{S}_\Sigma} \|u_\lambda\|_{\mathrm{C}([0,T],\mathbb{E})} < \eta$. By (iii) there is $l_\eta > 0$ such that

$$\|c_n(t)\| \leq \int_0^t l_\eta \|u_{\lambda^\infty}(s) - u_{\lambda^n}(s)\|\,ds$$

for all $t \in [0,T]$. Hence

$$a_n(t) \leq d_n(t) + \int_0^t l_\eta a_n(s)\,ds$$

for all $t \in [0,1]$. By Lemma 7.3.4, we have

$$\exp(-\int_0^t l_\eta\,ds)\int_0^t l_\eta\,a_n(s)\,ds \le \int_0^t \exp(-\int_0^t l_\eta\,ds)\,l_\eta(t)d_n(t)\,dt \tag{7.3.2}$$

for all $t \in [0,T]$ and for all $n \in \mathbb{N}$. We are going to check that $d_n(t) \to 0$ for every $t \in [0,T]$. Since (b_n) is relatively compact in $\mathrm{C}([0,T],\mathbb{E})$, it is sufficient to prove that

$$\lim_{n\to\infty}\langle x', b_n(t)\rangle = 0$$

for every $x' \in \mathbb{E}^*$ and for every $t \in [0,T]$. Let us set

$$f(s) : u \mapsto \langle x', g(s, u_{\lambda^\infty}(s), u)\rangle \quad (s \in [0,T],\ u \in \mathbb{U}).$$

By (iv), $f \in \mathrm{L}^1_{\mathrm{C}(\mathbb{U})}([0,T])$ and we have

$$\langle x', b_n(t)\rangle = \int_0^t \langle f(s), \lambda^\infty_s - \lambda^n_s\rangle\,ds.$$

Thus the second member tends to 0 because $\lambda^n \to \lambda^\infty$ stably. So

$$\lim_{n\to\infty}\int_0^t l_\eta\,d_n(s)\,ds = \int_0^t \lim_{n\to\infty} l_\eta\,d_n(s)\,ds = 0.$$

By (7.3.2) this implies that

$$\lim_{n\to\infty}\int_0^t l_\eta\,a_n(s)\,ds = 0.$$

Finally making use of (7.3.1) we get

$$\lim_{n\to\infty} a_n(t) \le \lim_{n\to\infty}\left[d_n(t) + \int_0^t l_\eta a_n(s)\,ds\right] = 0.$$

□

The preceding proof is somewhat traditional. We will present below a short and different proof using the fiber product of Young measures (Theorem 3.3.1). This fact will be used in other places.

Second proof of Proposition 7.3.3. We are going to prove that the graph of the mapping $\lambda \mapsto u_\lambda$ from S_Σ into the Banach space $\mathrm{C}([0,T],\mathbb{E})$ is compact. Recall that the set $\mathcal{S}_\mathcal{R}$ is relatively compact in $\mathrm{C}([0,T],\mathbb{E})$, Hence the set $K := \{u_\lambda(t);\ (\lambda,t) \in \mathrm{S}_\Sigma \times [0,T]\}$ is bounded (even relatively compact) in $\mathbb{E}$, so that, with the notations of the proof of Proposition 7.3.2, $\|g(t,x,z)\| \le (1+|K|)|\Phi(t)|$ for all $(t,x,z) \in [0,T] \times K \times \mathbb{U}$. Let (λ^n) be a sequence in S_Σ. By compactness we may assume that λ^n stably converges to λ^∞ and u_{λ^n} converges to $u_\infty \in \mathrm{C}([0,T],\mathbb{E})$. Let $x' \in \mathbb{E}^*$ and $t \in [0,T]$. Then we have $\langle x', u_\infty(t)\rangle = \lim_n \langle x', u_{\lambda^n}(t)\rangle$ and

$\underline{\delta}_{u_{\lambda^n}} \otimes \lambda^n$ stably converges to $\underline{\delta}_{u_\infty} \otimes \lambda^\infty$ by the Fiber Product Lemma (Theorem 3.3.1). As the integrand $(s,x,z) \mapsto \langle x', g(s,x,z)\rangle$ is L^1-bounded on $[0,T] \times K \times \mathbb{U}$, it follows that

$$\lim_n \int_0^t [\int_{\mathbb{U}} \langle x', g(s, u_{\lambda^n}(s), z)\rangle \lambda_s^n(dz)]ds = \int_0^t [\int_{\mathbb{U}} \langle x', g(s, u_\infty(s), z)\rangle \lambda_s^\infty(dz)]ds$$

So we deduce that $\dot{u}_\infty(t) = \int_{\mathbb{U}} g(s, u_\infty(s), z)\lambda_s^\infty(dz)$ a.e. By uniqueness, we have necessarily $u_\infty = u_{\lambda^\infty}$. □

7.4 Integral representation theorem via Young measures

When dealing with relaxed trajectories of the above differential inclusions we are concerned with trajectories of the evolution equation of the form

$$(P_{\mathcal{R}}) \begin{cases} \dot{u}_\lambda(t) \in A(t)u_\lambda(t) + \int_{\Gamma(t)} g(t, u_\lambda(t), u)\, d\lambda_t(u) & \text{a.e. } t \in [0,T], \\ u_\lambda(0) = x_0 \in \operatorname{dom} A(0), \end{cases}$$

where $A(t)$ is a maximal monotone operator and λ belongs to the set S_Σ of all relaxed controls, i.e. λ is the Lebesgue-measurable selection of the multifunction Σ defined by

$$\Sigma(t) := \{\nu \in \mathcal{M}^{+,1}(\mathbb{U});\ \nu(\Gamma(t)) = 1\}$$

for all $t \in [0,T]$. As

$$\int_{\Gamma(t)} g(t, u_\lambda(t), u)\, d\lambda_t(u) \in \overline{\text{co}}\ g(t, u_\lambda(t), \Gamma(t))$$

for almost all $t \in [0,T]$, u_λ is an absolutely continuous solution of the differential inclusion

$$(*) \qquad \dot{u}(t) \in A(t)u_\lambda(t) + \overline{\text{co}}\, F(t, u(t))$$

where

$$F(t,x) = g(t,x,\Gamma(t)) := \{g(t,x,u);\ u \in \Gamma(t)\}.$$

In the present context the relaxed trajectories for $(P_{\mathcal{R}})$ coincides with the trajectories of the differential inclusion (*). This consideration leads to some integral representation theorem involving Young measures. In the following the control space $\mathbb{U}$ is a Polish space and $\mathbb{E}$ is a separable Banach space.

Proposition 7.4.1 *Suppose that $\Gamma : [0,T] \to \mathcal{K}(\mathbb{U})$ is a compact valued measurable multifunction and $g : [0,T] \times \mathbb{U} \to \mathbb{E}$ is a Carathéodory mapping such that*

$t \mapsto \sup\{\|g(t,u)\|;\ u \in \Gamma(t)\}$ is integrable on $[0,T]$, and $v : [0,T] \to \mathbb{E}$ is an integrable function such that $v(t) \in \overline{\text{co}}\ g(t,\Gamma(t))$ for a.e. $t \in [0,T]$, then there exists a Young measure $\lambda \in \mathcal{Y}^1(\Omega, \mathcal{S}, \mathrm{P}; \mathbb{U})$ such that

$$v(t) = \int_{\Gamma(t)} g(t,u)\, d\lambda_t(u) \ \text{ a.e.}$$

Proof. We only sketch the proof. Let

$$h(t,\nu) := \int_{\Gamma(t)} g(t,u)\, d\nu(u).$$

Then h is a Carathéodory mapping from $[0,T] \times \mathcal{M}^{+,1}(\mathbb{U})$ to $\mathbb{E}_\sigma$. By our assumption we have

$$v(t) \in h(t, \Sigma(t)) \ \text{ a.e.}$$

where

$$\Sigma(t) = \{\nu \in \mathcal{M}^{+,1}(\mathbb{U});\ \nu(\Gamma(t)) = 1\}$$

for all $t \in [0,T]$. As Σ is measurable, applying a measurable selection theorem [CV77, Theorem III.22], we get the required result. □

Proposition 7.4.1 is a parametric version of Choquet's theorem. In the same spirit, using Young measures, we provide a characterization of measurable selections of a measurable multifunction of the form $\overline{\text{co}}(\Gamma)$, where Γ is a given measurable multifunction, below an epigraph valued one. See also [Ben91, Proposition 2.1].

Let $\mathbb{R}^{(\mathbb{N})}$ denote the set of real sequences with finite support, and for $k \in \mathbb{N}$

$$\Lambda_k := \Big\{(r_i)_{i\geq 0} \in \mathbb{R}^{(\mathbb{N})};\ r_i \geq 0,\ \sum_{i\geq 0} r_i = 1 \text{ and } r_i = 0 \text{ if } i < k\Big\}.$$

Let $(u_n)_n$ be a sequence in $\mathrm{L}^1_{\mathbb{E}^*}[\mathbb{E}]([0,T])$ and $u \in \mathrm{L}^1_{\mathbb{E}^*}[\mathbb{E}]([0,T])$. One says that u_n *Mazur converges a.e. to* u if there exist sequences $(t_i^k)_i$ with $(t_i^k)_i \in \Lambda_k$ such that $\left\|u(t) - \sum_{i\geq 0} t_i^k u_i(t)\right\| \to 0$ a.e. as $k \to \infty$.

Proposition 7.4.2 *Let $\mathbb{E}$ be separable Banach space and let $f : [0,T] \times \mathbb{E}^*_\sigma \to [0,+\infty]$ be a normal integrand. Let (u_n) (resp. (ψ_n)) be a bounded sequence in $\mathrm{L}^1_{\mathbb{E}^*}[\mathbb{E}]([0,T])$ (resp. $\mathrm{L}^1_{\mathbb{R}}([0,T])$) which satisfy:*

(i) $f(t, u_n(t)) \leq \psi_n(t)$ for all $t \in [0,T]$ and for all n.

(ii) (u_n) Mazur converges a.e. to $u \in \mathrm{L}^1_{\mathbb{E}^*}[\mathbb{E}]([0,T])$ and (ψ_n) Mazur converges a.e. to $\psi \in \mathrm{L}^1_{\mathbb{R}}([0,T])$ with the same coefficients, that is, there exist sequences $(t_i^k)_i \in \Lambda_k$ satisfying both

$$\Big\|u(t) - \sum_{i\geq 0} t_i^k u_i(t)\Big\| \to 0 \ \text{ a.e.}$$

and

$$\left|\psi(t) - \sum_{i\geq 0} t_i^k \psi_i(t)\right| \to 0 \ \text{a.e.}$$

Then there exists a Young measure $\lambda \in \mathcal{Y}^1(\Omega, \mathcal{S}, \mathrm{P}; \mathbb{E}_\sigma^*)$ *such that*

$$u(t) = \operatorname{bar}(\lambda_t) \quad \text{and} \quad \int_{\mathbb{E}} f(t,x)\, d\lambda_t(x) \leq \psi(t) \ \text{a.e.}$$

Proof. We will follow some arguments in [Ben91, Proposition 2.1]. We can consider as $\mathbb{R}^{(\mathbb{N})}$ as a topological subspace of the separable Banach space ℓ^1. For every positive integer k let us consider the multifunction Σ_k from $[0,T]$ to subsets of $\mathbb{R}^{(\mathbb{N})}$ defined by

$$\Sigma_k(t) = \Big\{(r_i)_{i\geq 0} \in \Lambda_k;\ \big\|u(t) - \sum_{i\geq 0} r_i\, u_i(t)\big\| \leq 1/k \text{ and } \big|\psi(t) - \sum_{i\geq 0} r_i\, \psi_i(t)\big| \leq 1/k\Big\}.$$

Then $\Sigma_k(t)$ is nonempty for all $t \in [0,T]$. Further the multifunction Σ_k is graph-measurable because setting, for $n \geq k$, $\Lambda_k^n := \{(r_i)_{i\geq 0} \in \Lambda_k;\ r_i = 0 \text{ if } i > n\}$, $\operatorname{gph}(\Sigma_k)$ equals

$$\bigcup_{n\geq k} \Big\{(t,(r_i)_{i\geq 0}) \in [0,T] \times \Lambda_k^n;$$
$$\big\|u(t) - \sum_{i\geq 0} r_i\, u_i(t)\big\| \leq 1/k \text{ and } \big|\psi(t) - \sum_{i\geq 0} r_i \psi_i(t)\big| \leq 1/k\Big\}$$

and $(t,(r_i)_{i\geq 0}) \mapsto \big\|u(t) - \sum_{i\geq 0} r_i\, u_i(t)\big\|$ and $(t,(r_i)_{i\geq 0}) \mapsto \big|\psi(t) - \sum_{i\geq 0} r_i\, \psi_i(t)\big|$ are measurable on $[0,T] \times \Lambda_k^n$. Pick a measurable selection $t^k : t \mapsto (t_i^k(t))_{i\geq 0}$ of Σ_k and let us set

$$\lambda_t^k := \sum_{i\geq 0} t_i^k(t) \delta_{u_i(t)}$$

for every $t \in [0,T]$. It is obvious that $(\lambda^k)_k$ is tight in $\mathcal{Y}^1(\Omega, \mathcal{S}, \mathrm{P}; \mathbb{E}_\sigma^*)$. By Theorem 4.3.5 we may suppose that $(\lambda^k)_k$ stably converges to a Young measure $\lambda \in \mathcal{Y}^1(\Omega, \mathcal{S}, \mathrm{P}; \mathbb{E}_\sigma^*)$. Let A be a fixed measurable subset in $[0,T]$. By the Portmanteau Theorem 2.1.3, we have

$$\begin{aligned}
\int_A \Big[\int_{\mathbb{E}} f(t,x)\, d\lambda_t(x)\Big]\, dt &\leq \liminf_{k\to\infty} \int_A \Big[\int_{\mathbb{E}} f(t,x)\, d\lambda_t^k(x)\Big]\, dt \\
&= \liminf_{k\to\infty} \int_A \Big[\sum_{i\geq 0} t_i^k(t) f(t, u_i(t))\Big]\, dt \\
&\leq \liminf_{k\to\infty} \int_A \Big[\sum_{i\geq 0} t_i^k(t)\, \psi_i(t)\Big]\, dt\,.
\end{aligned}$$

In view of the Biting Lemma, there is a nondecreasing sequence (T_p) of measurable subsets of $[0,T]$ such that $\cup_p T_p = [0,T]$ and a subsequence (ψ'_n) of (ψ_n) and a subsequence (u'_n) of (u_n) such that $(\mathbb{1}_{T_p}\psi'_n)$ (resp. $(\mathbb{1}_{T_p}u'_n)$) is uniformly integrable in $\mathrm{L}^1_{\mathbb{R}}(T_p)$ (resp. $\mathrm{L}^1_{\mathbb{E}^*}[\mathbb{E}](T_p)$). It follows that

$$\int_A \Big[\int_{\mathbb{E}} f(t,x)\,d\lambda_t(x)\Big]\,dt \le \lim_{k\to\infty}\int_A \Big[\sum_{i\ge 0} t_i^k(t)\,\psi_i(t)\Big]\,dt = \int_A \psi(t)\,dt,$$

for every measurable set A contained in T_p. So we conclude that

$$\int_{\mathbb{E}} f(t,x)\,d\lambda_t(x) \le \psi(t) \quad \text{a.e.}$$

on each T_p, thus the preceding inequality is true a.e. on $[0,T]$. Now consider the integrand $h := \mathbb{1}_A\langle x, .\rangle$ where A is a measurable set subset of T_p and $x \in \mathbb{E}$. Then h is a Carathéodory integrand such that $h(u'_n)$ is uniformly integrable in $\mathrm{L}^1_{\mathbb{R}}(T_p)$. From Theorem 6.3.5 and the stable convergence of $(\lambda^k)_k$ to λ, it follows that

$$\int_A \langle x, \mathrm{bar}\,(\lambda_t)\rangle\,dt = \lim_{k\to\infty}\int_A \langle x, \sum_{i\ge 0} t_i^k(t)\,u_i(t)\rangle\,dt = \int_A \langle x, u(t)\rangle\,dt,$$

for every measurable set $A \subset T_p$ and for every $x \in \mathbb{E}$. So we can conclude that $u(t) = \mathrm{bar}\,(\lambda_t)$ a.e. □

7.5 Relaxed trajectories of a differential inclusion governed by a nonconvex sweeping process

The material in this section is borrowed from [CST01]. Let us consider a compact metric space $\mathbb{U}$. Recall that the set S_Γ of all Lebesgue-measurable selections (alias original controls) of the Lebesgue-measurable multifunction $\Gamma : [0,T] \to \mathcal{K}(\mathbb{U})$ is dense for the above mentioned topologies in the set S_Σ of Lebesgue-measurable selections of Lebesgue-measurable multifunction $\Sigma : [0,T] \to \mathcal{M}^{+,1}(\mathbb{U})$ defined by

$$\Sigma(t) := \{\nu \in \mathcal{M}^{+,1}(\mathbb{U});\ \nu(\Gamma(t)) = 1\}$$

for all $t \in [0,T]$.

For each $t \in [0,T]$, let $C(t)$ be a nonempty closed subset in $\mathbb{R}^d$. We will asssume that

(H_1) For each $t \in [0,T]$, $C(t)$ is a nonempty closed subset in $\mathbb{R}^d$ and the sets $C(t)$ are ρ-proximal-regular (in the sense of Poliquin–Rockafellar–Thibault [PRT00]) for some fixed $\rho \in [0,+\infty]$,

(H_2) $C(t)$ varies in an absolute way, that is, there exists an absolutely continuous function $v : [0,T] \to \mathbb{R}$ such that

$$|d(x, C(t)) - d(y, C(s))| \leq \|x - y\| + |v(t) - v(s)|$$

for all $x, y \in \mathbb{R}^d$ and $s, t \in [0,T]$.

Recall (see [PRT00]) that, when a closed set $S \subset \mathbb{R}^d$ is ρ-proximal regular, the Clarke normal cone $N_S(.)$ is ρ-hypomonotone, that is, for all $x_i \in S$, $i = 1, 2$ and for all $v_1 \in N_S(u_i) \cap \overline{B}_{\mathbb{R}^d}(0, \rho)$, one has

$$\langle v_1 - v_2, u_1 - u_2 \rangle \geq - \|u_1 - u_2\|^2 .$$

Here, $\langle .,. \rangle$ denotes the usual inner product un $\mathbb{R}^d$ and $\|.\|$ is the Euclidean norm. The following is a uniqueness result.

Proposition 7.5.1 *Let $g : [0,1] \times \mathbb{R}^d \to \mathbb{R}^d$ be such that*

(i) for every $\eta > 0$ there exists a nonnegative Lebesgue-integrable function l_η defined on $[0,1]$ such that $\|g(t,x) - g(t,y)\| \leq l_\eta(t)\|x - y\|$ for all $t \in [0,1]$ and for all $(x,y) \in \overline{B}(0,\eta) \times \overline{B}(0,\eta)$,

(ii) there exists a nonnegative Lebesgue-integrable function c on $[0,1]$ such that $\|g(t,x)\| \leq c(t)(1 + \|x\|)$ for all $(t,x) \in [0,1] \times \mathbb{R}^d$.

Let $u_0 \in C(0)$. If u_1 and u_2 are absolutely continuous solutions to

$$\begin{cases} \dot{u}(t) \in -N_{C(t)}(u(t)) + g(t, u(t)), \text{ a.e. } t \in [0,1], \\ u(0) = u_0, \end{cases}$$

then $u_1 = u_2$.

Proof. Let u_1, u_2 be two absolutely continuous solutions to the problem under consideration whose existence is ensured by [CST01, Theorem 1.5 page 225]. Then we have

$$g(t, u_1(t)) - \dot{u}_1(t) \in N_{C(t)}(u_1(t)) \tag{7.5.1}$$

and

$$g(t, u_2(t)) - \dot{u}_2(t) \in N_{C(t)}(u_2(t)) \tag{7.5.2}$$

a.e. Let $m(t) = \|g(t, u_1(t))\| + \|g(t, u_2(t))\|$. Then by (7.5.1)–(7.5.2) and repeating the arguments given the proof of Proposition 1.2 in [CST01] via the hypomonotonicity of the normal cone, one has

$$\begin{aligned} &\langle g(t, u_1(t)) - \dot{u}_1(t) - (g(t, u_2(t)) - \dot{u}_2(t)), u_1(t) - u_2(t) \rangle \\ &\qquad \geq - \frac{\dot{v}(t) + m(t)}{\rho} \|u_1(t) - u_2(t)\|^2 \quad \text{a.e.} \end{aligned} \tag{7.5.3}$$

Consequently

$$\begin{aligned}\langle \dot{u}_1(t) - \dot{u}_2(t), u_1(t) - u_2(t)\rangle \leq \langle g(t, u_1(t)) - g(t, u_2(t)), u_1(t) - u_2(t)\rangle \\ + \frac{\dot{v}(t) + m(t)}{\rho} \|u_1(t) - u_2(t)\|^2.\end{aligned} \tag{7.5.4}$$

Pick $\eta > 0$ such that $\eta > \|u_1\|_\infty + \|u_2\|_\infty$. The Lipschitz property (i) allows us to derive from the inequality above that

$$\langle \dot{u}_1(t) - \dot{u}_2(t), u_1(t) - u_2(t)\rangle \leq \left(l_\eta(t) + \frac{\dot{v}(t) + m(t)}{\rho}\right) \|u_1(t) - u_2(t)\|^2$$

and hence

$$\frac{d}{dt}(\|u_1(t) - u_2(t)\|^2) \leq 2\left(l_\eta(t) + \frac{\dot{v}(t) + m(t)}{\rho}\right) \|u_1(t) - u_2(t)\|^2.$$

So we obtain for all $t \in [0, T]$

$$\frac{1}{2}\|u_1(t) - u_2(t)\|^2 \leq \int_0^t \left(l_\eta(t) + \frac{\dot{v}(t) + m(t)}{\rho}\right) \|u_1(s) - u_2(s)\|^2\, ds.$$

By applying Gronwall's lemma we conclude that $u_1 = u_2$. □

Let us consider a mapping $g : [0, 1] \times \mathbb{R}^d \times \mathbb{U} \to \mathbb{R}^d$ satisfying:

(*i*) For every fixed $t \in [0, 1]$, $g(t, ., .)$ is continuous on $\mathbb{R}^d \times \mathbb{U}$.

(*ii*) For every $(x, u) \in \mathbb{R}^d \times \mathbb{U}$, $g(., x, u)$ is Lebesgue-measurable on $[0, 1]$.

(*iii*) For every $\eta > 0$, there exists $l_\eta \in \mathrm{L}^1_{\mathbb{R}^+}([0, 1])$ such that $\|g(t, x, u) - g(t, y, u)\| \leq l_\eta(t)\|x - y\|$ for all $t \in [0, 1]$ and for all $(x, y) \in \overline{B}_{\mathbb{R}^d}(0, \eta) \times \overline{B}_{\mathbb{R}^d}(0, \eta)$.

(*iv*) There exists $c \in \mathrm{L}^1_{\mathbb{R}^+}([0, 1])$ such that $g(t, x, \mathbb{U}) \subset c(t)(1 + \|x\|)\overline{B}_{\mathbb{R}^d}$ for all $(t, x) \in [0, 1] \times \mathbb{R}^d$.

We aim to compare the solutions set of the following two differential inclusions:

$$(P_{\mathcal{O}}) \begin{cases} \dot{u}_\zeta(t) \in -N_{C(t)}(u_\zeta(t)) + g(t, u_\zeta(t), \zeta(t)) \text{ a.e. } t \in [0, 1], \\ u_\zeta(0) = x_0 \in C(0), \end{cases}$$

where ζ belongs to the set $\mathcal{U}$ of all original controls, i.e. ζ is a Lebesgue-measurable mapping from $[0, 1]$ into U with $\zeta(t) \in \Gamma(t)$ for a.e. $t \in [0, 1]$, and

$$(P_{\mathcal{R}}) \begin{cases} \dot{u}_\lambda(t) \in -N_{C(t)}(u_\lambda(t)) + \int_{\Gamma(t)} g(t, u_\lambda(t), u)\, d\lambda_t(u) \text{ a.e. } t \in [0, 1], \\ u_\lambda(0) = x_0 \in C(0), \end{cases}$$

where λ belongs to the set S_Σ of all relaxed controls, i.e. λ is a Lebesgue-measurable selection of the multifunction Σ defined by

$$\Sigma(t) := \{\nu \in \mathcal{M}^{+,1}(\mathbb{U});\ \nu(\Gamma(t)) = 1\}$$

for all $t \in [0,1]$. A nice example was given in [Jaw84] in which the author studies a similar problem in the case of an evolution equation governed by a convex sweeping process, namely when $C(.)$ is a closed convex absolutely continuous multifunction. For each $\lambda \in S_\Sigma$, let us set

$$h_\lambda(t,x) := \int_{\Gamma(t)} g(t,x,u)\, d\lambda_t(u),$$

for all $(t,x) \in [0,1] \times \mathbb{R}^d$. Then h_λ is Lebesgue-measurable on $[0,1]$ and satisfies:

(*) For every $\eta > 0$, there exists $l_\eta \in L^1_{\mathbb{R}+}$ such that $\|h_\lambda(t,x) - h_\lambda(t,y)\| \leq l_\eta(t)\|x-y\|$ for all $t \in [0,1]$ and for all $(x,y) \in \overline{B}_{\mathbb{R}^d}(0,\eta) \times \overline{B}_{\mathbb{R}^d}(0,\eta)$ thanks to (iii).

Moreover since $\lambda_t(\Gamma(t)) = 1$, we have

(**) $$h_\lambda(t,x) \in \overline{\text{co}}\; g(t,x,\Gamma(t)) \subset c(t)(1+\|x\|)\,\overline{B}_{\mathbb{R}^d},$$

for all $(t,x) \in [0,1] \times \mathbb{R}^d$.

So, in view of Proposition 7.5.1 and for each $\lambda \in S_\Sigma$, there is a unique solution u_λ for the problem

$$(P_{\mathcal{R}}) \begin{cases} -\dot{u}_\lambda(t) \in N_{C(t)}(u_\lambda(t)) + \int_{\Gamma(t)} g(t,u_\lambda(t),u)\, d\lambda_t(u) & \text{a.e. } t \in [0,1], \\ u_\lambda(0) = x_0 \in C(0). \end{cases}$$

By (**) and Theorem 1.5 in [CST01], we see that the solutions set $\{u_\lambda;\ \lambda \in S_\Sigma\}$ of $(P_{\mathcal{R}})$ is relatively compact in $C([0,1],\mathbb{R}^d)$. It is obvious that the solutions set $\{u_\rho;\ \rho \in \mathcal{U}\}$ is a nonempty subset of $\{u_\lambda;\ \lambda \in S_\Sigma\}$.

Theorem 7.5.2 *The set $\{u_\rho;\ \rho \in \mathcal{U}\}$ is dense for the uniform norm in the compact solutions set $\{u_\lambda;\ \lambda \in S_\Sigma\}$.*

The above denseness result is well-known in Optimal Control when $(P_{\mathcal{R}})$ and $(P_{\mathcal{O}})$ are reduced to clasical ordinary differential equations. Theorem 7.5.2 follows from

Lemma 7.5.3 *The mapping $\lambda \mapsto u_\lambda$ defined on the $\sigma(L^\infty_{C(\mathbb{U})^*}([0,1], L^1_{C(\mathbb{U})}([0,1]))$ compact set S_Σ has a closed graph.*

First proof. More precisely, let $\lambda^n \to \lambda^\infty$ for the $\sigma(L^\infty_{C(\mathbb{U})^*}([0,1]), L^1_{C(\mathbb{U})}([0,1]))$ topology and $u_{\lambda^n} \to u_\infty$ in $C([0,1],\mathbb{R}^d)$, where u_{λ^n} $(n \in \mathbb{N} \cup \{\infty\})$ is the unique absolutely continuous solution (cf. Proposition 7.5.1) of the equation

$$\begin{cases} \dot{u}_{\lambda^n}(t) \in -N_{C(t)}(u_{\lambda^n}(t)) + \int_{\Gamma(t)} g(t,u_{\lambda^n}(t),u)\, d\lambda^n_t(u) & \text{a.e. } t \in [0,1], \\ u_{\lambda^n}(0) = x_0 \in C(0), \end{cases}$$

then $u_\infty = u_{\lambda^\infty}$. Put $\eta := \sup_{\lambda\in S_\Sigma} \|u_\lambda\|_{C([0,1],\mathbb{R}^d)}$, then we have $\eta < +\infty$ remembering that the solutions set $\{u_\lambda;\ \lambda \in S_\Sigma\}$ of $(P_{\mathcal{R}})$ is relatively compact in $C([0,1],\mathbb{R}^d)$. We have

$$h_{\lambda^n}(t, u_{\lambda^n}(t)) - \dot{u}_{\lambda^n}(t) \in N_{C(t)}(u_{\lambda^n}(t)),$$

and

$$h_{\lambda^\infty}(t, u_{\lambda^\infty}(t)) - \dot{u}_{\lambda^\infty}(t) \in N_{C(t)}(u_{\lambda^\infty}(t)),$$

for a.e. $t \in [0,1]$. Notice that

$$\|h_{\lambda^n}(t, u_{\lambda^n}(t))\| \le c(t)(1 + \|u_{\lambda^n}(t)\|) \le m(t) := c(t)(1+\eta) \tag{7.5.5}$$

for all $n \in \mathbb{N} \cup \{\infty\}$ and for all $t \in [0,1]$. Using the hypomonotonicity of the normal cone and arguing as in the proof of [CST01, Proposition 1.2 page 222], one has

$$\begin{aligned}\langle h_{\lambda^n}(t, u_{\lambda^n}(t)) - \dot{u}_{\lambda^n}(t) - (h_{\lambda^\infty}(t, u_{\lambda^\infty}(t)) - \dot{u}_{\lambda^\infty}(t)), u_{\lambda^n}(t) - u_{\lambda^\infty}(t)\rangle \\ \ge -\gamma(t)\|u_{\lambda^n}(t) - u_{\lambda^\infty}(t)\|^2\end{aligned}$$

a.e. in $[0,1]$, where $\gamma(t) := \dot{v}(t) + \dfrac{m(t)}{\rho}$. So

$$\begin{aligned}\frac{1}{2}\frac{d}{dt}\|u_{\lambda^n}(t) - u_{\lambda^\infty}(t)\|^2 &\le \langle u_{\lambda^n}(t) - u_{\lambda^\infty}(t), h_{\lambda^n}(t, u_{\lambda^n}(t)) + \gamma(t)u_{\lambda^n}(t)\rangle \\ &\quad - \langle u_{\lambda^n}(t) - u_{\lambda^\infty}(t), h_{\lambda^\infty}(t, u_{\lambda^\infty}(t) + \gamma(t)u_{\lambda^\infty}(t)\rangle,\end{aligned}$$

a.e. in $[0,1]$. Integrating on $[0,t]$ gives

$$\begin{aligned}\frac{1}{2}\|u_{\lambda^n}(t) - u_{\lambda^\infty}(t)\|^2 &\le \int_0^t \langle u_{\lambda^n}(s) - u_{\lambda^\infty}(s), h_{\lambda^n}(s, u_{\lambda^n}(s)) + \gamma(s)u_{\lambda^n}(s)\rangle\, ds \\ &\quad - \int_0^t \langle u_{\lambda^n}(s) - u_{\lambda^\infty}(s), h_{\lambda^\infty}(s, u_{\lambda^\infty}(s)) + \gamma(s)u_{\lambda^\infty}(s)\rangle\, ds.\end{aligned}$$

Let us set

$$\begin{aligned}L_n(t) = &\int_0^t \langle u_{\lambda^n}(s) - u_{\lambda^\infty}(s), h_{\lambda^n}(s, u_{\lambda^n}(s)) + \gamma(s)u_{\lambda^n}(s)\rangle\, ds \\ &- \int_0^t \langle u_{\lambda^n}(s) - u_{\lambda^\infty}(s), h_{\lambda^\infty}(s, u_{\lambda^\infty}(s)) + \gamma(s)u_{\lambda^\infty}(s)\rangle\, ds.\end{aligned}$$

Then $L_n(t) = L_n^1(t) + L_n^2(t) + L_n^3(t)$ where

$$\begin{aligned}L_n^1(t) = &\int_0^t \langle u_{\lambda^n}(s) - u_{\lambda^\infty}(s), h_{\lambda^n}(s, u_{\lambda^n}(s)) + \gamma(s)u_{\lambda^n}(s)\rangle\, ds \\ &- \int_0^t \langle u_{\lambda^n}(s) - u_{\lambda^\infty}(s), h_{\lambda^n}(s, u_{\lambda^\infty}(s)) + \gamma(s)u_{\lambda^\infty}(s)\rangle\, ds,\end{aligned}$$

$$L_n^2(t) = \int_0^t \langle u_{\lambda^n}(s) - u_\infty(s), h_{\lambda^n}(s, u_{\lambda^\infty}(s)) - h_{\lambda^\infty}(s, u_{\lambda^\infty}(s))\rangle \, ds,$$

$$L_n^3(t) = \int_0^t \langle u_\infty(s) - u_{\lambda^\infty}(s), h_{\lambda^n}(s, u_{\lambda^\infty}(s)) - h_{\lambda^\infty}(s, u_{\lambda^\infty}(s))\rangle \, ds.$$

By (7.5.5), we have

$$|L_n^2(t)| \leq (2\int_0^1 m(s)\, ds)\|u_{\lambda^n} - u_\infty\|_{\mathrm{C}([0,1],\mathbb{R}^d)}.$$

Thus $L_n^2(t) \to 0$ when $n \to \infty$ uniformly in $[0,1]$. Similarly by (iv) the integrand

$$f(s,v) := \langle u_\infty(s) - u_{\lambda^\infty}(s), g(s, u_{\lambda^\infty}(s), v)\rangle$$

is estimated by

$$|f(s,v)| \leq c(t)(1+\eta)\|u_\infty - u_{\lambda^\infty}\|_{\mathrm{C}([0,1],\mathbb{R}^d)},$$

for all $(s,v) \in [0,1]\times\mathbb{U}$. Hence $f \in \mathrm{L}^1_{\mathrm{C}(\mathbb{U})}([0,1])$. As (λ^n) converges $\sigma(\mathrm{L}^\infty_{\mathrm{C}(\mathbb{U})^*}, \mathrm{L}^1_{\mathrm{C}(\mathbb{U})})$ to λ^∞, it is immediate that for every $t \in [0,1]$,

$$\int_0^t \Big[\int_{\mathbb{U}} f(s,v)\, d\lambda_s^n(v)\Big]\, ds \to \int_0^t \Big[\int_{\mathbb{U}} f(s,v)\, d\lambda_s^\infty(v)\Big]\, ds$$

when $n \to \infty$. So $\lim_{n\to\infty} L_n^3(t) \to 0$. Using (*), there is $l_\eta \in \mathrm{L}^1_{\mathbb{R}^+}([0,1])$ such that

$$|L_n^1(t)| \leq \int_0^t (l_\eta(s) + \gamma(s))\|u_{\lambda^n}(s) - u_{\lambda^\infty}(s)\|^2\, ds.$$

Finally we get

$$\frac{1}{2}\|u_{\lambda^n}(t) - u_{\lambda^\infty}(t)\|^2 \leq L_n^2(t) + L_n^3(t) + \int_0^t (l_\eta(s) + \gamma(s))\|u_{\lambda^n}(s) - u_{\lambda^\infty}(s)\|^2\, ds.$$

As $L_n^2(t) \to 0$ and $L_n^3(t) \to 0$, by Gronwall's inequality we obtain

$$u_{\lambda^n}(t) \to u_{\lambda^\infty}(t) = u_\infty(t),$$

for all $t \in [0,1]$ and so $\{u_\lambda;\ \lambda \in \mathrm{S}_\Sigma\}$ is compact in $\mathrm{C}\left([0,1],\mathbb{R}^d\right)$. □

Here is an alternative proof of the preceding lemma using the fiber product of Young measures.

Second proof. We are going to prove that the graph of the mapping $\lambda \mapsto u_\lambda$ from S_Σ into the Banach space $\mathrm{C}\left([0,T],\mathbb{R}^d\right)$ is compact. Let (λ^n) be a sequence in S_Σ. By compactness we may assume that λ^n stably converges to λ^∞, u_{λ^n} converges to $u_\infty \in \mathrm{C}\left([0,T],\mathbb{R}^d\right)$ and $(\dot u_{\lambda_n})$ weakly converges to $\dot u^\infty$ in $\mathrm{L}^1_{\mathbb{R}^d}([0,1])$. Let $w \in$

$L^\infty_{\mathbb{R}^d}([0,1])$. Then it is easy to see that the integrand $(t,x,z) \mapsto \langle w(t), g(t,x,z)\rangle$ defined on $[0,T] \times K \times \mathbb{U}$ is L^1-bounded, using the growth condition (iv) and the boundedness of $K := \{u_\lambda(t);\ (\lambda,t) \in S_\Sigma \times [0,T]\}$. Let us put, for all $t \in [0,1]$,

$$v^n(t) = \int_{\mathbb{U}} g(t, u^n_\lambda(t), z)\, d\lambda^n_t(z)],$$
$$v^\infty(t) = \int_{\mathbb{U}} g(t, u^\infty(t), z)\, d\lambda^\infty_t(z)].$$

By Theorem 3.3.1, $\underline{\delta}_{u^n_\lambda} \underline{\otimes}\, \lambda^n$ stably converges to $\underline{\delta}_{u_\infty} \underline{\otimes}\, \lambda^\infty$, hence

$$\lim_{n\to\infty} \int_0^1 \langle w(t), v^n(t)\rangle\, dt = \int_0^1 \langle w(t), v^\infty(t)\rangle\, dt.$$

Using the weak convergence in $L^1_{\mathbb{R}^d}([0,1])$ of $(\dot u_{\lambda_n})$ to $\dot u^\infty$ and the preceding limit, we conclude that the sequence $(\dot u_{\lambda_n} - v^n)$ weakly converges in $L^1_{\mathbb{R}^d}([0,1])$ to $\dot u^\infty - v^\infty$. As u_{λ_n} is the solution of the corresponding evolution inclusion, we have

$$\dot u_{\lambda_n} - v^n(t) \in -N(C(t); u_{\lambda_n}(t))\ a.e.\ t \in [0,1],$$

with $u_{\lambda_n}(0) = x_0$. In view of [Thi99], this inclusion is equivalent to

$$\dot u_{\lambda_n} - v^n(t) \in -\psi(t)\, \partial[d_{C(t)}](u_{\lambda_n}(t)) \text{ a.e. } t \in [0,1], \tag{7.5.6}$$

where $\psi(t) = 2c(t) + \dot v(t)$ for all $t \in [0,1]$, and $\partial[d_{C(t)}]$ denotes the subdifferential of the distance function $d_{C(t)} : x \mapsto d(x, C(t))$. Since u_{λ^n} converges uniformly to $u^\infty(.)$, by (7.5.6) and by virtue of a closure-type lemma [CV77, Theorem VI-4] we get

$$\dot u^\infty(t) \in -\psi(t)\, \partial[d_{C(t)}](u^\infty(t)) + \int_{\mathbb{U}} g(t, u^\infty(t), z)\, d\lambda^\infty_t(z)],$$

with $u^\infty(0) = x_0$, and $u^\infty(t) \in C(t)$ for all $t \in [0,1]$.

So, we have necessarily $u^\infty(.) = u_{\lambda^\infty}(.)$. □

Chapter 8

Semicontinuity of integral functionals using Young measures

In this chapter, $\mathbb{E}$ is a separable Banach space and, as usual, $(\Omega, \mathcal{S}, \mathrm{P})$ is a probability space.

8.1 Weak-strong lower semicontinuity of integral functionals

Recall that we have already proved a fiber product result (Theorem 3.3.1). Now we begin with a different proof of a particular case of this result.

Theorem 8.1.1 *Let $\mathbb{T}_1$ and $\mathbb{T}_2$ be metrizable Suslin spaces, ν^n and τ^n ($n \in \mathbb{N} \cup \{+\infty\}$) be Young measures in $\mathcal{Y}^1(\Omega, \mathcal{S}, \mathrm{P}; \mathbb{T}_1)$ and $\mathcal{Y}^1(\Omega, \mathcal{S}, \mathrm{P}; \mathbb{T}_2)$ respectively which are tight sequences and which satisfy $\nu^n \to \nu^\infty$ and $\tau^n \to \tau^\infty$. Let us define, for $n \in \mathbb{N} \cup \{+\infty\}$, $\theta^n \in \mathcal{Y}^1(\Omega, \mu; \mathbb{T}_1 \times \mathbb{T}_2)$ by*

$$\theta^n_\omega = \nu^n_\omega \otimes \tau^n_\omega.$$

If $\nu^\infty = \underline{\delta}_u$ where $u : \Omega \to \mathbb{T}_1$ is a measurable function, then θ^n converges stably to θ^∞ (with $\theta^\infty_\omega = \delta_{u(\omega)} \otimes \tau^\infty_\omega$).

Remark 8.1.2 In Theorem 8.1.6 the result will be applied with, for $n < +\infty$, $\nu^n = \underline{\delta}_{u_n}$ and $\tau^n = \underline{\delta}_{v_n}$ that is $\theta^n = \underline{\delta}_{(u_n, v_n)}$. Without the hypothesis "ν^∞ is associated with a function", the result is false: see Counterexample 3.3.3 page 72.

Proof. 1) If $\varphi : \Omega \times \mathbb{T}_1 \to \overline{\mathbb{R}}$ is an integrand which is either ≥ 0 or bounded and Carathéodory, the following holds

$$\begin{aligned}\int_{\Omega\times\mathbb{T}_1\times\mathbb{T}_2} \varphi(\omega,\xi)\, d\theta^n(\omega,\xi,\zeta) &= \int_\Omega \Big[\int_{\mathbb{T}_1\times\mathbb{T}_2} \varphi(\omega,\xi)\, d(\nu^n_\omega \otimes \tau^n_\omega)(\xi,\zeta)\Big]\, d\mathrm{P}(\omega)\\ &= \int_\Omega \Big[\int_{\mathbb{T}_2} \varphi(\omega,\xi)\, d\nu^n_\omega(\xi)\Big]\, d\mathrm{P}(\omega)\\ &= \int_{\Omega\times\mathbb{T}_1} \varphi\, d\nu^n.\end{aligned}$$

Similarly if φ' is an integrand on $\Omega \times \mathbb{T}_2$,

$$\int_{\Omega\times\mathbb{T}_1\times\mathbb{T}_2} \varphi'(\omega,\zeta)\, d\theta^n(\omega,\xi,\zeta) = \int_{\Omega\times\mathbb{T}_2} \varphi'\, d\tau^n.$$

2-a) The sequence $(\theta^n)_{n\in\mathbb{N}}$ is tight because, for $i = 1, 2$, there exists an integrand $h_i \geq 0$ inf-compact in the second variable, such that

$$\sup_{n\in\mathbb{N}} \int_{\Omega\times\mathbb{T}_1} h_1\, d\nu^n =: M_1 < +\infty \quad \text{and} \quad \sup_{n\in\mathbb{N}} \int_{\Omega\times\mathbb{T}_2} h_2\, d\tau^n =: M_2 < +\infty.$$

Then $h(\omega,\xi,\zeta) := h_1(\omega,\xi) + h_2(\omega,\zeta)$ defines an inf-compact integrand (details of this exercise are in [Val94, page 381]). By Part 1 it satisfies $\forall n \in \mathbb{N}$, $\int_{\Omega\times\mathbb{T}_1\times\mathbb{T}_2} h\, d\theta^n \leq M_1 + M_2$.

2-b) Let us process by contradiction. Suppose there exists a bounded Carathéodory integrand ψ_0 on $\Omega \times \mathbb{T}_1 \times \mathbb{T}_2$ satisfying

$$\int \psi_0\, d\theta^n \not\to \int \psi_0\, d\theta^\infty,$$

that is, such that, for some $\varepsilon > 0$ and for infinitely many n,

$$\Big|\int \psi_0\, d\theta^n - \int \psi_0\, d\theta^\infty\Big| > \varepsilon. \tag{8.1.1}$$

Extracting a subsequence one may suppose that θ^{n_k} converges to σ in the space $\mathcal{Y}^1(\Omega, \mathcal{S}, \mathrm{P}; \mathbb{T}_1 \times \mathbb{T}_2)$ and that (8.1.1) holds for all θ^{n_k}.

3) Let us prove that P-a.e. σ_ω is carried by $\{u(\omega)\} \times \mathbb{T}_2$. Let us set

$$\psi(\omega,\xi,\zeta) = \varphi(\omega,\xi) = \min[1, d(\xi, u(\omega))].$$

They are bounded Carathéodory integrands. Then

$$\int_{\Omega\times\mathbb{T}_1\times\mathbb{T}_2} \psi\, d\theta^{n_k} = \int_{\Omega\times\mathbb{T}_1} \varphi\, d\nu^{n_k} \to \int_{\Omega\times\mathbb{T}_1} \varphi\, d\underline{\delta}_u = 0.$$

Hence

$$\int_{\Omega\times\mathbb{T}_1\times\mathbb{T}_2} \psi\, d\sigma = 0.$$

So σ is carried by $\{(\omega,\xi,\zeta);\ \psi(\omega,\xi,\zeta)=0\}$ and P-a.e. σ_ω is carried by

$$\{(\xi,\zeta);\ \psi(x,\xi,\zeta)=0\} = \{(\xi,\zeta);\ \varphi(x,\xi)=0\} = \{u(\omega)\}\times\mathbb{T}_2.$$

Consequently

$$\sigma_\omega = \delta_{u(\omega)} \otimes \bar{\sigma}_\omega\,,$$

where $\bar{\sigma}_\omega$ is a probability on $\mathbb{T}_2$. Clearly $\omega \mapsto \bar{\sigma}_\omega$ is measurable (for any Borel subset B of $\mathbb{T}_2$, $\bar{\sigma}_\omega(B) = \sigma_\omega(\{u(\omega)\}\times B) = \sigma_\omega(\mathbb{T}_1\times B)$), and so is the disintegrated version of some $\bar{\sigma} \in \mathcal{Y}^1(\Omega,\mathcal{S},\mathrm{P};\mathbb{T}_2)$.

4) Now we prove $\bar{\sigma}_\omega = \tau^\infty_\omega$. This will end the proof since $\sigma_\omega = \theta^\infty_\omega$ implies $\sigma = \theta^\infty$ and $\theta^{n_k} \to \theta^\infty$, hence the contradiction to (8.1.1):

$$\int \psi_0\, d\theta^{n_k} \to \int \psi_0\, d\theta^\infty.$$

Let φ' a bounded Carathéodory integrand on $\Omega\times\mathbb{T}_2$. On one side

$$\begin{aligned}\int_{\Omega\times\mathbb{T}_1\times\mathbb{T}_2} \varphi'(\omega,\zeta)\, d\theta^{n_k}(\omega,\xi,\zeta) &\to \int_{\Omega\times\mathbb{T}_1\times\mathbb{T}_2} \varphi'(\omega,\zeta)\, d\sigma(\omega,\xi,\zeta)\\ &= \int_\Omega \Big[\int_{\mathbb{T}_2} \varphi'(\omega,\zeta)\, d\bar{\sigma}_\omega(\zeta)\Big]\, d\mu(\omega)\\ &= \int_{\Omega\times\mathbb{T}_2} \varphi'\, d\bar{\sigma}.\end{aligned}$$

On the other side, by Part 1,

$$\int_{\Omega\times\mathbb{T}_1\times\mathbb{T}_2} \varphi'(\omega,\zeta)\, d\theta^{n_k}(\omega,\xi,\zeta) = \int_{\Omega\times\mathbb{T}_2} \varphi'\, d\tau^{n_k} \to \int_{\Omega\times\mathbb{T}_2} \varphi'\, d\tau^\infty.$$

Thus the expected equality $\bar{\sigma} = \tau^\infty$ holds. □

Let us recall the following definition. Let $\mathcal{H}$ be a subset of $\mathrm{L}^1_{\mathbb{E}}(\Omega,\mathcal{S},\mathrm{P})$. One says that $\mathcal{H}$ is $\mathcal{R}\mathrm{w}\mathcal{K}(\mathbb{E})$-*tight* if, for any $\varepsilon > 0$, there exists an $\mathcal{R}\mathrm{w}\mathcal{K}(\mathbb{E})$-valued measurable multifunction, L_ε, such that

$$\forall u \in \mathcal{H} \quad \mathrm{P}(\{\omega\in\Omega;\ u(\omega) \notin L_\varepsilon(\omega)\}) \le \varepsilon.$$

Remarks 1) The set $\{\omega\in\Omega;\ u(\omega)\notin L_\varepsilon(\omega)\}$ belongs to $\mathcal{S}$ since it is the projection on Ω of $\mathrm{gph}\,(u)\setminus \mathrm{gph}\,(L_\varepsilon)$.

2) If $\mathbb{E}$ is a reflexive separable Banach space, $\mathcal{H}$ is necessarily $\mathcal{R}\mathrm{w}\mathcal{K}(\mathbb{E})$-tight because $L_\varepsilon(\omega) = \mathbb{E}$ is a possible choice.

3) If $\mathcal{H}$ is $\mathcal{R}\mathrm{w}\mathcal{K}$-tight and bounded in $\mathrm{L}^1_{\mathbb{E}}$, it is weakly flexibly tight (see page 124).

We need here the following result. Compare with Lemma 6.1.1.

Lemma 8.1.3 *Let $\mathbb{E}$ be a separable Banach space and let $\mathcal{H}$ be a bounded and $\mathcal{R}\mathrm{w}\mathcal{K}(\mathbb{E})$-tight subset of $\mathrm{L}^1_{\mathbb{E}}(\Omega,\mathcal{S},\mathrm{P})$. Then $\mathcal{H}$ is $\sigma(\mathbb{E},\mathbb{E}^*)$-tight,* i.e. *for any $\varepsilon>0$, there exists a measurable multifunction with weakly compact values, Φ, such that*

$$\forall u\in\mathcal{H}\quad \mathrm{P}(\{\omega\in\Omega;\ u(\omega)\notin\Phi(\omega)\})\le\varepsilon.$$

There exists a measurable ≥ 0 integrand h on $\Omega\times\mathbb{E}$ which is $\sigma(\mathbb{E},\mathbb{E}^*)$-inf-compact, such that $\sup\{\int_\Omega h(\omega,u(\omega))\,d\mathrm{P}(\omega);\ u\in\mathcal{H}\}$ is finite.

Proof. Let

$$K:=\sup\{\|u\|_{\mathrm{L}^1};\ u\in\mathcal{H}\}.$$

Let $\varepsilon>0$, $\alpha:=\frac{2L}{\varepsilon}$ and $\Omega^u:=\{\omega\in\Omega;\ u(\omega)\in L_{\varepsilon/2}(\omega)\}$. We set

$$\Phi(\omega):=L_{\varepsilon/2}(\omega)\cap\overline{B}\,(0,\alpha)\,.$$

The multifunction Φ is measurable with weakly compact values. Then, $\forall u\in\mathcal{H}$, $\omega\in\Omega^u\cap\{\|u(.)\|\le\alpha\}$ implies $u(\omega)\in\Phi(\omega)$. As

$$\mathrm{P}(\Omega\setminus\Omega^u)\le\frac{\varepsilon}{2}\ \text{and}\ \mathrm{P}(\{\|u(.)\|>\alpha\})\le\frac{\varepsilon}{2},$$

$\mathrm{P}(\{\omega\in\Omega;\ u(\omega)\notin\Phi(\omega)\})\le\varepsilon$, hence $\mathcal{H}$ is tight. The existence of h was proved in [Jaw84, Proposition 2.2 page 13.17] or [Bal89a, Remark 2.4 page 9.5]. □

We will need also the following result which extends to infinite dimension known results about the connection between weak convergence of functions and Young measures.

Theorem 8.1.4 *Let $\mathbb{E}$ be a separable Banach space and $(v_n)_{n\in\mathbb{N}}$ a bounded uniformly integrable sequence in $\mathrm{L}^1_{\mathbb{E}}(\Omega,\mathcal{S},\mathrm{P})$ which is $\mathcal{R}\mathrm{w}\mathcal{K}(\mathbb{E})$-tight. There exists a metric d on $\mathbb{E}$ whose topology is weaker than $\sigma(\mathbb{E},\mathbb{E}^*)$ satisfying the following: The sequence $(\tau^n)_{n\in\mathbb{N}}$, where $\tau^n=\underline{\delta}_{v_n}$, is tight in $\mathcal{Y}^1(\Omega,\mathcal{S},\mathrm{P};(\mathbb{E},d))$ and for any stably convergent subsequence $(\tau^{n_k})_{k\in\mathbb{N}}$, its limit, τ^∞, is of first order in the sense that $\int_{\Omega\times\mathbb{E}}\|\zeta\|\,d\tau^\infty(\omega,\zeta)<+\infty$. Moreover if the function v_∞ is defined by*

$$\mathrm{P}\text{-a.e.},\quad v_\infty(\omega)=\int_{\mathbb{E}}\zeta\,d\tau^\infty_\omega(\zeta)=:\mathrm{bar}\,(\tau^\infty_\omega)\,,$$

then $v_\infty\in\mathrm{L}^1_{\mathbb{E}}(\Omega,\mathcal{S},\mathrm{P})$ and the sequence $(u_{n_k})_{k\in\mathbb{N}}$ weakly converges (that is, with respect to $\sigma(\mathrm{L}^1_{\mathbb{E}},\mathrm{L}^\infty_{\mathbb{E}^*_\sigma})$) to v_∞.

Proof. 1) Let $(z'_n)_{n\in\mathbb{N}}$ be a sequence in $\mathbb{E}^*$ which separates the points of $\mathbb{E}$. Let

$$d(\zeta,\zeta')=\sum_{n\in\mathbb{N}}2^{-n}\,\frac{|\langle z'_n,\zeta-\zeta'\rangle|}{1+|\langle z'_n,\zeta-\zeta'\rangle|}\,.$$

The topology defined by d is the coarsest on $\mathbb{E}$ making continuous the maps $\zeta \mapsto \langle z_n', \zeta\rangle$. Recall that the Borel σ-algebra of the norm topology and the Borel σ-algebra of any Hausdorff coarser topology coincide. In particular this applies to $\sigma(\mathbb{E}, \mathbb{E}^*)$ and to the d-topology.

2) Let $\varepsilon > 0$. By Lemma 8.1.3, there exists a weakly compact valued multifunction Φ, such that

$$\forall n \in \mathbb{N} \quad \mathrm{P}(\{\omega \in \Omega;\, v_n(\omega) \notin \Phi(\omega)\}) \leq \varepsilon.$$

The sets $\Phi(\omega)$ are also d-compact. So $(v_n)_{n\in\mathbb{N}}$ is d-flexibly tight. Consequently there exists a subsequence $(\tau^{n_k})_{k\in\mathbb{N}}$ stably convergent (relatively to d) to τ^∞. Let h be a d-inf-compact integrand (whose existence is ensured by Lemma 8.1.3) and

$$M := \sup\Big\{\int_\Omega h(\omega, v_n(\omega))\, d\mathrm{P}(\omega);\, n \in \mathbb{N}\Big\}.$$

3) Let $\psi : \Omega \times \mathbb{E} \to [0, +\infty]$ be measurable and $\sigma(\mathbb{E}, \mathbb{E}^*)$-l.s.c. in the second variable. Following Balder ([Bal89a, formula (3.5) page 9.7], see also ([Bal86a, page 113]), we will prove the inequality

$$\int_{\Omega\times\mathbb{E}} \psi\, d\tau^\infty \leq \liminf_k \int_{\Omega\times\mathbb{E}} \psi\, d\tau^{n_k}. \tag{8.1.2}$$

Let

$$\mathcal{Y}_M := \Big\{\tau \in \mathcal{Y}^1(\Omega, \mathrm{P}; (\mathbb{E}, d));\, \int_{\Omega\times\mathbb{E}} h\, d\tau \leq M\Big\}.$$

Then, on $\mathcal{Y}_M$,

$$\int_{\Omega\times\mathbb{E}} \psi\, d\tau = \sup_{\varepsilon>0}\Big[\int_{\Omega\times\mathbb{E}} (\psi + \varepsilon h)\, d\tau - \varepsilon M\Big]. \tag{8.1.3}$$

As $\psi(\omega, .) + \varepsilon h(\omega, .)$ is $\sigma(\mathbb{E}, \mathbb{E}^*)$-inf-compact, hence d-inf-compact, hence d-l.s.c,

$$\tau \mapsto \int_{\Omega\times\mathbb{E}} (\psi + \varepsilon h)\, d\tau$$

is l.s.c. for the stable topology of d. By (8.1.3), $\tau \mapsto \int_{\Omega\times\mathbb{E}} \psi\, d\tau$ is also l.s.c. for the same topology on $\mathcal{Y}_M$, hence (8.1.2).

Applying the foregoing lower semicontinuity result to the integrand

$$(\omega, \zeta) \mapsto \|\zeta\|$$

gives $\int_{\Omega\times\mathbb{E}} \|\zeta\|\, d\tau^\infty(\omega, \zeta) \leq \sup\{\|v_n\|_{\mathrm{L}^1};\, n \in \mathbb{N}\} < +\infty$, that is, τ^∞ is of first order as said in the statement. Classically for P-almost every ω, τ^∞_ω is of first order in the sense of Probability Theory and has a barycenter $\mathrm{bar}\,(\tau^\infty_\omega)$.

4) Consider now an integrand φ such that $\varphi(\omega,.)$ is $\sigma(\mathbb{E},\mathbb{E}^*)$-l.s.c. and the sequence of negative parts $(\varphi(.,v_n(.))^-)_{n\in\mathbb{N}}$ is uniformly integrable. Using an idea of Ioffe [Iof77, bottom of page 530 and top of page 531], one can, introducing $\sup(-r,\varphi)+r$ ($r\in[0,+\infty[$), prove

$$\int_{\Omega\times\mathbb{E}}\varphi\,d\tau^\infty\leq\liminf_k\int_{\Omega\times\mathbb{E}}\varphi\,d\tau^{n_k}. \tag{8.1.4}$$

Indeed let $\varphi_r:=\sup(-r,\varphi)$. Since $\psi_r:=\varphi_r+r$ is ≥ 0, Part 3 implies

$$\int_{\Omega\times\mathbb{E}}\psi_r\,d\tau^\infty\leq\liminf_k\int_\Omega\psi_r(\omega,v_{n_k}(\omega))\,d\mathrm{P}(\omega)\,,$$

hence subtracting r,

$$\int_{\Omega\times\mathbb{E}}\varphi_r\,d\tau^\infty\leq\liminf_k\int_\Omega\varphi_r(\omega,v_{n_k}(\omega))\,d\mathrm{P}(\omega).$$

But thanks to the uniform integrability of negative parts, $\forall\varepsilon>0$, $\exists r$ such that

$$\forall n\in\mathbb{N}\quad\int_\Omega\varphi(\omega,v_n(\omega))\,d\mathrm{P}(\omega)\geq\int_\Omega\varphi_r(\omega,v_n(\omega))\,d\mathrm{P}(\omega)-\varepsilon.$$

Hence for any $\varepsilon>0$

$$\begin{aligned}\liminf_k\int_\Omega\varphi(\omega,v_{n_k}(\omega))\,d\mathrm{P}(\omega)&\geq\liminf_k\int_\Omega\varphi_r(\omega,v_{n_k}(\omega))\,d\mathrm{P}(\omega)-\varepsilon\\&\geq\int_{\Omega\times\mathbb{E}}\varphi_r\,d\tau^\infty-\varepsilon\geq\int_{\Omega\times\mathbb{E}}\varphi\,d\tau^\infty-\varepsilon.\end{aligned}$$

5) Let $z\in\mathrm{L}^\infty_{\mathbb{E}^*_\sigma}$ and $\varphi(\omega,\zeta):=\langle z(\omega),\zeta\rangle$. The sequence $(\langle z(.),v_n(.)\rangle)_{n\in\mathbb{N}}$ being uniformly integrable, inequality (8.1.4) applies to φ and to $-\varphi$, and we get

$$\int_{\Omega\times\mathbb{E}}\langle z(\omega),\zeta\rangle\,d\tau^\infty(\omega,\zeta)=\lim_k\int_{\Omega\times\mathbb{E}}\langle z(\omega),\zeta\rangle\,d\tau^{n_k}(\omega,\zeta)\,,$$

hence

$$\int_\Omega\Big\langle z(\omega),\int_\mathbb{E}\zeta\,d\tau^\infty_\omega(\zeta)\Big\rangle\,d\mathrm{P}(\omega)=\lim_k\int_\Omega\langle z(\omega),v_{n_k}(\omega)\rangle\,d\mathrm{P}(\omega)\,,$$

which proves the weak convergence $v_{n_k}\to v_\infty$. □

We shall use the following semicontinuity lemma for the fiber product, which we give in a more general form than necessary.

Lemma 8.1.5 *Let $\mathbb{S}$ and $\mathbb{T}$ be two completely regular spaces, let f be a normal integrand on $\Omega\times(\mathbb{S}\times\mathbb{T})$ with values in $[0,+\infty]$, let (u_n) be a sequence of measurable mappings from Ω to $\mathbb{S}$ which converges in measure to some $u_\infty\in\mathfrak{X}(\mathbb{S})$ and let (λ^n) be a sequence in $\mathcal{Y}^1(\mathbb{T})$ which W-stably converges to some $\lambda^\infty\in\mathcal{Y}^1(\mathbb{T})$. Assume furthermore that one of the following conditions is satisfied:*

(*i*) $\mathcal{M}^{+,1}_{\tau}(\mathbb{S}\times\mathbb{T}) = \mathcal{M}^{+,1}(\mathbb{S}\times\mathbb{T})$ *(e.g.* $\mathbb{S}\times\mathbb{T}$ *is Radon or hereditarily Lindelöf) and* (u_n) *and* (λ^n) *are both flexibly tight, or*

(*ii*) $\mathbb{S}$ *and* $\mathbb{T}$ *are sequentially Prohorov and their compact subsets are metrizable.*

Then

$$(8.1.5)\quad \liminf_n \int_\Omega [\int_\mathbb{T} f(\omega, u_n(\omega), z)\, d\lambda^n_\omega(z)]\, d\,\mathrm{P}(\omega) \geq \int_\Omega [\int_\mathbb{T} f(\omega, u_\infty(\omega), z)\, d\lambda^\infty_\omega(z)]\, d\,\mathrm{P}(\omega).$$

Proof. From Theorem 3.3.1, $(\underline{\delta}_{u_n} \otimes \lambda^n)$ W–stably converges to $\underline{\delta}_{u_\infty} \otimes \lambda^\infty$. But, in case (*i*), from Theorem 4.3.8, as $(\underline{\delta}_{u_n} \otimes \lambda^n)$ is flexibly tight, it also S–stably converges to $\underline{\delta}_{u_\infty} \otimes \lambda^\infty$. The same conclusion follows from Theorem 4.5.1 in case (*ii*). Then Theorem 2.1.12-(*d*) immediately yields (8.1.5). □

Now the weak-strong lower semicontinuity theorem which is fundamental in the Calculus of Variations follows easily.

Theorem 8.1.6 *Let* $\mathbb{E}$ *be a separable Banach space and* $\mathbb{S}$ *a regular cosmic space (e.g.* $\mathbb{S}$ *is separable metrizable or* $\mathbb{S}$ *is Suslin regular). Let* f *be a normal integrand on* $\Omega\times(\mathbb{S}\times\mathbb{E})$. *Let* (u_n) *be a sequence of measurable mappings from* Ω *to* $\mathbb{S}$ *which converges in measure to some* $u_\infty \in \mathfrak{X}(\mathbb{S})$ *and let* (v_n) *be a* $\mathcal{R}$w$\mathcal{K}$*–tight sequence in* $\mathrm{L}^1_\mathbb{E}(\Omega, \mathcal{S}, \mathrm{P})$ *which* $\sigma(\mathrm{L}^1_\mathbb{E}, \mathrm{L}^\infty_{\mathbb{E}^*})$*–converges to some* $v_\infty \in \mathrm{L}^1_\mathbb{E}$. *Assume that the sequence of negative parts* $(f(\omega, u_n(\omega), v_n(\omega))^-)_n$ *is uniformly integrable and that* $f(\omega, u_\infty(\omega), .)$ *is convex on* $\mathbb{E}$. *Then*

$$(8.1.6)\quad \liminf_n \int_\Omega f(\omega, u_n(\omega), v_n(\omega))\, d\,\mathrm{P}(\omega) \;\geq\; \int_\Omega f(\omega, u_\infty(\omega), v_\infty(\omega))\, d\,\mathrm{P}(\omega).$$

Proof. The sequence (v_n) is UI and bounded in $\mathrm{L}^1_\mathbb{E}$ because it is $\sigma(\mathrm{L}^1_\mathbb{E}, \mathrm{L}^\infty_{\mathbb{E}^*})$–convergent (see e.g. [DU77, Theorem 4 page 104]). Thus, as (v_n) is $\mathcal{R}$w$\mathcal{K}$–tight, it is weakly flexibly tight. From Theorem 8.1.4, each subsequence of (v_n) has a further subsequence (again denoted by (v_n)) such that $(\underline{\delta}_{v_n})_n$ S–stably converges in $\mathcal{Y}^1(\mathbb{E}_\sigma)$ to some Young measure $\nu^\infty \in \mathcal{Y}^1(\mathbb{E}_\sigma)$ (depending on the subsequence) with bar $(\nu^\infty_\omega) = v_\infty(\omega)$ a.e. Now, $\mathbb{S}\times\mathbb{E}_\sigma$ is cosmic regular (because $\mathbb{S}$ and $\mathbb{E}_\sigma$ are both cosmic regular), thus it is hereditarily Lindelöf.

Assume that f is bounded from below, that is, there exists $M \in \mathbb{R}$ such that $f \geq M$. Then Lemma 8.1.5-(*i*) gives

$$\liminf_n \int_\Omega f(\omega, u_n(\omega), v_n(\omega))\, d\,\mathrm{P}(\omega) \geq \int_\Omega [\int_\mathbb{E} f(\omega, u_\infty(\omega), z)\, d\nu^\infty_\omega(z)]\, d\,\mathrm{P}(\omega).$$

As $f(\omega, u_\infty(\omega), .)$ is convex, we then get (8.1.6) by Jensen inequality.

To prove (8.1.6) in the general case, we can apply the same method as in Part 4) of the proof of Theorem 8.1.4. □

Remarks 8.1.7 1) We can assume in Theorem 8.1.6 that $\mathbb{S}$ is simply a metric space (non–necessarily separable), because then the measurability of the functions v_n implies that they take their values in a separable part of $\mathbb{S}$.

2) Even if all $\psi(\omega, \xi, .)$ are convex, the topology on $\mathbb{E}$ with respect to which the functions $\psi(\omega, ., .)$ are l.s.c. is fundamental because of the parameter $\xi \in \mathbb{T}$. The lower semicontinuity on $\mathbb{T} \times (\mathbb{E}, \|.\|)$ does not imply the lower semicontinuity on $\mathbb{T} \times (\mathbb{E}, \sigma(\mathbb{E}, \mathbb{E}^*))$: see Example 8.1.8 below.

3) Actually we have

$$v_\infty(\omega) \in \overline{\text{co}}(\bigcap_{p \in \mathbb{N}} \text{w–sequ cl}\,\{v_m(\omega);\ m \geq p\}) \text{ a.e.}$$

This property can be proved by applying the Biting Lemma and [ACV92, Theorem 8]. We refer to [ACV92, Theorem 9] for details. See also Theorem 6.5.1 and Remarks 6.5.2.

4) As a few references see de Giorgi [dG69] (quoted by Buttazzo [But89]: see Th.2.3.1 p.46), Serrin [Ser59] quoted by Pedregal [Ped97], Ioffe [Iof77], Valadier [Val90a], Balder [Bal84a, Bal85, Bal86a, Bal00b].

5) We produce in the end of this section other lower semicontinuity-type examples for integral functionals, using the techniques developed in Castaing–Clauzure [CC82]. See also [MM89] for some applications of the strong-weak lower semicontinuity theorem in the problem of minimization of functionals of classes of Lipschitz domains.

Example 8.1.8 Let $\mathbb{E}$ be a separable Hilbert space, let $(e_n)_{n \in \mathbb{N}}$ be an orthonormal basis of $\mathbb{E}$ and let $\mathbb{T}$ be the closed unit ball of $\mathbb{E}$ with the topology $\sigma(\mathbb{E}, \mathbb{E})$. Define ψ by

$$\psi(\omega, \xi, \zeta) = \langle \xi, \zeta \rangle.$$

Then ψ is continuous on $\mathbb{T} \times (\mathbb{E}, \|.\|)$ and linear on $\mathbb{E}$, hence convex on $\mathbb{E}$. But it is not l.s.c. on $\mathbb{T} \times (\mathbb{E}, \sigma(\mathbb{E}, \mathbb{E}))$. Indeed, if $\xi_n = -e_n$, $\zeta_n = e_n$, $\xi_n \to 0$ in $\mathbb{T}$ and $\zeta_n \to 0$ in $(\mathbb{E}, \sigma(\mathbb{E}, \mathbb{E}))$, but $\forall n$, $\langle \xi_n, \zeta_n \rangle = -1$. With these data Theorem 8.1.6 would not hold if $\psi(\omega, ., .)$ was l.s.c. on $\mathbb{T} \times (\mathbb{E}, \|.\|)$. Indeed take $\Omega = [0, 1]$ with the Lebesgue measure, and $u_n \equiv -e_n$, $v_n \equiv e_n$.

Now we present an application of Theorem 8.1.1 to a Bolza type problem in Optimal Control.

The following result is a direct consequence of Theorem 3.3.1 (or Theorem 8.1.1), so its proof is omitted.

Proposition 8.1.9 *Let $\mathbb{S}$ be a Polish space. Let (u^n) be a sequence of $(\mathcal{S}, \mathcal{B}_{\mathbb{E}})$-measurable mappings from Ω to $\mathbb{E}$ which pointwise converges on Ω to a $(\mathcal{S}, \mathcal{B}_{\mathbb{E}})$-measurable mapping u^∞ and $(\nu^n)_n$ a sequence in $\mathcal{Y}^1(\Omega, \mathcal{S}, \mathrm{P}; \mathbb{S})$ which stably converges to a Young measure $\nu^\infty \in \mathcal{Y}^1(\Omega, \mathcal{S}, \mathrm{P}; \mathbb{S})$. Let $J : \Omega \times (\mathbb{E} \times \mathbb{S}) \to \mathbb{R}$ be an L^1-bounded Carathéodory integrand (here we mean that the continuity is over $\mathbb{E} \times \mathbb{S}$) such that there exists a nonnegative integrable function $\varphi \in \mathrm{L}^1_{\mathbb{R}}(\Omega, \mathcal{S}, \mathrm{P})$ satisfying $|J(\omega, x, s)| \leq \varphi(\omega)$ for all $(\omega, x, s) \in \Omega \times \mathbb{E} \times \mathbb{S}$. Then we have*

$$\lim_{n\to\infty} \int_\Omega \Big[\int_{\mathbb{S}} J(\omega, u^n(\omega), s)\, d\nu^n_\omega(s)\Big]\, d\mathrm{P}(\omega) = \int_\Omega \Big[\int_{\mathbb{S}} J(\omega, u^\infty(\omega), s)\, d\nu^\infty_\omega(s)\Big]\, d\mathrm{P}(\omega).$$

In the remainder of this section $\mathbb{U}$ is a compact metric space and $\mathbb{H}$ is a separable Hilbert space. We denote by S_Γ the set of all Lebesgue-measurable selections (alias original controls) of the Lebesgue-measurable multifunction $\Gamma : [0,1] \to \mathcal{K}(\mathbb{U})$ and by S_Σ the set of Lebesgue-measurable selections of the Lebesgue-measurable multifunction $\Sigma : [0,1] \to \mathcal{M}^{+,1}(\mathbb{U})$ defined by

$$\Sigma(t) := \{\nu \in \mathcal{M}^{+,1}(\mathbb{U});\ \nu(\Gamma(t)) = 1\}$$

for all $t \in [0,1]$.

Let $A(t) : \mathbb{H} \to 2^{\mathbb{H}}$ $(t \in [0,1])$ be a maximal monotone operator in $\mathbb{H}$ satisfying:

- (H_1) There exist a continuous function $\rho : [0,1] \to \mathbb{H}$ and a nondecreasing function $L : [0, +\infty[\to [0, +\infty[$ such that

$$\|J_\lambda A(t)x - J_\lambda A(s)x\| \leq \lambda \|\rho(t) - \rho(s)\| L(\|x\|)$$

  for all $\lambda \in\,]0,1]$, for all $(t,s) \in [0,1] \times [0,1]$, and for all $x \in \mathbb{H}$.

- (H_2)

  (*a*) For every $\mathrm{L}^2_{\mathbb{H}}([0,1])$-mapping $u : [0,1] \to \mathbb{H}$ satisfying $u(t) \in D(A(t))$ for all $t \in [0,1]$, the multifunction $t \mapsto A(t)u(t)$ is measurable,

  (*b*) for every $x \in \mathbb{H}$ and for every $\lambda > 0$ the function $t \mapsto (I_{\mathbb{H}} + \lambda A(t))^{-1}x$ is Lebesgue-measurable and,

  (*c*) there exists $\bar{g} \in \mathrm{L}^2_{\mathbb{H}}([0,1])$ such that $t \mapsto J_\lambda(t)g(t) := (I_{\mathbb{H}} + \lambda A(t))^{-1}\bar{g}(t)$ belongs to $\mathrm{L}^2_{\mathbb{H}}([0,1])$ for all $\lambda > 0$.

It is well-known that (H_1) implies that the domain $D(A(t)) = \{x \in \mathbb{H};\ A(t)x \neq \emptyset\}$ is constant. So we put $D(A(t)) = D$, $\forall t \in [0,1]$.

- (H_3) $\overline{D}$ is *ball-compact* (i.e., the intersection of $\overline{D}$ with any closed ball in $\mathbb{H}$ is norm compact).

- (H_4) For every $s > 0$, $\sup\{\|A_\lambda(0)x\|; \lambda \in\,]0,1],\ x \in \overline{D},\ \|x\| \leq s\} < +\infty$, where $A_\lambda(0) := \frac{I_{\mathbb{H}} - (I_{\mathbb{H}} + \lambda A(0))^{-1}}{\lambda}$.

As for every $t \in [0,1]$,

$$\frac{1}{\lambda}\|J_\lambda(t)x - x\| = \|A_\lambda(t)x\| \leq |A(t)x|_0 := \inf_{y\in A(t)x} ||y||, \ \forall x \in D(A(t)),$$

where $A_\lambda(t) = \dfrac{I_{\mathbb{H}} - J_{\lambda(t)}}{\lambda}$, (H_4) is satisfied if $0 \in D(A(0)) = \overline{D(A(0))}$ and $A(0)$ satisfies the following boundedness type condition, namely, for any closed ball $\overline{B}_{\mathbb{H}}(0,\eta)$ of center 0 with radius η, the set $\{|A(0)x|_0 : x \in D(A(0)) \cap \overline{B}_{\mathbb{H}}(0,\eta)\}$ is bounded in $\mathbb{R}$. In particular, (H_4) is satisfied if $A(0) : D(A(0)) \rightrightarrows \mathbb{H}$ is cw$\mathcal{K}(\mathbb{H})$-valued and scalarly upper semicontinuous, because, by (H_3), $D(A(0)) \cap \overline{B}_{\mathbb{H}}(0,\eta)$ is compact for any closed ball $\overline{B}_{\mathbb{H}}(0,\eta)$ of center 0 with radius η, and so, the sets $\{A(0)x : x \in D(A(0)) \cap \overline{B}_{\mathbb{H}}(0,\eta)\}$ are weakly compact.

Let us consider a mapping $g : [0,1] \times \mathbb{H} \times \mathbb{U} \to \mathbb{H}$ satisfying:

(*i*) For every fixed $t \in [0,1]$, $g(t,.,.)$ is continuous on $\mathbb{H} \times \mathbb{U}$.

(*ii*) For every $(x,u) \in \mathbb{H} \times \mathbb{U}$, $g(.,x,u)$ is Lebesgue-measurable on $[0,1]$.

(*iii*) For every $\eta > 0$, there exists $l(\eta) > 0$ such that $\|g(t,x,u) - g(t,y,u)\| \leq l(\eta)\|x-y\|$ for all $t \in [0,1]$ and for all $(x,y) \in \overline{B}_{\mathbb{H}}(0,\eta) \times \overline{B}_{\mathbb{H}}(0,\eta)$.

(*iv*) There exists $c > 0$ such that $g(t,x,\mathbb{U}) \subset c\overline{B}_{\mathbb{H}}$ for all $(t,x) \in [0,1] \times \mathbb{H}$.

We aim to compare the solutions set of the two following evolution equations:

$$(P_{\mathcal{O}}) \begin{cases} \dot{u}_\rho(t) \in -A(t)u_\rho(t) + g(t,u_\rho(t),\rho(t)) \ \text{ a.e. } t \in [0,1], \\ u_\rho(0) = x_0 \in \overline{D}, \end{cases}$$

where ρ belongs to the set $\mathcal{U}$ of all original controls, i.e. ρ is a Lebesgue-measurable mapping from $[0,1]$ into $\mathbb{U}$ with $\rho(t) \in \Gamma(t)$ for a.e. $t \in [0,1]$, and

$$(P_{\mathcal{R}}) \begin{cases} \dot{u}_\lambda(t) \in -A(t)u_\lambda(t) + \int_{\Gamma(t)} g(t,u_\lambda(t),u)\, d\lambda_t(u) \ \text{ a.e. } t \in [0,1], \\ u_\lambda(0) = x_0 \in \overline{D}, \end{cases}$$

where λ belongs to the set S_Σ of all relaxed controls, i.e. λ is the Lebesgue-measurable selection of the multifunction Σ defined by

$$\Sigma(t) := \{\nu \in \mathcal{M}^{+,1}(\mathbb{U}); \nu(\Gamma(t)) = 1\}$$

for all $t \in [0,1]$.

Now is a Bolza-type example for an optimal control problem associated with the preceding evolution equations. As we deal here with evolution inclusions, the proofs we present below need some unusual techniques via the fiber product of Young measures. These techniques have been already given in Chapter 7 dealing with evolution equations governed by nonconvex proximal closed sweeeping process and m-accretive operators and allow to study the properties of the associated value function and its link with the viscosity solution of the associated Hamiton-Jacobi-Bellman equation arisen in theses evolution equations. See Theorem 8.3.12 for the case of ordinary differential equation and the forthcoming works by Castaing, Jofre, Raynaud de Fitte, Salvadori for the case of the evolution inclusions under consideration dealing with two controls Young measures.

Theorem 8.1.10 *Let $J : [0,1] \times (\mathbb{H} \times \mathbb{U}) \to \mathbb{R}$ be an L^1-bounded Carathéodory integrand, (here we mean that the continuity is over $\mathbb{H} \times \mathbb{U}$) such that there is a positive integrable function $\varphi \in \mathrm{L}^1([0,1], dt)$ with $|J(t,x,u)| \leq \varphi(t)$ for all $(t,x,u) \in [0,1] \times \mathbb{H} \times \mathbb{U}$. Let us consider the control problem*

$$\inf_{\lambda \in \mathrm{S}_\Sigma} \int_0^1 \Big[\int_{\mathbb{U}} J(t, u_\lambda(t), u)\, d\lambda_t(u)\Big]\, dt$$

where u_λ is the unique absolutely continuous solution to

$$(P_{\mathcal{R}}) \begin{cases} \dot{u}_\lambda(t) \in -A(t)u_\lambda(t) + \int_{\Gamma(t)} g(t, u_\lambda(t), u)\, d\lambda_t(u) & \text{a.e. } t \in [0,1], \\ u_\lambda(0) = x_0 \in \overline{D}. \end{cases}$$

Then one has $\inf(\mathcal{P}_O) = \min(\mathcal{P}_R)$.

We will give two different proofs.

First proof.

First Step: The graph of the single-valued mapping $\lambda \mapsto u_\lambda$ defined on the $\sigma(\mathrm{L}^\infty_{\mathrm{C}(\mathbb{U})^*}([0,1]), \mathrm{L}^1_{\mathrm{C}(\mathbb{U})}([0,1]))$ compact set S_Σ is closed.

Here we will proceed as in [CI03, Lemma 3.3]. First, by [CI03, Proposition 2.10 b], the solutions set $\{u_\lambda; \lambda \in \mathrm{S}_\Sigma\}$ is compact in $\mathrm{C}([0,1], \mathbb{H})$ with $\sup_{\lambda \in \mathrm{S}_\Sigma} ||\dot{u}_\lambda(.)|| \leq M$ for some constant $M > 0$. Secondly, let $\lambda^n \to \lambda^\infty$ for the $\sigma(\mathrm{L}^\infty_{\mathrm{C}(\mathbb{U})^*}[0,1], \mathrm{L}^1_{\mathrm{C}(\mathbb{U})}[0,1])$ topology and $u_{\lambda^n} \to u_\infty$ in $\mathrm{C}([0,1], \mathbb{H})$, where u_{λ^n} ($n \in \mathbb{N} \cup \{\infty\}$) is the unique absolutely continuous solution of the equation

$$\begin{cases} \dot{u}_{\lambda^n}(t) \in -A(t)u_{\lambda^n}(t) + \int_{\Gamma(t)} g(t, u_{\lambda^n}(t), z)\, d\lambda^n_t(z) \text{ a.e } t \in [0,1], \\ u_{\lambda^n}(0) = x_0 \in \overline{D}, \end{cases}$$

then $u_\infty = u_{\lambda^\infty}$. For simplicity we set

$$h_\lambda(t,x) = \int_{\Gamma(t)} g(t,x,z)\, d\lambda_t(z) = \int_Z g(t,x,z)\, d\lambda_t(z) = h(t,x,\lambda_t)$$

for all $(t,x,\lambda) \in [0,1] \times \mathbb{H} \times \mathrm{S}_\Sigma$. We have

$$h_{\lambda^n}(t, u_{\lambda^n}(t)) - \dot{u}_{\lambda^n}(t) \in A(t)u_{\lambda^n}(t),$$

and

$$h_{\lambda^\infty}(t, u_{\lambda^\infty}(t)) - \dot{u}_{\lambda^\infty}(t) \in A(t)u_{\lambda^\infty}(t),$$

for a.e $t \in [0,1]$. Since $A(t)$ is monotone,

$$\langle u_{\lambda^n}(t) - u_{\lambda^\infty}(t), h_{\lambda^n}(t, u_{\lambda^n}(t)) - \dot{u}_{\lambda^n}(t) - h_{\lambda^\infty}(t, u_{\lambda^\infty}(t)) + \dot{u}_{\lambda^\infty}(t) \geq 0,$$

a.e in $[0,1]$. So

$$\frac{1}{2}\frac{d}{dt}||u_{\lambda^n}(t) - u_{\lambda^\infty}(t)||^2 \leq \langle u_{\lambda^n}(t) - u_{\lambda^\infty}(t), h_{\lambda^n}(t, u_{\lambda^n}(t)) - h_{\lambda^\infty}(t, u_{\lambda^\infty}(t))\rangle,$$

a.e in $[0,1]$. Integrating on $[0,t]$ gives

$$\frac{1}{2}||u_{\lambda^n}(t) - u_{\lambda^\infty}(t)||^2$$
$$\leq \int_0^t \langle u_{\lambda^n}(s) - u_{\lambda^\infty}(s), h_{\lambda^n}(s, u_{\lambda^n}(s)) - h_{\lambda^\infty}(s, u_{\lambda^\infty}(s))\rangle \, ds.$$

Let us set

$$L_n(t) = \int_0^t \langle u_{\lambda^n}(s) - u_{\lambda^\infty}(s), h_{\lambda^n}(s, u_{\lambda^n}(s)) - h_{\lambda^\infty}(s, u_{\lambda^\infty}(s))\rangle \, ds.$$

Then $L_n(t) = L_n^1(t) + L_n^2(t) + L_n^3(t)$ where

$$L_n^1(t) = \int_0^t \langle u_{\lambda^n}(s) - u_{\lambda^\infty}(s), h_{\lambda^n}(s, u_{\lambda^n}(s)) - h_{\lambda^n}(s, u_{\lambda^\infty}(s))\rangle \, ds,$$

$$L_n^2(t) = \int_0^t \langle u_{\lambda^n}(s) - u_\infty(t), h_{\lambda^n}(s, u_{\lambda^\infty}(s)) - h_{\lambda^\infty}(s, u_{\lambda^\infty}(s))\rangle \, ds,$$

$$L_n^3(t) = \int_0^t \langle u_\infty(s) - u_{\lambda^\infty}(s), h_{\lambda^n}(s, u_{\lambda^\infty}(s)) - h_{\lambda^\infty}(s, u_{\lambda^\infty}(s))\rangle \, ds.$$

As $||h_\lambda(t,x)|| \leq c$ for all $\lambda \in \mathrm{S}_\Sigma$ and for all $(t,x) \in [0,1] \times H$, using (iv), we get the estimation

$$|L_n^2(t)| \leq 2c||u_{\lambda^n} - u_\infty||_{\mathrm{C}([0,1],\mathbb{H})}.$$

Thus $L_n^2(t) \to 0$ uniformly in $[0,1]$, when $n \to \infty$. Similarly by (iv) the integrand

$$f(s,z) := \langle u_\infty(s) - u_{\lambda^\infty}(s), g(s, u_{\lambda^\infty}(s), z)\rangle$$

is estimated by

$$|f(s,z)| \leq c||u_\infty - u_{\lambda^\infty}||_{\mathrm{C}([0,1],\mathbb{H})},$$

for all $(s,z) \in [0,1] \times Z$. Hence $f \in \mathrm{L}^1_{\mathrm{C(U)}}([0,1])$. As (λ^n) converges $\sigma(\mathrm{L}^\infty_{\mathrm{C(U)}^*}, \mathrm{L}^1_{\mathrm{C(U)}})$ to λ^∞, it is immediate that for every $t \in [0,1]$,

$$\int_0^t [\int_Z f(s,z)\lambda_s^n(dz)] \, ds \to \int_0^t [\int_Z f(s,z)\lambda_s^\infty(dz)] \, ds$$

when $n \to \infty$. So $\lim_{n\to\infty} L_n^3(t) = 0$. Further there is $\eta > 0$ such that

$$\sup\{||u_\lambda||_{C([0,1],\mathbb{H})};\ \lambda \in S_\Sigma\} < \eta < \infty.$$

Using $(iv)'$, there is $l(\eta) > 0$ such that

$$|L_n^1(t)| \leq \int_0^t l(\eta)||u_{\lambda^n}(s) - u_{\lambda^\infty}(s)||^2\, ds.$$

Finally we get

$$\frac{1}{2}||u_{\lambda^n}(t) - u_{\lambda^\infty}(t)||^2 \leq L_n^2(t) + L_n^3(t) + \int_0^t l(\eta)||u_{\lambda^n}(s) - u_{\lambda^\infty}(s)||^2\, ds.$$

As the functions $L_n^2(.)$ and $L_n^3(.)$ are continuous with $L_n^2(t) \to 0$ and $L_n^3(t) \to 0$, by Gronwall's lemma we conclude that

$$u_{\lambda^n}(t) \to u_{\lambda^\infty}(t)$$

for all $t \in [0,1]$. So, $u_{\lambda^\infty} = u_\infty$. and so the set $\{u_\lambda;\ \lambda \in S_\Sigma\}$ is compact in $C([0,1],\mathbb{H})$.

Second Step: $\inf(\mathcal{P})_\mathcal{O} = \min(\mathcal{P})_\mathcal{R}$ *and conclusion.*

As a consequence of Step 1, the solutions set $\{u_\lambda;\ \lambda \in S_\Sigma\}$ is compact in $C([0,1],\mathbb{H})$ for the uniform convergence. As $\mathcal{O}$ is dense in $\mathcal{R} := S_\Sigma$ for the stable topology according to Lemma 7.1.1, it suffices to prove that the mapping $\Psi : \lambda \mapsto \int_0^1 [\int_\mathbb{U} J(t, u_\lambda(t), z)\, d\lambda_t(z)]\, dt$ is continuous on S_Σ. Let (λ_n) be a sequence in $\mathcal{R}$ stably converging to λ. Applying Step 1 shows that (u_λ^n) converges uniformly to u_λ that is solution of

$$\dot{u}_\lambda(t) \in -A(t)u_\lambda(t) + \int_{\Gamma(t)} g(t, u_\lambda(t), z)\, d\lambda_t(z); \quad u_\lambda(0) = x.$$

This implies that $\underline{\delta}_{u_{\lambda^n}} \otimes \lambda^n$ stably converges to $\underline{\delta}_{u_\lambda} \otimes \lambda$. As J is an L^1-bounded Carathéodory integrand, by virtue of Proposition 8.1.9, we get

$$\lim_{n\to\infty} \int_0^1 [\int_\mathbb{U} J(t, u_{\lambda^n}(t), z)\, d\lambda_t^n(z)]\, dt = \int_0^1 [\int_\mathbb{U} J(t, u_\lambda(t), z)\, d\lambda_t(z)]\, dt.$$

The first proof is therefore complete. □

Second proof.

It is clear that the main point is to prove that the mapping $\lambda \mapsto u_\lambda$ is continuous on $\mathcal{S}_\Sigma$. Here we will produce a different proof which can be applied to evolution inclusions governed by m–accretive operators [CI03]. Let (λ_n) be a sequence in $\mathcal{R}$ stably converging to λ. Let $h \in L^2_\mathbb{H}([0,1])$. Let us set $L_h(t,x,z) := \langle h(t), g(t,x,z)\rangle$

for all $(t, x, z) \in [0, 1] \times \mathbb{H} \times \mathbb{U}$. Then L_h is an L^1-bounded integrand, because $|\langle h(t), g(t, x, z)\rangle| \leq c||h(t)||$ for all $(t, x, z) \in [0, 1] \times \mathbb{H} \times \mathbb{U}$ with $c|h| \in \mathrm{L}^2_{\mathbb{R}}([0, 1])$. By compactness, we may assume that (u_{λ^n}) uniformly converges to an absolutely continuous function u^∞ and $(\dot{u}_{\lambda^n})$ converges weakly in $\mathrm{L}^2_{\mathbb{H}}([0, 1])$ to $\dot{u}^\infty$ with $u^\infty(0) = x$. Hence $\underline{\delta}_{u_{\lambda^n}} \underline{\otimes} \lambda^n$ stably converges to $\underline{\delta}_{u^\infty} \underline{\otimes} \lambda$. Applying Proposition 8.1.9 or Theorem 3.3.1 gives

$$\lim_{n\to\infty} \int_0^1 [\int_{\Gamma(t)} L_h(t, u_{\lambda^n}(t), z)\, d\lambda_t^n(z)]\, dt = \int_0^1 [\int_{\Gamma(t)} L_h(t, u^\infty(t), z)\, d\lambda_t(z)]\, dt.$$

Hence the sequence (v_n) in $\mathrm{L}^2_{\mathbb{H}}([0, 1])$ given by

$$v_n(t) := \int_{\Gamma(t)} g(t, u_{\lambda^n}(t), z)\, d\lambda_t^n(z), \quad \forall t \in [0, 1]$$

converges weakly in $\mathrm{L}^2_{\mathbb{H}}([0, 1])$ to the function v given by

$$v(t) := \int_{\Gamma(t)} g(t, u^\infty(t), z)\, d\lambda_t(z), \quad \forall t \in [0, 1].$$

As

$$-\dot{u}_{\lambda^n}(t) - v_n(t) \in A(t)u_{\lambda^n}(t) \quad \text{a.e.}$$

$-\dot{u}_{\lambda^n} - v_n$ weakly converges in $\mathrm{L}^2_{\mathbb{H}}([0, 1])$ to $-\dot{u}^\infty - v$, u_{λ^n} converges uniformly to u^∞, by invoking a closure type lemma [CI03, Lemma 2.3], we get

$$-\dot{u}^\infty(t) - v(t) \in A(t)u^\infty(t) \quad \text{a.e.}$$

The preceding inclusion shows that u^∞ is the unique absolutely continuous solution u_λ associated with the control λ of the evolution equation

$$\dot{u}_\lambda(t) \in -A(t)u_\lambda(t) + \int_{\Gamma(t)} g(t, u_\lambda(t), z) d\lambda_t(z); \quad u_\lambda(0) = x.$$

Hence we can conclude that the mapping $\lambda \mapsto u_\lambda$ is continuous on S_Σ. The remainder of the proof is straightforward. □

Example 8.1.11 Suppose that $\mathbb{E}$ is a reflexive separable Banach space and that $(\Omega, \mathcal{S}, \mathrm{P})$ is $([0, 1], \mathcal{L}_{[0,1]}, dt)$, where $\mathcal{L}_{[0,1]}$ denotes the σ–algebra of Lebesgue measurable sets of $[0, 1]$. Let $\mathbb{T}$ be a topological Hausdorff space. Let $(u_n)_{n\in\mathbb{N}}$ be a sequence of Borel functions from Ω into $\mathbb{T}$ which pointwise converges to a Borel function u_∞. Let Γ be a convex weakly compact subset in $\mathbb{E}$ and let $(v_n)_{n\in\mathbb{N}}$ be a sequence in $\mathrm{L}^1_{\mathbb{E}}(\Omega, \mathcal{S}, \mathrm{P})$ such that $v_n(\omega) \in \Gamma$ for all $n \in \mathbb{N}$ and for all $\omega \in \Omega$ and that v_n $\sigma(\mathrm{L}^1_{\mathbb{E}}, \mathrm{L}^\infty_{\mathbb{E}^*})$ converges to v_∞. If $\psi : \Omega \times \mathbb{T} \times \mathbb{E} \to [0, +\infty[$ is lower semicontinuous on $\Omega \times \mathbb{T} \times (\mathbb{E}, \sigma(\mathbb{E}, \mathbb{E}^*))$, if $\psi(\omega, u_\infty(\omega), .)$ is convex for all $\omega \in \Omega$, then

$$\int_\Omega \psi(\omega, u_\infty(\omega), v_\infty(\omega))\, d\mathrm{P}(\omega) \leq \liminf_{n\to+\infty} \int_\Omega \psi(\omega, u_n(\omega), v_n(\omega))\, d\mathrm{P}(\omega).$$

Remark 8.1.12 Here $\mathbb{T}$ is only assumed to be a Hausdorff topological space. But other assumptions are stronger than those given in Theorem 8.1.6. Apart from this fact, the proof is quite different from the techniques developed above. Namely the use of Young measures in unnecessary.

To prove Example 8.1.11, we will need the two following results.

Lemma 8.1.13 *Let $\mathbb{T}$ be a topological Hausdorff space and let $\mathbb{S}$ be a topological Suslin space. Let $\Gamma : \Omega \to \mathcal{K}(\mathbb{S})$ be a compact valued multifunction such that its graph belongs to $\mathcal{S} \otimes \mathcal{B}_{\mathbb{S}}$, let (u_n) be a sequence of $(\mathcal{S}, \mathcal{B}_{\mathbb{T}})$-measurable mappings from Ω into $\mathbb{T}$ pointwise converging to a $(\mathcal{S}, \mathcal{B}_{\mathbb{T}})$-measurable mapping u_∞. Let $\psi : \Omega \times \mathbb{T} \times S \to [0, +\infty]$ be an $\mathcal{S} \otimes \mathcal{B}_{\mathbb{T}} \otimes \mathcal{B}_{\mathbb{S}}$-measurable integrand on $\Omega \times \mathbb{T} \times \mathbb{S}$, lower semicontinuous on $\mathbb{T} \times \mathbb{S}$ for all $\omega \in \Omega$ such that $\psi(\omega, u_\infty(\omega), .)$ is finite and continuous on $\Gamma(\omega)$ for all $\omega \in \Omega$. Then, for every $\varepsilon > 0$, the following holds*

$$\lim_{n\to\infty} \mathrm{P}\left[\{\omega \in \Omega;\ \inf_{x\in\Gamma(\omega)} [\psi(\omega, u_n(\omega), x) - \psi(\omega, u_\infty(\omega), x)] \le -\varepsilon\}\right] = 0.$$

Proof. For every $n \in \mathbb{N}$ and for every $\omega \in \Omega$, the function $x \mapsto \psi(\omega, u_n(\omega), x) - \psi(\omega, u_\infty(\omega), x)$ is well defined and lower semicontinuous on $\Gamma(\omega)$. So there exists $\sigma_n(\omega) \in \Gamma(\omega)$ such that

$$\inf_{x\in\Gamma(\omega)} [\psi(\omega, u_n(\omega), x) - \psi(\omega, u_\infty(\omega), x)] = \psi(\omega, u_n(\omega), \sigma_n(\omega)) - \psi(\omega, u_\infty(\omega), \sigma_n(\omega)).$$

Notice that the function

$$r_n : \omega \mapsto \inf_{x\in\Gamma(\omega)} [\psi(\omega, u_n(\omega), x) - \psi(\omega, u_\infty(\omega), x)]$$

is $\mathcal{S}$-measurable by [CV77, Lemma II.39]. Now we claim that

$$\forall\omega \in \Omega \quad \liminf_n r_n(\omega) \ge 0.$$

Let $\omega \in \Omega$ be fixed. There is a subsequence $r_{\varphi(n)}(\omega)$ such that

$$\liminf_n r_n(\omega) = \lim_n r_{\varphi(n)}(\omega).$$

As $\Gamma(\omega)$ is compact, there is a subsequence $(\sigma_{\theta\circ\varphi(n)}(\omega))$ of $(\sigma_{\varphi(n)}(\omega))$ and $\eta(\omega) \in \Gamma(\omega)$ such that $\lim_n \sigma_{\theta\circ\varphi(n)}(\omega) = \eta(\omega)$. Thus we have

$$\lim_n (u_{\theta\circ\varphi(n)}(\omega), \sigma_{\theta\circ\varphi(n)}(\omega)) = (u_\infty(\omega), \eta(\omega)).$$

As $\psi(\omega, ., .)$ is lower semicontinuous on $\mathbb{T} \times \mathbb{S}$ and $\psi(\omega, u_\infty(\omega), .)$ is finite and continuous on $\Gamma(\omega)$

$$\liminf_n [\psi(\omega, u_{\theta\circ\varphi(n)}(\omega), \sigma_{\theta\circ\varphi(n)}(\omega)) - \psi(\omega, u_\infty(\omega), \sigma_{\theta\circ\varphi(n)}(\omega))] \ge 0.$$

But

$$\liminf_n r_{\theta\circ\varphi(n)}(\omega) = \lim_n r_{\theta\circ\varphi(n)}(\omega) = \lim_n r_{\varphi(n)}(\omega) = \liminf_n r_n(\omega).$$

Hence $\liminf_n r_n(\omega) \geq 0$. It follows that

$$\Omega = \{\omega \in \Omega;\ \liminf_n r_n(\omega) \geq 0\} = \bigcap_{\varepsilon>0} \bigcup_{n\in\mathbb{N}} \bigcap_{m\geq n} \{\omega \in \Omega;\ r_m(\omega) > -\varepsilon\}.$$

So, for all $\varepsilon > 0$, we have

$$\mathrm{P}\,[\bigcap_{n\in\mathbb{N}} \bigcup_{m\geq n} \{\omega \in \Omega;\ r_m(\omega) \leq -\varepsilon\}] = 0.$$

Consequently

$$\forall \varepsilon > 0 \quad \lim_n \mathrm{P}\,[\bigcup_{m\geq n} \{\omega;\ r_m(\omega) \leq -\varepsilon\}] = 0. \quad \square$$

References. [Ole77, CC82]. The preceding lemma is extracted from [CC82]. The lower semicontinuity of $\psi(\omega,.,.)$ on $\mathbb{T} \times \mathbb{S}$ is essential. In most applications, $\mathbb{S} = (\mathbb{E}, \sigma(\mathbb{E}, \mathbb{E}^*))$, or $\mathbb{S} = (\mathbb{E}^*, \sigma(\mathbb{E}^*, \mathbb{E}))$ or $\mathbb{S} = (\mathbb{E}, \|.\|)$.

Here is an approximation result that is an analogous of Hausdorff–Baire's approximation for lower semicontinuous function defined on a metric space.

Theorem 8.1.14 *Let $\mathbb{X}$ be a topological locally convex Suslin space. Let Γ be a multifunction on Ω with convex circled compact values such that its graph belongs to $\mathcal{S} \otimes \mathcal{B}_{\mathbb{X}}$ and let $g : \Omega \times \mathbb{X} \to [0, +\infty]$ be a l.s.c. integrand such that, $\forall \omega \in \Omega$, $g(\omega, .)$ is finite on $\Gamma(\omega)$. Then there exists a nondecreasing sequence of nonnegative integrands $(g_k)_k$ on $\Omega \times \mathbb{X}$ which satisfies the following properties:*

(a) $\forall \omega \in \Omega$, $g_k(\omega, .)$ *is convex on* X.

(b) $\forall x \in \mathbb{X}$, $g_k(., x)$ *is* $\mathcal{S}$*-measurable.*

(c) $\forall \omega \in \Omega$, $g_k(\omega, .)$ *finite and continuous on* $\Gamma(\omega)$.

(d) $\forall \omega \in \Omega$, $\forall x \in \mathbb{X}$, $g(\omega, x) = \lim_k \uparrow g_k(\omega, x)$.

Proof. Let (e'_n) be a sequence in the topological dual $\mathbb{X}^*$ of $\mathbb{X}$ which separates the points of $\mathbb{X}$. Let us set

$$(8.1.7) \qquad \varphi(\omega, x) = \begin{cases} \sum_{n=1}^{\infty} 2^{-n} \dfrac{|\langle e'_n, x\rangle|}{1 + \delta^*(e'_n, \Gamma(\omega))} & \text{if } x \in \Gamma(\omega), \\ +\infty \quad \text{if } x \notin \Gamma(\omega). \end{cases}$$

Then φ is a convex normal integrand. Further, for each $\omega \in \Omega$, the function $\varphi(\omega, .)$ is subadditive, convex and continuous on $\Gamma(\omega)$. For each $k \in \mathbb{N}$ let us set

$$(8.1.8) \qquad g_k(\omega, x) = \begin{cases} \inf_{y\in\Gamma(\omega)} \left[k\varphi(\omega, \frac{x-y}{2}) + g(\omega, y)\right] & \text{if } x \in \Gamma(\omega), \\ +\infty \quad \text{if } x \notin \Gamma(\omega). \end{cases}$$

Using the Projection Theorem, it is not difficult to see that $g_k(.,x)$ is $\mathcal{S}$-measurable for all $x \in \mathbb{X}$. By its definition, $g_k(\omega,.)$ is convex lower semicontinuous on $\mathbb{X}$, for all $\omega \in \Omega$. It is obvious that

$$\forall k \in \mathbb{N} \quad \forall \omega \in \Omega \quad \forall x \in \Gamma(\omega) \quad 0 \leq g_k(\omega,x) \leq g_{k+1}(\omega,x) \leq g(\omega,x).$$

So $\sup_k g_k(\omega,x) \leq g(\omega,x)$ for all $\omega \in \omega$ and for all $x \in \Gamma(\omega)$. Now let $(\omega,x) \in \mathrm{gph}\,(\Gamma)$ be fixed. Let $r \in \mathbb{R}$ such that $g(\omega,x) > r$. We claim that there is $k_0 \in \mathbb{N}$ such that $g_{k_0}(\omega,x) \geq r$. As $g(\omega,.)$ is lower semicontinuous on $\Gamma(\omega)$ there is an open neigbourhood $V(x)$ of x such that $g(\omega,y) > r$ for all $y \in V(x) \cap \Gamma(\omega)$. This implies

$$k\varphi(\omega,\frac{x-y}{2}) + g(\omega,y) > r$$

for all $k \in \mathbb{N}$ and for all $y \in V(x) \cap \Gamma(\omega)$. So it is enough to find $k_0 \in \mathbb{N}$ such that $k_0\varphi(\omega,\frac{x-y}{2}) \geq r - \min_{y\in V(x)\setminus\Gamma(\omega)} g(\omega,y)$. Notice that

$$m := \min_{y\in V(x)\setminus\Gamma(\omega)} \varphi(\omega,\frac{x-y}{2})$$

is strictly positive. Pick $k_0 \in \mathbb{N}$ such that $k_0\, m \geq r - \min_{y\in V(x)\setminus\Gamma(\omega)} g(\omega,y)$, then $g_{k_0}(\omega,x) \geq r$. Consequently $\sup_k g_k(\omega,x) = g(\omega,x)$. Thanks to the subadditivity of φ we have

$$g_k(\omega,x) - g_k(\omega,y) \leq k\varphi(\omega,\frac{x-y}{2})$$

for all $\omega \in \Omega$ and x, y in $\Gamma(\omega)$. This proves the continuity of $g_k(\omega,.)$ on $\Gamma(\omega)$.

□

Remark 8.1.15 If Ω is a K_σ-topological space, $\mathbb{X} = (\mathbb{E}, \sigma(\mathbb{E},\mathbb{E}^*))$ where $\mathbb{E}$ is a reflexive separable space, Γ is a fixed convex compact circled set in $\mathbb{X}$, $g : \Omega \times \mathbb{X} \to [0,+\infty]$ is lower semicontinuous, convex on $\mathbb{X}$ and finite on Γ, then (8.1.7) is reduced to

$$\varphi(x) = \begin{cases} \sum_{n=1}^{\infty} 2^{-n} \dfrac{|\langle e'_n, x\rangle|}{1+\delta^*(e'_n,\Gamma)} & \text{if } x \in \Gamma, \\ +\infty \quad \text{if } x \notin \Gamma, \end{cases}$$

and thus the integrand g_k given by the formula (8.1.8) which becomes

$$g_k(\omega,x) = \begin{cases} \inf_{y\in\Gamma} \left[k\varphi(\frac{x-y}{2}) + g(\omega,y)\right] & \text{if } x \in \Gamma, \\ +\infty \quad \text{if } x \notin \Gamma, \end{cases}$$

is Borel on $\Omega \times \mathbb{X}$. Indeed, since $\mathbb{X} = \mathbb{E}_\sigma$ is a K_σ-space, we deduce using the preceding formulas that, for every $r \in \mathbb{R}$, the set $\{g_k \leq r\}$ is a K_σ subset in $\Omega \times \mathbb{X}$. To see this, let us denote by δ_Γ the indicator of Γ. Then the function

$$\varphi_k : (\omega,x,y) \mapsto k\varphi\left(\frac{x-y}{2}\right) + g(\omega,y) + \delta_\Gamma(y)$$

is lower semicontinuous on $\Omega \times \mathbb{E}_\sigma \times \mathbb{E}_\sigma$. Hence $\{(\omega, x) \in \Omega \times \mathbb{E}_\sigma;\ g_k(\omega, x) \leq r\}$ is equal to

$$(\Omega \times \Gamma) \cap \pi_{\Omega \times \mathbb{E}_\sigma}[\{(\omega, x, y) \in \Omega \times \mathbb{E}_\sigma \times \mathbb{E}_\sigma;\ \varphi_k(\omega, x, y) \leq r\}],$$

where $\pi_{\Omega \times \mathbb{E}_\sigma}$ is the canonical projection onto $\Omega \times \mathbb{E}_\sigma$. As $\Omega \times \mathbb{E}_\sigma$ is a K_σ-set and the projection of a K_σ-set is a K_σ-set, the result follows. In particular, this shows that g_k is Borel on $\Omega \times \mathbb{E}_\sigma$.

Reference: [CC82, Théorèmes 1–2].

Proof of Example 8.1.11. *Step 1* Let us focus our attention to the integrand $(\omega, x) \mapsto \psi(\omega, u_\infty(\omega), x)$. We may suppose that Γ is circled. Now apply Remark 8.1.15 to the space $\mathbb{X} = \mathbb{E}_\sigma := (\mathbb{E}, \sigma(\mathbb{E}, \mathbb{E}^*))$, the integrand $\psi(\omega, u_\infty(\omega), x)$ with $\omega \in \Omega = [0, 1]$ and $x \in \mathbb{E}$, and the convex compact set Γ in $\mathbb{E}_\sigma$. Then there exists a nondecreasing sequence of nonnegative integrands $(g_k)_k$ on $[0, 1] \times \mathbb{E}_\sigma$ which satisfies the properties (a)–(d) of the theorem and is Borel on $[0, 1] \times \mathbb{E}_\sigma$. Let us set

$$\psi_k(\omega, s, x) = \begin{cases} g_k(\omega, x) \text{ if } s = u_\infty(\omega), \\ \hat{\psi}(\omega, s, x) \text{ if } s \neq u_\infty(\omega), \end{cases} \tag{8.1.9}$$

where

$$\hat{\psi}(\omega, s, x) = \begin{cases} \psi(\omega, s, x) \text{ if } (s, x) \in \mathbb{T} \times \Gamma, \\ = +\infty \text{ if } (s, x) \in \mathbb{T} \times (\mathbb{E} \setminus \Gamma). \end{cases}$$

Then $\psi(\omega, s, x) = \lim_k \uparrow \psi_k(\omega, s, x)$, $\forall (\omega, s, x) \in \Omega \times \mathbb{T} \times \mathbb{E}$. Further, for all $k \in \mathbb{N}$, $\forall n \in \mathbb{N}$, $\forall \omega \in \Omega$, we have

$$\psi_k(\omega, u_n(\omega), v_n(\omega)) \leq \psi(\omega, u_n(\omega), v_n(\omega)) \tag{8.1.10}$$

and

$$\psi_k(\omega, u_\infty(\omega), v_n(\omega)) = g_k(\omega, v_n(\omega)) \leq \psi(\omega, u_\infty(\omega), v_n(\omega)) \tag{8.1.11}$$

using the formula (8.1.9). As ψ is lower semicontinuous on $[0, 1] \times \mathbb{T} \times \mathbb{E}_\sigma$ by hypothesis, so $\psi_k(\omega, ., .)$ is lower semicontinuous on $\mathbb{T} \times \mathbb{E}_\sigma$ thanks to formula (8.1.9) (see [CC82, Lemme 3]). Again by (8.1.9), the integrands $\psi_k(\omega, u_\infty(\omega), x)$ are finite and $\sigma(\mathbb{E}, \mathbb{E}^*)$-continuous on Γ for every fixed $\omega \in \Omega$. Further ψ_k is Borel on $[0, 1] \times \mathbb{T} \times \mathbb{E}_\sigma$ since g_k is Borel on $[0, 1] \times \mathbb{E}_\sigma$ (see [CC82, Lemme 2]).

Step 2 We may suppose that $a := \liminf_n \int_\Omega \psi(\omega, u_n(\omega), v_n(\omega))\, d\,\mathrm{P}(\omega)$ is finite and by extracting a subsequence that $a = \lim_n \int_\Omega \psi(\omega, u_n(\omega), v_n(\omega))\, d\,\mathrm{P}(\omega)$. Let $k \in \mathbb{N}$, $\varepsilon > 0$ and $n \in \mathbb{N}$. Let us consider the truncated integrand $\psi_k(\omega, s, x)$ associated with the integrand $\psi(\omega, s, x)$. We claim that

$$\liminf_n \int_\Omega \psi_k(\omega, u_\infty(\omega), v_n(\omega))\, d\,\mathrm{P}(\omega) \geq \int_\Omega \psi_k(\omega, u_\infty(\omega), v_\infty(\omega))\, d\,\mathrm{P}(\omega). \tag{8.1.12}$$

Similarly we may suppose that

$$b_k := \liminf_n \int_\Omega \psi_k(\omega, u_\infty(\omega), v_n(\omega))\, d\,\mathrm{P}(\omega) \in \mathbb{R}$$

and by extracting a subsequence that

$$b_k = \lim_n \int_\Omega \psi_k(\omega, u_\infty(\omega), v_n(\omega))\, d\,\mathrm{P}(\omega).$$

By Mazur's lemma, there is a sequence $(\widetilde{v}_n)_n$ where $\widetilde{v}_n = \sum_{j=n}^{\nu_n} \lambda^{n_j} v_{n_j}$ with $0 \le \lambda^{n_j} \le 1$ and $\sum_{j=n}^{\nu_n} \lambda^{n_j} = 1$ which converges a.e. to $v_\infty(\omega)$. By convexity, we have

$$\forall\omega \quad \forall n \quad \psi_k(\omega, u_\infty(\omega), \widetilde{v}_n(\omega)) \le \sum_{j=n}^{\nu_n} \lambda^{n_j} \psi_k(\omega, u_\infty(\omega), v_{n_j}(\omega)).$$

Hence

$$\limsup_n \int_\Omega \psi_k(\omega, u_\infty(\omega), \widetilde{v}_n(\omega))\, d\,\mathrm{P}(\omega) \le b_k.$$

By the lower semicontinuity of $\psi_k(\omega, u_\infty(\omega), .)$ and by Fatou's lemma, we get

$$b_k \ge \liminf_n \int_\Omega \psi_k(\omega, u_\infty(\omega), \widetilde{v}_n(\omega))\, d\,\mathrm{P}(\omega) \ge \int_\Omega \psi_k(\omega, u_\infty(\omega), v_\infty(\omega))\, d\,\mathrm{P}(\omega),$$

thus proving the claim. At this point let us observe that (8.1.12) holds if we replace the sequence (v_n) by any sequence which $\sigma(\mathrm{L}^1_{\mathbb{E}}, \mathrm{L}^\infty_{\mathbb{E}^*})$-converges to v_∞. By the properties of the integrand ψ_k obtained in Step 1 we may consider the set

$$\Omega^k_{n,\varepsilon} = \{\omega \in \Omega;\ \inf_{x\in\Gamma} [\psi_k(\omega, u_n(\omega), x) - \psi_k(\omega, u_\infty(\omega), x)] \le -\varepsilon\}$$

and apply Lemma 8.1.13 to the integrand ψ_k by noting that the vector space $\mathbb{E}$ equipped with the $\sigma(\mathbb{E}, \mathbb{E}^*)$ topology is a Lusin space. So we have

$$\lim_{n\to\infty} \mathrm{P}(\Omega^k_{n,\varepsilon}) = 0. \tag{8.1.13}$$

By what has been demonstrated, there exists $\bar{v}$ with $\bar{v}(\omega) \in \Gamma$ for all ω such that

$$\int_\Omega \psi_k(\omega, u_\infty(\omega), \bar{v}(\omega))\, d\,\mathrm{P}(\omega) < +\infty. \tag{8.1.14}$$

Let us set

$$v^k_{n,\varepsilon} = \mathbb{1}_{\Omega^k_{n,\varepsilon}} \bar{v} + \mathbb{1}_{\Omega\setminus\Omega^k_{n,\varepsilon}} v_n. \tag{8.1.15}$$

Using (8.1.13) and(8.1.15) it is easily seen that $(v_{n,\varepsilon}^k - v_n)_n$ weakly converges to 0 and thus $(v_{n,\varepsilon}^k)_n$ weakly converges to v_∞. We have

$$\int_\Omega \psi_k(\omega, u_\infty(\omega), v_{n,\varepsilon}^k(\omega))\, d\,\mathrm{P}(\omega) = \int_{\Omega_{n,\varepsilon}^k} \psi_k(\omega, u_\infty(\omega), \bar{v}(\omega))\, d\,\mathrm{P}(\omega) + \int_{\Omega\setminus\Omega_{n,\varepsilon}^k} \psi_k(\omega, u_\infty(\omega), v_n(\omega))\, d\,\mathrm{P}(\omega).$$

Using (8.1.13) and (8.1.14) we pick $N_\varepsilon \in \mathbb{N}$ such that $n > N_\varepsilon$ implies

$$\int_\Omega \psi_k(\omega, u_\infty(\omega), v_{n,\varepsilon}^k(\omega))\, d\,\mathrm{P}(\omega) \le \varepsilon + \int_{\Omega\setminus\Omega_{n,\varepsilon}^k} \psi_k(\omega, u_\infty(\omega), v_n(\omega))\, d\,\mathrm{P}(\omega).$$

Thus, for $n > N_\varepsilon$ we get

$$\begin{aligned}
&\int_\Omega \psi_k(\omega,u_\infty(\omega), v_{n,\varepsilon}^k(\omega))\, d\,\mathrm{P}(\omega) \\
&\qquad \le \varepsilon + \int_{\Omega\setminus\Omega_{n,\varepsilon}^k} [\psi_k(\omega, u_\infty(\omega), v_n(\omega)) - \psi_k(\omega, u_n(\omega), v_n(\omega))]\, d\,\mathrm{P}(\omega) \\
&\qquad\quad + \int_\Omega \psi_k(\omega, u_n(\omega), v_n(\omega))\, d\,\mathrm{P}(\omega) \\
&\qquad < \varepsilon + \varepsilon + \int_\Omega \psi_k(\omega, u_n(\omega), v_n(\omega))\, d\,\mathrm{P}(\omega).
\end{aligned}$$

Consequently, there is $N'_\varepsilon \in \mathbb{N}$ such that $n > N'_\varepsilon$ implies

$$\int_\Omega \psi_k(\omega, u_\infty(\omega), v_{n,\varepsilon}^k(\omega))\, d\,\mathrm{P}(\omega) < 3\varepsilon + a.$$

So we get

$$\limsup_n \int_\Omega \psi_k(\omega, u_\infty(\omega), v_{n,\varepsilon}^k(\omega))\, d\,\mathrm{P}(\omega) \le a.$$

Since

$$\liminf_n \int_\Omega \psi_k(\omega, u_\infty(\omega), v_{n,\varepsilon}^k(\omega))\, d\,\mathrm{P}(\omega) \ge \int_\Omega \psi_k(\omega, u_\infty(\omega), v_\infty(\omega))\, d\,\mathrm{P}(\omega)$$

as we have already proved in the first step, we finally get

$$\int_\Omega \psi_k(\omega, u_\infty(\omega), v_\infty(\omega))\, d\,\mathrm{P}(\omega) \le a = \liminf_n \int_\Omega \psi(\omega, u_n(\omega), v_n(\omega))\, d\,\mathrm{P}(\omega).$$

As $\psi_k(\omega, u_\infty(\omega), v_\infty(\omega)) \uparrow \psi(\omega, u_\infty(\omega), v_\infty(\omega))$, by monotone convergence theorem, we deduce that

$$\int_\Omega \psi(\omega, u_\infty(\omega), v_\infty(\omega))\, d\,\mathrm{P}(\omega) \le a = \liminf_n \int_\Omega \psi(\omega, u_n(\omega), v_n(\omega))\, d\,\mathrm{P}(\omega).$$

The proof is therefore complete. □

References Castaing–Clauzure [CC82, Theorem 4], Olech [Ole77]. Example 8.1.11 holds for any topological compact space $(\Omega, \mathcal{T}_{\mathrm{P}}, \mathrm{P})$ equipped with a probability Radon measure P, $\mathcal{T}_{\mathrm{P}}$ being the σ-algebra of all P-measurable sets. A variant of Example 8.1.11 is:

Example 8.1.16 Suppose that $\mathbb{E}^*_\sigma$ is the weak dual of a separable Banach space $\mathbb{E}$ and (Ω, P) is a topological compact space equipped with a probability Radon measure P. Let $\mathbb{T}$ be a topological Hausdorff space. Let $(u_n)_{n\in\mathbb{N}}$ be a sequence of Borel functions from Ω into $\mathbb{T}$ which pointwise converges to a Borel function u_∞. Let Γ be convex weakly compact in $\mathbb{E}^*_\sigma$ and let $(v_n)_{n\in\mathbb{N}}$ be a sequence in $\mathrm{L}^\infty_{\mathbb{E}^*_\sigma}(\Omega, \mathrm{P})$ such that $v_n(\omega) \in \Gamma$ for all $n \in \mathbb{N}$ and for all $\omega \in \Omega$ and that v_n $\sigma(\mathrm{L}^\infty_{\mathbb{E}^*_\sigma}, \mathrm{L}^1_{\mathbb{E}})$-converges to v_∞. If $\psi : \Omega \times \mathbb{T} \times \mathbb{E} \to [0, +\infty[$ is lower semicontinuous on $\Omega \times \mathbb{T} \times \mathbb{E}^*_\sigma$, if $\psi(\omega, u_\infty(\omega), .)$ is convex for all $\omega \in \Omega$, then

$$\int_\Omega \psi(\omega, u_\infty(\omega), v_\infty(\omega))\, d\,\mathrm{P}(\omega) \le \liminf_{n\to+\infty} \int_\Omega \psi(\omega, u_n(\omega), v_n(\omega))\, d\,\mathrm{P}(\omega).$$

Proof. Since $\mathbb{E}^*_\sigma$ is a K_σ Suslin space, the proof follows the same line as Step 1 of the proof of Example 8.1.11. Using the approximated integrand ψ_k one can prove that the Fatou property holds in this case via the duality of convex integral functionals [CV77]. Indeed we only need to check that

$$c_k := \liminf_n \int_\Omega \psi_k(\omega, u_\infty(\omega), w_n(\omega))\, d\,\mathrm{P}(\omega) \ge \int_\Omega \psi_k(\omega, u_\infty(\omega), v_\infty(\omega))\, d\,\mathrm{P}(\omega),$$

for any sequence (w_n) of measurable selections of Γ which $\sigma(\mathrm{L}^\infty_{\mathbb{E}^*_\sigma}, \mathrm{L}^1_{\mathbb{E}})$-converges to v_∞. We may suppose that $c_k \in \mathbb{R}$ and by extracting a subsequence that

$$c_k = \lim_n \int_\Omega \psi_k(\omega, u_\infty(\omega), w_n(\omega))\, d\,\mathrm{P}(\omega).$$

Pick $\bar v$ with $\bar v(\omega) \in \Gamma$ a.e. such that $\psi_k(\omega, u_\infty(\omega), \bar v(\omega))$ is integrable. Hence the convex dual function g_k of the convex integrand $h_k(\omega, x) := \psi_k(\omega, u_\infty(\omega), x))$ satisfies

$$0 \le -g_k(\omega, 0) \le h_k(\omega, \bar v(\omega)).$$

Thus the convex integral functionals

$$I_{h_k}(v) = \int_\Omega h_k(\omega, v(\omega))\, d\,\mathrm{P}(\omega), \;\; v \in \mathrm{L}^\infty_{\mathbb{E}^*_\sigma}$$

and

$$I_{g_k}(u) = \int_\Omega g_k(\omega, u(\omega))\, d\,\mathrm{P}(\omega), \;\; u \in \mathrm{L}^1_{\mathbb{E}}$$

are dual functionals. In particular I_{h_k} is convex lower semicontinuous for the $\sigma(\mathrm{L}^\infty_{\mathbb{E}^*_\sigma}, \mathrm{L}^1_{\mathbb{E}})$ topology, thus proving the desired lower semicontinuity property. The remainder of the proof is mutatis mutandis the same as in the proof of Example 8.1.11. □

To finish this section let us mention another example (compare with the example just after Theorem 8.1.6) which arises from the theory of sweeping process (see [Mor77, CV77, MM93]).

Example 8.1.17 Suppose that $(\mathbb{H}, \|.\|_{\mathbb{H}})$ is a Hilbert space. Let C be a lower semicontinuous multifunction on $[0,1]$ with closed convex values in $\mathbb{H}$. Let (u_n) be a uniformly bounded sequence in $\mathrm{L}^\infty_{\mathbb{H}}([0,1])$ which converges in measure with respect to the norm $\|.\|_{\mathbb{H}}$ to a function $u_\infty \in \mathrm{L}^\infty_{\mathbb{H}}([0,1])$ and (v_n) be a uniformly integrable sequence in $\mathrm{L}^1_{\mathbb{H}}([0,1], dt)$ which $\sigma(\mathrm{L}^1_{\mathbb{H}}, \mathrm{L}^\infty_{\mathbb{H}})$ converges to a function $v_\infty \in \mathrm{L}^1_{\mathbb{H}}([0,1], dt)$. Let ψ be the integrand defined on $[0,1] \times \mathbb{H} \times \mathbb{H}$ by

$$\psi(t, y, x) := \delta^*(x, C(t)) + \langle y, x \rangle$$

for $(t, x, y) \in [0,1] \times \mathbb{H} \times \mathbb{H}$. Then

$$\int_0^1 \psi(t, u_\infty(t), v_\infty(t))\, dt \leq \liminf_{n \to +\infty} \int_0^1 \psi(t, u_n(t), v_n(t))\, dt.$$

Hint Firstly, use the fact that the topology of the convergence in measure on bounded subsets of $\mathrm{L}^\infty_{\mathbb{H}}$ coincides with the topology of uniform convergence on uniformly integrable subsets of $\mathrm{L}^1_{\mathbb{H}}$ [Cas80], to get

$$\lim_{n \to +\infty} \int_0^1 \langle u_n(t), v_n(t) \rangle\, dt = \int_0^1 \langle u_\infty(t), v_\infty(t) \rangle\, dt.$$

Secondly, since C is lower semicontinuous, the integrand $(t, x) \mapsto \delta^*(x, C(t))$ is lower semicontinuous on $[0,1] \times \mathbb{H}$, using Michael's selection theorem, and is convex lower semicontinuous on $\mathbb{H}$. Thus

$$\int_0^1 \delta^*(v_\infty(t)), C(t))\, dt \leq \liminf_{n \to +\infty} \int_0^1 \delta^*(v_n(t)), C(t))\, dt.$$

The result follows. □

8.2 Reshetnyak–type theorems for Banach-valued measure

Most parts of the material is this section is borrowed from [CJ95]. Let $\mathbb{T}$ be a Polish space and $\mathcal{B}_{\mathbb{T}}$ its Borel σ-algebra. Let $\mathbb{E}$ be a separable reflexive Banach space with norm $\|\cdot\|$, $\mathbb{E}^*$ be its strong dual space and $\langle ., . \rangle$ be the duality bilinear form between $\mathbb{E}$ and $\mathbb{E}^*$. The closed unit ball $\overline{B}_{\mathbb{E}^*}$ of $\mathbb{E}^*$ is equipped with the weak* topology $\sigma(\mathbb{E}^*, \mathbb{E})$; it is a metrizable compact space.

We denote by $\mathrm{C}_b(\mathbb{T}, \mathbb{E})$ the Banach space of all bounded continuous $\mathbb{E}$-valued functions on $\mathbb{T}$ equipped with the sup-norm.

An $\mathbb{E}^*$-valued measure on $\mathbb{T}$ is a σ-additive set-function m from $\mathcal{B}_{\mathbb{T}}$ into $\mathbb{E}^*$. The variation of the measure m is the non-negative real-valued measure $|m|$ on $\mathbb{T}$ defined, for all $A \in \mathcal{B}_{\mathbb{T}}$, by

$$|m|(A) = \sup\{\sum_{i \in I} \|m(A_i)\|;\ (A_i)_{i \in I} \text{ finite } \mathcal{B}_{\mathbb{T}}\text{-partition of } A\}.$$

We denote by $\mathcal{M}(\mathbb{T}, \mathbb{E}^*)$ the space of all $\mathbb{E}^*$-valued measures m on $\mathbb{T}$ with bounded variation (i.e. $|m|$ is a bounded Radon measure on $\mathbb{T}$). We set $\|m\| = |m|(\mathbb{T})$.

For every $m \in \mathcal{M}(\mathbb{T}, \mathbb{E}^*)$, there is a $|m|$-measurable function $(dm/d|m|) : \mathbb{T} \longrightarrow \mathbb{E}^*$ such that $m = (dm/d|m|)|m|$, that is,

$$m(A) = \int_A \frac{dm}{d|m|}(t)\, d|m|(t), \quad \forall A \in \mathcal{B}_{\mathbb{T}}.$$

For every $f \in \mathrm{C}_b(\mathbb{T}, \mathbb{E})$ and every $m \in \mathcal{M}(\mathbb{T}, \mathbb{E}^*)$, we define the integral of f with respect to m by

$$\int f\, dm = \int \langle f(t), \frac{dm}{d|m|}(t) \rangle\, d|m|(t).$$

Thus, the space $\mathcal{M}(\mathbb{T}, \mathbb{E}^*)$ is identified with a subspace of the topological dual space of $\mathrm{C}_b(\mathbb{T}, \mathbb{E})$. It is equipped with the weak* topology $\sigma(\mathcal{M}(\mathbb{T}, \mathbb{E}^*), \mathrm{C}_b(\mathbb{T}, \mathbb{E})$ usually called *weak* (or *narrow*) topology.

For further details on vector measures, we refer to [DU77].

A subset $\mathcal{H}$ of $\mathcal{M}(\mathbb{T}, \mathbb{E}^*)$ is *bounded* if

$$\sup_{m \in \mathcal{H}} \|m\| < +\infty.$$

It is *tight* (or it satisfies *Prohorov's condition*) if for every $\varepsilon > 0$, there is a compact subset K_ε of $\mathbb{T}$ such that

$$|m|(\mathbb{T} \setminus K_\varepsilon) \leq \varepsilon \quad \forall m \in \mathcal{H}.$$

We recall a classical statement of Prohorov's theorem which is, for positive measures, a particular case of Theorem 4.3.5.

References: [Bou69, §5 Théorème 1], [DM75, Théorème III.59].

Theorem 8.2.1 *Let $\mathcal{H}$ be a bounded tight subset of $\mathcal{M}(\mathbb{T}, \mathbb{R})$. Then $\mathcal{H}$ is relatively narrowly compact in $\mathcal{M}(\mathbb{T}, \mathbb{R})$.*

The following result is the generalization to Banach valued measures defined on a Polish space of a well-known result of Y. Reshetnyak ([Res68, Theorem 2]).

Theorem 8.2.2 *Let $\phi : \mathbb{T} \times \overline{B}_{\mathbb{E}^*} \longrightarrow [0, +\infty]$ be a lower semicontinuous function on $\mathbb{T} \times \overline{B}_{\mathbb{E}^*}$ such that for all $t \in \mathbb{T}$, $\phi(t, .)$ is convex and positively 1-homogeneous on $\overline{B}_{\mathbb{E}^*}$ i.e. for all $x \in \overline{B}_{\mathbb{E}^*}$ and all $\lambda \in [0,1]$, $\phi(t, \lambda x) = \lambda\phi(t,x)$. Let $(m_k)_k$ be a bounded tight sequence in $\mathcal{M}(\mathbb{T}, \mathbb{E}^*)$ which narrowly converges to $m \in \mathcal{M}(\mathbb{T}, \mathbb{E}^*)$. Then we have*

$$\liminf_k \int_{\mathbb{T}} \phi(t, \frac{dm_k}{d|m_k|}(t))\, d|m_k|(t) \geq \int_{\mathbb{T}} \phi(t, \frac{dm}{d|m|}(t))\, d|m|(t).$$

Proof. This proof follows some steps developed by Y. Reshetnyak [Res68, Cas87] in the case when $\mathbb{T}$ is locally compact, with some necessary modifications.

Extracting a subsequence, we can suppose that

$$a = \liminf_k \int_{\mathbb{T}} \phi(t, \frac{dm_k}{d|m_k|}(t))\, d|m_k|(t) = \lim_k \int_{\mathbb{T}} \phi(t, \frac{dm}{d|m|})(t))\, d|m_k|(t).$$

For each $k \in \mathbb{N}$, we consider the measure $\nu_k \in \mathcal{M}^+(\mathbb{T} \times \overline{B}_{\mathbb{E}^*}, \mathbb{R})$, image of $|m_k|$ by the map $t \mapsto (t, (dm_k/d|m_k|(t)), \mathbb{T} \to \mathbb{T} \times \overline{B}_{\mathbb{E}^*}$. We have ([DM75, III(73) page 128]),

$$\nu_k = \int_{\mathbb{T}} \delta_t \otimes \delta_{dm_k/d|m_k|(t)}\, d|m_k|(t).$$

Since $(m_k)_k$ is bounded and since $\|\nu_k\| = \int d|m_k| = \|m_k\|$, the sequence $(\nu_k)_k$ is bounded too. Moreover, since the sequence $(m_k)_k$ is tight, for every $\varepsilon > 0$, there exists a compact subset K of $\mathbb{T}$ such that $\sup_k |m_k|(\mathbb{T} \setminus K) \leq \varepsilon$. It follows that

$$\nu_k([\mathbb{T} \times \overline{B}_{\mathbb{E}^*}] \setminus [K \times \overline{B}_{\mathbb{E}^*}]) = \nu_k([\mathbb{T} \setminus K] \times \overline{B}_{\mathbb{E}^*}) = |m_k|(\mathbb{T} \setminus K).$$

And so $\sup_k \nu_k([\mathbb{T} \times \overline{B}_{\mathbb{E}^*}] \setminus [K \times \overline{B}_{\mathbb{E}^*}]) \leq \varepsilon$. Since $K \times \overline{B}_{\mathbb{E}^*}$ is compact, this proves that $(\nu_k)_k$ is tight. Thus, by Prohorov's theorem (Theorem 8.2.1), it is relatively weakly compact in the space $\mathcal{M}^+(\mathbb{T} \times \overline{B}_{\mathbb{E}^*}, \mathbb{R})$ of non-negative bounded Radon measures on $\mathbb{T} \times \overline{B}_{\mathbb{E}^*}$; observe that $\mathcal{M}^+(\mathbb{T} \times \overline{B}_{\mathbb{E}^*}, \mathbb{R})$ equipped with the weak topology is a Polish space: see [Bou69, §5 Proposition 10]. Therefore, there exist a subsequence $(\nu_{k_p})_p$ of $(\nu_k)_k$ and a measure $\nu \in \mathcal{M}^+(\mathbb{T} \times \overline{B}_{\mathbb{E}^*}, \mathbb{R})$ such that $(\nu_{k_p})_p$ converges weakly to ν. Since ϕ is non-negative and lower semicontinuous on $\mathbb{T} \times \overline{B}_{\mathbb{E}^*}$, the map $\tau \longmapsto \int \phi\, d\tau$ defined on $\mathcal{M}^+(\mathbb{T} \times \overline{B}_{\mathbb{E}^*}, \mathbb{R})$ is narrowly lower semicontinuous [DM75, Théorème III.55]. Then we obtain

$$\begin{aligned} a = \lim_k \int_{\mathbb{T}} \phi(t, \frac{dm_k}{d|m_k|}(t))\, d|m_k|(t) &= \lim_p \int_{\mathbb{T} \times \overline{B}_{\mathbb{E}^*}} \phi(t,x)\, d\nu_{k_p}(t,x) \\ &\geq \int_{\mathbb{T} \times \overline{B}_{\mathbb{E}^*}} \phi(t,x)\, d\nu(t,x). \end{aligned} \tag{8.2.1}$$

Let μ be the projection of ν onto $\mathbb{T}$ defined by

$$\mu = \int_{\mathbb{T}\times\overline{B}_{\mathbb{E}^*}} \delta_t \, d\nu(t,x),$$

that is,

$$\mu(A) = \nu(A \times \overline{B}_{\mathbb{E}^*})$$

for every $A \in \mathcal{B}_{\mathbb{T}}$. Thanks to a result on disintegration of measures ([SP75], [Val72, Théorème 9], [Val73, Théorème 2], [CV77, VII.17], [IT69]), there is a μ-measurable function $\lambda : t \longmapsto \lambda_t$ from $\mathbb{T}$ into the space $\mathcal{M}^{+,1}(\overline{B}_{\mathbb{E}^*})$ of Radon probabilities on $\overline{B}_{\mathbb{E}^*}$ equipped with the narrow topology, such that

$$\nu = \int_{\mathbb{T}} \delta_t \otimes \lambda_t \, d\mu(t).$$

Coming back to (8.2.1), we find

$$a \geq \int_{\mathbb{T}\times\overline{B}_{\mathbb{E}^*}} \phi(t,x)\, d\nu(t,x) = \int_{\mathbb{T}}\int_{\overline{B}_{\mathbb{E}^*}} \phi(t,x)\, d\lambda_t(x)\, d\mu(t). \tag{8.2.2}$$

Let now $\operatorname{bar}(\lambda_t) \in \overline{B}_{\mathbb{E}^*}$ be the barycenter of λ_t defined by

$$\operatorname{bar}(\lambda_t) = \int_{\overline{B}_{\mathbb{E}^*}} x \, d\lambda_t(x).$$

Since $\operatorname{bar}(\lambda) : \mathbb{T} \longrightarrow \overline{B}_{\mathbb{E}^*}, t \longmapsto \operatorname{bar}(\lambda_t)$ is measurable and for every $t \in \mathbb{T}$, the function $\phi(t,.)$ is convex and lower semicontinuous on $\overline{B}_{\mathbb{E}^*}$, by Jensen's inequality we have

$$\int_{\overline{B}_{\mathbb{E}^*}} \phi(t,x)\, d\lambda_t(x) \geq \phi(t, \operatorname{bar}(\lambda_t)), \quad \forall t \in \mathbb{T}.$$

It follows from (8.2.2) that

$$a \geq \int_{\mathbb{T}} \phi(t, \operatorname{bar}(\lambda_t))\, d\mu(t). \tag{8.2.3}$$

Let us show now that $m = (\operatorname{bar}(\lambda))\,\mu$. Let $f \in \mathrm{C}_b(\mathbb{T},\mathbb{E})$. For each $k \in \mathbb{N}$, we have

$$\int_{\mathbb{T}} f(t)\, dm_k(t) = \int_{\mathbb{T}} \langle f(t), \frac{dm_k}{d|m_k|}(t)\rangle \, d|m_k|(t) = \int_{\mathbb{T}\times\overline{B}_{\mathbb{E}^*}} \langle f(t), x\rangle \, d\nu_k(t,x).$$

On the other hand, since $(m_{k_p})_p$ narrowly converges to m, we have

$$\int_{\mathbb{T}} f(t)\, dm(t) = \lim_p \int_{\mathbb{T}} f(t)\, dm_{k_p}(t).$$

Moreover, since $\langle f(.), .\rangle \in C_b(\mathbb{T} \times \overline{B}_{\mathbb{E}^*}, \mathbb{R})$ and the sequence $(\nu_{k_p})_p$ narrowly converges to ν, we have

$$\int_{\mathbb{T}\times\overline{B}_{\mathbb{E}^*}} \langle f(t), x\rangle\, d\nu(t,x) = \lim_p \int_{\mathbb{T}\times\overline{B}_{\mathbb{E}^*}} \langle f(t), x\rangle\, d\nu_{k_p}(t,x).$$

Then, we put together the three previous equalities to obtain

$$\begin{aligned}\int_{\mathbb{T}} f(t)\, dm(t) &= \int_{\mathbb{T}\times\overline{B}_{\mathbb{E}^*}} \langle f(t), x\rangle\, d\nu(t,x) = \int_{\mathbb{T}}\int_{\overline{B}_{\mathbb{E}^*}} \langle f(t), x\rangle\, d\lambda_t(x)\, d\mu(t)\\ &= \int_{\mathbb{T}} \langle f(t), \int_{\overline{B}_{\mathbb{E}^*}} x\, d\lambda_t(x)\rangle\, d\mu(t) = \int_{\mathbb{T}} \langle f(t), \mathrm{bar}\,(\lambda_t)\rangle\, d\mu(t)\\ &= \int_{\mathbb{T}} f(t)\, d(\mathrm{bar}\,(\lambda)\,\mu)(t).\end{aligned}$$

This proves that $m = \mathrm{bar}\,(\lambda)\,\mu$ and thus $\dfrac{dm}{d\mu} = \mathrm{bar}\,(\lambda)$. We deduce now from (8.2.3) that

$$a \geq \int_{\mathbb{T}} \phi(t, \mathrm{bar}\,(\lambda_t))\, d\mu(t) = \int_{\mathbb{T}} \phi(t, \frac{dm}{d\mu}(t))\, d\mu(t). \tag{8.2.4}$$

Let $|\,\mathrm{bar}\,(\lambda)\,| : \mathbb{T} \longrightarrow [0,1], t \longmapsto \|\,\mathrm{bar}\,(\lambda_t)\,\|$. Since μ is a non-negative measure on $\mathbb{T}$, we have $d|m| = |\,\mathrm{bar}\,(\lambda)\,|\, d\mu$ and so

$$dm = \frac{dm}{d\,|m|}\frac{d\,|m|}{d\mu}\, d\mu = \frac{dm}{d\,|m|}|\,\mathrm{bar}\,(\lambda)\,|\, d\mu.$$

We obtain for almost every $t \in \mathbb{T}$,

$$\phi(t, \frac{dm}{d\mu}(t)) = \phi(t, \|\,\mathrm{bar}\,(\lambda_t)\,\|\frac{dm}{d|m|}(t)) = \|\,\mathrm{bar}\,(\lambda_t)\,\|\phi(t, \frac{dm}{d|m|}(t)).$$

Finally, with (8.2.4), we find

$$\begin{aligned}a &\geq \int_{\mathbb{T}} \phi(t, \frac{dm}{d\mu}(t))\, d\mu(t) = \int_{\mathbb{T}} \phi(t, \frac{dm}{d|m|}(t))\,\|\,\mathrm{bar}\,(\lambda_t)\,\|\, d\mu(t)\\ &= \int_{\mathbb{T}} \phi(t, \frac{dm}{d|m|}(t))\, d(|\,\mathrm{bar}\,(\lambda)\,|\mu)(t) = \int_{\mathbb{T}} \phi(t, \frac{dm}{d|m|}(t))\, d|m|(t).\end{aligned}$$

This proves the result. □

Remark 8.2.3 1) Reshetnyak's theorem shows that the functional

$$I_\phi : \mathcal{M}(\mathbb{T}, \mathbb{E}^*) \longrightarrow \mathbb{R}^+,\ m \longmapsto \int_{\mathbb{T}} \phi(t, \frac{dm}{d|m|}(t))\, d|m|(t)$$

is narrowly sequentially lower semicontinuous on the bounded tight subsets of $\mathcal{M}(\mathbb{T}, \mathbb{E}^*)$.

2) When $\mathbb{T}$ is a locally compact topological space, the first author proves in [Cas87] the same result that we present below.

In the following $\mathbb{T}$ is a locally compact space, $C_{\mathcal{K}}(\mathbb{T}, \mathbb{E})$ is the space of all continuous mappings from $\mathbb{T}$ into $\mathbb{E}$ with compact support, $\mathcal{M}^+(\mathbb{T})$ is the space of positive Radon measures on $\mathbb{T}$ equipped with the *vague topology* $\sigma(\mathcal{M}^+(\mathbb{T}), C_{\mathcal{K}}(\mathbb{T}, \mathbb{R}))$ where $C_{\mathcal{K}}(\mathbb{T}, \mathbb{R})$ is the space of all real valued continuous functions defined on $\mathbb{T}$ with compact support. We endow the space $\mathcal{M}(\mathbb{T}, \mathbb{E}^*)$ of all $\mathbb{E}^*$-valued measures m on $\mathbb{T}$ with bounded variation with the *vague topology* $\sigma(\mathcal{M}(\mathbb{T}, \mathbb{E}^*), C_{\mathcal{K}}(\mathbb{T}, \mathbb{E}))$. Now we proceed to a second type of Reshetnyak's theorem in the locally compact framework. Although the proof follows some lines in Reshetnyak's theorem, we need to be a bit more careful since we work with the vague topology, in contrast to Theorem 8.2.2 dealing with the narrow topology. So, in order to avoid any risk of confusion, we will produce the proof of this variant with full details.

Theorem 8.2.4 *Let $\phi : \mathbb{T} \times \overline{B}_{\mathbb{E}^*} \longrightarrow [0, +\infty]$ be a lower semicontinuous function on $\mathbb{T} \times \overline{B}_{\mathbb{E}^*}$ such that for all $t \in \mathbb{T}$, $\phi(t, .))$ is convex and positively 1-homogeneous on $\overline{B}_{\mathbb{E}^*}$ i.e. for all $x \in \overline{B}_{\mathbb{E}^*}$ and all $\lambda \in [0,1]$, $\phi(t, \lambda x) = \lambda \phi(t, x)$. Let $(m_k)_k$ be a bounded sequence in $\mathcal{M}(\mathbb{T}, \mathbb{E}^*)$ which vaguely converges to $m \in \mathcal{M}(\mathbb{T}, \mathbb{E}^*)$. Then we have*

$$\liminf_k \int_{\mathbb{T}} \phi(t, \frac{dm_k}{d|m_k|}(t))\, d|m_k|(t) \geq \int_{\mathbb{T}} \phi(t, \frac{dm}{d|m|}(t))\, d|m|(t).$$

Proof. Extracting a subsequence, we can suppose that

$$(8.2.5) \qquad a = \liminf_k \int_{\mathbb{T}} \phi(t,, \frac{dm_k}{d|m_k|}(t))\, d|m_k|(t) = \lim_k \int_{\mathbb{T}} \phi(t, \frac{dm}{d|m|}(t))\, d|m_k|(t).$$

For each $k \in \mathbb{N}$, we consider the measure $\nu_k \in \mathcal{M}^+(\mathbb{T} \times \overline{B}_{\mathbb{E}^*}, \mathbb{R})$, image of $|m_k|$ by the map $t \mapsto (t, (dm_k/d|m_k|(t)), \mathbb{T} \to \mathbb{T} \times \overline{B}_{\mathbb{E}^*}$. We have ([DM75, III(73)]),

$$\nu_k = \int_{\mathbb{T}} \delta_t \otimes \delta_{dm_k/d|m_k|(t)}\, d|m_k|(t).$$

Since $(m_k)_k$ is bounded and since $\|\nu_k\| = \int d|m_k| = \|m_k\|$, the sequence $(\nu_k)_k$ is bounded too. Hence $(\nu_k)_k$ is relatively vaguely compact in the space $\mathcal{M}^+(\mathbb{T} \times \overline{B}_{\mathbb{E}^*})$ of positive bounded Radon measures on the locally compact space $\mathbb{T} \times \overline{B}_{\mathbb{E}^*}$. So, there exist a filter $\mathcal{F}$ finer than the Fréchet filter and a positive Radon measure $\nu \in \mathcal{M}^+(\mathbb{T} \times \overline{B}_{\mathbb{E}^*})$ such that

$$\sigma(\mathcal{M}^+(\mathbb{T} \times \overline{B}_{\mathbb{E}^*}), C_{\mathcal{K}}(\mathbb{T} \times \overline{B}_{\mathbb{E}^*}, \mathbb{R}))\text{-}\lim_{\mathcal{F}} \nu_k = \nu.$$

Using the lower semicontinuity of ϕ and (8.2.5) we get

$$a = \lim_k \int_{\mathbb{T}} \phi(t, , \frac{dm_k}{d|m_k|}(t))\, d|m_k|(t) = \lim_k \int_{\mathbb{T}\times \overline{B}_{\mathbb{E}^*}} \phi(t,s)\, d\nu_k(t,s)$$
$$\geq \int_{\mathbb{T}\times \overline{B}_{\mathbb{E}^*}} \phi(t,s)\, d\nu_k(t,s).$$

Let μ be the projection of ν onto $\mathbb{T}$ defined by

$$\mu = \int_{\mathbb{T}\times \overline{B}_{\mathbb{E}^*}} \delta_t\, d\nu(t,x).$$

As in the proof of Theorem 8.2.2 thanks to a result on disintegration of measures, there is a μ-measurable function $\lambda : t \longmapsto \lambda_t$ from $\mathbb{T}$ into the space $\mathcal{M}^{+,1}(\overline{B}_{\mathbb{E}^*})$ of Radon probabilities on $\overline{B}_{\mathbb{E}^*}$ equipped with the vague topology $\sigma(\mathcal{M}^{+,1}(\overline{B}_{\mathbb{E}^*}), \mathrm{C}(\overline{B}_{\mathbb{E}^*}))$ such that

$$\nu = \int_{\mathbb{T}} \delta_t \otimes \lambda_t\, d\mu(t).$$

Here $\mathrm{C}(\overline{B}_{\mathbb{E}^*})$ is the space of all real-valued continous function defined on the $\sigma(\mathbb{E}^*, \mathbb{E})$ compact set $\overline{B}_{\mathbb{E}^*}$. Coming back to (8.2.5) we find

$$a \geq \int_{\mathbb{T}\times \overline{B}_{\mathbb{E}^*}} \phi(t,x)\, d\nu(t,x) = \int_{\mathbb{T}} \int_{\overline{B}_{\mathbb{E}^*}} \phi(t,x)\, d\lambda_t(x)\, d\mu(t). \tag{8.2.6}$$

Let now $\mathrm{bar}\,(\lambda_t) \in \overline{B}_{\mathbb{E}^*}$ be the barycenter of λ_t defined by

$$\mathrm{bar}\,(\lambda_t) = \int_{\overline{B}_{\mathbb{E}^*}} x\, d\lambda_t(x).$$

Since the mapping $t \longmapsto \mathrm{bar}\,(\lambda_t)$ is measurable and for every $t \in \mathbb{T}$, the function $\phi(t, .)$ is convex and lower semicontinuous on $\overline{B}_{\mathbb{E}^*}$, by Jensen's inequality we have

$$\int_{\overline{B}_{\mathbb{E}^*}} \phi(t,x)\, d\lambda_t(x) \geq \phi(t, \mathrm{bar}\,(\lambda_t)), \ \forall t \in \mathbb{T}.$$

It follows from (8.2.6) that

$$a \geq \int_{\mathbb{T}} \phi(t, \mathrm{bar}\,(\lambda_t))\, d\mu(t). \tag{8.2.7}$$

Let us show now that $m = \mathrm{bar}\,(\lambda)\,\mu$. Let $f \in \mathrm{C}_{\mathcal{K}}(\mathbb{T}, \mathbb{E})$. For each $k \in \mathbb{N}$, we have

$$\int_{\mathbb{T}} f(t)\, dm_k(t) = \int_{\mathbb{T}} \langle f(t), \frac{dm_k}{d|m_k|}(t)\rangle\, d|m_k|(t) = \int_{\mathbb{T}\times \overline{B}_{\mathbb{E}^*}} \langle f(t), x\rangle\, d\nu_k(t,x).$$

On the other hand, since $(m_{k_p})_p$ vaguely converges to m, we have

$$\int_{\mathbb{T}} f(t)\, dm(t) = \lim_p \int_{\mathbb{T}} f(t)\, dm_{k_p}(t).$$

Moreover, since $\langle f(.), . \rangle \in \mathrm{C}_{\mathcal{K}}(\mathbb{T} \times \overline{B}_{\mathbb{E}^*}, \mathbb{R})$ and the sequence $(\nu_{k_p})_p$ vaguely converges to ν, we have

$$\int_{\mathbb{T}\times \overline{B}_{\mathbb{E}^*}} \langle f(t), x \rangle \, d\nu(t,x) = \lim_p \int_{\mathbb{T}\times \overline{B}_{\mathbb{E}^*}} \langle f(t), x \rangle \, d\nu_{k_p}(t,x).$$

Then, we put together the three previous equalities to obtain

$$\begin{aligned}
\int_{\mathbb{T}} f(t)\, dm(t) &= \int_{\mathbb{T}\times \overline{B}_{\mathbb{E}^*}} \langle f(t), x \rangle \, d\nu(t,x) = \int_{\mathbb{T}} \int_{\overline{B}_{\mathbb{E}^*}} \langle f(t), x \rangle \, d\lambda_t(x)\, d\mu(t) \\
&= \int_{\mathbb{T}} \langle f(t), \int_{\overline{B}_{\mathbb{E}^*}} x\, d\lambda_t(x) \rangle \, d\mu(t) = \int_{\mathbb{T}} \langle f(t), \mathrm{bar}\,(\lambda_t) \rangle \, d\mu(t) \\
&= \int_{\mathbb{T}} f(t)\, d(\mathrm{bar}\,(\lambda)\, \mu)(t).
\end{aligned}$$

This proves that $m = \mathrm{bar}\,(\lambda)\,\mu$ and thus $\dfrac{dm}{d\mu} = \mathrm{bar}\,(\lambda)$. We deduce now from (8.2.7) that

$$a \geq \int_{\mathbb{T}} \phi(t, \mathrm{bar}\,(\lambda_t))\, d\mu(t) = \int_{\mathbb{T}} \phi(t, \frac{dm}{d\mu}(t))\, d\mu(t). \tag{8.2.8}$$

Let $|\,\mathrm{bar}\,(\lambda)\,| : \mathbb{T} \longrightarrow [0,1], t \longmapsto \|\,\mathrm{bar}\,(\lambda_t)\,\|$. Since μ is a non-negative measure on $\mathbb{T}$, we have $d|m| = |\,\mathrm{bar}\,(\lambda)\,|\, d\mu$ and so

$$dm = \frac{dm}{d|m|}\frac{d|m|}{d\mu}\, d\mu = \frac{dm}{d|m|} |\,\mathrm{bar}\,(\lambda)\,|\, d\mu.$$

We obtain for almost every $t \in \mathbb{T}$,

$$\phi(t, \frac{dm}{d\mu}(t)) = \phi(t, \|\,\mathrm{bar}\,(\lambda_t)\,\| \frac{dm}{d|m|}(t)) = \|\,\mathrm{bar}\,(\lambda_t)\,\|\, \phi(t, \frac{dm}{d|m|}(t)).$$

Finally, with (8.2.8), we find

$$\begin{aligned}
a &\geq \int_{\mathbb{T}} \phi(t, \frac{dm}{d\mu}(t))\, d\mu(t) = \int_{\mathbb{T}} \phi(t, \frac{dm}{d|m|}(t)) \|\,\mathrm{bar}\,(\lambda_t)\,\|\, d\mu(t) \\
&= \int_{\mathbb{T}} \phi(t, \frac{dm}{d|m|}(t))\, d(|\,\mathrm{bar}\,(\lambda)\,|\mu)(t) = \int_{\mathbb{T}} \phi(t, \frac{dm}{d|m|}(t))\, d|m|(t).
\end{aligned}$$

□

Remark 8.2.5 We refer to [CJ95] for the usefulness of the Reshetnyak-type theorems presented above in the statement of epiconvergence of some integral functionals and the sweeping process. In this vein, using the techniques developed therein, we give below some examples of epiconvergence integral functionals involving the fiber product type limit theorem for Young measures.

Example 8.2.6 Let $(\mathbb{T}, \mathcal{T}_{\mathrm{P}}, \mathrm{P}) = ([0,1], \mathcal{L}_{[0,1]}, dt)$, $\widehat{\mathbb{N}} = \mathbb{N} \cup \{\infty\}$ and $\mathbb{S}$ and Y be two Polish spaces. Let $\{\varphi_k;\, k \in \widehat{\mathbb{N}}\}$ be lower semicontinuous functions on $\mathbb{T}\times\mathbb{S}\times Y$ with values in $[0,+\infty]$ such that, for every sequence $((t_k, x_k, y_k))_k$ in $\mathbb{T}\times\mathbb{S}\times Y$ which converges to (t,x,y), we have

$$\liminf_k \varphi_k(t_k, x_k, y_k) \geq \varphi_\infty(t,x,y). \tag{8.2.9}$$

Let (λ^k) be a tight sequence of Young measures in $\mathcal{Y}^1(\mathbb{T}, \mathcal{T}_{\mathrm{P}}, \mathrm{P}; \mathbb{S})$ which stably converges to $\lambda^\infty \in \mathcal{Y}^1(\mathbb{T}, \mathcal{T}_{\mathrm{P}}, \mathrm{P}; \mathbb{S})$ and let (u_k) be a sequence of continuous functions on $\mathbb{T}$ with values in Y which converges uniformly on $\mathbb{T}$ to a continuous function u_∞. Then we have

$$\liminf_k \int_{\mathbb{T}} \Big[\int_{\mathbb{S}} \varphi_k(t,s,u_k(t))\, d\lambda_t^k(s)\Big]\, dt \geq \int_{\mathbb{T}} \Big[\int_{\mathbb{S}} \varphi_\infty(t,s,u_\infty(t))\, d\lambda_t^\infty(s)\Big]\, dt.$$

Proof. Remark that the (Alexandroff) one point compactification $\hat{\mathbb{N}}$ of $\mathbb{N}$ is a compact metric space. We define a function $\psi : \mathbb{T} \times \hat{\mathbb{N}} \times S \longrightarrow [0,+\infty]$, by $\psi(t,k,x) = \varphi_k(t,x,u_k(t))$. Let us show that ψ is lower semicontinuous on $\mathbb{T} \times \hat{\mathbb{N}} \times \mathbb{S}$. Let $(t_k, p_k, x_k)_k$ be a sequence in $\mathbb{T} \times \hat{\mathbb{N}} \times \mathbb{S}$ converging to (t,p,x). We set $a = \liminf_k \psi(t_k, p_k, x_k)$. If $p \in \mathbb{N}$, for k large enough, $p_k = p$. Since u_p is continuous, the sequence $(u_p(t_k))_k$ converges to $u_p(t)$ in Y. It follows from the lower semicontinuity of φ_p that

$$\begin{aligned}\liminf_k \varphi(t_k, p_k, x_k) = \liminf_k \varphi_{p_k}(t_k, x_k, u_{p_k}(t_k)) &= \liminf_k \varphi_p(t_k, x_k, u_p(t_k)) \\ &\geq \varphi_p(t,x,u_p(t)) = \psi(t,p,x).\end{aligned}$$

Consider now the case $p = \infty$. Extracting a subsequence, we may assume that $a = \lim_k \psi(t_k, p_k, x_k)$ and that the sequence $(p_k)_k$ is increasing. For each $n \in \mathbb{N}$, we define

$$s_n = \begin{cases} t_1, & \text{if } n < p_1\,, \\ t_k, & \text{if } p_k \leq n < p_{k+1}\,, \end{cases} \quad \text{and} \quad y_n = \begin{cases} x_1, & \text{if } n < p_1\,, \\ x_k, & \text{if } p_k \leq n < p_{k+1}\,. \end{cases}$$

Then, we have

$$t = \lim_n s_n\,, \quad (s_{p_k})_k = (t_k)_k\,, \tag{8.2.10}$$

$$x = \lim_n y_n\,, \quad (y_{p_k})_k = (x_k)_k\,. \tag{8.2.11}$$

Moreover, since u is continuous and $(u_n)_n$ converges uniformly to u on the compact set $\{t, s_n;\ n \in \mathbb{N}\}$, the sequence $(u_n(s_n))_n$ converges to $u(t)$ in Y. Hence, by (8.2.9), we get

$$\begin{aligned}\liminf_k \psi(t_k, p_k, x_k) \geq \liminf_n \psi(s_n, n, y_n) &= \liminf_n \phi_n(s_n, y_n, u_n(s_n)) \\ &\geq \phi_\infty(t, x, u(t)) = \psi(t, \infty, x).\end{aligned}$$

This proves the lower semicontinuity of ψ. For each $k \in \widehat{\mathbb{N}}$, consider the Young measure $\nu^k = \underline{\delta}_k \otimes \lambda^k$. Then $\nu^k \in \mathcal{Y}^1(\mathbb{T}; \widehat{\mathbb{N}} \times \mathbb{S})$. Since the sequence $(\underline{\delta}_k)$ is tight and stably converges to $\underline{\delta}_\infty$ in $\mathcal{M}^b(\widehat{\mathbb{N}}, \mathbb{R})$, the sequence $(\nu^k) = (\underline{\delta}_k \otimes \lambda^k)$ stably converges to $\nu^\infty = \underline{\delta}_\infty \otimes \lambda^\infty$, using Theorem 8.1.1 or Theorem 3.3.1. Applying Portmanteau Theorem 2.1.3 to (ν^k) and the function ψ gives

(8.2.12)

$$\liminf_k \int_{\mathbb{T}} \Big[\int_{\widehat{\mathbb{N}} \times S} \psi(t, n, s)\, d\nu_t^k((n, s))\Big]\, dt \geq \int_{\mathbb{T}} \Big[\int_{\widehat{\mathbb{N}} \times S} \psi(t, n, s)\, d\nu_t^\infty((n, s))\Big]\, dt.$$

Since for each $k \in \widehat{\mathbb{N}}$, we have

$$\int_{\mathbb{T}} \Big[\int_{\widehat{\mathbb{N}} \times S} \psi(t, n, s)\, d\nu_t^k((n, s))\Big]\, dt = \int_{\mathbb{T}} \Big[\int_{\mathbb{S}} \psi(t, k, s)\, d\lambda_t^k(s)\Big]\, dt \tag{8.2.13}$$

So the result follows from (8.2.12) and (8.2.13). □

Example 8.2.7 Let $(\mathbb{T}, \mathcal{T}_{\mathrm{P}}, \mathrm{P}) = ([0,1], \mathcal{L}_{[0,1]}, dt)$, $\widehat{\mathbb{N}} = \mathbb{N} \cup \{\infty\}$ and $\mathbb{S}$ be a compact metric space. Let $\{\varphi_k;\ k \in \widehat{\mathbb{N}}\}$ be lower semicontinuous functions on $\mathbb{T} \times \mathbb{S}$ with values in $[0, +\infty]$. Assume that there is $\bar{s} \in \mathbb{S}$ such that $\varphi_k(t, \bar{s}) = 0$ for all $(t, k) \in \mathbb{S} \times \widehat{\mathbb{N}}$. And assume that for every $(t, \nu) \in \mathbb{T} \times \mathcal{M}^{+,1}(\mathbb{S})$ ($\mathcal{M}^{+,1}(\mathbb{S})$ being the space of all Probability Radon measures on $\mathbb{S}$ equipped with the vague topology), there is a sequence (ν^k) in $\mathcal{M}^{+,1}(\mathbb{S})$ such that

$$\limsup_k \int_{\mathbb{S}} \varphi_k(t, s)\, d\nu^k(s) \leq \int_{\mathbb{S}} \varphi_\infty(t, s)\, d\nu(s). \tag{8.2.14}$$

Then for every $\lambda \in \mathcal{Y}^1(\mathbb{T}, \mathcal{T}_{\mathrm{P}}, \mathrm{P}; \mathbb{S})$ there exists a sequence of Young measures (λ^k) in $\mathcal{Y}^1(\mathbb{T}, \mathcal{T}_{\mathrm{P}}, \mathrm{P}; \mathbb{S})$ which pointwise vaguely converges to $\lambda_t \in \mathcal{Y}^1(\mathbb{T}, \mathcal{T}_{\mathrm{P}}, \mathrm{P}; \mathbb{S})$ (that is, for every $t \in \mathbb{T}$, λ_t^k vaguely converges to λ_t) and satisfies

$$\limsup_k \int_{\mathbb{T}} \Big[\int_{\mathbb{S}} \varphi_k(t, s)\, d\lambda_t^k(s)\Big]\, dt \leq \int_{\mathbb{T}} \Big[\int_{\mathbb{S}} \varphi_\infty(t, s)\, d\lambda_t(s)\Big]\, dt.$$

Proof. Let $\lambda \in \mathcal{Y}^1(\mathbb{T}, \mathcal{T}_{\mathrm{P}}, \mathrm{P}; \mathbb{S})$. We may assume that

$$\int_{\mathbb{S}} \varphi_\infty(t, s)\, d\lambda_t(s) < +\infty$$

for all $t \in \mathbb{T}$. For each $k \in \widehat{\mathbb{N}}$, let us consider the integrand $\psi_k : \mathbb{T} \times \mathcal{M}^{+,1}(\mathbb{S}) \to [0, +\infty]$ defined by

$$\psi_k(t, \mu) = \Big[\int_{\mathbb{S}} \varphi_k(t, s)\, d\mu(s) - \int_{\mathbb{S}} \varphi_\infty(t, s)\, d\lambda_t(s)\Big]^+.$$

It is clear that ψ_k is $\mathcal{T}_{\mathrm{P}} \otimes \mathcal{B}_{\mathcal{M}^{+,1}(\mathbb{S})}$-measurable on $\mathbb{T} \times \mathcal{M}^{+,1}(\mathbb{S})$ and lower semicontinuous on $\mathcal{M}^{+,1}(\mathbb{S})$. Let us denote by d a distance compatible with the vaguely compact metrizable topology on $\mathcal{M}^{+,1}(\mathbb{S})$. Consider now the multifunction Γ_k on $\mathbb{T}$ with values in the set of all subsets of $\mathcal{M}^{+,1}(\mathbb{S})$ defined by

$$\Gamma_k(t) = \{\nu \in \mathcal{M}^{+,1}(\mathbb{S});\ d(\lambda_t, \nu) + \psi_k(t, \nu) = \min_{\nu \in \mathcal{M}^{+,1}(\mathbb{S})} \{d(\lambda_t, \nu) + \psi_k(t, \nu)\}\}.$$

Since $\mathcal{M}^{+,1}(\mathbb{S})$ is vaguely compact set, it is clear that for all $t \in \mathbb{T}$, $\Gamma_k(t)$ is non empty. Then, by [CV77, III.39], there exists a $(\mathcal{T}_{\mathrm{P}}, \mathcal{B}_{\mathcal{M}^{+,1}(\mathbb{S})})$-measurable selection λ^k of Γ_k. For every $t \in \mathbb{T}$, by (8.2.14), there exists a sequence $(\nu_k)_k$ in $\mathcal{M}^{+,1}(\mathbb{S})$ vaguely converging to λ_t such that $\lim_k \psi_k(t, \nu_k) = 0$. Since

$$d(\lambda_t, \lambda_t^k) + \psi_k(t, \lambda_t^k) \le d(\lambda_t, \nu_k) + \psi_k(t, \nu_k),$$

we get

$$\lim_k d(\lambda_t, \lambda_t^k) = 0 \text{ and } \lim_k \psi_k(t, \lambda_t^k) = 0.$$

On the other hand, since $\varphi_k(t, \bar{s}) = 0$ by our assumption, we have $\psi_k(t, \delta_{\bar{s}}) = 0$ and thus

$$d(\lambda_t, \lambda_t^k) + \psi_k(t, \lambda_t^k) \le d(\lambda_t, \delta_{\bar{s}}) + \psi_k(t, \delta_{\bar{s}}) \le M,$$

where $M = \sup\{d(\mu, \nu);\ (\mu, \nu) \in \mathcal{M}^{+,1}(\mathbb{S}) \times \mathcal{M}^{+,1}(\mathbb{S})\}$. So by Lebesgue's dominated convergence theorem, we deduce that $(\psi_k(., \lambda_t^k))_k$ converges to 0 in the space $\mathrm{L}^1([0,1], \mathcal{L}_{[0,1]}, dt)$. Thus we have

$$\lim_k \int_{\mathbb{T}} \Big[\int_{\mathbb{S}} \varphi_k(t, s)\, d\lambda_t^k(s) - \int_{\mathbb{S}} \varphi_\infty(t, s)\, d\lambda_t(s)\Big]^+ dt = 0,$$

hence

$$\limsup_k \int_{\mathbb{T}} \Big[\int_{\mathbb{S}} \varphi_k(t, s)\, d\lambda_t^k(s)\Big]\, dt \le \int_{\mathbb{T}} \Big[\int_{\mathbb{S}} \varphi_\infty(t, s)\, d\lambda_t(s)\Big]\, dt.$$

That completes the proof. □

8.3 Some new applications of the Fiber Product Lemma for Young measures

This section is essentially taken from [CRdF04]. It is devoted to the study of the value functions of a control problem where the dynamic is governed by an ordinary differential equation (ODE) where the controls are Young measures. Here the stable convergence for the fiber product of Young measures is crucial in the statement of the variational properties of the value functions in the control problems under consideration and the developments of Mathematical Economics (see e.g. [Tat02]). References for control problems are e.g. [EK72, Ell87, ES84, KS88, BJ91].

Here, $\mathbb{E} = \mathbb{R}^d$ is a finite dimensional space and $[0,1]$ is equipped with the Lebesgue measure.

8.3.A The value function of a control problem governed by a first order ordinary differential equation

In this section we present a study of the value function of a control problem where the controls are Young measures. As the proofs are rather long, we do not make weak assumptions on the Control spaces but we only focus on the main ideas in order to present some sharp applications of the fiber product for Young measures presented above. Namely we assume here that $\mathbb{S}$ and $\mathbb{Z}$ are metric compact spaces. Let $k(\mathbb{Z})$ be the set of all compact subsets of $\mathbb{Z}$, $\Gamma : [0,1] \to k(\mathbb{Z})$ be a compact valued Lebesgue measurable multifunction from $[0,1]$ to $\mathbb{Z}$. It is well-known that $\mathcal{M}^{+,1}(\mathbb{S})$ (resp. $\mathcal{M}^{+,1}(\mathbb{Z})$ is a compact metrizable space for the $\sigma(\mathrm{C}^*(\mathbb{S}), \mathrm{C}(\mathbb{S}))$ (resp. $\sigma(\mathrm{C}(\mathbb{Z})^*, \mathrm{C}(\mathbb{Z}))$)–topology.

Let us consider a mapping $f : [0,1] \times \mathbb{E} \times \mathbb{S} \times \mathbb{Z} \to \mathbb{E}$ satisfying:

(i) For every fixed $t \in [0,1]$, $f(t,.,.,.)$ is continuous on $\mathbb{E} \times \mathbb{S} \times \mathbb{Z}$;

(ii) For every $(x,s,z) \in \mathbb{E} \times \mathbb{S} \times \mathbb{Z}$, $f(.,x,s,z)$ is Lebesgue-measurable on $[0,1]$;

(iii) There is a positive Lebesgue integrable function c such that $f(t,x,s,z) \in c(t)\overline{B}_{\mathbb{E}}$ for all (t,x,s,z) in $[0,1] \times \mathbb{E} \times \mathbb{S} \times \mathbb{Z}$;

(iv) There exists a Lipschitz constant λ such that

$$||f(t,x_1,s,z) - f(t,x_2,s,z)|| \leq \lambda ||x_1 - x_2||$$

for all $(t,x_1,s,z), (t,x_2,s,z) \in [0,1] \times \mathbb{E} \times \mathbb{S} \times \mathbb{Z}$.

We consider the absolutely continuous solutions set of the following ordinary differential equations (ODE)

$$(\mathcal{I}_{M,\mathcal{H},\mathcal{O}}) \begin{cases} \dot{u}_{x,\mu,\zeta}(t) = \int_{\mathbb{S}} f(t, u_{x,\mu,\zeta}(t), s, \zeta(t))\, d\mu_t(s) \\ u_{x,\mu,\zeta}(0) = x \in M \subset \mathbb{E} \end{cases}$$

where ζ belongs to the set S_Γ of all original controls, which means that ζ is a Lebesgue-measurable mapping from $[0,1]$ into $\mathbb{Z}$ with $\zeta(t) \in \Gamma(t)$ for a.e. $t \in [0,1]$, and M is a compact subset of $\mathbb{E}$, $\mu \in \mathcal{H}$, where $\mathcal{H}$ is a subset in the space of Young measures $\mathcal{Y}^1([0,1];\mathbb{S})$ defined on $\mathbb{S}$, and

$$(\mathcal{I}_{M,\mathcal{H},\mathcal{R}}) \begin{cases} \dot{u}_{x,\mu,\nu}(t) = \int_{\Gamma(t)} [\int_{\mathbb{S}} f(t, u_{x,\mu,\nu}(t), s, z)\, d\mu_t(s)]\, d\nu_t(z) \\ u_{x,\mu,\nu}(0) = x \in M \subset \mathbb{E} \end{cases}$$

where ν belongs to the set $\mathcal{R}$ of all relaxed controls, which means that ν is a Lebesgue-measurable selection of the multifunction Σ defined by

$$\Sigma(t) := \{\sigma \in \mathcal{M}^{+,1}(\mathbb{Z});\ \sigma(\Gamma(t)) = 1\}$$

for all $t \in [0,1]$. These assumptions are sufficient to guarantee that for each $(x, \mu, \nu) \in M \times \mathcal{H} \times \mathcal{R}$ there is a unique absolutely continuous solution $u_{x,\mu,\nu}$ for the ODE under consideration on the interval $[0,1]$ with $u_{x,\mu,\nu}(0) = x \in M \subset \mathbb{E}$.

Theorem 8.3.1 *Assume that $J : [0,1] \times \mathbb{E} \times \mathbb{S} \times \mathbb{Z} \to \mathbb{R}$ is an L^1-bounded Carathéodory integrand, (that is, $J(t,.,.,.)$ is continuous on $\mathbb{E}\times\mathbb{S}\times\mathbb{Z}$ for every $t \in [0,1]$ and $J(.,x,s,z)$ is Lebesgue-measurable on $[0,1]$, for every $(x,s,z) \in \mathbb{E}\times\mathbb{S}\times\mathbb{Z}$) which satisfies the condition: There is an integrable function $\varphi \in \mathrm{L}^1_{\mathbb{R}^+}([0,1]$ such that $|J(t,x,s,z)| \le \varphi(t)$ for all $(t,x,s,z) \in [0,1] \times \mathbb{E} \times \mathbb{S} \times \mathbb{Z}$. Assume further that $\mathcal{H}$ is compact for the convergence in probability. Let us consider the control problems*

$$(P_{M,\mathcal{H},\mathcal{O}}) : \inf_{(x,\mu,\zeta)\in M\times\mathcal{H}\times \mathrm{S}_\Gamma} \int_0^1 [\int_{\mathbb{S}} J(t, u_{x,\mu,\zeta}(t), s, \zeta(t))\, d\mu_t(s)]\, dt$$

and

$$(P_{M,\mathcal{H},\mathcal{R}}) : \inf_{(x,\mu,\nu)\in M\times\mathcal{H}\times\mathcal{R}} \int_0^1 [\int_{\mathbb{Z}} [\int_{\mathbb{S}} J(t, u_{x,\mu,\nu}(t), s, z)\, d\mu_t(s)]\, d\nu_t(z)]\, dt$$

where $u_{x,\mu,\zeta}$ (resp. $u_{x,\mu,\nu}$) is the unique solution associated with (x,μ,ζ) (resp. (x,μ,ν) to the ODE $(\mathcal{I}_{M,\mathcal{H},\mathcal{O}})$ (resp. $(\mathcal{I}_{M,\mathcal{H},\mathcal{R}})$. Then one has $\inf(P_{M,\mathcal{H},\mathcal{O}}) = \min(P_{M,\mathcal{H},\mathcal{R}})$.

Proof.

Claim 1. The graph of the single valued mapping $(x,\mu,\nu) \mapsto u_{x,\mu,\nu}$ defined on the compact space $M \times \mathcal{H} \times \mathcal{R}$ with value in the Banach space $\mathrm{C}_{\mathbb{E}}([0,1])$ endowed with the topology of the sup-norm is compact.

It is obvious that the solution $u_{x,\mu,\nu}$ for the ODE under consideration is given explicitely by

$$u_{x,\mu,\nu}(t) = x + \int_0^t [\int_{\mathbb{Z}} [\int_{\mathbb{S}} f(\tau, u_{x,\mu,\nu}(\tau), s, z)\, d\mu_\tau(s)]\, d\nu_\tau(z)]dt$$

for each $t \in [0,1]$. Let (x^n, μ^n, ν^n) be a sequence in $M \times \mathcal{H} \times \mathcal{R}$ and let u_{x^n,μ^n,ν_n} be the unique absolutely continuous solution to

$$\begin{cases} \dot{u}_{x^n,\mu^n,\nu^n}(t) = \int_{\mathbb{Z}}[\int_{\mathbb{S}} f(t, u_{x^n,\mu^n,\nu^n}(t), s, z)\, d\mu^n_t(s)]\, d\nu^n_t(z) \text{ a.e.} \\ u_{x^n,\mu^n,\nu^n}(0) = x^n \in M. \end{cases}$$

Since M is compact we may suppose that (x^n) converges to a point $x^\infty \in M$. Taking into account the assumption on f, it is easily seen that the sequence

(u_{x^n,μ^n,ν^n}) is relatively compact in $C_{\mathbb{E}}([0,1])$. We may suppose, by extracting subsequences, that (u_{x^n,μ^n,ν^n}) converges uniformly to an absolutely continuous function $u^\infty(.)$ with $u^\infty(0) = x^\infty$ and $(\dot{u}_{x^n,\mu^n,\nu^n})$ converges $\sigma(L^1_{\mathbb{E}}, L^\infty_{\mathbb{E}})$ to $\dot{u}^\infty$. We may also assume that (μ^n) converges in probability to $\mu^\infty \in \mathcal{H}$. Further the sequence (ν^n) of Young measures is relatively compact for the stable topology on the space $\mathcal{Y}^1([0,1];\mathbb{Z})$ of Young measures, and hence by extracting a subsequence, we may suppose that (ν^n) stably converges to a Young measure ν^∞ with $\nu^\infty_t(\Gamma(t)) = 1$ a.e. So, in view of Theorem 3.3.1, $\underline{\delta}_{u_{x^n,\mu^n,\nu^n}} \underline{\otimes}\, \mu^n \,\underline{\otimes}\, \nu^n$ stably converges to $\underline{\delta}_{u^\infty} \underline{\otimes}\, \mu^\infty \,\underline{\otimes}\, \nu^\infty$. Let $h \in \mathrm{L}^\infty_{\mathbb{E}}([0,1])$. It is clear that the function $L : (t,x,s,z) \mapsto \langle h(t), f(t,x,s,z)\rangle$ is an L^1- bounded Carathéodory integrand defined on the compact space $[0,1] \times M \times \mathbb{S} \times \mathbb{Z}$, namely, $|L(t,x,s,z)| \leq h(t)c(t)$ for all $(t,x,s,z) \in [0,1] \times M \times \mathbb{S} \times \mathbb{Z}$, using condition (iii). Consequently, by the stable convergence of $(\underline{\delta}_{u_{x^n,\mu^n,\nu^n}} \underline{\otimes}\, \mu^n \,\underline{\otimes}\, \nu^n)$ to $\underline{\delta}_{u^\infty} \underline{\otimes}\, \mu^\infty \,\underline{\otimes}\, \nu^\infty$, we get

$$\lim_{n\to\infty} \int_0^1 [\int_{\mathbb{Z}} [\int_{\mathbb{S}} \langle h(t), f(t, u_{x^n,\mu^n,\nu^n}(t), s, z)\rangle\, \mu^n_t(s)]\, d\nu^n_t(z)]\, dt$$
$$= \lim_{n\to\infty} \int_0^1 \langle h(t), v_n(t)\rangle\, dt = \int_0^1 [\int_{\mathbb{Z}} [\int_{\mathbb{S}} \langle h(t), f(t, u^\infty(t), s, z)\rangle\, d\mu^\infty_t(s)]\, d\nu^\infty_t(z)]\, dt$$
$$= \int_0^1 \langle h(t), v_\infty(t)\rangle\, dt$$

where, for notational convenience,

$$v^n(t) = \int_{\mathbb{Z}} [\int_{\mathbb{S}} f(t, u_{x^n,\mu^n,\nu^n}(t), s, z)\, d\mu^n_t(s)]\, d\nu^n_t(z),\ \forall t \in [0,1],$$

and

$$v^\infty(t) = \int_{\mathbb{Z}} [\int_{\mathbb{S}} f(t, u^\infty(t), s, z)\, d\mu^\infty_t(s)]\, d\nu^\infty_t(z),\ \forall t \in [0,1].$$

Hence (v^n) weakly converges in $\mathrm{L}^1_{\mathbb{E}}([0,1])$ to v^∞. Using the weak convergence in $\mathrm{L}^1_{\mathbb{E}}([0,1])$ of $(\dot{u}_{x^n,\mu^n,\nu^n})$ to $\dot{u}^\infty$, and the preceding limit, we get

$$\dot{u}^\infty(t) = \int_{\mathbb{Z}} [\int_{\mathbb{S}} f(t, u^\infty(t), s, z)\, d\mu^\infty_t(s)]\, d\nu^\infty_t(z) \text{ a.e..}$$

So, we have necessarily $u^\infty(.) = u_{x^\infty,\mu^\infty,\nu^\infty}(.)$, where $u_{x^\infty,\mu^\infty,\nu^\infty}$ is the unique absolutely continuous solution of the ODE $(\mathcal{I}_{M,\mathcal{H},\mathcal{R}})$ associated with $(x^\infty, \mu^\infty, \nu^\infty)$. Hence Claim 1 is proved.

Claim 2. $\inf(P_{M,\mathcal{H},\mathcal{O}}) = \min(P_{M,\mathcal{H},\mathcal{R}})$.

As a consequence of Claim 1, the solutions set $\{u_{x,\mu,\nu} : (x,\mu,\nu) \in M \times \mathcal{H} \times \mathcal{R}\}$ is compact for the topology of uniform convergence. Since $\mathcal{O}$ is dense in $\mathcal{R}$ for the stable topology, it suffices to prove that the mapping

$$\Psi : (x,\mu,\nu) \to \int_0^1 [\int_{\mathbb{Z}} [\int_{\mathbb{S}} J(t, u_{x,\mu,\nu}(t), s, z)\, d\mu_t(s)]\, d\nu_t(z)]\, dt$$

is continuous on $M \times \mathcal{H} \times \mathcal{R}$. Let (x^n, μ^n, ν^n) be a sequence in $M \times \mathcal{H}, \times \mathcal{R}$ such that (x^n) converges to $x \in M$ and (ν^n) converges in probability to $\mu \in \mathcal{H}$ and (ν^n) stably converges to $\nu \in \mathcal{R}$. Applying the result in Claim 1, shows that (u_{x^n,μ^n,ν^n}) converges uniformly to $u_{x,\mu,\nu}$ that is a solution of our ODE

$$\begin{cases} \dot{u}_{x,\mu,\nu}(t) = \int_{\mathbb{Z}}[\int_{\mathbb{S}} f(t, u_{x,\mu,\nu}(t), s, z)\, d\mu_t(s)]\, d\nu_t(z) \text{ a.e.} \\ u_{x,\mu,\nu}(0) = x \in M. \end{cases}$$

This implies that $(\underline{\delta}_{u_{x^n,\mu^n,\nu^n}} \underline{\otimes}\, \mu^n \underline{\otimes}\, \nu^n)$ stably converges to $(\underline{\delta}_{u_{x,\mu,\nu}} \underline{\otimes}\, \mu \underline{\otimes}\, \nu)$. As J is an L^1-bounded Carathéodory integrand, it follows that

$$\lim_{n\to\infty} \int_0^1 [\int_{\mathbb{Z}} [\int_{\mathbb{S}} J(t, u_{x^n,\mu^n,\nu^n}(t), s, z)\, d\mu_t^n(s)]\, d\nu_t^n(z)]\, dt$$
$$= \int_0^1 [\int_{\mathbb{Z}} [\int_{\mathbb{S}} J(t, u_{x,\mu,\nu}(t), s, z)\, d\mu_t(s)]\, d\nu_t(z)]\, dt.$$

The proof is therefore complete. □

Remark 8.3.2 In the course of the proof of Theorem 8.3.1, we have proven a significant property, namely, the continuous dependence of the trajectories $u_{x,\mu,\nu}$ of the dynamic under consideration with respect to the data $(x, \mu, \nu) \in M \times \mathcal{H} \times \mathcal{R}$. At this point it is worth to mention that this continuity property is valid under more general conditions.

First, we may assume that $\mathbb{E}$ is a separable Banach space, $\mathbb{S}$ and $\mathbb{Z}$ are topological Lusin spaces, and the dynamic f is a mapping from $[0,1] \times \mathbb{E} \times \mathbb{S} \times \mathbb{Z}$ into $\mathbb{E}$ satisfying (i), (ii), (iv) and

(iii)' There exists a measurable and integrably bounded convex compact valued multifunction $\Gamma : [0,1] \rightrightarrows \mathbb{E}$ such that $f(t, x, s, z) \in (1 + ||x||)\Gamma(t)$ for all $(t, x, x, s, z) \in [0,1] \times \mathbb{E} \times \mathbb{S} \times \mathbb{Z}$.

Secondly, we may assume that $\mathcal{H}$ is sequentially compact for the convergence in probability and $\mathcal{R}$ is compact metrizable for the stable topology.

Indeed (i)- (ii)-(iii)'-(iv) are sufficient to guarantee the measurability of the mappings $t \mapsto \int_{\mathbb{Z}}[\int_{\mathbb{S}} f(t, x, s, z)\, \mu_t(ds)]\, \nu_t(dz)$ for all $(\mu, \nu) \in \mathcal{H} \times \mathcal{R}$ and that, for each $(x, \mu, \nu) \in M \times \mathcal{H} \times \mathcal{R}$, there exists a unique absolutely continuous solution and the solutions set $\{u_{x,\mu\nu};\ (x, \mu, \nu) \in M \times \mathcal{H} \times \mathcal{R}\}$ is relatively compact for the topology of uniform convergence (see Proposition 6.2.3, Example 7.3.1 and Proposition 7.3.2).

To illustrate this fact and before going further we present below a min-max type result for Young measures.

Proposition 8.3.3 *Let $\mathbb{S}$ be a Polish space and $\mathcal{H}$ be a convex subset in $\mathcal{Y}^1([0,1]; \mathbb{S})$ which is closed for the convergence in probability (or for the stable topology, see*

Corollary 4.5.9). Let $\mathbb{T}$ be a Polish space and let $\mathcal{K}$ be a $\tau^*_{\mathcal{Y}^1}$-compact (metrizable) subset of $\mathcal{Y}^1([0,1];\mathbb{T})$. Let us consider a real-valued function $\Phi : \mathcal{H} \times \mathcal{K}$ such that, for every fixed $\mu \in \mathcal{H}, \Phi(\mu,.)$ is upper-semicontinuous on $\mathcal{K}$ and for every fixed $\nu \in \mathcal{K}, \Phi(.,\nu)$ is convex lower-semicontinuous on $\mathcal{H}$. Then there exist a pair $(\widetilde{\mu},\widetilde{\nu}) \in \mathcal{H} \times \mathcal{K}$ such that

$$\max_{\nu\in\mathcal{K}} \min_{\mu\in\mathcal{H}} \Phi(\mu,\nu) \le \Phi(\widetilde{\mu},\widetilde{\nu}) \le \min_{\mu\in\mathcal{H}} \max_{\nu\in\mathcal{K}} \Phi(\mu,\nu).$$

Proof. Let us set

$$p(\mu) := \max_{\nu\in\mathcal{K}} \Phi(\mu,\nu),\ \forall \mu \in \mathcal{H},$$

$$q(\nu) = \inf_{\mu\in\mathcal{H}} \Phi(\mu,\nu),\ \forall \nu \in \mathcal{K}.$$

Then $p(.)$ is convex lower semicontinuous on $\mathcal{H}$ and $q(.)$ is upper semicontinuous on $\mathcal{K}$. As $\mathcal{H}$ has the Mazur property (see Theorem 4.5.8), it is not difficult to provide an element $\widetilde{\mu} \in \mathcal{H}$ such that

$$p(\widetilde{\mu}) = \min_{\mu\in\mathcal{H}} p(\mu).$$

As $q(.)$ is upper semicontinuous for the stable topology on the compact set $\mathcal{K}$ there exists $\widetilde{\nu} \in \mathcal{K}$ such that

$$q(\widetilde{\nu}) = \max_{\nu\in\mathcal{K}} q(\nu).$$

So we get

$$q(\widetilde{\nu}) \le \Phi(\widetilde{\mu},\widetilde{\nu}) \le p(\widetilde{\mu}).$$

□

Proposition 8.3.4 *Assume that the hypotheses of Theorem 8.3.1 are satisfied. Let $l : \mathbb{E} \to \mathbb{E}$ be a continuous mapping. Let $x \in M$ and $\theta \in]0,1]$ be fixed. Then the control problem*

$$\Psi_x(\mu,\nu) := \int_0^1 [\int_{\mathbb{Z}} [\int_{\mathbb{S}} J(t, u_{x,\mu,\nu}(t), s, z)\, d\mu_t(s)]\, d\nu_t(z)]\, dt + l(u_{x,\mu,\nu}(\theta))$$

subject to

$$\begin{cases} \dot{u}_{x,\mu,\nu}(t) = \int_{\Gamma(t)}[\int_{\mathbb{S}} f(t, u_{x,\mu,\nu}(t), s, z)\, d\mu_t(s)]\, d\nu_t(z) \ a.e.\ t \in [0,1]; \\ u_{x,\mu,\nu}(0) = x \in M \end{cases}$$

admits at least a quasi-saddle point $(\widetilde{\mu},\widetilde{\nu}) \in \mathcal{H} \times \mathcal{R}$, that is

$$\max_{\nu\in\mathcal{R}} \min_{\mu\in\mathcal{H}} \Psi_x(\mu,\nu) \le \Psi_x(\widetilde{\mu},\widetilde{\nu}) \le \min_{\mu\in\mathcal{H}} \max_{\nu\in\mathcal{R}} \Psi_x(\mu,\nu).$$

Proof. Taking Theorem 8.3.1 and Remark 8.3.2 into account, we see that the function Ψ_x is continuous on $\mathcal{H} \times \mathcal{R}$. Repeating the arguments of Proposition 8.3.3 and using the compactness assumption on $\mathcal{H}$ gives the result. □

Remark 8.3.5 The conclusions of Theorem 8.3.1 and Proposition 8.3.4 may fail if one only assumes that $\mathcal{H}$ is compact for the stable topology (instead of the topology of convergence in probability), because the Fiber Product Lemma is not valid: By compactness, the sequence $\underline{\delta}_{u_{x^n,\mu^n,\nu^n}} \underline{\otimes}\, \mu^n \underline{\otimes}\, \nu^n$ has limit points in $\mathcal{Y}^1([0,1]; C_{\mathbb{E}}([0,1]) \times \mathbb{S} \times \mathbb{Z})$, but they do not necessarily have the form $\underline{\delta}_{u^\infty} \underline{\otimes}\, \mu^\infty \underline{\otimes}\, \nu^\infty$ (see Counterexample 3.3.3).

In the preceding dynamical system, we focus essentially on the continuous dependence of the solutions on the data $(x, \mu, \nu) \in M \times \mathcal{H} \times \mathcal{R}$. In order to illustrate our techniques, we develop some new properties of the value function $V_g(x, \mu, t)$ associated with the dynamical system under consideration as follows:

$$V_g(x, \mu, t) = \inf_{\nu \in \mathcal{R}} g(u_{x,\mu,\nu}(t))$$

with the data $(x, \mu, \nu) \in M \times \mathcal{H} \times \mathcal{R}$ where g is a bounded lower semicontinuous function defined on $\mathbb{E}$ and $t \in [0,1]$. Recall that $u_{x,\mu,\nu}(t)$ is the value of the solution $u_{x,\mu,\nu}$ at t of the dynamic $(\mathcal{I}_{M,\mathcal{H},\mathcal{R}})$. The following result provides a lower semicontinuity property for the integrand

$$J : (g, x, \mu, t) \mapsto V_g(x, \mu, t)$$

defined on $\mathcal{I}_{\mathbb{E}} \times M \times \mathcal{H} \times [0,1]$ where $\mathcal{I}_{\mathbb{E}}$ denotes the set of all bounded lower semicontinuous functions defined on $\mathbb{E}$.

Proposition 8.3.6 *Assume the same hypothesis as in Theorem 8.3.1. Let $g \in \mathcal{I}_{\mathbb{E}}$, let $(x, \mu) \in M \times \mathcal{H}$ and let $t \in [0,1]$. Then for any an increasing sequence (g_i) of bounded Lipschitz functions defined on $\mathbb{E}$ converging pointwisely to g, for any sequence (t^i) in $[0,1]$ converging to t, for any a sequence (x^i) in M converging to $x \in M$ and for any sequence (μ^i) in $\mathcal{H}$ converging in measure to μ,*

$$\liminf_i V_{g^i}(x^i, \mu^i, t^i) \geq V_g(x, \mu, t).$$

Proof. Note that, for every $s \in [0,1]$, the functions $(x, \mu, \nu) \to g_i(u_{x,\mu,\nu}(s))$ are continuous (thus uniformly continuous) on the compact space $M \times \mathcal{H} \times \mathcal{R}$ because the mapping $(x, \mu, \nu) \to (u_{x,\mu,\nu})$ is continuous on $M \times \mathcal{H} \times \mathcal{R}$ as we have already proved in Theorem 8.3.1. Therefore, for each i, the mapping $(s, x, \mu, \nu) \to g_i(u_{x,\mu,\nu}(s))$ is continuous. On the other hand, for each i, there is $\nu^i \in \mathcal{R}$ with

$$V_{g^i}(x^i, \mu^i, t^i) = g_i(u_{x^i,\mu^i,\nu^i}(t^i))$$

where u_{x^i,μ^i,ν^i} is the solution of the dynamic $(\mathcal{I}_{M,\mathcal{H},\mathcal{R}})$, that is

$$\dot{u}_{x^i,\mu^i,\nu^i}(\tau) = \int_{\Gamma(\tau)} [\int_{\mathbb{S}} f(\tau, u_{x^i,\mu^i,\nu^i}(\tau), s, z)\, d\mu^i_\tau(s)]\, d\nu^i_\tau(z) \text{ a.e. } \tau \in [0,1],$$
$$u_{x^i,\mu^i,\nu^i}(0) = x^i.$$

Then, there is a subsequence of (μ^i, ν^i) with the same notation such that $(\mu^i \underline{\otimes} \nu^i)$ stably converges to $\mu \underline{\otimes} \nu$ and such that (u_{x^i,μ^i,ν^i}) converges uniformly to $u_{x,\mu,\nu}$ which is the solution of this dynamic, using Theorem 8.3.1, that is

$$\dot{u}_{x,\mu,\nu}(\tau) = \int_{\Gamma(\tau)} [\int_{\mathbb{S}} f(\tau, u_{x,\mu,\nu}(\tau), s, z)\, d\mu_\tau(s)]\, d\nu_\tau(z) \text{ a.e. } \tau \in [0,1],$$
$$u_{x,\mu,\nu}(0) = x.$$

So, for $i \geq k$, we immediately have the estimate

$$V_{g_i}(x^i, \mu^i, t^i) = g_i(u_{x^i,\mu^i,\nu^i}(t^i)) \geq g_k(u_{x^i,\mu^i,\nu^i}(t^i)) - g(u_{x,\mu,\nu}(t)) + V_g(x, \mu, t).$$

Let $\varepsilon > 0$. There is $k > 0$ such that $0 \leq g(u_{x,\mu,\nu}(t)) - g_k(u_{x,\mu,\nu}(t)) \leq \varepsilon$. As (g_i) is increasing, we have

$$\liminf_{i\to\infty} g_i(u_{x^i,\mu^i,\nu^i}(t^i)) \geq \liminf_{i\to\infty} g_k(u_{x^i,\mu^i,\nu^i}(t^i)) = g_k(u_{x,\mu,\nu}(t)) \geq g(u_{x,\mu,\nu}(t)) - \varepsilon.$$

The preceding inequality implies that

$$\liminf_{i\to\infty} V_{g_i}(x^i, \mu^i, t^i) = \liminf_{i\to\infty} g_i(u_{x^i,\mu^i,\nu^i}(t^i)) \geq V_g(x, \mu, t) - \varepsilon.$$

□

In the preceding result the data g is of simple nature. Now we give an extension to Proposition 8.3.6 concerning a relaxation property for the relaxed value function, when the data is the integral functional associated with a positive, bounded normal integrand. Namely, let $h : [0,1] \times \mathbb{E} \times \mathbb{S} \times \mathbb{Z} \to \mathbb{R}^+$ be a bounded normal $\mathcal{L}_{[0,1]}$–$\mathcal{B}(\mathbb{E} \times \mathbb{S} \times \mathbb{Z})$–measurable integrand, that is, h is measurable and $h(t, ., ., ., .)$ is lower-semicontinuous on $\mathbb{E} \times \mathbb{S} \times \mathbb{Z}$ for every fixed $t \in [0,1]$, and let I_h be the integral functional on $\mathcal{H} \times \mathcal{R}$ given by

$$I_h(x, \mu, \nu) = \int_0^1 [\int_{\mathbb{Z}} [\int_{\mathbb{S}} h(t, u_{x,\mu,\nu}(t), s, z)\, \mu_t(s)]\, d\nu_t(z)]\, dt,$$

where $u_{x,\mu,\nu}$ is the unique solution to

$$\dot{u}_{x,\nu,\mu}(t) = \int_{\Gamma(t)} [\int_{\mathbb{S}} f(t, u_{x,\mu,\nu}(t), s, z)\, d\mu_t(s)]\, d\nu_t(z) \text{ a.e. } t \in [0,1],$$
$$u_{x,\mu,\nu}(0) = x.$$

Proposition 8.3.7 *Let h be as above, and let $(x,\mu) \in M \times \mathcal{H}$. Let (h^i) be an increasing sequence of bounded, Carathéodory integrands defined on $[0,1]\times\mathbb{E}\times\mathbb{S}\times\mathbb{Z}$ such that $h = \sup_i h^i$. Let (x^i) be a sequence in M which converges to $x \in M$, let (μ^i) be a sequence in $\mathcal{H}$ which converges in probability to μ. Let us consider the value function*

$$W_h(x,\mu) = \inf_{\nu\in\mathcal{R}} I_h(x,\mu,\nu) = \inf_{\nu\in\mathcal{R}} \int_0^1 [\int_{\mathbb{Z}} [\int_{\mathbb{S}} h(t, u_{x,\mu,\nu}(t), s, z)\, d\mu_t(s)]\, d\nu_t(z)]\, dt,$$

where $u_{x,\mu,\nu}$ is the solution of the dynamic $(\mathcal{I}_{M,\mathcal{H},\mathcal{R}})$. We then have

$$\liminf_i W_{h^i}(x^i,\mu^i) \geq W_h(x,\mu),$$

where, for all i,

$$W_{h^i}(x^i,\mu^i) := \inf_{\zeta\in \mathrm{S}_\Gamma} I_{h^i}(x^i,\mu^i,\zeta) = \inf_{\zeta\in \mathrm{S}_\Gamma} \int_0^1 [\int_{\mathbb{S}} h^i(t, u_{x^i,\mu^i,\zeta}(t), s, \zeta(t))\, d\mu_t^i(s)]\, dt.$$

Remark 8.3.8 It is well–known that a sequence such as (h^i) always exists. It is enough to combine the Baire-Hausdorff approximations with the Projection Theorem to obtain such approximates (h^i).

Proof of Proposition 8.3.7. By Theorem 8.3.1, for each i, we have

$$\begin{aligned}\inf_{\nu\in \mathrm{S}_\Sigma} I_{h^i}(x,\mu,\nu) &= \inf_{\nu\in \mathrm{S}_\Sigma} \int_0^1 [\int_{\mathbb{Z}} [\int_{\mathbb{S}} h^i(t, u_{x,\mu,\nu}(t), s, z)\, d\mu_t(s)]\, d\nu_t(z)]\, dt \\ &= \inf_{\zeta\in \mathrm{S}_\Gamma} \int_0^1 [\int_{\mathbb{S}} h^i(t, u_{x,\mu,\zeta}(t), s, \zeta(t))\, d\mu_t(s)]\, dt = W_{h^i}(x,\mu).\end{aligned}$$

We have already observed that the mappings $(x,\mu,\nu) \mapsto u_{x,\mu,\nu}$ and $(x,\mu,\nu) \mapsto I_{h^i}(x,\mu,\nu)$ are continuous on the compact space $M \times \mathcal{H} \times \mathcal{R}$. For each i, there is $\nu^i \in \mathrm{S}_\Sigma$ with

$$W_{h^i}(x^i,\mu^i) = I_{h^i}(x^i,\mu^i,\nu^i),$$

where u_{x^i,μ^i,ν^i} is the trajectory solution of $(\mathcal{I}_{M,\mathcal{H},\mathcal{R}})$. Then, as in the proof of Proposition 8.3.6, there is a subsequence with the same notation such that $(\mu^i \underline{\otimes} \nu^i)$ stably converges to $\mu \underline{\otimes} \nu$ and (u_{x^i,μ^i,ν^i}) converges uniformly to $u_{x,\mu,\nu}$ which is a trajectory solution with the initial condition $u_{x,\mu,\nu}(0) = x$. So, for $i \geq k$, we immediately have the estimate

$$W_{h^i}(x^i,\mu^i) = I_{h^i}(x^i,\mu^i,\nu^i) \geq I_{h^k}(x^i,\mu^i,\nu^i) - I_h(x,\mu,\nu) + W_h(x,\mu).$$

For every fixed $(x,\mu,\nu) \in M\times\mathcal{H}\times\mathcal{R}$, we have, by using the monotone convergence theorem,

$$\begin{aligned}\sup_i I_{h^i}(x,\mu,\nu) &= \sup_i \int_0^1 \int_{\mathbb{Z}} [\int_{\mathbb{S}} h^i(t, u_{x,\mu,\nu}(t), s, z)\, d\mu_t(s)]\, d\nu_t(z)]\, dt \\ &= \int_0^1 \int_{\mathbb{Z}} [\int_{\mathbb{S}} h(t, u_{x,\mu,\nu}(t), s, z)\, d\mu_t(s)]\, d\nu_t(z)]\, dt = I_h(x,\mu,\nu).\end{aligned}$$

Let $\varepsilon > 0$. There is $k > 0$ such that $0 \le I_h(x,\mu,\nu) - I_{h^k}(x,\mu,\nu) \le \varepsilon$. As (I_{h^i}) is increasing, we have

$$\liminf_{i\to\infty} I_{h^i}(x^i,\mu^i,\nu^i) \ge \liminf_{i\to\infty} I_{h^k}(x^i,\mu^i,\nu^i) = I_{h^k}(x,\mu,\nu) \ge I_h(x,\mu,\nu) - \varepsilon.$$

The preceding inequality implies that

$$\liminf_{i\to\infty} W_{h^i}(x^i,\mu^i) = \liminf_{i\to\infty} I_{h^i}(x^i,\mu^i) \ge W_h(x,\mu) - \varepsilon.$$

□

8.3.B Dynamic programming

In the following we aim to present other types of value functions which occur in the problem of viscosity solutions of the Hamilton–Jacobi–Bellman equation associated with the relaxed upper Hamiltonian H^+ defined by

$$H^+(t,x,\rho) = \min_{\mu\in\mathcal{M}^{+,1}(\mathbb{S})} \max_{\nu\in\mathcal{M}^{+,1}(\mathbb{Z})} \{\langle \rho, \int_{\mathbb{Z}}[\int_{\mathbb{S}} f(t,x,s,z)\,d\mu(s)]\,d\nu(z)\rangle + \int_{\mathbb{Z}}[\int_{\mathbb{S}} J(t,x,s,z)\,d\mu(s)]\,d\nu(z)\}$$

where the cost function has the form

$$\int_\tau^1 [\int_{\mathbb{Z}}[\int_{\mathbb{S}} J(t,u_{x,\mu,\nu}(t),s,z)\,d\mu_t(s)]\,d\nu_t(z)]\,dt + g(u_{x,\mu,\nu}(1)),$$

where $J : [0,1]\times\mathbb{E}\times\mathbb{S}\times\mathbb{Z}\to\mathbb{R}$ is an L^1-bounded Carathéodory integrand and $u_{x,\mu,\nu}$ is the trajectory solution starting at position x at intermediate time $\tau\in[0,1[$, associated with the control $(\mu,\nu)\in\mathcal{H}\times\mathcal{R}$, where $\mathcal{H}$ is the set of all Lebesgue-measurable mappings $\mu : [0,1]\to\mathcal{M}^{+,1}(\mathbb{S})$, namely

$$\dot u_{x,\mu,\nu}(t) = \int_{\mathbb{Z}}[\int_{\mathbb{S}} f(t,u_{x,\mu,\nu}(t),s,z)\,d\mu_t(s),]\,d\nu_t(z),\ u_{x,\mu,\nu}(\tau) = x$$

and g is an upper-semicontinuous continuous function defined on $\mathbb{E}$ (see Remark 8.3.10-1).

In the remainder of this section, we assume that $\mathcal{H}$ is a *decomposable* subset of the space of Young measures $\mathcal{Y}^1([0,1];\mathbb{S})$, that is, for every Lebesgue-measurable set $A\subset[0,1]$ and for every μ^1, μ^2 in $\mathcal{H}$, $1_A\mu^1 + 1_{A^c}\mu^2 \in \mathcal{H}$, in particular, $\mathcal{H}$ is the set of all Lebesgue-measurable mappings $\zeta : [0,1]\to S$. Here only decomposability assumption on $\mathcal{H}$ is used by contrast to the results obtained above.

Theorem 8.3.9 (of dynamic programming) *Assume that* $J : [0,1] \times \mathbb{E} \times \mathbb{S} \times \mathbb{Z} \to \mathbb{R}$ *is an* L^1*-bounded Carathéodory integrand. Let* $x \in \mathbb{E}$*, let* $\tau \in [0,1[$ *and let* $\sigma > 0$ *such that* $\tau + \sigma < 1$*. Assume further that* $\mathcal{H}$ *is a decomposable subset of the space of Young measures* $\mathcal{Y}^1([0,1];\mathbb{S})$*. Let us consider the* upper value function

$$V_J(\tau, x) := \inf_{\mu \in \mathcal{H}} \max_{\nu \in \mathcal{R}} \{ \int_\tau^1 [\int_{\mathbb{Z}} [\int_{\mathbb{S}} J(t, u_{x,\mu,\nu}(t), s, z)\, d\mu_t(s)]\, d\nu_t(z)]\, dt \}.$$

Here $u_{x,\mu,\nu}$ *denotes the solution trajectory determined by the relaxed dynamic associated with* f *and the control* $(\mu, \nu) \in \mathcal{H} \times \mathcal{R}$ *starting at position* x *at time* $\tau \in [0,1]$*, namely*

$$\dot{u}_{x,\mu,\nu}(t) = \int_{\mathbb{Z}} [\int_{\mathbb{S}} f(t, u_{x,\mu,\nu}(t), s, z)\, \mu_t(s)]\, d\nu_t(z);\ u_{x,\mu,\nu}(\tau) = x.$$

Then the following hold:

$$V_J(\tau, x) = \inf_{\mu \in \mathcal{H}} \max_{\nu \in \mathcal{R}} \{ \int_\tau^{\tau+\sigma} [\int_{\mathbb{Z}} [\int_{\mathbb{S}} J(t, u_{x,\mu,\nu}(t), s, z)\, d\mu_t(s)]\, d\nu_t(z)]\, dt$$
$$+ V_J(\tau + \sigma, u_{x,\mu,\nu}(\tau + \sigma)) \}. \tag{8.3.1}$$

Here

$$V_J(\tau+\sigma, u_{x,\mu,\nu}(\tau+\sigma)) := \inf_{\beta \in \mathcal{H}} \max_{\gamma \in \mathcal{R}} \{ \int_{\tau+\sigma}^1 [\int_{\mathbb{Z}} [\int_{\mathbb{S}} J(t, v_{x,\beta,\gamma}(t), y, z)\, d\beta_t(s)]\, d\gamma_t(z)]\, dt \}.$$

$v_{x,\beta,\gamma}$ *denotes the trajectory on* $[\tau + \sigma, 1]$ *associated with* $(\beta, \gamma) \in \mathcal{H} \times \mathcal{R}$ *with the initial condition* $v_{x,\beta,\gamma}(\tau + \sigma) = u_{x,\mu,\nu}(\tau + \sigma)$*.*

Proof. We will use the continuity results obtained above. Let $W_J(\tau, x)$ denote the right hand side of (8.3.1). Take $\varepsilon > 0$ and $\mu^1 \in \mathcal{H}$ such that

$$W_J(\tau, x) \geq \max_{\nu \in \mathcal{R}} \{ \int_\tau^{\tau+\sigma} [\int_{\mathbb{Z}} [\int_{\mathbb{S}} J(t, u_{x,\mu^1,\nu}(t), s, z)\, d\mu_t^1(s)]\, d\nu_t(z)]\, dt$$
$$+ V_J(\tau + \sigma, u_{x,\mu^1,\nu}(\tau + \sigma)) \} - \varepsilon.$$

There is $\mu^2 \in \mathcal{H}$ such that

$$V_J(\tau+\sigma, u_{x,\mu^1,\nu}(\tau+\sigma)) > \max_{\gamma \in \mathcal{R}} \int_{\tau+\sigma}^1 [\int_{\mathbb{Z}} [\int_{\mathbb{S}} J(t, v_{x,\mu^2,\gamma}(t), s, z)\, d\mu_t^2(s)]\, d\gamma_t(z)] dt - \varepsilon.$$

Here $v_{x,\mu^2,\gamma}$ denotes the solution trajectory on $[\tau + \sigma, 1]$ associated with $(\mu^2, \gamma) \in \mathcal{H} \times \mathcal{R}$ with the initial condition $v_{x,\mu^2,\gamma}(\tau + \sigma) = u_{x,\mu^1,\nu}(\tau + \sigma)$. By compactness of $\mathcal{R}$ and by the continuity of

$$\nu \mapsto \int_\tau^{\tau+\sigma} [\int_{\mathbb{Z}} [\int_{\mathbb{S}} J(t, u_{x,\mu,\nu}(t), s, z)\, d\mu_t(s)]\, d\nu_t(z)]\, dt$$

on $\mathcal{R}$, we can choose $\nu^1 \in \mathcal{R}$ such that

$$\max_{\nu\in\mathcal{R}} \int_\tau^{\tau+\sigma} [\int_{\mathbb{Z}} [\int_{\mathbb{S}} J(t, u_{x,\mu^1,\nu}(t), s, z)\, d\mu_t^1(s)]\, d\nu_t(z)]\, dt$$
$$= \int_\tau^{\tau+\sigma} [\int_{\mathbb{Z}} [\int_{\mathbb{S}} J(t, u_{x,\mu^1,\nu^1}(t), s, z)\, d\mu_t^1(s)]\, d\nu_t^1(z)]\, dt$$

and similarly there is $\nu^2 \in \mathcal{R}$ such that

$$\max_{\gamma\in\mathcal{R}} \int_{\tau+\sigma}^{1} [\int_{\mathbb{Z}} [\int_{\mathbb{S}} J(t, v_{x,\mu^2,\gamma}(t), s, z)\, d\mu_t^2(s)]\, d\gamma_t(z)]\, dt$$
$$= \int_{\tau+\sigma}^{1} [\int_{\mathbb{Z}} [\int_{\mathbb{S}} J(t, v_{x,\mu^2,\nu^2}(t), s, z)\, \mu_t^2(s)]\, d\nu_t^2(z)]\, dt.$$

Here v_{x,μ^2,ν^2} is the trajectory solution defined on $[\tau+\sigma, 1]$ associated with $(\mu^2, \nu^2) \in \mathcal{H} \times \mathcal{R}$ with the initial condition $v_{x,\mu^2,\nu^2}(\tau+\sigma) = u_{x,\mu^1,\nu}(\tau+\sigma)$. Let us set

$$\overline{\mu} := 1_{[\tau,\tau+\sigma]}\mu^1 + 1_{[\tau+\sigma,1]}\mu^2$$

and

$$\overline{\nu} := 1_{[\tau,\sigma]}\nu^1 + 1_{[\tau+\sigma,1]}\nu^2.$$

By the decomposability of $\mathcal{H}$ and $\mathcal{R}$ (recall that $\mathcal{R}$ is the set of measurable selections of a multifunction), we have that $\overline{\mu} \in \mathcal{H}$ and $\overline{\nu} \in \mathcal{R}$. Let $\overline{w}_{x,\overline{\mu},\overline{\nu}}$ be the trajectory on $[\tau, 1]$ associated with $(\overline{\mu}, \overline{\nu}) \in \mathcal{H} \times \mathcal{R}$ with the initial condition x, that is, $\overline{w}_{x,\overline{\mu},\overline{\nu}}(t) = u_{x,\mu^1,\nu^1}(t)$ for $t \in [\tau, \tau+\sigma]$ and $\overline{w}_{x,\overline{\mu},\overline{\nu}}(t) = v_{x,\mu^2,\nu^2}(t)$ for $t \in [\tau+\sigma, 1]$. Coming back to the definition of $V_J(\tau, x)$, we have

$$V_J(\tau, x) \le \sup_{\nu\in\mathcal{R}} \int_\tau^{1} [\int_{\mathbb{Z}} [\int_{\mathbb{S}} J(t, u_{x,\overline{\mu},\nu}, s, z)\, d\overline{\mu}_t(s)]\, d\nu_t(z)]\, dt$$
$$= \int_\tau^{\tau+\sigma} [\int_{\mathbb{Z}} [\int_{\mathbb{S}} J(t, u_{x,\mu^1,\nu^1}, s, z)\, d\mu_t^1(s)]\, d\nu_t^1(z)]\, dt$$
$$+ \int_{\tau+\sigma}^{1} [\int_{\mathbb{Z}} [\int_{\mathbb{S}} J(t, v_{x,\mu^2,\nu^2}, s, z)\, d\mu_t^2(s)]\, d\nu_t^2(z)]\, dt$$
$$= \int_\tau^{1} [\int_{\mathbb{Z}} [\int_{\mathbb{S}} J(t, \overline{w}_{x,\overline{\mu},\overline{\nu}}, s, z)\, d\overline{\mu}_t(s)]\, d\overline{\nu}_t(z)]\, dt \le W_J(\tau, x) + 2\varepsilon.$$

On the other hand, there is $\widetilde{\mu} \in \mathcal{H}$ such that

$$V_J(\tau, x) \ge \max_{\nu\in\mathcal{R}}\{\int_\tau^{1} [\int_{\mathbb{Z}} [\int_{\mathbb{S}} J(t, u_{x,\widetilde{\mu},\nu}(t), s, z)\, \widetilde{\mu}_t(s)]\, d\nu_t(z)]\, dt\} - \varepsilon.$$

Then

$$W_J(\tau, x) \le \max_{\nu\in\mathcal{R}}\{\int_\tau^{\tau+\sigma} [\int_{\mathbb{Z}} [\int_{\mathbb{S}} J(t, u_{x,\widetilde{\mu},\nu}(t), s, z)\, d\widetilde{\mu}_t(s)]\, d\nu_t(z)]\, dt$$
$$+V_J(\tau+\sigma, u_{x,\widetilde{\mu},\nu}(\tau+\sigma)\},$$

with

$$V_J(\tau+\sigma, u_{x,\widetilde{\mu},\nu}(\tau+\sigma) \le \max_{\nu\in\mathcal{R}} \int_{\tau+\sigma}^{1} [\int_{\mathbb{Z}}[\int_{\mathbb{S}} J(t, u_{x,\widetilde{\mu},\nu}(t), s, z)\, d\widetilde{\mu}_t(s)]\, d\nu_t(z)]dt.$$

Here $v_{x,\widetilde{\mu},\nu}$ is the trajectory solution associated with the controls $(\widetilde{\mu}, \nu) \in \mathcal{H} \times \mathcal{R}$ with the initial condition $v_{x,\widetilde{\mu},\nu}(\tau+\sigma) = u_{x,\widetilde{\mu},\nu}(\tau+\sigma)$. By continuity of the integral functionals under consideration and by compactness of $\mathcal{R}$, there exists $\overline{\nu^1} \in \mathcal{R}$ such that

$$\begin{aligned}\max_{\nu\in\mathcal{R}} \int_{\tau}^{\tau+\sigma} &[\int_{\mathbb{Z}}[\int_{\mathbb{S}} J(t, u_{x,\widetilde{\mu},\nu}, s, z)\, d\widetilde{\mu}_t(s)]\, d\nu_t(z)]\, dt \\ &= \int_{\tau}^{\tau+\sigma} [\int_{\mathbb{Z}}[\int_{\mathbb{S}} J(t, u_{x,\widetilde{\mu},\overline{\nu^1}}, s, z)\, d\widetilde{\mu}_t(s)]\, d\overline{\nu^1}_t(z)]\, dt\end{aligned}$$

and similarly there exists $\overline{\nu}^2 \in \mathcal{R}$ such that

$$\begin{aligned}\int_{\tau+\sigma}^{1} & J(t, v_{x,\widetilde{\mu},\overline{\nu}^2}(t), s, z)\, \widetilde{\mu}_t(s)]\, d\overline{\nu}^2_t(z)]\, dt \\ &= \max_{\nu\in\mathcal{R}} \int_{\tau+\sigma}^{1} J(t, v_{x,\widetilde{\mu},\nu}(t), s, z)\, d\widetilde{\mu}_t(s)]\, d\nu_t(z)]\, dt.\end{aligned}$$

Here $v_{x,\widetilde{\mu},\overline{\nu}^2}$ is the trajectory on $[\tau+\sigma, 1]$ associated with $(\widetilde{\mu}, \overline{\nu}^2) \in \mathcal{H}\times\mathcal{R}$ with the initial condition $v_{x,\widetilde{\mu},\overline{\nu}^2}(\tau+\sigma) = u_{x,\widetilde{\mu},\nu}(\tau+\sigma)$. Let $\widetilde{\nu} = 1_{[\tau,\tau+\sigma]}\overline{\nu^1} + 1_{[\tau+\sigma,1]}\overline{\nu}^2 \in \mathcal{R}$ and $\widetilde{u}_{x,\widetilde{\mu},\widetilde{\nu}}(t) = u_{x,\widetilde{\mu},\nu^1}(t)$ if $t \in [\tau, \tau+\sigma]$ and $\widetilde{u}_{x,\widetilde{\mu},\widetilde{\nu}}(t) = v_{x,\widetilde{\mu},\overline{\nu}^2}(t)$ if $t \in [\tau+\sigma, 1]$. Taking the above estimate of $W_J(\tau, x)$ into account, we get

$$\begin{aligned}W_J(\tau, x) \le & \int_{\tau}^{1} [\int_{\mathbb{Z}}[\int_{\mathbb{S}} J(t, \widetilde{u}_{x,\widetilde{\mu},\widetilde{\nu}}(t), s, z)\, \widetilde{\mu}_t(s)]\, d\widetilde{\nu}_t(z)]\, dt \\ \le & \max_{\nu\in\mathcal{R}} \int_{\tau}^{1} [\int_{\mathbb{Z}}[\int_{\mathbb{S}} J(t, u_{x,\widetilde{\mu},\nu}(t), s, z)\, d\widetilde{\mu}_t(s)]\, d\nu_t(z)]\, dt \\ \le & V_J(\tau, x) + \varepsilon.\end{aligned}$$

□

Remarks 8.3.10 1) The preceding result still holds when the upper value functions has the form

$$V_J^g(\tau, x) = \inf_{\mu\in\mathcal{H}} \max_{\nu\in\mathcal{R}} \{ \int_{\tau}^{1} [\int_{\mathbb{Z}}[\int_{\mathbb{S}} J(t, u_{x,\mu,\nu}(t), y, z)\, d\mu_t(s)]\, d\nu_t(z)]\, dt + g(u_{x,\mu,\nu}(1))\},$$

where g is a bounded upper semicontinuous function defined on $\mathbb{E}$. Here the decomposability properties of $\mathcal{H}$ and $\mathcal{R}$ and the compactness of $\mathcal{R}$ for the stable topology in the space of Young measures $\mathcal{Y}^1([0,1]; \mathbb{Z})$ are very useful in the proof of Theorem 8.3.9. Similarly, one can give an analogous result for the lower value function

$$U_J^g(\tau, x) = \sup_{\mathcal{R}} \inf_{\mathcal{H}} \{ \int_\tau^{\tau+\sigma} [\int_{\mathbb{Z}} [\int_{\mathbb{S}} J(t, u_{x,\mu,\nu}(t), s, z)\, d\mu_t(s)]\, d\nu_t(z)]\, dt + g(u_{x,\mu,\nu}(1)\}$$

by permuting the role of $\mathcal{R}$ and $\mathcal{H}$ and by assuming that g is a bounded lower semicontinous function defined on $\mathbb{E}$.

2) It is worth to mention that the preceding result holds in the particular case where $\mathcal{H}$ is the set of all Lebesgue-measurable mappings from $[0,1]$ to $\mathcal{M}^{+,1}(\mathbb{S})$ endowed with the vague topology $\sigma(\mathcal{C}(\mathbb{S})^*, C(\mathbb{S}))$.

3) Taking the remarks of Theorem 8.3.1 into account, let us mention that Theorem 8.3.9 is valid if we assume that $\mathbb{S}$ and $\mathbb{Z}$ are topological Lusin spaces, and the dynamic f satisfies the conditions (i)- (ii)-(iii)'-(iv) of Remark 8.3.2, $\mathcal{H}$ is a decomposable space of $\mathcal{Y}^1([0,1];\mathbb{S})$ and $\mathcal{R}$ is compact metrizable for the stable topology.

In the sequel of this section, we will make the following assumptions.

(H_1) $\mathcal{H} = \mathcal{Y}^1([0,1];\mathbb{S})$ and $\mathcal{K} = \mathcal{Y}^1([0,1];\mathbb{Z})$. In particular, the sets $\mathcal{H}$ and $\mathcal{K}$ are decomposable and compact for the stable topology.

(H_2) $f : [0,1] \times \mathbb{E} \times \mathbb{S} \times \mathbb{Z} \to \mathbb{E}$ is bounded continuous, f is uniformly Lipschitz with respect to the variable $x \in \mathbb{E}$ and the family $(f(.,.,s,z))_{(s,z)\in\mathbb{S}\times\mathbb{Z}}$ is equicontinuous; $J : [0,1] \times \mathbb{E} \times \mathbb{S} \times \mathbb{Z} \to \mathbb{R}$ is bounded continuous and the family $(J(.,.,s,z))_{(s,z)\in\mathbb{S}\times\mathbb{Z}}$ is equicontinuous.

It is obvious that the mappings $\widetilde{f} : [0,1] \times \mathbb{E} \times \mathcal{M}^{+,1}(\mathbb{S}) \times \mathcal{M}^{+,1}(\mathbb{Z}) \to \mathbb{E}$ and $\widetilde{J} : [0,1] \times \mathbb{E} \times \mathcal{M}^{+,1}(\mathbb{S}) \times \mathcal{M}^{+,1}(\mathbb{Z}) \to \mathbb{R}$ defined by

$$\widetilde{f}(t, x, \mu, \nu) = \int_{\mathbb{Z}} [\int_{\mathbb{S}} f(t, x, s, z)\, d\mu(s)]\, d\nu(z)$$

and

$$\widetilde{J}(t, x, \mu, \nu) = \int_{\mathbb{Z}} [\int_{\mathbb{S}} J(t, x, s, z)\, d\mu(s)]\, d\nu(z)$$

inherit the properties of f and J respectively.

The remainder is a careful adaptation of the techniques given in [Ell87] and [ES84]. Before going further let us mention a useful lemma.

Lemma 8.3.11 *Let* $(t_0, x_0) \in [0,1] \times \mathbb{E}$*, and let* $\Lambda : [t_0, 1] \times \mathbb{E} \times \mathcal{M}^{+,1}(\mathbb{S}) \times \mathcal{M}^{+,1}(\mathbb{Z}) \to \mathbb{R}$ *be a continuous mapping such that the family*

$$(\Lambda(.,.,\mu,\nu))_{(\mu,\nu)\in\mathcal{M}^{+,1}(\mathbb{S})\times\mathcal{M}^{+,1}(\mathbb{Z})}$$

is equicontinuous.

(a) *If* $\min_{\mu\in\mathcal{M}^{+,1}(\mathbb{S})}\max_{\nu\in\mathcal{M}^{+,1}(\mathbb{Z})}\Lambda(t_0,x_0,\mu,\nu) < -\eta < 0$ *for some* $\eta > 0$, *then there exist* $\overline{\mu}\in\mathcal{H}$ *and* $\sigma > 0$ *such that*

$$\max_{\nu\in\mathcal{K}}\int_{t_0}^{t_0+\sigma}\Lambda(t,u_{x_0,\overline{\mu},\nu}(t),\overline{\mu},\nu_t)\,dt < -\frac{\sigma\eta}{2},$$

where $u_{x_0,\overline{\mu},\nu}$ *denotes the trajectory solution of the relaxed dynamic associated with* f *and the controls* $\overline{\mu}$ *and* ν *with the initial condition* $u_{x_0,\overline{\mu}_t,\nu}(t_0) = x_0$, *that is,*

$$\dot{u}_{x_0,\overline{\mu},\nu}(t) = \int_{\mathbb{Z}}[\int_{\mathbb{S}} f(t,u_{x,\overline{\mu},\nu}(t),s,z)\,d\overline{\mu}_t(s)]\,d\nu_t(z) \ a.e.\ t\in[0,1],$$
$$u_{x_0,\overline{\mu},\nu}(t_0) = x_0.$$

(b) *If* $\min_{\mu\in\mathcal{M}^{+,1}(\mathbb{S})}\max_{\nu\in\mathcal{M}^{+,1}(\mathbb{Z})}\Lambda(t_0,x_0,\mu,\nu) > \eta > 0$ *for some* $\eta > 0$, *then there exists* $\sigma > 0$ *such that, for each* $\mu\in\mathcal{H}$, *we have*

$$\max_{\nu\in\mathcal{K}}\int_{t_0}^{t_0+\sigma}\Lambda(t,u_{x_0,\mu,\nu}(t),\mu_t,\nu_t)\,dt > \frac{\sigma\eta}{2}.$$

Proof.

(a) By hypothesis, there is $\overline{\mu}\in\mathcal{M}^{+,1}(\mathbb{S})$ such that $\max_{\nu\in\mathcal{M}^{+,1}(\mathbb{Z})}\Lambda(t_0,x_0,\overline{\mu},\nu) < -\eta < 0$. Also, by the equicontinuity hypothesis, there exists $\xi > 0$ such that $\max_{\nu\in\mathcal{M}^{+,1}(\mathbb{Z})}\Lambda(t,x,\overline{\mu},\nu) < -\eta/2$ for $0\le t-t_0\le\xi$ and $\|x-x_0\|\le\xi$. Take $\sigma > 0$ such that $\int_{t_0}^{t_0+\sigma}c(t)\,dt\le\xi$ so that $\|u_{x_0,\overline{\mu},\nu}(t)-u_{x_0,\overline{\mu},\nu}(t_0)\|\le\int_{t_0}^{t_0+\sigma}c(t)\,dt\le\xi$ for all $t\in[t_0,t_0+\sigma]$ and for all $\nu\in\mathcal{K}$ (we also denote by $\overline{\mu}$ the constant Young measure $t\mapsto\overline{\mu}_t=\overline{\mu}$). Then, by integrating,

$$\int_{t_0}^{t_0+\sigma}\Lambda(t,u_{x_0,\overline{\mu},\nu}(t),\overline{\mu},\nu_t)\,dt \le \int_{t_0}^{t_0+\sigma}\max_{\nu'\in\mathcal{M}^{+,1}(\mathbb{Z})}\Lambda(t,u_{x_0,\overline{\mu},\nu}(t),\overline{\mu},\nu')\,dt < -\frac{\sigma\eta}{2}$$

for all $\nu\in\mathcal{K}$ and the result follows.

(b) Let $\xi > 0$ such that, for all $(\mu,\nu)\in\mathcal{M}^{+,1}(\mathbb{S})\times\mathcal{M}^{+,1}(\mathbb{Z})$, if $0\le t-t_0<\xi$ and $\|x-x_0\|\le\xi$, then $|\Lambda(t,x,\mu,\nu)-\Lambda(t_0,x_0,\mu,\nu|<\eta/2$. Let $\mu:[0,1]\to\mathcal{M}^{+,1}(\mathbb{S})$ be a Lebesgue-measurable mapping. By virtue of the Sainte Beuve–von Neumann–Aumann Selection Theorem, there exists a Lebesgue-measurable mapping $\nu^\mu:[0,1]\to\mathcal{M}^{+,1}(\mathbb{Z})$ such that

$$\Lambda(t_0,x_0,\mu_t,\nu_t^\mu) = \max_{\nu\in\mathcal{M}^{+,1}(\mathbb{Z})}\Lambda(t_0,x_0,\mu_t,\nu)$$

for all $t\in[0,1]$, because the nonempty compact valued multifunction

$$t\mapsto\{\nu\in\mathcal{M}^{+,1}(\mathbb{Z});\ \Lambda(t_0,x_0,\mu_t,\nu) = \max_{\nu'\in\mathcal{M}^{+,1}(\mathbb{Z})}\Lambda(t_0,x_0,\mu_t,\nu')\}$$

has its graph in $\mathcal{L}_{[0,1]} \otimes \mathcal{B}(\mathcal{M}^{+,1}(\mathbb{Z}))$. Take $\sigma > 0$ such that $\int_{t_0}^{t_0+\sigma} c(t)\,dt \leq \xi$ so that $\|u_{x_0,\mu,\nu}(t) - u_{x_0,\mu,\nu}(t_0)\| \leq \int_{t_0}^{t_0+\sigma} c(t)\,dt \leq \xi$ for all $t \in [t_0, t_0+\sigma]$ and for all $\nu \in \mathcal{K}$. As in (a), we then have, by integrating,

$$\begin{aligned}\int_{t_0}^{t_0+\sigma} \Lambda(t, u_{x_0,\mu,\nu^\mu}(t), \mu_t, \nu_t^\mu)\,dt &\geq \int_{t_0}^{t_0+\sigma} [\Lambda(t, x_0, \mu_t, \nu_t^\mu) - \frac{\eta}{2}]\,dt \\ &> \int_{t_0}^{t_0+\sigma} \frac{\eta}{2}\,dt = \frac{\sigma\eta}{2}.\end{aligned}$$

□

Theorem 8.3.12 (of viscosity solution) *Assume that (H_1) and (H_2) are satisfied. Let us consider the* lower value function

$$U_J(\tau, x) := \max_{\nu\in\mathcal{K}} \min_{\mu\in\mathcal{H}} \{\int_\tau^1 [\int_\mathbb{Z} [\int_\mathbb{S} J(t, u_{x,\mu,\nu}(t), s, z)\,d\mu_t(s)]\,d\nu_t(z)]\,dt\}.$$

Here $u_{x,\mu,\nu}$ is the trajectory solution to

$$\dot{u}_{x,\mu,\nu}(t) = \int_\mathbb{Z} [\int_\mathbb{S} f(t, u_{x,\mu,\nu}(t), s, z)\,d\mu_t(s)]\,d\nu_t(z),\ u_{x,\mu,\nu}(\tau) = x.$$

Let us consider the relaxed upper Hamiltonian

$$\begin{aligned}H^+(t, x, \rho) := \min_{\mu\in\mathcal{M}^{+,1}(\mathbb{S})} \max_{\nu\in\mathcal{M}^{+,1}(\mathbb{Z})} \{\langle \rho, \int_\mathbb{Z} [\int_\mathbb{S} f(t, x, s, z)\,d\mu(s)]\,d\nu(z)\rangle \\ + \int_\mathbb{Z} [\int_\mathbb{S} J(t, x, s, z)\,d\mu(s)]\,d\nu(z)\}.\end{aligned}$$

Then U_J is a viscosity solution to the Hamilton–Jacobi–Bellman equation $\frac{\partial U}{\partial t} + H^+(t, x, \nabla U) = 0$, that is, for any $\varphi \in \mathrm{C}^1([0,1] \times \mathbb{E})$ for which $U_J - \varphi$ attains a local maximum at (t_0, x_0), we have

$$\frac{\partial \varphi}{\partial t}(t_0, x_0) + H^+(t_0, x_0, \nabla\varphi(t_0, x_0)) \geq 0,$$

and for any $\varphi \in \mathrm{C}^1([0,1]) \times \mathbb{E})$ for which $U_J - \varphi$ attains a local minimum at (t_0, x_0), we have

$$\frac{\partial \varphi}{\partial t}(t_0, x_0) + H^+(t_0, x_0, \nabla\varphi(t_0, x_0)) \leq 0.$$

Proof. Assume by contradiction that there exists a $\varphi \in \mathrm{C}^1([0,1] \times \mathbb{E})$ and a point (t_0, x_0) for which

$$\frac{\partial \varphi}{\partial t}(t_0, x_0) + H^+(t_0, x_0, \nabla\varphi(t_0, x_0)) \leq -\eta < 0 \text{ for } \eta > 0.$$

Applying Lemma 8.3.11-(a), by taking $\Lambda = \widetilde{J} + \langle \nabla\varphi, \widetilde{f}\rangle + \frac{\partial \varphi}{\partial t}$, where, for all $(t, x, \mu, \nu) \in [0,1] \times \mathbb{E} \times \mathcal{M}^{+,1}(\mathbb{S}) \times \mathcal{M}^{+,1}(\mathbb{Z})$,

$$\widetilde{J}(t, x, \mu, \nu) = \int_{\mathbb{Z}} [\int_{\mathbb{S}} J(t, x, s, z)\, d\mu(s)]\, d\nu(z)$$

and

$$\widetilde{f}(t, x, \mu, \nu) = \int_{\mathbb{Z}} [\int_{\mathbb{S}} f(t, x, s, z)\, d\mu(s)]\, d\nu(z),$$

provides a control $\overline{\mu} \in \mathcal{M}^{+,1}(\mathbb{S})$ and $\sigma > 0$ such that

$$\begin{aligned} \max_{\nu\in\mathcal{K}}\{ & \int_{t_0}^{t_0+\sigma} [\int_{\mathbb{Z}} [\int_{\mathbb{S}} J(t, u_{x_0,\overline{\mu},\nu}(t), s, z)\, d\overline{\mu}(s)]\, d\nu_t(z)]\, dt \\ & + \int_{t_0}^{t_0+\sigma} [\int_{\mathbb{Z}} [\int_{\mathbb{S}} \langle \nabla\varphi(t, u_{x_0,\overline{\mu},\nu}(t)), f(t, u_{x_0,\overline{\mu},\nu}(t), s, z)\rangle\, d\overline{\mu}(s)]\, d\nu_t(z)]\, dt \\ & + \int_{t_0}^{t_0+\sigma} \frac{\partial \varphi}{\partial t}(t, u_{x_0,\overline{\mu},\nu}(t))\, dt\} \\ & \leq -\sigma\eta/2. \end{aligned}$$

Thus

$$\begin{aligned} \max_{\nu\in\mathcal{K}} \min_{\mu\in\mathcal{H}}\{ & \int_{t_0}^{t_0+\sigma} [\int_{\mathbb{Z}} [\int_{\mathbb{S}} J(t, u_{x_0,\mu,\nu}(t), s, z)\, d\mu(s)]\, d\nu_t(z)]\, dt \\ & + \int_{t_0}^{t_0+\sigma} [\int_{\mathbb{Z}} [\int_{\mathbb{S}} \langle \nabla\varphi(t, u_{x_0,\mu,\nu}(t)), f(t, u_{x_0,\mu,\nu}(t), s, z)\rangle\, d\mu_t(s)]\, d\nu_t(z)]\, dt \\ & + \int_{t_0}^{t_0+\sigma} \frac{\partial \varphi}{\partial t}(t, u_{x_0,\mu,\nu}(t))\, dt\} \end{aligned}$$

$$\leq -\sigma\eta/2. \tag{8.3.2}$$

From the dynamic programming identity (see Remark 8.3.10-1),

$$\begin{aligned} U_J(t_0, x_0) = \max_{\nu\in\mathcal{K}} \min_{\mu\in\mathcal{H}}\{ & \int_{t_0}^{t_0+\sigma} [\int_{\mathbb{Z}} [\int_{\mathbb{S}} J(t, u_{x,\mu,\nu}(t), s, z)\, d\mu_t(s)]\, d\nu_t(z)]\, dt \\ & + U_J(t_0 + \sigma, u_{x_0,\mu,\nu}(t_0 + \sigma))\}. \end{aligned} \tag{8.3.3}$$

Here $u_{x_0,\mu,\nu}$ is the trajectory solution of the relaxed dynamic $\widetilde{f}$ associated with $\mu \in \mathcal{H}$ and $\nu \in \mathcal{K}$ with the initial condition $u_{x_0,\zeta,\nu}(t_0) = x_0$. Since $U_J - \varphi$ has a local maximum at (t_0, x_0), so for small enough σ

$$U_J(t_0, x_0) - \varphi(t_0, x_0) \geq U_J(t_0+\sigma, u_{x_0,\mu,\nu}(t_0+\sigma)) - \varphi(t_0+\sigma, u_{x_0,\mu,\nu}(t_0+\sigma)). \tag{8.3.4}$$

From (8.3.3) and (8.3.4),

$$\begin{aligned}\max_{\nu\in\mathcal{K}}\min_{\mu\in\mathcal{H}}\{&\int_{t_0}^{t_0+\sigma}[\int_{\mathbb{Z}}[\int_{\mathbb{S}}J(t,u_{x,\mu,\nu}(t),s,z)\,d\mu_t(s)]\,d\nu_t(z)]\,dt\\&+\varphi(t_0+\sigma,u_{x_0,\mu,\nu}(t_0+\sigma)-\varphi(t_0,x_0))\}\\&\geq 0.\end{aligned}\tag{8.3.5}$$

As φ is C^1

$$\begin{aligned}\varphi(t_0+\sigma,&u_{x_0,\mu,\nu}(t_0+\sigma))-\varphi(t_0,x_0)\\&=\int_{t_0}^{t_0+\sigma}\langle\nabla\varphi(t,u_{x_0,\mu,\nu}(t)),[\int_{\mathbb{Z}}[\int_{\mathbb{S}}f(t,u_{x_0,\mu,\nu}(t),s,z)\,d\mu_t(s)]\,d\nu_t(z)]\,\rangle dt\\&+\int_{t_0}^{t_0+\sigma}\frac{\partial\varphi}{\partial t}(t,u_{x_0,\mu,\nu}(t))\,dt.\end{aligned}\tag{8.3.6}$$

Substituting (8.3.6) in (8.3.5) we have a contradiction to (8.3.2). Therefore we must have

$$\frac{\partial\varphi}{\partial t}(t_0,x_0)+H^+(t_0,x_0,\nabla\varphi(t_0,x_0))\geq 0.$$

The verification for the min point is similar, using Lemma 8.3.11-(b). □

Remark 8.3.13 Theorem 8.3.12 remains valid under the hypothesis (H_1) and (H_2) if $\mathbb{E}$ is a separable Banach space with additional assumptions and the dynamic f satisfies the conditions (i)-(ii)-(iii)'-(iv) of Remark 8.3.2. See e.g. Example 7.3.1.

Now we give a variant of the dynamic programming identity with strategies. Let $\mathcal{A}\subset\mathcal{Y}^1([0,1];\mathbb{S})$ be compact for the convergence in probability (note that $\mathcal{A}$ is then compact for the stable topology) and let $\mathcal{B}\subset\mathcal{Y}^1([0,1];\mathbb{Z})$ be compact (metrizable) for the stable topology, e.g. $\mathcal{B}=\mathcal{R}$. We denote by Δ the set of all continuous mappings (strategies) $\alpha:\mathcal{A}\to\mathcal{B}$. Then the dynamic programming identity becomes as follows.

Theorem 8.3.14 (The dynamic programming identity with strategies) *Let $x\in\mathbb{E}$ and let $\sigma,\tau\in[0,1[$ such that $\tau+\sigma<1$. Let $J:[0,1]\times E\times S\times Z\to\mathbb{R}$ be an L^1-bounded Carathéodory integrand. Let us consider the* lower value function

$$H_J(\tau,x):=\inf_{\alpha\in\Delta}\max_{\mu\in\mathcal{A}}\{\int_\tau^1[\int_{\mathbb{Z}}[\int_{\mathbb{S}}J(t,u_{x,\mu,\alpha(\mu)}(t),s,y,z)\,d\mu_t(s)]\,d\alpha(\mu)_t(z)]\,dt.\}$$

Here $u_{x,\mu,\alpha(\mu)}$ denotes the solution trajectory determined by the relaxed dynamic associated with f and the controls $\alpha(\mu)$, $(\alpha\in\Delta)$ and $\mu\in\mathcal{A}$ starting at position x at time $\tau\in[0,1[$, that is,

$$\dot u_{x,\mu,\alpha(\mu)}(t)=\int_{\mathbb{Z}}[\int_{\mathbb{S}}f(t,u_{x,\mu,\alpha(\mu)}(t),s,z)\,d\mu_t(s)]\,d\alpha(\mu)_t(z);\ u_{x,\mu,\alpha(\mu)}(\tau)=x.$$

Then the following holds:

$$H_J(\tau, x) = \inf_{\alpha\in\Delta} \max_{\mu\in\mathcal{A}}\{ \int_\tau^{\tau+\sigma} [\int_{\mathbb{Z}}[\int_{\mathbb{S}} J(t, u_{x,\mu,\alpha(\mu)}(t), s, z)\, d\mu_t(s)]\, d\alpha(\mu)_t(z)]\, dt$$

$$+ H_J(\tau+\sigma, u_{x,\mu,\alpha(\mu),\nu}(\tau+\sigma)\}. \tag{8.3.7}$$

Here

$$H_J(\tau+\sigma, u_{x,\mu,\alpha(\mu),}(\tau+\sigma))$$
$$:= \inf_{\gamma\in\Delta} \sup_{\mu\in\mathcal{A}}\{ \int_{\tau+\sigma}^{1} [\int_{\mathbb{Z}}[\int_{\mathbb{S}} J(t, v_{x,\mu,\gamma(\mu)}(t), y, z)\, d\mu_t(s)]\, d\gamma(\mu)_t(z)]\, dt\}$$

where $v_{x,\mu,\gamma(\mu)}$ *denotes the trajectory on* $[\tau+\sigma, 1]$ *associated with* $(\gamma(\mu), \mu)$ $(\gamma \in \Delta, \mu \in \mathcal{A})$ *with the initial condition* $v_{x,\mu,\gamma(\mu)}(\tau+\sigma) = u_{x,\mu,\alpha(\mu)}(\tau+\sigma)$.

Proof. Let $K_J(\tau, x)$ be the second member of (8.3.7). Take $\varepsilon > 0$ and $\gamma^1 \in \Delta$ such that

$$K_J(\tau, x) \geq \max_{\mu\in\mathcal{A}}\{ \int_\tau^{\tau+\sigma} [\int_{\mathbb{Z}}[\int_{\mathbb{S}} J(t, u_{x,\mu,\gamma^1(\mu)}(t), s, z)\, d\mu_t(s)]\, d\gamma^1(\mu)_t(z)]\, dt$$
$$+ H_J(\tau+\sigma, u_{x,\mu,\gamma^1(\mu)}(\tau+\sigma)\} - \varepsilon.$$

There is $\gamma^2 \in \Delta$ such that

$$H_J(\tau+\sigma, u_{x,\mu,\gamma^1(\mu)}(\tau+\sigma))$$
$$> \sup_{\mu\in\mathcal{A}} \int_{\tau+\sigma}^{1} [\int_{\mathbb{Z}}[\int_{\mathbb{S}} J(t, v_{x,\mu,\gamma^2(\mu)}(t), s, z)\, d\mu_t(s)]\, d\gamma^2(\mu)_t(z)]dt - \varepsilon.$$

Here $v_{x,\mu,\gamma^2(\mu)}$ denotes the trajectory of the dynamic on $[\tau+\sigma, 1]$ associated with $(\gamma^2(\mu), \mu)$ $(\gamma^2 \in \Delta, \mu \in \mathcal{A})$ and the initial condition $v_{x,\mu,\gamma^2(\mu)}(\tau+\sigma) = u_{x,\mu,\gamma^1(\mu)}(\tau+\sigma)$. In view of the continuity of γ^1 and the Fiber Product Lemma (Theorem 3.3.1), we see that the mapping

$$\mu \mapsto \int_\tau^{\tau+\sigma} [\int_{\mathbb{Z}}[\int_{\mathbb{S}} J(t, u_{x,\mu,\gamma^1(\mu)}(t), s, z)\, \mu_t(s)]\, d\gamma^1(\mu)_t(z)]\, dt$$

is continuous on the compact (for the convergence in probability) set $\mathcal{A}$. Hence we can choose $\mu^1 \in \mathcal{A}$ such that

$$\max_{\mu\in\mathcal{A}} \int_\tau^{\tau+\sigma} [\int_{\mathbb{Z}}[\int_{\mathbb{S}} J(t, u_{x,\mu,\gamma^1(\mu)}(t), s, z)\, d\mu_t(s)]\, d\gamma^1(\mu)_t(z)]\, dt$$
$$= \int_\tau^{\tau+\sigma} [\int_{\mathbb{Z}}[\int_{\mathbb{S}} J(t, u_{x,\mu^1,\gamma^1(\mu^1)}(t), s, z)\, d\mu_t^1(s)]\, d\gamma^1(\mu^1)_t(z)]\, dt$$

where $u_{x,\mu^1,\gamma^1(\mu^1)}$ is the solution trajectory defined on $[\tau, \tau+\sigma]$ by $\gamma^1(\mu^1), (\gamma^1 \in \Delta), \mu^1 \in \mathcal{A}$. Similarly there is $\mu^2 \in \mathcal{A}$ such that

$$\max_{\mu\in\mathcal{A}} \int_{\tau+\sigma}^1 [\int_{\mathbb{Z}} [\int_{\mathbb{S}} J(t, v_{x,\mu,\gamma^2(\mu)}(t), s, z)\, d\mu_t(s)]\, d\gamma^2(\mu)_t(z)]\, dt$$
$$= \int_{\tau+\sigma}^1 [\int_{\mathbb{Z}} [\int_{\mathbb{S}} J(t, v_{x,\mu^2,\gamma^2(\mu^2)}(t), s, z)\, d\mu_t^2(s)]\, d\gamma^2(\mu^2)_t(z)]\, dt.$$

Let us define $\overline{\gamma} \in \Delta$ by setting, for all $\mu \in \mathcal{A}$: $\overline{\gamma}(\mu)_t = \gamma^1(\mu)_t$ for $t \in [\tau, \sigma]$ and $\overline{\gamma}(\mu)_t = \gamma^2(\mu)_t$ for $t \in [\tau+\sigma, 1]$. Coming back to the definition of $H_J(\tau, x)$

$$H_J(\tau, x) \le \sup_{\mu\in\mathcal{A}} \int_{\tau}^1 [\int_{\mathbb{Z}} [\int_{\mathbb{S}} J(t, u_{x,\mu,\overline{\gamma}(\mu)}, s, z)\, d\mu_t(s)]\, d\overline{\gamma}(\mu)_t(z)]\, dt$$
$$= \int_{\tau}^{\tau+\sigma} [\int_{\mathbb{Z}} [\int_{\mathbb{S}} J(t, u_{x,\mu^1,\gamma^1(\mu^1)}, s, z)\, d\mu_t^1(s)]\, d\gamma^1(\mu^1)_t(z)]\, dt$$
$$+ \int_{\tau+\sigma}^1 [\int_{\mathbb{Z}} [\int_{\mathbb{S}} J(t, v_{x,\mu^2,\gamma^2(\mu)}, s, z)\, d\mu_t^2(s)]\, d\gamma^2(\mu^2)_t(z)]\, dt$$
$$\le K_J(\tau, x) + 2\varepsilon.$$

On the other hand, there is $\widetilde{\gamma} \in \Delta$ such that

$$H_J(\tau, x) \ge \sup_{\mu\in\mathcal{A}} \{ \int_{\tau}^1 [\int_{\mathbb{Z}} [\int_{\mathbb{S}} J(t, u_{x,\mu,\widetilde{\gamma}(\mu)}(t), s, z)\, d\mu_t(s)]\, d\widetilde{\gamma}(\mu)_t(z)]\, dt\} - \varepsilon.$$

Then

$$K_J(\tau, x) \le \max_{\mu\in\mathcal{A}} \{ \int_{\tau}^{\tau+\sigma} [\int_{\mathbb{Z}} [\int_{\mathbb{S}} J(t, u_{x,\mu,\widetilde{\gamma}(\mu)}(t), s, z)\, d\mu_t(s)]\, d\widetilde{\gamma}(\mu)_t(z)]\, dt$$
$$+ H_J(\tau+\sigma, u_{x,\mu,\widetilde{\gamma}(\mu)}(\tau+\sigma)\}$$

with

$$H_J(\tau+\sigma, u_{x,\mu,\widetilde{\gamma}(\mu)}(\tau+\sigma)$$
$$\le \sup_{\mu\in\mathcal{A}} \int_{\tau+\sigma}^1 [\int_{\mathbb{Z}} [\int_{\mathbb{S}} J(t, v_{x,\mu,\widetilde{\gamma}(\mu)}(t), s, z)\, \mu_t(s)]\, d\widetilde{\gamma}(\mu)_t(z)] dt.$$

Here $v_{x,\mu,\widetilde{\gamma}}(\mu)$ is the trajectory on $[\tau+\sigma, 1]$ associated with $(\widetilde{\gamma}(\mu^2), \mu^2)$ with the initial condition $v_{x,\mu,\widetilde{\gamma}(\mu)}(\tau+\sigma) = u_{x,\mu,\widetilde{\gamma}(\mu)}(\tau+\sigma)$. As above, by continuity of the integral functionals under consideration and by compactness of $\mathcal{A}$, there exists $\mu^1 \in \mathcal{A}$ such that

$$\max_{\mu\in\mathcal{A}} \int_{\tau}^{\tau+\sigma} [\int_{\mathbb{Z}} [\int_{\mathbb{S}} J(t, u_{x,\mu,\widetilde{\gamma}(\mu)}, s, z)\, \mu_t(s)]\, d\widetilde{\gamma}(\mu)_t(z)]\, dt$$
$$= \int_{\tau}^{\tau+\sigma} [\int_{\mathbb{Z}} [\int_{\mathbb{S}} J(t, u_{x,\mu^1,\widetilde{\gamma}(\mu^1)}, s, z)\, d\mu_t^1(s)]\, d\widetilde{\gamma}(\mu^1)_t(z)]\, dt.$$

Here $u_{x,\widetilde{\gamma}(\overline{\mu^1}),\overline{\mu^1}}$ is the trajectory on $[\tau, \tau+\sigma]$ associated with $(\widetilde{\gamma}(\overline{\mu^1}), \overline{\mu^1})$ and similarly there is $\mu^2 \in \mathcal{A}$ such that

$$\int_{\tau+\sigma}^1 [\int_{\mathbb{Z}} [\int_{\mathbb{S}} J(t, v_{x,\mu^2,\widetilde{\gamma}(\mu^2)}(t), s, z)\, d\widetilde{\gamma}(\overline{\mu}^2)_t(s)]\, d\widetilde{\gamma}(\mu^2)_t(z)]\, dt$$
$$= \max_{\mu\in\mathcal{A}} \int_{\tau+\sigma}^1 [\int_{\mathbb{Z}} [\int_{\mathbb{S}} J(t, v_{x,\mu,\widetilde{\gamma}(\mu)}(t), s, z)\, d\widetilde{\mu}_t(s)]\, d\gamma(\mu)_t(z)]\, dt.$$

Let

$$w_{x,\widetilde{\gamma}(\mu),\mu}(t) = \begin{cases} u_{x,\widetilde{\gamma}(\overline{\mu^1}),\overline{\mu^1}}(t) & \text{for } t \in [\tau, \tau+\sigma] \\ v_{x,\widetilde{\gamma}(\mu^2),\overline{\mu}^2}(t) & \text{for } t \in [\tau+\sigma, 1]. \end{cases}$$

Taking the expression of $K_J(\tau, x)$ into account, we get

$$\begin{aligned} K_J(\tau, x) &\le \int_\tau^1 [\int_{\mathbb{Z}} [\int_{\mathbb{S}} J(t, w_{x,\widetilde{\gamma}(\mu),\mu}(t), s, z)\, d\widetilde{\gamma}(\mu)_t(s)]\, d\mu_t(z)]\, dt \\ &\le \sup_{\mu\in\mathcal{A}} \int_\tau^1 [\int_{\mathbb{Z}} [\int_{\mathbb{S}} J(t, u_{x,\widetilde{\gamma}(\mu),\mu}(t), s, z)\, d\widetilde{\gamma}(\mu)_t(s)]\, d\mu_t(z)]\, dt \\ &\le H_J(\tau, x) + \varepsilon. \end{aligned}$$

□

Remark 8.3.15 The preceding techniques can be applied to the case when $\mathcal{A}$ is equal to $\mathcal{M}^{+,1}(\mathbb{S})$, Γ is the set of all Carathéodory mappings $\alpha : [0,1]\times\mathcal{M}^{+,1}(\mathbb{S}) \to \mathcal{M}^{+,1}(\mathbb{Z})$ and Δ is the set of all mappings $\hat{\alpha} : \mathcal{M}^{+,1}(\mathbb{S}) \to \mathcal{Y}^1([0,1], \mathbb{Z})$ given by $\hat{\alpha}(\mu)_t = \alpha_t(\mu)$ for all $\alpha \in \Gamma, \mu \in \mathcal{M}^{+,1}(\mathbb{S})$ and for all $t \in [0,1]$.

Chapter 9

Stable convergence in limit theorems of probability theory

A. Rényi [Rén66, Rén70] was the first to observe that, in most limit theorems of probability theory, when weak convergence is obtained, it is actually possible to prove stable convergence, which is a much stronger result. We shall illustrate this fact here in locally convex spaces, and give an application of the Fiber Product Lemma to limit theorems with a random number of random vectors.

9.1 Weak limit theorems in locally convex spaces

In order to make further developments readable, we need to recall some definitions and results.

Measures and random elements on locally convex spaces By *LCS* we shall always mean "real locally convex Hausdorff topological vector space with metrizable compact subsets". This implies in particular that a measure on an LCS is Radon if and only if it is tight (see page 12), and that, in this case, it has a separable support (see page 96).

Let $\mathbb{E}$ be an LCS. Let $\mathrm{Cyl}(\mathbb{E})$ denote the algebra of sets of the form

$$\{x \in \mathbb{E};\ (\langle x_1{}^*, x\rangle, \dots, \langle x_n{}^*, x\rangle) \in B\}$$

where $n \geq 1$, $(x_1{}^*, \dots, x_n{}^*) \in (\mathbb{E}^*)^n$ and $B \in \mathcal{B}_{\mathbb{R}^n}$. The elements of $\mathrm{Cyl}(\mathbb{E})$ are called *cylinder sets*. Let $\widehat{\mathrm{Cyl}}(\mathbb{E})$ denote the σ–algebra generated by $\mathrm{Cyl}(\mathbb{E})$. From [VTC87, Theorem I.3.4], every tight measure on $\widehat{\mathrm{Cyl}}(\mathbb{E})$ has a unique extension to a Radon measure on $\mathcal{B}_{\mathbb{E}}$. In particular, if two Radon measures on $\mathcal{B}_{\mathbb{E}}$ coincide on $\mathrm{Cyl}(\mathbb{E})$, they are equal. Note that, if $\mathbb{E}$ is a countable union of metrizable compact subsets, it is not difficult to prove that $\widehat{\mathrm{Cyl}}(\mathbb{E}) = \mathcal{B}_{\mathbb{E}}$ (see [Bad70, page 44] or [VTC87, Proposition I.1.6]).

Let $p \geq 0$. We say that a Radon measure μ on an LCS $\mathbb{E}$ has *strong p–th order* if, for each continuous seminorm $\|.\|$ on $\mathbb{E}$, we have $\int_{\mathbb{E}} \|x\|^p \, d\mu(x) < +\infty$.

We say that μ has *weak p–th order* if, for every $x^* \in \mathbb{E}^*$, the measure $(x^*)_\sharp \mu$ on $\mathbb{R}$ has a moment of order p.

We call *mean* of a Radon measure $\mu \in \mathcal{M}_R^{+,1}(\mathbb{E})$ (or of a random vector X such that $\mathcal{L}(X) = \mu$) the Pettis integral, if it exists, of μ, that is, the only element m of $\mathbb{E}$ such that $\langle x^*, m\rangle = \int_{\mathbb{E}} \langle x^*, x\rangle \, d\mu(x)$ for every $x^* \in \mathbb{E}^*$. The vector m is then denoted by $E(\mu)$ (or $E(X)$). If $\mathbb{E}$ is reflexive and $\mu \in \mathcal{M}_R^{+,1}(\mathbb{E})$ has weak first order, then μ has a mean. The same conclusion holds true if $\mathbb{E}$ is a Fréchet space and μ has strong first order, see e.g. [CRdF00, Proposition 1] or [VT78, Theorem 4]. If $E(\mu) = 0$, we say that μ (or X) is *centered.*

The *covariance operator* of a law $\mu \in \mathcal{M}_R^{+,1}(\mathbb{E})$ with weak second order and with mean m is the mapping

$$\mathrm{Cov}_\mu : \left\{ \begin{array}{ccc} \mathbb{E}^* \times \mathbb{E}^* & \rightarrow & \mathbb{R} \\ (x^*, y^*) & \mapsto & \int_{\mathbb{E}} \langle x^*, x - m\rangle \langle y^*, x - m\rangle \, d\mu(x). \end{array} \right.$$

We say that a Radon law $\mu \in \mathcal{M}_R^{+,1}(\mathbb{E})$ is *Gaussian* if, for every $x^* \in \mathbb{E}^*$, the measure $(x^*)_\sharp \mu$ on $\mathbb{R}$ is Gaussian (see [Bog98a]). Then μ is characterized by its mean $E(\mu)$ and its covariance operator, and we denote it by $\mathcal{N}(m, \mathrm{Cov}_\mu)$.

In this chapter, we are interested in limits of weighted sums of random vectors of an LCS $\mathbb{E}$. Note that, if X and Y are random elements of $\mathbb{E}$, then $X + Y$ is not necessarily $\mathcal{B}_{\mathbb{E}}$–measurable (see [Sto76]), because we do not necessarily have $\mathcal{B}_{\mathbb{E}\times\mathbb{E}} = \mathcal{B}_{\mathbb{E}} \otimes \mathcal{B}_{\mathbb{E}}$ (see Theorem 2.1.13.C about this equality). However, $X + Y$ is obviously $\widehat{\mathrm{Cyl}}(\mathbb{E})$–measurable. If X and Y have Radon laws, then it is easy to check that $X + Y$ is measurable, with Radon law. Indeed, there exists an increasing sequence $(K_m)_{m\geq 1}$ of compact subsets of $\mathbb{E}$ such that, for each m, the set $\Omega_m = \{X \in K_m \text{ and } Y \in K_m\}$ satisfies $\mathrm{P}(\Omega_m) \geq 1 - 1/m$. Let $A_1 = \Omega_1$ and, for $m > 1$, let $A_m = \Omega_m \setminus \Omega_{m-1}$. For each $m \geq 1$, the restriction to A_m of $X + Y$ is measurable, thus $X + Y$ is measurable.

Another advantage of Radon laws is that, on an LCS, weak convergence of Radon laws amounts to narrow convergence of these laws (see page 13).

For the sake of simplicity, we shall in the sequel consider only random vectors which have Radon laws. If $\mathbb{E}$ is a Fréchet space and contains a dense subspace with non–measurable cardinal, then every Borel measure on $\mathbb{E}$ is separable (see Section 1.4), thus Radon.

Weak limit theorems in locally convex spaces Let $\mathbb{E}$ be an LCS. Let $(X_n)_{n\geq 1}$ be a sequence of centered random elements of $\mathbb{E}$, with Radon laws, and let $(C_n)_{n\geq 1}$ be a sequence of positive real numbers. For every $n \geq 1$, let

$S_n = \sum_{k=1}^n X_k$. We say that $(X_n)_{n\geq 1}$ satisfies a *weak limit theorem* with *norming coefficients* $(C_n)_{n\geq 1}$ if there exists a Radon measure $\gamma \in \mathcal{M}^{+,1}(\mathbb{E})$ such that $\mathcal{L}(S_n/C_n)$ weakly (thus narrowly) converges to γ (for the clarity of the presentation, we shall not consider limit theorems for triangular arrays). If the limit γ is Gaussian, we say that $(X_n)_{n\geq 1}$ satisfies the *Central Limit Theorem* (for short *CLT*). When the norming coefficients C_n are not specified, this means that $C_n = \sqrt{n}$.

It is well-known that a Banach space $\mathbb{E}$ has type 2 if and only if, for every i.i.d. sequence $(X_n)_{n\geq 1}$ with common centered law $\mu \in \mathcal{M}_R^{+,1}(\mathbb{E})$ of strong order 2, $(X_n)_{n\geq 1}$ satisfies the CLT; conversely, $\mathbb{E}$ has cotype 2 if and only if, for every i.i.d. sequence $(X_n)_{n\geq 1}$ with common centered law $\mu \in \mathcal{M}_R^{+,1}(\mathbb{E})$ which satisfies the CLT, μ has strong second order; a Banach space has type 2 and cotype 2 if and only if it is isomorphic to a Hilbert space (see [AG80, LT91]).

A Banach space in which every i.i.d. sequence $(X_n)_{n\geq 1}$ of centered random vectors with common law $\mu \in \mathcal{M}_R^{+,1}(\mathbb{E})$ of weak order 2 satisfies the CLT is necessarily finite dimensional; however, in some non–normable infinite dimensional spaces, in particular in duals of nuclear Fréchet spaces, the CLT holds for all i.i.d. sequences with weak second order [Bog86]. We say that an LCS $\mathbb{F}$ is *nuclear* if there exists a net $(\langle .,. \rangle_\alpha)_{\alpha\in\mathbb{A}}$ of pre-Hilbertian scalar products on $\mathbb{F}$ which defines the topology of $\mathbb{F}$ and such that, for each $\alpha \in \mathbb{A}$, there exists $\beta \in \mathbb{A}$ such that the associated norm $\|.\|_\beta$ is finer than $\|.\|_\alpha$ and such that the canonical mapping $(\widehat{\mathbb{F}}_\beta, \langle .,. \rangle_\beta) \to (\widehat{\mathbb{F}}_\alpha, \langle .,. \rangle_\alpha)$ is Hilbert-Schmidt, where $\widehat{\mathbb{F}}_\alpha$ denotes the completion of $\mathbb{F}/\{\|.\|_\alpha = 0\}$ for $\|.\|_\alpha$ [Sch73, Proposition 5 page 220].

Every Fréchet nuclear space $\mathbb{F}$ is a Montel space [Trè67, Corollary 3 to Proposition 50.2], thus it is reflexive [Trè67, Corollary to Proposition 36.9] and its strong dual $\mathbb{G} = \mathbb{F}_b^*$ is a complete barrelled nuclear space. We have seen in Example 4.4.1 that $\mathbb{G}$ is a submetrizable k_ω–space, and, in particular, it is Prohorov and Radon. An important example of dual of a nuclear Fréchet space is the Schwartz space $\mathcal{S}'$ of tempered distributions (see e.g. [Trè67]).

Recall that the characteristic function (or Fourier transform) Φ_μ of a law $\mu \in \mathcal{M}^{+,1}(\mathbb{G})$ is the mapping

$$\Phi_\mu : \begin{cases} \mathbb{G}^* & \to \quad \mathbb{C} \\ x^* & \mapsto \quad \int_{\mathbb{G}} e^{i\langle x^*, x\rangle}\, d\mu(x). \end{cases}$$

Here, as $\mathbb{G}$ is reflexive, its dual space $\mathbb{G}^*$ is $\mathbb{F}$. Let $(\mu_n)_n$ be a sequence in $\mathcal{M}^{+,1}(\mathbb{G})$. Fernique [Mey66], extending Lévy's continuity theorem, proved that

1. if $(\Phi_{\mu_n})_n$ is equicontinuous (equivalently, if $(\Phi_{\mu_n})_n$ is equicontinuous at 0), then $(\mu_n)_n$ is tight;

2. if $(\Phi_{\mu_n})_n$ is pointwise convergent to some continuous function $\Phi : \mathbb{F} \to \mathbb{C}$, then Φ is the characteristic function of some Radon probability law μ on $\mathbb{G}$ and $(\mu_n)_n$ narrowly converges to μ.

To show the equicontinuity at 0 of $(\Phi_{\mu_n})_n$, one can estimate the weak second moments of $(\mu_n)_n$. Indeed, it is easy to check that, for any $t \in \mathbb{R}$, we have

$$\left|1 - e^{it}\right|^2 = 4\sin^2\left(\frac{t}{2}\right) \le t^2,$$

thus, for any $n \ge 1$ and any $x^* \in \mathbb{F}$, using Jensen inequality,

$$\left|1 - \Phi_{\mu_n}(x^*)\right|^2 \le \int_{\mathbb{G}} \left|1 - e^{i\langle x^*, x\rangle}\right|^2 \, d\mu_n(x) \le \int_{\mathbb{G}} \langle x^*, x\rangle^2 \, d\mu_n(x).$$

For each $n \ge 1$, let X_n be a random vector of $\mathbb{G}$ such that $\mathcal{L}(X_n) = \mu_n$ and consider the operator

$$T_n : \begin{cases} \mathbb{F} & \rightarrow \quad L^2_{\mathbb{R}}(\Omega, \mathcal{S}, P) \\ x^* & \mapsto \quad \langle x^*, X_n\rangle . \end{cases}$$

The space $\mathbb{F}$ is barrelled, thus, from [VT78, Corollary 2], each T_n is continuous. Furthermore, the sequence $(T_n)_n$ is equicontinuous if and only if it is pointwise bounded (see [Bou81, Theorem 1 page III.25] or [Sch66, Theorem 4.2 page 83]). We thus have the following lemma.

Lemma 9.1.1 *Let $\mathbb{F}$ be a Fréchet nuclear LCS. Let $\mathbb{G}$ be the strong dual of $\mathbb{F}$. Let $(\mu_n)_n$ be a sequence of laws with weak second order on $\mathbb{G}$. If, for each $x^* \in \mathbb{F}$, the sequence $\left(\int_{\mathbb{G}} \langle x^*, x\rangle^2 \, d\mu_n(x)\right)_n$ is bounded, then $(\mu_n)_n$ is tight.*

The following result is a variant of [Bog86, Theorem 2].

Theorem 9.1.2 (Central limit theorem in the dual of a Fréchet nuclear space) *Assume that $\mathbb{F}$ is a Fréchet nuclear LCS. Let $\mathbb{G}$ be the strong dual of $\mathbb{F}$. Let $(X_n)_{n\ge 1}$ be a sequence of random vectors of $\mathbb{G}$, with (necessarily Radon) laws of weak order 2. For each $n \ge 1$, let $S_n = X_1 + \cdots + X_n$. Let $(C_n)_{n\ge 1}$ be a sequence of positive real numbers. Assume that*

1. *for each finite sequence $({x_1}^*, \ldots, {x_m}^*)$ of elements of $\mathbb{F}$, the sequence $((\langle {x_1}^*, X_n\rangle, \ldots, \langle {x_m}^*, X_n\rangle)_n$ satisfies the CLT in $\mathbb{R}^m$ with norming coefficients $(C_n)_{n\ge 1}$,*
2. *for each $x^* \in \mathbb{F}$, the sequence $\left(E\left(\langle x^*, S_n/C_n\rangle^2\right)\right)_n$ is bounded.*

Then there exists a Gaussian Radon probability law γ on $\mathbb{G}$ such that

$$\mathcal{L}\left(\frac{1}{C_n} S_n\right) \xrightarrow{\text{narrow}} \gamma.$$

Proof. For each $n \ge 1$, let

$$\mu^{*n} = \mathcal{L}\left(\frac{1}{C_n} S_n\right).$$

From Lemma 9.1.1, the sequence $(\mu^{*n})_n$ is tight. Now, from the finite dimensional CLT in Condition 1, all limits of $(\mu^{*n})_n$ coincide on $\mathrm{Cyl}(\mathbb{G})$, thus they are equal. Thus $(\mu^{*n})_n$ weakly (thus narrowly) converges to some $\gamma \in \mathcal{M}^{+,1}(\mathbb{G})$. Furthermore, by Condition 1, each $(x^*)_\sharp \mu$ is Gaussian, thus γ is Gaussian. □

Corollary 9.1.3 ([Bog86]) *Assume that $\mathbb{F}$ is a Fréchet nuclear LCS. Let $\mathbb{G}$ be the strong dual of $\mathbb{F}$. Let $(X_n)_{n\geq 1}$ be a centered i.i.d. sequence with common law $\mu \in \mathcal{M}^{+,1}(\mathbb{G})$ of weak order 2. Then $(X_n)_{n\geq 1}$ satisfies the CLT, that is,*

$$\mathcal{L}\left(\frac{1}{\sqrt{n}}(X_1+\cdots+X_n)\right) \xrightarrow{narrow} \mathcal{N}(0, \mathrm{Cov}_\mu)$$

(this proves in particular that there exists a Gaussian law on $\mathbb{E}$ with same covariance as μ!).

Proof. Condition 1 of Theorem 9.1.2 is obviously satisfied. Furthermore, with the notations of the proof of Theorem 9.1.2, we have

$$\int_{\mathbb{G}} \langle x^*, x\rangle^2 \, d\mu^{*n}(x) = E\left(\left\langle x^*, \frac{1}{\sqrt{n}}S_n\right\rangle^2\right) = \int_{\mathbb{G}} \langle x^*, x\rangle^2 \, d\mu(x)$$

which proves Condition 2. Thus $(\mu^{*n})_n$ converges to some Gaussian law $\gamma \in \mathcal{M}^{+,1}(\mathbb{G})$. From the finite dimensional CLT, we have

$$\gamma = \mathcal{N}(0, \mathrm{Cov}_\mu).$$

□

Remark 9.1.4 The strong dual of an inductive limit of locally convex Fréchet spaces is the projective limit of the strong duals of these spaces, see [Sch66, pages 140 and 146]. Thus Theorem 9.1.2 and Corollary 9.1.3 have obvious extensions to the case when $\mathbb{F}$ is the inductive limit of a sequence $(\mathbb{F}_n)_n$ of Fréchet nuclear locally convex spaces.

9.2 More on stable convergence

Portmanteau Theorem revisited We shall need in this chapter the following easy reduction lemma, which holds without any hypothesis on $\mathbb{T}$. The technique of the proof has already been used in Lemma 4.1.1)

Lemma 9.2.1 *Let $(\mu^\alpha)_{\alpha\in\mathbb{A}}$ be a net in $\mathcal{Y}^1_{\mathrm{dis}}$ and let $\mu^\infty \in \mathcal{Y}^1_{\mathrm{dis}}$. Let $\mathcal{S}_0$ be the σ-algebra generated by the random laws $\mu^\alpha_{.}$, $\alpha \in \mathbb{A}\cup\{\infty\}$. The following conditions are equivalent.*

(a) $\mu^\alpha \xrightarrow{W\text{-stably}} \mu^\infty$.

(b) $\mu^\alpha \xrightarrow{\mathcal{S}_0\text{-}W\text{-stably}} \mu^\infty$.

Proof. The implication (a)⇒(b) is obvious.

Conversely, assume (b) and let $g : \Omega \to [0,1]$ be measurable and $f : \mathbb{T} \to [0,1]$ be continuous. The conditional expectation $E^{\mathcal{S}_0}(g)$ of g w.r.t. $\mathcal{S}_0$ is a bounded measurable function thus we have, from Part D of Theorem 2.1.3,

$$\mu^\infty(g \otimes f) = \int_\Omega g(\omega)\mu^\infty_\omega(f)\,d\,\mathrm{P}(\omega) = \int_\Omega \left[E^{\mathcal{S}_0}(g)(\omega)\right]\mu^\infty_\omega(f)\,d\,\mathrm{P}(\omega)$$
$$= \lim_\alpha \int_\Omega \left[E^{\mathcal{S}_0}(g)(\omega)\right]\mu^\alpha_\omega(f)\,d\,\mathrm{P}(\omega) = \lim_\alpha \mu^\alpha(g \otimes f).$$

From Theorem 2.1.3, this proves (a). □

The next result is an easy consequence of results of Chapters 2, 3 and 4. For any $A \in \mathcal{S}$ such that $\mathrm{P}(A) > 0$, we denote by P^A the probability on A defined by $\mathrm{P}^A(B) = (1/\mathrm{P}(A))\,\mathrm{P}(B)$ for every measurable $B \subset A$, and, for any $X \in \mathfrak{X}(\mathbb{E})$, we denote by $X|_A$ the restriction to A of X. If X is a random element, we denote by $\mathcal{L}^{(\mathrm{P}^A)}(X)$ its law under P^A.

Proposition 9.2.2 (Portmanteau Theorem for stable convergence of random vectors) *Let $\mathbb{T}$ be a completely regular topological space with metrizable compact subsets. Let $(X_n)_n$ be a* strictly tight *sequence of random elements of $\mathbb{T}$. Let $\mathcal{S}_0$ be a sub-σ-algebra of $\mathcal{S}$ such that each X_n is $\mathcal{S}_0$-measurable. Let $\mathcal{C}$ be a subset of $\mathcal{S}_0$ which is stable under finite intersections and generates $\mathcal{S}_0$. Let $\mathcal{H}$ be a set of bounded continuous functions on $\mathbb{T}$ which separate the elements of $\mathcal{M}^{+,1}_R(\mathbb{T})$ (i.e. for all $\mu, \nu \in \mathcal{M}^{+,1}_R(\mathbb{T})$ such that $\mu \neq \nu$, there exists $f \in \mathcal{H}$ such that $\mu(f) \neq \nu(f)$). Finally, let $\mathbb{S}$ be a regular cosmic topological space which contains at least two elements. The following conditions are equivalent:*

1. *$(X_n)_n$ is $\mathcal{S}$-stably convergent.*
2. *For each $A \in \mathcal{C}$ and for each $f \in \mathcal{H}$, the sequence $(E(\mathbb{1}_A f \circ X_n))_n$ is convergent.*
3. *For each $A \in \mathcal{C}$ such that $\mathrm{P}(A) > 0$, the sequence $\left(\mathcal{L}^{(\mathrm{P}^A)}\left(X_n|_A\right)\right)_n$ is narrowly (equivalently, weakly) convergent.*
4. *For any sequence $(Z_n)_n$ of random elements of $\mathbb{S}$ with Radon laws which converges in probability to a random element Z of $\mathbb{T}$, the sequence $(\mathcal{L}(X_n, Z_n))_n$ is narrowly (equivalently, weakly) convergent.*
5. *For any $\mathcal{S}_0$-measurable random element Z of $\mathbb{S}$ with Radon law, the sequence $(\mathcal{L}(X_n, Z))_n$ is narrowly (equivalently, weakly) convergent.*

If these conditions are satisfied, the limit $\mu^{\infty}_{.}$ of $(X_n)_n$ is $\mathcal{S}_0$-measurable. Moreover, with the notations of Condition 4, we have

$$\lim_{n\to+\infty} \mathcal{L}(X_n, Z_n) = \mu^{\infty}_{.} \underline{\otimes}\, \underline{\delta}_Z. \tag{9.2.1}$$

Proof. With the help of Lemma 9.2.1, the equivalence $1 \Leftrightarrow 2 \Leftrightarrow 3$ is immediate, following the lines of Theorem 4.3.8.

Now, as $(X_n)_n$ is strictly tight, by the Change of Topology Lemma 4.4.3, there exists a K_σ subset $\mathbb{T}_0$ of $\mathbb{T}$ and a topology τ_0' on $\mathbb{T}_0$ which is Lusin regular and finer than the topology induced by $\mathbb{T}$ and has the same Borel subsets, such that all X_n take their values in $\mathbb{T}_0$ with probability 1 and the topology $\mathcal{Y}^1(\mathbb{T}_0, \tau_0')$ coincides with $\mathcal{Y}^1(\mathbb{T})$ on the closure in $\mathcal{Y}^1(\mathbb{T})$ of $(X_n)_n$ (we have identified each X_n with $\underline{\delta}_{X_n}$). Thus we can consider without loss of generality that $\mathbb{T}$ is Lusin regular. Then the space $\mathbb{T} \times \mathbb{S}$ is hereditarily Lindelöf regular.

The implication $1 \Rightarrow 4$ and Formula (9.2.1), are obvious from Corollary 3.3.5, or from Theorem 3.3.1 and the observation that $\underline{\delta}_X \underline{\otimes}\, \underline{\delta}_Z = \underline{\delta}_{(X,Z)}$ for all $X \in \mathfrak{X}(\mathbb{E})$ and all $Z \in \mathfrak{X}(\mathbb{S})$.

The implication $4 \Rightarrow 5$ is clear.

Now, we only need to prove $5 \Rightarrow 2$. Let $A \in \mathcal{S}$ and let f be a bounded continuous function on $\mathbb{T}$. Let a and b be two distinct elements of $\mathbb{S}$ and let $Z \in \mathfrak{X}(\mathbb{S})$ be defined by $Z(\omega) = a$ if $\omega \in A$ and $Z(\omega) = b$ if $\omega \in B$. Let $g : \mathbb{S} \to [0, 1]$ be a continuous function such that $g(a) = 1$ and $g(b) = 0$. Let $h = f \otimes g$. Assuming 5, the sequence $(X_n, Z)_n$ converges in law to some probability law ν on $\mathbb{T} \times \mathbb{S}$. We thus have

$$\lim_n \underline{\delta}_{X_n} (\mathbb{1}_A \otimes f) = \lim_n E(\mathbb{1}_A f(X_n)) = \lim_n E(h(X_\alpha, Z)) = \nu(h),$$

which proves 2.

Finally, the $\mathcal{S}_0$-measurability of the limit $\mu^{\infty}_{.}$ can be deduced from e.g. Lemma 4.5.4. □

Remark 9.2.3 The equivalence $5 \Leftrightarrow 1$ appears in [AE78].

Corollary 9.2.4 *Let $\mathbb{E}$ be an LCS. Let $(X_n)_n$ be a* strictly tight *sequence of random elements of $\mathbb{E}$. Let $\mathcal{S}_0$ be a sub-σ-algebra of $\mathcal{S}$ such that each X_n is $\mathcal{S}_0$-measurable. Let $\mathcal{C}$ be a subset of $\mathcal{S}_0$ which is stable under finite intersections and generates $\mathcal{S}_0$. The following conditions are equivalent:*

1. *$(X_n)_n$ is $\mathcal{S}$-stably convergent.*
2. *For each $x^* \in \mathbb{E}^*$ and for each $A \in \mathcal{C}$, the sequence $\left(E\left(\mathbb{1}_A e^{i\langle x^*, X_n\rangle}\right)\right)_n$ is convergent.*

Proof. The implication $1 \Rightarrow 2$ is obvious. Now, the functions $\phi_{x^*} : x \mapsto \exp(i\langle x^*, x\rangle)$ separate the measures on $\widehat{\mathrm{Cyl}}(\mathbb{E})$, thus they separate Radon measures on $\mathbb{E}$. Thus $2 \Rightarrow 1$ follows from the corresponding implication $2 \Rightarrow 1$ in Proposition 9.2.2. □

Rényi–mixing Let $(X_\alpha)_\alpha$ be a net in $\mathfrak{X}(\mathbb{T})$. We say that $(X_\alpha)_\alpha$ is *Rényi–mixing* if $(X_\alpha)_\alpha$ W–stably converges to a homogeneous limit $\mu^\infty \in \mathcal{Y}^1_{\text{dis}}$ (that is, μ^∞ has the form $\mu = \mathrm{P} \otimes \nu$ for some $\nu \in \mathcal{M}^{+,1}(\mathbb{T})$). The following lemma is easily deduced from Parts D and E of Theorem 2.1.3 and from Lemma 9.2.1. We skip its demonstration.

Lemma 9.2.5 *Let $(X_\alpha)_\alpha$ be a net in $\mathfrak{X}(\mathbb{T})$ and let $\nu \in \mathcal{M}^{+,1}(\mathbb{T})$. Let $\mathcal{S}_0$ be the σ–algebra generated by $(X_\alpha)_\alpha$. Let $\mathcal{C}$ be a set of nonnegative $\mathcal{S}$–measurable bounded functions which is stable under multiplication of two elements, which contains the constant function $\mathbb{1}_\Omega$ and which generates $\mathcal{S}_0$. The following conditions are equivalent:*

(a) $(X_\alpha)_\alpha$ is Rényi–mixing with limit $\mathrm{P} \otimes \nu$.

(b) For any $A \in \mathcal{C}$ and for any $f \in \mathrm{C}_b(\mathbb{T})$, we have $\lim_\alpha E\left(f(X_\alpha)\,\mathbb{1}_A\right) = \mathrm{P}(A)\nu(f)$.

If furthermore $\mathbb{T}$ is is hereditarily Lindelöf and regular (thus completely regular) and $\mathcal{D}$ is a σ–upwards filtering family of semidistances which defines the topology of $\mathbb{T}$, then Conditions (a) and (b) are equivalent to

(c) For every $A \in \mathcal{C}$, for every $d \in \mathcal{D}$ and for any $f \in \mathrm{BL}_1(\mathbb{T}, d)$, we have $\lim_\alpha E\left(f(X_\alpha)\,\mathbb{1}_A\right) = \mathrm{P}(A)\nu(f)$.

W–Stable convergence of random elements generalizes both Rényi–mixing and convergence in probability, as illustrated in the following diagram, assuming that $\mathbb{T}$ is separable completely regular (see Theorem 3.1.2).

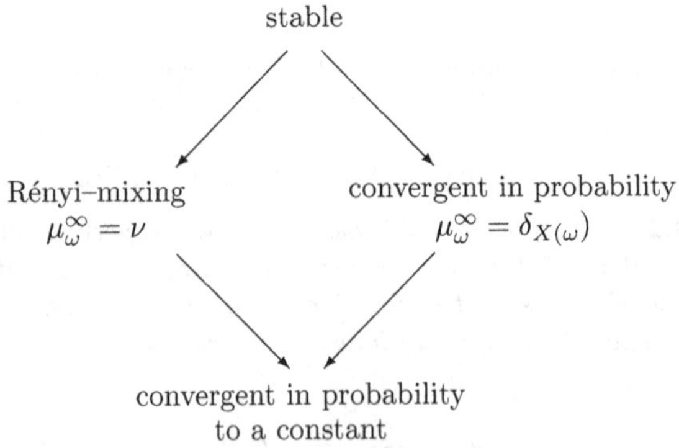

Change of probability Let us start with a simple observation. Let Q be a probability on $(\Omega, \mathcal{S})$ which is absolutely continuous w.r.t. P. Let $(X_\alpha)_\alpha$ be a net of random elements of a topological space $\mathbb{T}$ which $*$–stably converges in $\mathcal{Y}^1(\Omega, \mathcal{S}, \mathrm{P})$ to a Young measure $\mu \in \mathcal{Y}^1(\Omega, \mathcal{S}, \mathrm{P})$, $* = \mathrm{S}, \mathrm{M}, \mathrm{N}, \mathrm{W}$. The following results are straightforward:

1. $(X_\alpha)_\alpha$ $*$–stably converges in $\mathcal{Y}^1(\Omega, \mathcal{S}, \mathrm{Q})$ to the measure $\mu^{\mathrm{Q}} = \left(\frac{d\,\mathrm{Q}}{d\,\mathrm{P}} \otimes \mathbb{1}_{\mathbb{T}}\right).\mu$;

2. If furthermore $(X_\alpha)_\alpha$ is Rényi–mixing w.r.t. P, then it is Rényi–mixing w.r.t. Q.

This makes the difference with convergence in law. For example, let $\Omega = [0,1]$ and $\mathrm{P} = dt$ be the Lebesgue measure. Let $\mathrm{Q} = 2t\,dt$. Let $(a_n)_{n\in\mathbb{N}}$ be a sequence of elements of $[0, 1/2]$. For each $n \in \mathbb{N}$, set $X_n = \mathbb{1}_{[a_n, a_n+1/2]}$. The law of X_n under P is

$$\mathcal{L}^{(\mathrm{P})}(X_n) = \frac{1}{2}\delta_0 + \frac{1}{2}\delta_1,$$

thus $(X_n)_n$ is convergent in law under P. But we have

$$\mathcal{L}^{(\mathrm{Q})}(X_n) = c_n\delta_0 + (1 - c_n)\delta_1,$$

with

$$c_n = \int_{[0,a_n]\cup[a_n+1/2,1]} 2t\,dt = \frac{3}{4} - a_n,$$

thus, if $(a_n)_n$ is not convergent, $(X_n)_n$ is not convergent in law under Q.

Proposition 9.2.6 *Assume that* $\mathbb{T}$ *is completely regular separable and that each compact subset of* $\mathbb{T}$ *is metrizable. Let* (X_α) *be a* strictly tight *net of random elements of* $\mathbb{T}$. *The following assertions are equivalent:*

(*a*) $(X_\alpha)_\alpha$ *W–stably converges in* $\mathcal{Y}^1(\Omega, \mathcal{S}, \mathrm{P})$.

(*b*) $(X_\alpha)_\alpha$ *S–stably converges in* $\mathcal{Y}^1(\Omega, \mathcal{S}, \mathrm{P})$.

(*c*) *For any probability* Q *on* $(\Omega, \mathcal{S})$ *which is absolutely continuous w.r.t.* P, $(X_\alpha)_\alpha$ *converges in law under* Q *to some probability* ν^{Q} *on* $\mathbb{T}$.

(*d*) *For any probability* Q *on* $(\Omega, \mathcal{S})$ *which is equivalent to* P, $(X_\alpha)_\alpha$ *converges in law under* Q *to some probability* ν^{Q} *on* $\mathbb{T}$.

If these conditions are satisfied, $(X_\alpha)_\alpha$ *S–stably converges to a disintegrable Young measure* μ *such that, for any probability* Q *on* $(\Omega, \mathcal{S})$ *which is absolutely continuous w.r.t.* P,

$$\nu^{\mathrm{Q}} = \int_\Omega \mu_\omega\, d\,\mathrm{Q}(\omega). \tag{9.2.2}$$

Proof. The equivalence $(a)\Leftrightarrow(b)$ is contained in Theorem 4.3.8. Furthermore, from Lemma 4.3.1, we know that the limit μ of $(X_\alpha)_\alpha$ for $\tau^{\mathrm{S}}_{\mathcal{Y}^1}$ will be disintegrable. The implications $(a)\Rightarrow(c)\Rightarrow(d)$ are obvious. Assume now (d). Let $A \in \mathcal{S}$ such that $\mathrm{P}(A) > 0$. For each $\varepsilon \in \,]0,1[$, let Q^ε be the probability on $(\Omega, \mathcal{S})$ with density

$$g^\varepsilon = \frac{1-\varepsilon}{\mathrm{P}(A)}\,\mathbb{1}_A + \frac{\varepsilon}{\mathrm{P}(A^c)}\,\mathbb{1}_{A^c}.$$

The probability Q^ε is equivalent to P, thus $(X_\alpha)_\alpha$ converges in law under Q^ε to some probability ν^ε on $\mathbb{T}$, that is, for any continuous $f : \mathbb{T} \to [0,1]$, the limit along α of

$$a_\alpha^\varepsilon := \frac{1-\varepsilon}{P(A)} \int_A f \circ X_\alpha \, dP + \frac{\varepsilon}{P(A^c)} \int_{A^c} f \circ X_\alpha \, dP$$

exists. Let a^ε be this limit. On the other hand, when ε goes to 0, a_α^ε converges, uniformly w.r.t. α, to

$$a_\alpha := \frac{1}{P(A)} \int_A f \circ X_\alpha \, dP\,.$$

It is classical and easy to check that both $(a_\alpha)_\alpha$ and $(a^\varepsilon)_\varepsilon$ are Cauchy and that they converge to the same limit. As $(X_\alpha)_\alpha$ is strictly tight in $\mathcal{Y}^1(\Omega, \mathcal{S}, P)$, this proves (a), using the last part of Theorem 4.3.8.

To prove (9.2.2), if $\mu \in \mathcal{Y}^1_{\text{dis}}$ and Q is a probability on $(\Omega, \mathcal{S})$ which is equivalent to P, the mixing

$$\nu^Q := \int_\Omega \mu_\omega \, dQ(\omega)$$

satisfies for any Borel $f : \mathbb{T} \to [0,1]$ the following formula:

$$\int_{\mathbb{T}} f \, d\nu^Q = \int_\Omega \Big[\int_\Omega f(x) \, d\mu_\omega(x) \Big] \, dQ(\omega) = \int_{\Omega\times\mathbb{T}} f(x) \frac{dQ}{dP}(\omega) \, d\mu(\omega, x)\,.$$

Now Formula (9.2.2) is easily explained: If moreover f is continuous, we have

$$\int_\Omega f(X_\alpha(\omega)) \, dQ(\omega) = \int_{\Omega\times\mathbb{T}} f(x) \frac{dQ}{dP}(\omega) \, d\underline{\delta}_{X_\alpha}(\omega, x) \to \int_{\Omega\times\mathbb{T}} f(x) \frac{dQ}{dP}(\omega) \, d\mu(\omega, x).$$

□

Remark 9.2.7 If $\mathbb{T}$ is Prohorov, and if $(X_\alpha)_\alpha$ is a sequence, we can drop the assumption of tightness in Theorem 9.2.6: Indeed, this assumption is automatically satisfied in each of Conditions (a), (b), (c) and (d).

This remark also applies to Corollary 9.2.8 below.

The equivalence $(a)\Leftrightarrow(e)$ in the following theorem is a generalization of a result of Padmanabhan [Pad70, Theorem 2.1].

Corollary 9.2.8 (Padmanabhan Criterion) *Assume that $\mathbb{T}$ is completely regular and that the compact subsets of $\mathbb{T}$ are metrizable. Let $(X_\alpha)_\alpha$ be a* strictly tight *net of random elements of $\mathbb{T}$. Let X be a random element of $\mathbb{T}$. The following propositions are equivalent:*

(a) *For any probability Q on $(\Omega, \mathcal{S})$ which is absolutely continuous w.r.t. P, $(X_\alpha)_\alpha$ converges in law under Q to X.*

(b) *For any probability Q on $(\Omega, \mathcal{S})$ which is equivalent to P, $(X_\alpha)_\alpha$ converges in law under Q to X.*

(c) $(X_\alpha)_\alpha$ *W–stably converges to* X.

(d) $(X_\alpha)_\alpha$ *S–stably converges to* X.

(e) $(X_\alpha)_\alpha$ *converges in* P*–probability to* X.

Proof. From Theorem 9.2.6, we have (a)⇔(b)⇔(c)⇔(d).

The equivalence (c)⇔(e) comes from the fact that, from Theorem 3.1.2, the topologies $\tau_{\text{prob}}\left(\mathcal{Y}^1_{\text{dis}}\right)$ and $\tau^*_{\mathcal{Y}^1}$ coincide on $\mathfrak{X}$ (in Chapter 3, and in particular in the proof of Theorem 3.1.2, we assumed that $\mathbb{T}$ contains a dense subspace with non–measurable cardinal, e.g. $\mathbb{T}$ is separable, see Section 1.4; here, from strict tightness, all Young measures considered here are supported by $\Omega \times \mathbb{T}_0$, where $\mathbb{T}_0$ is a countable union of compact metrizable spaces, thus Theorem 3.1.2 applies without any restriction). □

9.3 Rényi–mixing Central Limit Theorem for α–mixing sequences

Let $(X_n)_n$ be a sequence of random elements of some measurable space $(\mathbb{T}, \mathcal{B})$. Let $\alpha = (\alpha_n)_n$ be a sequence of nonnegative real numbers which converges to 0. Assume that, for all $k, n \geq 1$, if $A \in \sigma(X_1, \ldots, X_k)$ and $B \in \sigma(X_{k+n}, X_{k+n+1}, \ldots)$ we have

$$|\mathrm{P}(A \cap B) - \mathrm{P}(A)\,\mathrm{P}(B)| \leq \alpha_n. \tag{9.3.1}$$

Then we say that $(X_n)_n$ is *α–mixing* (here the letter α denotes both the kind of mixing condition satisfied by $(X_n)_n$ and the sequence $(\alpha_n)_n$). Note that α–mixing is often also called *strong mixing* since [Ros56]. There are some stronger notions of mixing in the literature, that we shall not detail here, such as φ–mixing or ρ–mixing. The reader is referred to [Dou94] or [Bra93, PU97]: We only need to know that they are stronger than "strong mixing".

We see that α–mixing is a kind of asymptotic independence. Inequality 9.3.1 passes to random variables in the following way. With the same hypothesis than in the above definition, let Y and Z be real random variables defined on the same probability space as $(X_n)_n$. Assume that Y and Z are essentially bounded and that Y is $\sigma(X_1, \ldots, X_k)$–measurable and Z is $\sigma(X_{k+n}, X_{k+n+1}, \ldots)$–measurable. We then have

$$|E(YZ) - E(Y)\,E(Z)| \leq 4\,\|Y\|_\infty\,\|Z\|_\infty\,\alpha_n \tag{9.3.2}$$

(this result appears in [VR59], with constant 16, it is given without proof in [Ibr62, Lemma 1.2] with constant 4, a proof can be found in e.g. [HH80, theorem A.5] or in [Bil95, Lemma 2, page 365]). It is easy to see that if $(X_n)_{n \in \mathbb{N}}$ is identically distributed and α–mixing, then it is Rényi–mixing: Indeed, using (9.3.2), we can apply Lemma 9.2.5, taking for $\mathcal{C}$ the set of elements A of $\mathcal{S}$ which are in $\sigma(X_1, \ldots, X_n)$ for some $n \in \mathbb{N}$.

There is a huge literature on the Central Limit Theorem for α–mixing (or φ–mixing or ρ–mixing) stationary sequences of real random variables since [Ros56]: see e.g. [Ibr62], [Bil95, Theorem 27.4], a survey is given in [Dou94]; for more recent results, see [PU97, Rio00]. There are less results about the Central Limit Theorem for α– or φ–mixing nonindependent random vectors of a separable Banach space: see e.g. [KP80, Sam84, Ute92, PS99].

We shall see that, actually, the weighted sums involved in all central limit theorems for stationnary α–mixing sequences are Rényi–mixing, as was already observed by Rényi himself [Rén66, Rén70] in the independent case. The proof is easy and natural, and in the same spirit as that of Rényi.

Let $(X_n)_{n\geq 1}$ be a sequence of random elements (with Radon laws) of an LCS. For all $k, n \geq 1$, let $S_k = X_1 + \cdots + X_k$ and $S_{n,n+k} = X_{n+1} + \cdots + X_{n+k}$. We say that $(X_n)_n$ *has stationary sums* if, for any $k \geq 1$ and for any $n \geq 1$, $(S_{n,n+1}, \ldots, S_{n,n+k})$ has the same law as $(S_1, \ldots, S_k)$. This condition is obviously less restrictive than that of stationarity of $(X_n)_n$.

Theorem 9.3.1 (Rényi–mixing Central Limit Theorem for α–mixing stationary sequences) *Let $\mathbb{E}$ be a hereditarily Lindelöf LCS and let $(X_n)_{n\geq 1}$ be a sequence of random elements of $\mathbb{E}$ with Radon laws. For each $n \geq 1$, let $S_n = X_1 + \cdots + X_n$. Let $(C_n)_n$ be a sequence of positive real numbers. Let ν be a Radon probability on $\mathbb{E}$. Assume that*

(i) $\lim_n C_n = +\infty$ *and* $\lim_n \left(1 - \dfrac{C_{n-m}}{C_n}\right) = 0$ *for each fixed* $m \geq 1$,

(ii) $(X_n)_{n\geq 1}$ *is α–mixing and has stationary sums,*

(iii)

$$\frac{S_n}{C_n} \xrightarrow{\text{weakly}} \nu.$$

Then $(S_n/C_n)_n$ is Rényi–mixing, that is,

$$\frac{S_n}{C_n} \xrightarrow{W\text{-stably}} \mathrm{P} \otimes \nu.$$

Note that Condition (i) is satisfied in the classical case $C_n = \sqrt{n}$.

The proof of Theorem 9.3.1 will use the following lemma. In this result, the space does not need to be hereditarily Lindelöf, because we deal with narrow convergence instead of stable convergence.

Lemma 9.3.2 *Let $\mathbb{E}$ be a LCS. Let $(X_n)_n$ be a sequence of random elements of a $\mathbb{E}$, which converges in law to a Radon law $\nu \in \mathcal{M}_R^{+,1}(\mathbb{E})$. Let $(\beta_n)_n$ be a sequence of real numbers which converges to 1. Then $(\beta_n X_n)_n$ converges in law to ν.*

Proof. Let $(\|.\|_\alpha)_{\alpha\in\mathbb{A}}$ be an upwards filtering family of seminorms which defines the topology of $\mathbb{E}$. From Lemma 1.3.3, we only need to prove that, for every $\alpha \in \mathbb{A}$ and every bounded $\|.\|_\alpha$–Lipschitz function f on $\mathbb{E}$, we have

$$\lim_n E\left(f(\beta_n X_n)\right) = \nu(f). \tag{9.3.3}$$

Let $\varepsilon > 0$. We can find a compact subset K_ε of $\mathbb{E}$ such that $\nu(K_\varepsilon) \geq 1 - \varepsilon/2$. Let

$$K_\varepsilon^\varepsilon = \{x \in \mathbb{E};\ d_\alpha(x, K_\varepsilon) < \varepsilon\},$$

where d_α is the distance associated with $\|.\|_\alpha$. As $K_\varepsilon^\varepsilon$ is open, there exists an integer N_ε such that

$$\forall n \geq N_\varepsilon \quad \mathrm{P}\{X_n \in K_\varepsilon^\varepsilon\} \geq 1 - \varepsilon.$$

Let $R_\varepsilon = \sup\{\|x\|_\alpha\,;\ x \in K_\varepsilon^\varepsilon\}$. Let C denote the Lipschitz constant of f for $\|.\|_\alpha$. We have

$$\begin{aligned}
\lim_n |E(f(\beta_n X_n)) - \nu(f)| &= \lim_n \left(|E(f(\beta_n X_n) - f(X_n))| + |E(f(X_n)) - \nu(f)|\right) \\
&\leq \lim_n \left|E\left((f(\beta_n X_n) - f(X_n))\, \mathbb{1}_{\{X_n \in K_\varepsilon^\varepsilon\}}\right)\right| + 2\varepsilon \|f\|_\infty \\
&\leq \lim_n C R_\varepsilon |1 - \beta_n| + 2\varepsilon \|f\|_\infty \\
&= 2\varepsilon \|f\|_\infty .
\end{aligned}$$

As ε is arbitrary, this proves (9.3.3). □

Proof of Theorem 9.3.1. Let $(\|.\|_\alpha)_{\alpha\in\mathbb{A}}$ be a σ–upwards filtering family of seminorms which defines the topology of $\mathbb{E}$. Let $\mathcal{C}$ be the set of elements A of $\mathcal{S}$ which are in $\sigma(X_1, \ldots, X_{n_0})$ for some $n_0 \geq 1$. From Lemma 9.2.5 we only need to show that, for any $A \in \mathcal{C}$, any $\alpha \in \mathbb{A}$ and any bounded $\|.\|_\alpha$–Lipschitz function f on $\mathbb{E}$, we have

$$\lim_n E\left(f(S_n/C_n)\, \mathbb{1}_A\right) = \mathrm{P}(A)\nu(f).$$

Let $A \in \mathcal{C}$ (we assume w.l.g. that $\mathrm{P}(A) > 0$) and let f be a bounded $\|.\|_\alpha$–Lipschitz function on $\mathbb{E}$. For $n > m$, let $S_{m,n} = X_{m+1} + \cdots + X_n$. As $(X_n)_n$ has stationary sums, we have

$$E\left(f\left(\frac{S_{m,n}}{C_{n-m}}\right)\right) = E\left(f\left(\frac{S_{n-m}}{C_{n-m}}\right)\right) \to \nu(f) \text{ when } n \to +\infty. \tag{9.3.4}$$

Furthermore, as the random variable $f\left(\dfrac{S_{m,n}}{C_{n-m}}\right)$ is measurable w.r.t. $\sigma(X_{m+1}, X_{m+2}, \ldots)$, we have, for $m > n_0$, using (9.3.2),

$$\left\| E\left(\mathbb{1}_A f\left(\frac{S_{m,n}}{C_{n-m}}\right)\right) - \mathrm{P}(A) E\left(f\left(\frac{S_{m,n}}{C_{n-m}}\right)\right)\right\|_\alpha \leq 4 \|f\|_\infty \alpha_{m-n_0}. \tag{9.3.5}$$

Let us denote by E^A the expectation with respect to the probability P^A (see page 276). From (9.3.4) and (9.3.5), we have

$$\lim_n E^A f\left(\frac{S_{m,n}}{C_{n-m}}\right) = \frac{1}{\mathrm{P}(A)} \lim_n E\left(\mathbb{1}_A f\left(\frac{S_{m,n}}{C_{n-m}}\right)\right) = \nu(f),$$

thus $(S_{m,n}/C_{n-m})_n$ converges in law to ν under P^A. From Lemma 9.3.2, we thus have

$$\lim_{n\to+\infty} E\left(\mathbb{1}_A f\left(\frac{S_{m,n}}{C_n}\right)\right) = \mathrm{P}(A) \lim_{n\to+\infty} E^A f\left(\frac{S_{m,n}}{C_n}\right) = \mathrm{P}(A)\nu(f). \tag{9.3.6}$$

On the other hand, we have

$$\left|f\left(\frac{S_n}{C_n}\right) - f\left(\frac{S_{m,n}}{C_n}\right)\right| \leq C \left\|\frac{S_m}{C_n}\right\|_\alpha \wedge 2\left\|f\right\|_\infty,$$

where C is the Lipschitz constant of f for $\|.\|_\alpha$. Note that the right hand side is bounded by $2\|f\|_\infty$ and that $(S_m/C_n)_n$ converges pointwise to 0. From Lebesgue's Dominated Convergence Theorem, we thus have, for any $m \geq 1$,

$$\lim_{n\to+\infty} E\left(\left|\mathbb{1}_A f\left(\frac{S_n}{C_n}\right) - \mathbb{1}_A f\left(\frac{S_{m,n}}{C_n}\right)\right|\right) = 0. \tag{9.3.7}$$

The result then follows from (9.3.6) and (9.3.7). □

Remark 9.3.3 This method does not apply without any extra hypothesis to triangular arrays such as in [Sam84]. For more sophisticated stable limit theorems for triangular arrays, (with $\mathbb{E} = \mathbb{R}$), see [LP96, Let98, Xue91]. For Rényi–mixing or stable theorems for martingale differences, see [AE78], and [Xue91] for arrays of martingale differences. Other very general stable theorems for stationary sequences of real variables can be found in [DM02], where actually $\tau^{\mathrm{W}}_{p,|.|}$–convergence is proved (see Section 2.4).

9.4 Stable Central Limit Theorem for a random number of random vectors

Let $(X_n)_n$ satisfy the Central Limit Theorem. We are interested here in the convergence of (S_{η_n}/C_{η_n}), where (η_n) is a sequence of random positive integers which converges in probability to $+\infty$. This topic is extensively studied and we only present here, in a general setting, some well-known results which illustrate our techniques.

We start with a presentation of the case when $(\eta_n)_n$ is independent of $(X_n)_n$. This case is easy and can be solved with the tools of $\mathcal{U}$–stable convergence (see the definition page 22): If $Z_n = S_n/C_n$ converges in law, then Z_{η_n} converges in law to the same limit. If Z_n converges stably, so does Z_{η_n}.

In the second case, no assumption is made on the interdependence between $(\eta_n)_n$ and $(X_n)_n$, and stable convergence plays an essential role.

Case when the number of random vectors is independent of the sequence of random vectors It is easy to see that, with the hypothesis and notations of Theorem 9.3.1, if $(\eta_n)_n$ is a sequence of random positive integers which is independent of $(X_n)_n$ and converges to $+\infty$ in probability, then

$$\frac{S_{\eta_n}}{C_{\eta_n}} \xrightarrow{\text{stably}} \mathrm{P} \otimes \nu.$$

This can be deduced from the following general result.

Theorem 9.4.1 *Let $\mathcal{U}$ be a sub–σ–algebra of $\mathcal{S}$. Let $(Z_n)_n$ be a sequence of random elements of $\mathbb{T}$ which $\mathcal{U}$–W–stably converges to a disintegrable Young measure $\mu \in \mathcal{Y}^1_{\text{dis}}$. Let $(\eta_n)_n$ be a sequence of random positive integers which is independent of $(Z_n)_n$ and of $\mu_.$ and converges to $+\infty$ in probability. Let $\mathcal{S}_1$ be the σ–algebra generated by $(Z_n)_n$ and $\mu_.$ and let $\mathcal{S}_2$ be the σ–algebra generated by $(\eta_n)_n$. Assume that $\mathcal{U}$ is generated by a sub–σ–algebra $\mathcal{S}_1'$ of $\mathcal{S}_1$ and a sub–σ–algebra of $\mathcal{S}_2'$ of $\mathcal{S}_2$. Then $(Z_{\eta_n})_n$ $\mathcal{U}$–W–stably converges to μ.*

Proof. Using the Monotone Class Theorem, the same reasoning as in the proof of Part C of the Portmanteau Theorem 2.1.3 shows that, to prove Theorem 9.4.1, we only need to prove

(9.4.1)
$$\forall B \in \mathcal{S}_1' \quad \forall C \in \mathcal{S}_2' \quad \forall f \in \mathrm{C}(\mathbb{T},[0,1]) \quad \lim_n E\left(\mathbb{1}_{B\cap C}\, f \circ Z_{\eta_n}\right) = \mu(\mathbb{1}_{B\cap C} \otimes f).$$

Let $B \in \mathcal{S}_1$, let $C \in \mathcal{S}_2$ and let $f \in \mathrm{C}(\mathbb{T},[0,1])$. We have

$$\mu(\mathbb{1}_{B\cap C} \otimes f) = E\left(E^{\mathcal{S}_1}(\mathbb{1}_C)\, \mathbb{1}_B\, \mu_.(f)\right) = \mathrm{P}(C)\mu(\mathbb{1}_B \otimes f). \tag{9.4.2}$$

Let $\varepsilon > 0$. There exists $N \geq 1$ such that

$$\forall n \geq N \quad \left[|E(\mathbb{1}_B\, f \circ Z_n) - \mu(\mathbb{1}_B \otimes f)| < \varepsilon \text{ and } |E(f \circ Z_n) - \mu(\mathbb{1}_\Omega \otimes f)| < \varepsilon \right].$$

Furthermore, there exists $N_1 > 0$ such that

$$\forall n \geq N_1 \quad \mathrm{P}(\eta_n < N) < \varepsilon. \tag{9.4.3}$$

We thus have, for $n \geq N_1$, using the independence assumption, (9.4.2) and (9.4.3),

$$\begin{aligned}
&|E(\mathbb{1}_{B\cap C}\, f\circ Z_{\eta_n}) - \mu(\mathbb{1}_{B\cap C}\otimes f)| \\
&= \Big|\sum_{k\geq 1} E(\mathbb{1}_{\{\eta_n=k\}}\,\mathbb{1}_{B\cap C}\, f\circ Z_k) - \mu(\mathbb{1}_{B\cap C}\otimes f)\Big| \\
&\leq \mathrm{P}\{\eta_n < N\} + \Big|\sum_{k\geq N} E(\mathbb{1}_{\{\eta_n=k\}}\,\mathbb{1}_{B\cap C}\, f\circ Z_k) - \mathrm{P}\{\eta_n\geq N\}\,\mu(\mathbb{1}_{B\cap C}\otimes f)\Big| \\
&\leq \varepsilon + \Big|\sum_{k\geq N} E(\mathbb{1}_{\{\eta_n=k\}}\,\mathbb{1}_C)\, E(\mathbb{1}_B\, f\circ Z_k) - \mathrm{P}\{\eta_n\geq N\}\,\mu(\mathbb{1}_{B\cap C}\otimes f)\Big| \\
&\leq 2\varepsilon + \Big|\sum_{k\geq N} E(\mathbb{1}_{\{\eta_n=k\}}\,\mathbb{1}_C)\,\mu(\mathbb{1}_B\otimes f) - \mathrm{P}\{\eta_n\geq N\}\,\mu(\mathbb{1}_{B\cap C}\otimes f)\Big| \\
&\leq 4\varepsilon + |\mathrm{P}(C)\,\mu(\mathbb{1}_B\otimes f) - \mu(\mathbb{1}_{B\cap C}\otimes f)| \\
&= 4\varepsilon.
\end{aligned}$$

As ε is arbitrary, this proves (9.4.1). □

Corollary 9.4.2 *The preceding result also holds true in the case when $\mathcal{U} = \mathcal{S}$.*

Proof. Let $A \in \mathcal{U}$ and let $f \in \mathrm{C}(\mathbb{T}, [0,1])$. Let us show that

$$\lim_n E(\mathbb{1}_A\, f\circ Z_{\eta_n}) = \mu(\mathbb{1}_A\otimes f). \tag{9.4.4}$$

Let $\mathcal{S}_0$ be the σ–algebra generated by $(Z_n)_n$, $\mu_.$ and $(\eta_n)_n$. Let $U = E^{\mathcal{S}_0}(\mathbb{1}_A)$. Then (9.4.4) is equivalent to

$$\lim_n E(U\, f\circ Z_{\eta_n}) = \mu(U\otimes f), \tag{9.4.5}$$

because we have

$$\begin{aligned}
\forall n\geq 1\quad E(\mathbb{1}_A\, f\circ Z_{\eta_n}) &= E(E^{\mathcal{S}_0}(\mathbb{1}_A\, f\circ Z_{\eta_n})) = E(U\, f\circ Z_{\eta_n}),\\
\mu(\mathbb{1}_A\otimes f) &= \int_\Omega U(\omega)\mu_\omega(f)\, d\mathrm{P}(\omega) = \mu(U\otimes f).
\end{aligned}$$

But (9.4.5) follows from Theorem 9.4.1. □

Principle of invariance for a random sum of random of vectors If we drop the condition of independence between $(\eta_n)_n$ and $(X_n)_n$, we have to replace it by some other hypothesis. The most studied one seems to be Anscombe's condition [Ans52] and its generalizations. The work of Aldous [Ald78] is fundamental in this topic, see also [KR98] for a recent work with many references. Anscombe's condition is satisfied in particular when $(X_n)_n$ satisfies a functional limit theorem (see the definition below). We shall restrict ourselves to this case, using the elegant ideas of Billingsley [Bil68, Section 17], developped by Fischler [Fis76] and Aldous [Ald78]. In our general topological setting, this is an application of the results of Chapters 2 and 3, in particular the Fiber Product Lemma.

Let $\mathbb{T}$ be a completely regular topological space and let $\mathcal{D}$ be a set of semidistances which defines the topology of $\mathbb{T}$. Let I be the interval $[0, T]$ for some $T > 0$ or the interval $\mathbb{R}^+$. We denote by $\mathbf{D}(I, \mathbb{T})$ the space of mappings from I to $\mathbb{T}$ which are right continuous and have left limits everywhere. We shall be mainly interested in $\mathbf{D}(\mathbb{R}^+, \mathbb{T})$. When no confusion should arise about the space $\mathbb{T}$, we shall denote simply $\mathbf{D} = \mathbf{D}(\mathbb{R}^+, \mathbb{T})$. The space $\mathbf{D}(I, \mathbb{T})$ is called a *Skorokhod space*. We endow $\mathbf{D}(I, \mathbb{T})$ with Skorokhod's J_1 topology. Actually, the space $\mathbf{D}$ considered by Skorokhod [Sko56] was $\mathbf{D}([0,1], \mathbb{R})$ (see also [Bil68]). The interval $[0, 1]$ was enlarged to $\mathbb{R}^+$ by [Lin73] (see also [Jac85]). Then $\mathbf{D}([0,1], \mathbb{R})$ and $\mathbf{D}(\mathbb{R}^+, \mathbb{R})$ were generalized by Mitoma [Mit83] to $\mathbf{D}([0,1], \mathbb{T})$ and $\mathbf{D}(\mathbb{R}^+, \mathbb{T})$ with $\mathbb{T}$ completely regular, and extensively studied by Jakubowski [Jak86] in this general case. When studying limit theorems with random indexes, it is more convenient to use $\mathbf{D}(\mathbb{R}^+, \mathbb{T})$ instead of $\mathbf{D}([0,1], \mathbb{T})$, as was noticed by Aldous [Ald78].

We shall not give a precise definition of the J_1 topology on $\mathbf{D}(I, \mathbb{T})$, we refer for this to [Jak86]. What we mainly need to know is that, with each $d \in \mathcal{D}$, we can associate a semidistance $\widetilde{d}$ on $\mathbf{D}$ in such a way that

1. the topology of $\mathbf{D}$ is defined by the set $\widetilde{\mathcal{D}} = \{\widetilde{d};\ d \in \mathcal{D}\}$,

2. for each sequence $(F_n)_{n\in\mathbb{N}}$ in $\mathbf{D}$ and for each $d \in \mathcal{D}$, the sequence $(F_n)_{n\in\mathbb{N}}$ converges in $(\mathbf{D}, \widetilde{d})$ to some $F \in \mathbf{D}$ if and only if there exists a sequence $(u_n)_n$ of continuous one–to–one mappings from $\mathbb{R}^+$ to itself which converges uniformly on bounded subsets of $\mathbb{R}^+$ to the identity mapping Id_I and such that $(F_n \circ u_n)$ converges to F d–uniformly on bounded subsets of I.

In particular, if F is continuous at t and $(F_n)_n$ converges to F in $\mathbf{D}$, then $F_n(t) \to F(t)$. Note that the uniform convergence of (u_n) on bounded sets is automatically satisfied if (u_n) converges pointwise, see Lemma 9.4.6.

If $\mathbb{T} = \mathbb{E}$ is a topological vector space, $\mathbf{D}$ is the projective limit of the spaces $\mathbf{D}([0, n], \mathbb{E})$ through the projections $\phi_n : \mathbf{D}(\mathbb{R}^+, \mathbb{E}) \to \mathbf{D}([0, n+1], \mathbb{E})$ defined by

$$\phi_n(F)(t) = \begin{cases} F(t) & \text{if } 0 \le t \le n \\ (-t + n + 1)F(t) & \text{if } n \le t \le n + 1. \end{cases}$$

If B is a subset of $\mathbb{T}$, the topology of $\mathbf{D}(I, B)$ is induced by the topology of $\mathbf{D}(I, \mathbb{T})$. If B is closed (resp. open) subset of $\mathbb{T}$, then $\mathbf{D}(I, B)$ is a closed (open) subset of $\mathbf{D}(I, \mathbb{T})$. If the compact subsets of $\mathbb{T}$ are metrizable, then the same holds for $\mathbf{D}(I, \mathbb{T})$.

We denote by $\mathbf{C}(\mathbb{R}^+, \mathbb{T})$ (or simply $\mathbf{C}$) the subspace of continuous mappings from $\mathbb{R}^+$ to $\mathbb{T}$. The space $\mathbf{C}$ is dense in $\mathbf{D}$ for the topology J_1, and closed in $\mathbf{C}$ for the topology of uniform convergence on bounded subsets of $\mathbb{R}^+$. Both topologies coincide on $\mathbf{C}$.

In general, measurability of $\mathbf{D}$–valued mappings can be a delicate matter (see [Jak86]). But the $\mathbf{D}$–valued mappings we shall consider here have a very special form and are constructed from a sequence of random vectors (in particular, they have at most a countable number of jumps).

Let $\mathbb{E}$ be an LCS, let $t_0 \in I$, let X be a random element of $\mathbb{E}$ with Radon law, and, for all $(\omega, t) \in \Omega \times I$, set $H(\omega)(t) = 0_{\mathbb{E}}$ if $0 \le t < t_0$ and $H(\omega)(t) = X(\omega)$ if $t_0 \le t$. Let us prove that H is a random element of $\mathbf{D}(I, \mathbb{E})$. There exists an increasing sequence $(K_m)_{m \ge 1}$ of compact subsets of $\mathbb{E}$ such that, for each m, the set $\Omega_m = \{X \in K_m\}$ satisfies $\mathrm{P}(\Omega_m) \ge 1 - 1/m$. We only need to prove that the restriction to each Ω_m of H is measurable, thus we can assume w.l.g. that X takes its values in a compact set K. As K is metrizable, the conclusion immediately follows by approximating uniformly X by finitely valued random elements, or by applying a well-known criterion of measurability for $\mathbf{D}(\mathbb{R}^+, \mathbb{T})$ when $\mathbb{T}$ is Polish [Jak86, Propositions 3.4 and 4.4]. Furthermore, the law of H is tight (see [Jak86, Theorems 3.1 and 4.6]), thus it is Radon. Note that the same holds if $I = [0, n+1]$ and if we replace H by $\phi_n \circ H$, because the convex hull of $\{0\} \cup K$ is compact.

Now, let $(Y_n)_{n \ge 1}$ be a sequence of random elements of $\mathbb{E}$, with Radon laws, and let $\theta > 0$. For each $\omega \in \Omega$ and each $t \in \mathbb{R}^+$, let $G(\omega, t) = Y_{[\theta t]}$, where $[\theta t]$ denotes the integer part of θt. For each $\omega \in \Omega$, $G(\omega, .)$ is in $\mathbf{D}(\mathbb{R}^+, \mathbb{E})$. Furthermore, for each integer $n \ge 0$, $\phi_n \circ G$ is a finite sum of random elements of $\mathbf{D}([0, n+1], \mathbb{E})$ with Radon laws, thus $\phi_n \circ G$ is a random element of $\mathbf{D}([0, n+1], \mathbb{E})$ with Radon law. Thus G is measurable. As the projective limit of compact sets is compact, it is straightforward to check that the law of G is tight, thus Radon.

Let $\mathbb{E}$ be an LCS, let $(X_n)_{n \ge 1}$ be a sequence of random elements of $\mathbb{E}$, with Radon laws, and let $(C_n)_n$ be a sequence of positive real numbers. Set, for any $n \ge 1$, any $\omega \in \Omega$ and any $t \in \mathbb{R}^+$,

(9.4.6)
$$S_n(\omega) = \sum_{i=1}^{n} X_i(\omega), \quad Z_n(\omega) = \frac{S_n(\omega)}{C_n} \quad \text{and} \quad G_n(\omega, t) = G_n(\omega)(t) = \frac{S_{[nt]}(\omega)}{C_n},$$

where $[nt]$ denotes the integer part of nt. Each G_n is a random element of $\mathbf{D}(\mathbb{R}^+, \mathbb{E})$, with Radon law. We say that $(X_n)_n$ satisfies *a functional limit theorem*, if $(G_n)_n$ converges in law in the space $\mathbf{D}(\mathbb{R}^+, \mathbb{E})$ to a probability law $\nu \in \mathcal{M}^{+,1}(\mathbf{D}(\mathbb{R}^+, \mathbb{E}))$. If furthermore ν is Gaussian (that is, if $\nu(\mathbf{C}) = 1$ and ν is a Gaussian Radon measure on $\mathbf{C}$), then we say that $(X_n)_n$ satisfies *the invariance principle*, or *the functional central limit theorem*. We then have

$$\mathcal{L}(Z_n) = \mathcal{L}(G_n(., 1)) \xrightarrow{\text{narrow}} \nu_1,$$

where ν_1 is the image of ν by the projection

$$\pi_1 : \left\{ \begin{array}{lcl} \mathbf{D}(\mathbb{R}^+, \mathbb{E}) & \to & \mathbb{E} \\ H & \mapsto & H(1), \end{array} \right.$$

thus the invariance principle implies the Central Limit Theorem.

There exist functional limit theorems fore more general **D**–valued random elements than the processes G_n generated by $(Z_n)_n$, for example one can consider semimartingales, see e.g. [Jac97, Fie90].

If the convergence of $(G_n)_n$ to ν is stable, that is, if there exists a Young measure $\gamma \in \mathcal{Y}^1(\mathbf{D}(\mathbb{R}^+, \mathbb{E}))$ (with Radon margin $\nu = \gamma(\Omega \times .)$) such that

$$G_n \xrightarrow{\text{W-stably}} \gamma, \tag{9.4.7}$$

we say that $(X_n)_n$ satisfies a *stable functional limit theorem*. Let us say that γ is a *Gaussian Young measure* if $\gamma(A \times .)$ is Gaussian for each $A \in \mathcal{S}$. If γ is Gaussian, we say that $(X_n)_n$ satisfies *the stable invariance principle*, or *the stable functional central limit theorem*. Of course, if furthermore the limit γ has the form $\gamma = \mathrm{P} \otimes \nu$, we say that $(X_n)_n$ satisfies *the (Rényi-)mixing invariance principle*, or *the (Rényi-)mixing functional central limit theorem*.

Theorem 9.4.3 (Invariance principle for a random number of random vectors) *Let $\mathbb{E}$ be an LCS. Assume that $\mathbb{E}$ is Lusin. Let $(X_n)_{n\geq 1}$ be a sequence of random elements of $\mathbb{E}$. Let $\alpha > 0$ and set $C_n = n^\alpha$ for each $n \geq 1$. Define $(S_n)_n$, $(Z_n)_n$ and $(G_n)_n$ as in (9.4.6). Let θ be a real random variable such that $\theta > 0$ a.e. and let $(a_n)_n$ be a sequence of positive real numbers which converges to $+\infty$. Assume that*

(i) $(X_n)_{n\geq 1}$ satisfies a stable *functional limit theorem:*

$$G_n \xrightarrow{W\text{-stably}} \gamma$$

for some $\gamma \in \mathcal{Y}^1_{\text{dis}}(\mathbf{D}(\mathbb{R}^+, \mathbb{E}))$,

(ii)

$$\frac{\eta_n}{a_n} \xrightarrow{prob} \theta.$$

Then we have

$$G_{\eta_n} \xrightarrow{W\text{-stably}} \gamma. \tag{9.4.8}$$

To prove Theorem 9.4.3, we need some auxilliary results. The Lusin condition on $\mathbb{E}$ will allow us to use the following lemma.

Lemma 9.4.4 *Under the conditions of Theorem 9.4.3, the space $\mathbf{D}(\mathbb{R}^+, \mathbb{E})$ is regular Suslin.*

Proof. As $\mathbb{E}$ is Lusin, there exists a Polish topology $\widetilde{\tau}$ on $\mathbb{E}$ which is finer than the original topology of $\mathbb{E}$. Let us denote by $\widetilde{\mathbb{E}}$ the space $\mathbb{E}$ endowed with $\widetilde{\tau}$. It is well-known that $\mathbf{D}([0,1], \widetilde{\mathbb{E}})$ is Polish. Thus $\mathbf{D}([0,1], \mathbb{E})$ is Lusin, thus Suslin. Similarly, for each integer n, $\mathbf{D}([0,n], \mathbb{E})$ is Suslin. Thus the space $\mathbf{D}(\mathbb{R}^+, \mathbb{E})$ is a countable projective limit of Suslin spaces, and therefore it is Suslin. □

Remark 9.4.5 We can replace the hypothesis that $\mathbb{E}$ is Lusin by assuming that the sequence $(Z_n)_n$ is tight. Then, with probability 1, the random vectors Z_n take their values in a (Lusin) K_σ subspace $\mathbb{T}_0$ of $\mathbb{E}$. Furthermore, the topology of $\mathbf{D}(\mathbb{R}^+, \mathbb{T}_0)$ is the topology induced by $\mathbf{D}(\mathbb{R}^+, \mathbb{E})$.

Let $\mathbf{C}_0^\infty$ be the set of one-to-one continuous (increasing) elements of $\mathbf{D}(\mathbb{R}^+, \mathbb{R}^+)$. We endow $\mathbf{C}_0^\infty$ with the topology induced by the Suslin space $\mathbf{D}(\mathbb{R}^+, \mathbb{R}^+)$.

Lemma 9.4.6 *The topology of $\mathbf{C}_0^\infty$ coincides with the topologies of pointwise convergence and of uniform convergence. Furthermore, if $(F_n)_{n\in\mathbb{N}}$ converges in $\mathbf{C}_0^\infty$ to an element F of $\mathbf{C}_0^\infty$, the sequence (F_n^{-1}) of converse mappings converges in $\mathbf{C}_0^\infty$ to F^{-1}.*

Proof. As $\mathbf{C}_0^\infty \subset \mathbf{C}$, the topology of $\mathbf{C}_0^\infty$ is the topology of uniform convergence on bounded subsets of $[0, +\infty[$. So the first part of the lemma will be proved if we show that, on $\mathbf{C}_0^\infty$, the topologies of pointwise convergence and of uniform convergence coincide.

Note that the elements of $\mathbf{C}_0^\infty$ can be seen as continuous increasing mappings from $[0, +\infty]$ to $[0, +\infty]$. Recall that, if $(F_n)_{n\in\mathbb{N}}$ is a sequence of distribution functions which pointwise converges to a continuous distribution function F, then this convergence is uniform (see e.g. the proof of Glivenko–Cantelli Theorem in [Chu74]). In particular, this holds if the (F_n) and F are increasing surjective mappings from $[0, +\infty]$ to $[0, 1]$, with F continuous. As $[0, +\infty]$ is homeomorphic to $[0, 1]$ through an increasing mapping (e.g. the function $x \mapsto \dfrac{x}{1+x}$), this result can be transposed to $\mathbf{C}_0^\infty$, thus the topologies of pointwise convergence and uniform convergence coincide on $\mathbf{C}_0^\infty$.

Now, let $(F_n)_{n\in\mathbb{N}}$ be a sequence in $\mathbf{C}_0^\infty$ which converges to $F \in \mathbf{C}_0^\infty$. We only need to prove that $(F_n^{-1})_{n\in\mathbb{N}}$ converges pointwise to F^{-1}. Let $y \in [0, +\infty]$. For each $n \in \mathbb{N}$, let $x_n = F_n^{-1}(y)$. Let $(x_{n'})$ be a subsequence of (x_n). From compactness of $[0, +\infty]$, there exists a subsequence $(x_{n''})$ of $(x_{n'})$ which converges to an $x \in [0, +\infty]$. We then have

$$|F_{n''}(x_{n''}) - F(x)| \le \|F_{n''} - F\|_\infty + |F(x_{n''}) - F(x)|,$$

thus $|F_{n''}(x_{n''}) - F(x)|$ converges to 0. But we have $F_{n''}(x_{n''}) = y$ for each n'', thus $F(x) = y$, that is, $x = F^{-1}(y)$. We have proved that, for any subsequence of $(F_n^{-1}(y))_n$, there is a further subsequence which converges to $F^{-1}(y)$, thus $(F_n^{-1}(y))_n$ converges to $F^{-1}(y)$. □

Lemma 9.4.7 *Let $\mathbb{E}$ be an LCS. The mapping*

$$\Phi : \begin{cases} \mathbf{C}_0^\infty \times \mathbf{D}(\mathbb{R}^+, \mathbb{E}) \times \mathbb{R} & \to \mathbf{D}(\mathbb{R}^+, \mathbb{E}) \\ (F, G, \rho) & \mapsto \rho G \circ F \end{cases}$$

is continuous.

Proof. We only need to check that, for any continuous semidistance d on $\mathbb{E}$, Φ is continuous for the associated semidistance $\widetilde{d}$ on $\mathbf{D}(\mathbb{R}^+, \mathbb{E})$. Let $(F_n)_{n\in\mathbb{N}}$ be a sequence in $\mathbf{C}_0^\infty$ which converges to $F \in \mathbf{C}_0^\infty$, let $(G_n)_{n\in\mathbb{N}}$ be a sequence in $\mathbf{D}$ which converges to $G \in \mathbf{D}$ for $\widetilde{d}$ and let $(\rho_n)_n$ be a sequence of real numbers which converges to $\rho \in \mathbb{R}$. Let $(u_n)_{n\in\mathbb{N}}$ be a sequence in $\mathbf{C}_0^\infty$ which converges in $\mathbf{C}_0^\infty$ to $\mathrm{Id}_{\mathbb{R}^+}$ and such that $(G_n \circ u_n)$ converges to G uniformly on the bounded subsets of $\mathbb{R}^+$. For each $n \in \mathbb{N}$, the mapping $F_n^{-1} \circ u_n \circ F$ is in $\mathbf{C}_0^\infty$ and, from Lemma 9.4.6, $(F_n^{-1} \circ u_n \circ F)_n$ converges in $\mathbf{C}_0^\infty$ to $\mathrm{Id}_{\mathbb{R}^+}$. Furthermore, $\rho_n\,(G_n \circ F_n)\left(F_n^{-1} \circ u_n \circ F\right) = \rho_n\, G_n \circ u_n \circ F$ converges to $\rho\, G \circ F$ uniformly on bounded subsets of $\mathbb{R}^+$. Thus $(\rho_n\, G_n \circ F_n)_n$ converges to $\rho\, G \circ F$ in $(\mathbf{D}, \widetilde{d})$.

□

The next lemma relies on the particular shape of the normalizing sequence $(C_n)_n$ in Theorem 9.4.3. It will ensure that the limit of G_{η_n} is γ.

Lemma 9.4.8 *Assume the hypothesis of Theorem 9.4.3. Let $\rho > 0$. Define a mapping $\Phi_\rho : \mathbf{D}(\mathbb{R}^+, \mathbb{E}) \to \mathbf{D}(\mathbb{R}^+, \mathbb{E})$ by*

$$\forall t \in \mathbb{R}^+ \quad \Phi_\rho(G)(t) = \frac{1}{\rho^\alpha}\, G(\rho t).$$

The Young measure γ is invariant by Φ_ρ, that is,

$$(\Phi_\rho)_\sharp\, \gamma_\omega = \gamma_\omega \quad a.e. \tag{9.4.9}$$

Proof. Let $F_\rho \in \mathbf{C}_0^\infty$ be defined by $F_\rho(t) = \rho t$ for every $t \in \mathbb{R}^+$. With the notations of Lemma 9.4.7, we have $\Phi_\rho(G) = \Phi(F_\rho, G, 1/\rho^\alpha)$ for every $G \in \mathbf{D}(\mathbb{R}^+, \mathbb{E})$. Thus Φ_ρ is continuous. Let

$$\underline{\Phi}_\rho : \begin{cases} \Omega \times \mathbf{D}(\mathbb{R}^+, \mathbb{E}) & \to \quad \Omega \times \mathbf{D}(\mathbb{R}^+, \mathbb{E}) \\ (\omega, G) & \mapsto \quad (\omega, \Phi_\rho(G)). \end{cases}$$

The mapping $\left(\underline{\Phi}_\rho\right)_\sharp : \mathcal{Y}^1(\mathbf{D}(\mathbb{R}^+, \mathbb{E})) \to \mathcal{Y}^1(\mathbf{D}(\mathbb{R}^+, \mathbb{E}))$ is continuous for W–stable convergence.

Assume that $\rho = N$ is an integer. We have, for every $(\omega, t) \in \Omega \times \mathbb{R}^+$ and for every $n \in \mathbb{N}$,

$$\Phi_N(G_n(\omega))(t) = \frac{1}{N^\alpha}\,\frac{S_{[nNt]}(\omega)}{n^\alpha} = G_{Nn}(\omega)(t).$$

Thus

$$\gamma = \lim_n \underline{\delta}_{G_n} = \lim_n \underline{\delta}_{G_{Nn}} = \lim_n\, (\underline{\Phi}_N)_\sharp\, \underline{\delta}_{G_n} = (\underline{\Phi}_N)_\sharp\, \gamma.$$

where the limits are in W–stable convergence. Thus (9.4.9) is satisfied when $\rho \in \mathbb{N}^*$.

Now, for every $N \in \mathbb{N}^*$ and every $n \in \mathbb{N}^*$, we have $\Phi_{1/N}(G_{nN}) = G_n$, thus the limit of $(\Phi_{1/N}(G_n))_n$ can be only γ. Thus (9.4.9) is equally satisfied for $\rho = 1/N$, with $N \in \mathbb{N}^*$. This implies that (9.4.9) is satisfied when ρ is rational. By continuity of $\rho \mapsto \Phi_\rho$, we thus have (9.4.9) for any $\rho > 0$. $\square$

Proof of Theorem 9.4.3. For every $(\omega, t) \in \Omega \times \mathbb{R}^+$, and for every $n \geq 1$, set

$$\theta_n(\omega) = \frac{\eta_n(\omega)}{a_n} \text{ and } F_n(\omega, t) = F_n(\omega)(t) = \theta_n(\omega)\, t.$$

Let $(k_n)_n$ be a sequence of integers such that $\lim_n k_n/a_n = 1$. We have

$$G_{\eta_n}(\omega)(t) = \frac{S_{[\eta_n(\omega)t](\omega)}}{\eta_n{}^\alpha(\omega)} \overset{\mathbf{D}}{\sim} \frac{S_{[\frac{\eta_n(\omega)}{a_n} k_n t](\omega)}}{\left(\frac{\eta_n(\omega)}{a_n}\right)^\alpha k_n^\alpha} = \Phi(F_n(\omega), G_{k_n}(\omega), 1/\theta_n^\alpha(\omega)),$$

where Φ has been defined in Lemma 9.4.7 and $\overset{\mathbf{D}}{\sim}$ means equivalence of sequences in $\mathbf{D}(\mathbb{R}^+, \mathbb{E})$. From (ii), the sequence $(\theta_n)_n$ converges in probability to θ and $(F_n)_n$ converges in probability to the random element F of $\mathbf{C}_0^\infty \subset \mathbf{D}(\mathbb{R}^+, \mathbb{R}^+)$ defined by

$$F(\omega, t) = \theta(\omega)t.$$

Furthermore, from Lemma 9.4.4, $\mathbf{D}(\mathbb{R}^+, \mathbb{R}^+) \times \mathbf{D}(\mathbb{R}^+, \mathbb{E})$ is Suslin regular. Thus, from (i), by Lemma 9.4.7 and Corollary 3.3.5, the sequence $(\Phi(F_n, G_{k_n}, 1/\theta_n^\alpha))_{n\geq 1}$ W–stably converges to the Young measure γ' such that, for almost every $\omega \in \Omega$,

$$\gamma'_\omega = \Phi_\sharp(\delta_{F(\omega)} \otimes \gamma_\omega \otimes \delta_{\theta^{-\alpha}(\omega)}) = \left(\Phi_{\theta(\omega)}\right)_\sharp \gamma_\omega,$$

where $\Phi_{\theta(\omega)}$ has been defined in Lemma 9.4.8. But Lemma 9.4.8 implies that we have $\gamma' = \gamma$ a.e. $\square$

Remark 9.4.9 The hypothesis (i) of stable convergence in Theorem 9.4.3 is essential in our reasoning, in order to apply the Fiber Product Lemma (see Counterexample 3.3.4).

On the other hand, Condition (ii) in Theorem 9.4.3 can easily be weakened in the following direction: Let θ' and θ'' be random real variables such that $\theta' > 0$ and $\theta'' > 0$ a.e., let $(a_n)_n$ and $(b_n)_n$ be two sequences of positive real numbers which converge to $+\infty$. Let $A \in \mathcal{S}$. Then Theorem 9.4.3 still holds true if we replace (ii) by

(ii)'

$$\frac{\eta_n}{a_n} \mathbb{1}_B + \frac{\eta_n}{b_n} \mathbb{1}_{B^c} \xrightarrow{\text{prob}} \theta := (\theta' \mathbb{1}_B + \theta'' \mathbb{1}_{B^c}).$$

Indeed, we can apply separately on B and B^c the reasoning of the proof of Theorem 9.4.3: Let $(k'_n)_n$ and $(k''_n)_n$ be two sequences of integers such that $\lim_n k'_n/a_n =$

$\lim_n k''_n/b_n = 1$. Then the sequence $(G_{k'_n} \mathbb{1}_B + G_{k''_n} \mathbb{1}_{B^c})_n$ stably converges to γ, thus we only need to replace θ_n in the proof of Theorem 9.4.3 by

$$\theta_n = \frac{\eta_n}{a_n} \mathbb{1}_B + \frac{\eta_n}{b_n} \mathbb{1}_{B^c}.$$

Aldous has proposed a more general condition to replace (ii):

(ii)" ([Ald78, Condition (3.4)]) For each $\varepsilon > 0$ and each $\delta > 0$, there exist a measurable partition $(A_1, \ldots, A_N)$ of Ω and constants $c_{n,i}$ $(n \geq 1,\ 1 \leq i \leq N)$ such that

$$\forall i = 1, \ldots, N \quad \lim_{n \to +\infty} c_{n,i} = +\infty$$

$$\limsup_{n \to +\infty} \sum_{i=1}^{N} \mathrm{P}\left(\{|\eta_n - c_{n,i}| > \delta c_{n,i}\} \cap A_i\right) \leq \varepsilon.$$

It is not difficult to check that Theorem 9.4.3 also holds true if we replace (ii) by (ii)". Furthermore, a simple argument given by Aldous [Ald78, Lemma 4] shows that the convergence of $(Z_n)_n$ to γ_1 must be stable in order that $\mathcal{L}(Z_{\eta_n})$ weakly converges to γ_1 for all sequences $(\eta_n)_n$ satisfying Condition (ii)' (or (ii)"). Indeed assume that the convergence of $(Z_n)_n$ to γ_1 is only $\mathcal{U}$–W–stable, for some proper sub–σ–algebra $\mathcal{U}$ of $\mathcal{S}$. There exists $A \in \mathcal{S} \setminus \mathcal{U}$, with $\mathrm{P}(A) > 0$, and a bounded continuous $f : \mathbb{E} \to \mathbb{R}$ such that the sequence $(E(\mathbb{1}_A f \circ Z_n))_n$ does not converge. We can thus find two sequences $(a_n)_n$ and $(b_n)_n$ of integers such that the sequences $(E(\mathbb{1}_A f \circ Z_{a_n}))_n$ and $(E(\mathbb{1}_A f \circ Z_{b_n}))_n$ converge to different limits. Then set $\eta_n = a_n$ and $\eta'_n = a_n \mathbb{1}_A + b_n \mathbb{1}_{A^c}$. Clearly, $(\eta_n)_n$ and $(\eta'_n)_n$ satisfy (ii)', but $(Z_{\eta_n})_n$ and $(Z_{\eta'_n})_n$ cannot converge to the same limit.

References

[AB97] J. J. Alibert and G. Bouchitté, *Non-uniform integrability and generalized Young measures*, J. Convex Anal. **4** (1997), no. 1, 129–147.

[ABM] Hedy Attouch, Giuseppe Buttazzo, and Gérard Michaille, *Variational Analysis in Sobolev and BV spaces: Applications to PDE and Optimization*, SIAM, Philadelphia, to appear.

[AC97] Allal Amrani and Charles Castaing, *Weak compactness in Pettis integration*, Bull. Pol. Acad. Sci., Math. **45** (1997), no. 2, 139–150.

[ACV92] Allal Amrani, Charles Castaing, and Michel Valadier, *Méthodes de troncature appliquées à des problèmes de convergence faible ou forte dans L^1*, Arch. Rational Mech. Anal. **117** (1992), 167–191.

[ACV98] ———, *Convergence in Pettis norm under extreme point condition*, Vietnam J. of Math. **26** (1998), no. 4, 323–335.

[AE78] David J. Aldous and G. K. Eagleson, *On mixing and stability of limit theorems*, Ann. Probab. **6** (1978), 325–331.

[AG80] Aloisio Araujo and Evarist Giné, *The Central Limit Theorem for real and Banach valued random variables*, John Wiley & Sons, Inc., New York, 1980.

[Ald78] David J. Aldous, *Weak convergence of randomly indexed sequences of random variables*, Math. Proc. Camb. Phil. Soc. **83** (1978), 117–126.

[All92] Grégoire Allaire, *Homogenization and two scale convergence*, SIAM J. Math. Anal. **23** (1992), no. 6, 1482–1518.

[All96] Boualem Alleche, *Quelques résultats sur la consonance, les multi-applications et la séquentialité*, Ph.D. thesis, Université de Rouen, 1996.

[Ans52] F. J. Anscombe, *Large-sample theory of sequential estimation*, Math. Proc. Camb. Phil. Soc. **48** (1952), 600–607.

[AP03] Zvi Artstein and C. C. Popa, *Convexity and the natural best approximation in spaces of integrable Young measures*, J. Convex Anal. **10** (2003), no. 1, 169–184.

[Art01a] Zvi Artstein, *Compact convergence of σ-fields and relaxed conditional expectation*, Probab. Theory Related Fields **120** (2001), no. 3, 369–394.

[Art01b] ———, *Projections on convex sets in the relaxed limit*, Set-Valued Anal. **9** (2001), no. 1-2, 13–34, Wellposedness in optimization and related topics (Gargnano, 1999).

[Bad70] Albert Badrikian, *Séminaire sur les fonctions aléatoires linéaires et les mesures cylindriques*, Lecture Notes in Math., no. 139, Springer Verlag, Berlin, 1970.

[Bal84a] Erik J. Balder, *A general approach to lower semicontinuity and lower closure in optimal control theory*, SIAM J. Control and Optimization **22** (1984), 570–598.

[Bal84b] ———, *A general denseness result for relaxed control theory*, Bull. Austral. Math. Soc. **30** (1984), 463–475.

[Bal85] ———, *An extension of Prohorov's theorem for transition probabilities with applications to infinite-dimensional lower closure problems*, Rend. Circ. Mat. Palermo II. Ser. (Suppl.) **34** (1985), 427–447.

[Bal86a] ———, *On seminormality of integral functionals and their integrands*, SIAM J. Control Optim. **24** (1986), no. 1, 95–121.

[Bal86b] ———, *On weak convergence implying strong convergence in L_1-spaces*, Bull. Austral. Math. Soc. **33** (1986), no. 3, 363–368.

[Bal88] ———, *Generalized equilibrium results for games with incomplete information*, Math. Oper. Res. **13** (1988), no. 2, 265–276.

[Bal89a] Erik Balder, *On Prohorov's theorem for transition probabilities*, Sém. Anal. Convexe **19** (1989), 9.1–9.11.

[Bal89b] Erik J. Balder, *Infinite-dimensional extension of a theorem of Komlós*, Probab. Theory Relat. Fields **81** (1989), 185–188.

[Bal89c] J. M. Ball, *A version of the fundamental theorem for Young measures*, PDEs and continuum models of phase transitions (Nice 1988) (Berlin, New York) (D. Serre, ed.), Lecture Notes in Phys., no. 344, Springer Verlag, 1989, pp. 207–215.

[Bal90] E. J. Balder, *New sequential compactness results for spaces of scalarly integrable functions*, J. Math. Anal. Appl. **151** (1990), 1–16.

[Bal91] ———, *On equivalence of strong and weak convergence in L^1-spaces under extreme point condition*, Israel J. Math. **75** (1991), 21–47.

[Bal95] Erik J. Balder, *Lectures on Young measures*, Cahiers de mathématiques de la décision 9517, CEREMADE, Université Paris-Dauphine, 1995.

[Bal00a] E. J. Balder, *New fundamentals of Young measure convergence*, Calculus of Variations and Optimal Control (Haifa 1998) (Boca Raton, FL), Chapman & Hall, 2000, pp. 24–48.

[Bal00b] Erik J. Balder, *Lectures on Young measure theory and its applications in economics*, Rend. Istit. Mat. Univ. Trieste **31, suppl.** (2000), 1–69, Workshop di Teoria della Misura et Analisi Reale Grado, 1997 (Italia).

[Bal01] ———, *On ws-convergence of product measures*, Math. Oper. Res. **26** (2001), no. 3, 494–518.

[BBR79] A. Bozorgnia and M. Bhaskara Rao, *A strong law of large numbers for subsequences of random elements in separable Banach spaces*, Ann. Probab. **7** (1979), no. 1, 156–158.

[BC77] J. R. Baxter and R. V. Chacon, *Compactness of stopping times*, Z. Wahrsch. Verw. Gebiete **40** (1977), 169–181.

[BC93] Ahmed Bouziad and Jean Calbrix, *Théorie de la mesure et de l'intégration*, Publications de l'Université de Rouen, Rouen, France, 1993.

[BC97] Houcine Benabdellah and Charles Castaing, *Weak compactness criteria and convergences in $L^1_E(\mu)$.*, Collect. Math. **48** (1997), no. 4–6, 423–448, Fourth International Conference on Function Spaces (Zielona Góra, 1995).

[BC01] ———, *Weak compactness and convergences in $L^1_{E'}[E]$.*, Adv. Math. Econ. **3** (2001), 1–44.

[BCG99a] Abdelhamid Bourras, Charles Castaing, and Mohamed Guessous, *Olech–types lemma and Visintin–types theorem in Pettis integration and $L^1_{E'}[E]$*, NLA98 Convex Analysis and Chaos (Sakado, 1998) (Sakado, Japan) (Kiyoko Nishizawa, ed.), Josai Math. Monogr., no. 1, Josai Univ., 1999, pp. 1–26.

[BCG99b] ———, *Olech-types lemma and Visintin-types theorem in Pettis integration and $L^1_{E'}[E]$*, NLA98: Convex analysis and chaos (Sakado, 1998), Josai Univ., Sakado, 1999, pp. 1–26.

[Bee93] Gerald Beer, *Topologies on closed and closed convex sets*, Kluwer Academic Publishers Group, Dordrecht, 1993.

[Ben91] H. Benabdellah, *Extrémalité et stricte convexité dans L^1_E*, Sém. Anal. Convexe **21** (1991), 4.1–4.18.

[Beš83] Amer Bešlagić, *Embedding cosmic spaces in Lusin spaces*, Proc. Amer. Math. Soc. **89** (1983), no. 3, 515–518.

[BGJ94] Erik J. Balder, Maria Girardi, and Vincent Jalby, *From weak to strong types of L^1_E-convergence by the Bocce criterion*, Studia Math. **111** (1994), no. 3, 241–262.

[BH95] Erik J. Balder and Christian Hess, *Fatou's lemma for multifunctions with unbounded values*, Math. Oper. Res. **20** (1995), 21–48.

[BH96] ———, *Two generalizations of Komlós' theorem with lower closure-type applications*, J. Convex Anal. **1** (1996), 25–44.

[Bil68] Patrick Billingsley, *Convergence of probability measures*, J. Wiley, New York, London, 1968.

[Bil95] ———, *Probability and measure*, third ed., Wiley Series in Probability and Mathematical Statistics, John Wiley and Sons, Inc., New York, 1995.

[BJ91] E. N. Barron and R. Jensen, *Optimal control and semicontinuous viscosity solutions*, Proc. Amer. Math. Soc. **113** (1991), no. 2, 397–402.

[BK83] V. V. Buldygin and A. B. Kharazishvili, *Borel measures in nonseparable metric spaces*, Ukrainian Math. J. **35** (1983), no. 5, 465–470, translated from the Russian: Ukrain. Mat. Zh. **35** (1983) no. 5, 552–556.

[BL71] Henri Berliocchi and Jean-Michel Lasry, *Sur le contrôle optimal de systèmes gouvernés par des équations aux dérivées partielles*, C. R. Acad. Sci. Paris Sér. A-B **273** (1971), A1222–A1225.

[BL73] ———, *Intégrandes normales et mesures paramétrées en calcul des variations*, Bull. Soc. Math. France **101** (1973), 129–184.

[Bog86] Vladimir I. Bogachev, *Locally convex spaces with the property of Central Limit Theorem and measures supports*, Vestnik Moskov. Univ. Ser. I Mat. Mekh. (1986), no. 6, 16–20, 86, (in Russian) English translation: Moskow Univ. Math. Bull. **41** (1986) no 6, 19–23.

[Bog98a] ———, *Gaussian measures*, Mathematical Surveys and Monographs, vol. 62, American Mathematical Society, Providence, RI, 1998.

[Bog98b] ———, *Measures on topological spaces*, J. Math. Sci. **91** (1998), no. 4, 3033–3156, Functional analysis, 1.

[Bou] Ahmed Bouziad, *Change of Topology Lemma*, private communication.

[Bou69] N. Bourbaki, *Intégration, chapitre 9*, second ed., Diffusion C.C.L.S., Paris, 1969.

[Bou71] ———, *Topologie générale, chapitres 1 à 4*, second ed., Diffusion C.C.L.S., Paris, 1971.

[Bou74] ———, *Topologie générale, chapitres 5 à 10*, second ed., Diffusion C.C.L.S., Paris, 1974.

[Bou79] J. Bourgain, *The Komlós theorem for vector-valued functions*, Wrije Universiteit Brussel (1979), 9 pages.

[Bou81] N. Bourbaki, *Espaces vectoriels topologiques*, second ed., Masson, Paris, 1981.

[Bou83] Richard D. Bourgin, *Geometric aspects of convex sets with the Radon–Nikodým property*, Lecture Notes in Math., no. 993, Springer Verlag, Berlin, 1983.

[Bou96] Ahmed Bouziad, *Borel measures in consonant spaces*, Topology Appl. **70** (1996), no. 2–3, 125–132.

[Bou98] ———, *A note on consonance of G_δ subsets*, Topology Appl. **87** (1998), no. 1, 53–61.

[Bou02] ———, *Coincidence of the upper Kuratowski topology with the co-compact topology on compact sets, and the Prohorov property*, Topology Appl. **120** (2002), 283–299.

[Bra93] Richard C. Bradley, *Equivalent mixing conditions for random fields*, Ann. Probab. **21** (1993), no. 4, 1921–1926.

[BS03] Erik J. Balder and Anna Rita Sambucini, *A note on strong convergence for Pettis integrable functions*, Vietnam J. Math. **31** (2003), no. 3, 341–347.

[But89] G. Buttazzo, *Semicontinuity, relaxation and integral representation in the calculus of variations*, Pitman Research Notes in Math., no. 207, Longman, Harlow, 1989.

[Cal82] Jean Calbrix, *Une propriété des espaces topologiques réguliers, images continues d'espaces métrisables séparables*, C. R. Acad. Sci. Paris Sér. I **295** (1982), 81–82.

[Cal84] ———, *Plongement dans les espaces lusiniens (d'après Amer Bešlagić) et sur un théorème d'Hurewicz*, Pub. Math. Univ. Pierre et Marie Curie, Sémin. Initiation Anal. 23ème année - 1983/84 **66** (1984), 11.1–11.10.

[Cas80] Charles Castaing, *Topologie de la convergence uniforme sur les parties uniformément intégrables de L^1 et théorèmes de compacité dans certains espaces du type Köthe–Orlicz*, Sém. Anal. Convexe **10** (1980), 5.1–5.27.

[Cas85] ______, *Compacité dans l'espace des mesures de probabilités de transition*, Atti Sem. Mat Fis. Modena **34** (1985), 337–352.

[Cas87] ______, *Validité du théorème de Reshetnyak dans les espaces hilbertiens*, Sém. Anal. Convexe **17** (1987), 8.1–8.9.

[Cas96] ______, *Weak compactness and convergence in Bochner and Pettis integration*, Vietnam J. Math. **24** (1996), no. 3, 241–286.

[CC82] Charles Castaing and Paulette Clauzure, *Semicontinuité des fonctionnelles intégrales*, Acta Math. Vietnam. **7** (1982), 139–170, (first published in Séminaire d'Analyse Convexe **11** (1981, 15.1–15.45, Montpellier, France).

[CC85] ______, *Compacité dans l'espace L^1_E et dans l'espace des multifonctions intégralement bornées, et minimisation*, Ann. Mat. Pura Appl. **140** (1985), no. 4, 345–364, (first published in Séminaire d'Analyse Convexe **14** (1984, 4.1–4.29, Montpellier, France).

[CE98] Charles Castaing and Fatima Ezzaki, *Convergences for weakly compact random sets in B–convex reflexive Banach spaces*, Atti Sem. Mat Fis. Modena (supplemento) **46** (1998), 123–149, Dedicated to Prof. C. Vinti (Perugia 1996).

[CFS00] C. Castaing, L. Aicha Faik, and A. Salvadori, *Evolution equations governed by m-accretive and subdifferential operators with delay*, Int. J. Appl. Math. **2** (2000), no. 9, 1005–1026.

[CG99] Charles Castaing and Mohamed Guessous, *Convergences in $L^1_X(\mu)$*, Adv. Math. Econ. **1** (1999), 17–37.

[Chr74] J. P. R. Christensen, *Topology and Borel structure*, North-Holland, Amsterdam, London, 1974.

[Chu74] K.L. Chung, *A course in probability theory*, second ed., Acad. Press, New york, 1974.

[CI03] Charles Castaing and M. G. Ibrahim, *Functional evolution governed by m–accretive operators*, Advances in mathematical economics, Vol. 5, Adv. Math. Econ., vol. 5, Springer, Tokyo, 2003, pp. 23–54.

[CJ95] Charles Castaing and Vincent Jalby, *Epi-convergence of integral functionals defined on the space of measures. Appplications to the sweeping process*, Atti Sem. Mat Fis. Modena **43** (1995), 113–157.

[CM97] Pilar Cembranos and José Mendoza, *Banach spaces of vector–valued functions*, Lecture Notes in Math., no. 1676, Springer Verlag, Berlin, 1997.

[CRdF00] Charles Castaing and Paul Raynaud de Fitte, $\mathfrak{S}$–*Uniform scalar integrability and strong laws of large numbers for Pettis integrable functions with values in a separable locally convex space*, J. Theor. Probab. **13** (2000), no. 1, 93–134.

[CRdF04] ———, *On the fiber product of Young measures with application to a control problem with measures*, Adv. Math. Econ. **6** (2004), 1–38.

[CS00] Charles Castaing and Mohammed Saadoune, *Dunford-Pettis-types theorem and convergences in set-valued integration*, J. Nonlinear Convex Anal. **1** (2000), no. 1, 37–71.

[CST01] Charles Castaing, Anna Salvadori, and Lionel Thibault, *Functional evolution equations governed by nonconvex sweeping process*, J. Nonlinear Convex Anal. **2** (2001), no. 2, 217–241, Special issue for Professor Ky Fan.

[CV77] Charles Castaing and Michel Valadier, *Convex analysis and measurable multifunctions*, Lecture Notes in Math., no. 580, Springer Verlag, Berlin, 1977.

[CV98] ———, *Weak convergence using Young measures*, Funct. Approximatio Comment. Math. **26** (1998), 7–17.

[Deb66] Gérard Debreu, *Integration of correspondences*, Proc. Fifth Berkeley Symposium on Mathematical Statistics and Probability Vol II, Part I (Berkeley), University of California Press, 1966, pp. 351–372.

[Del78] Claude Dellacherie, *Convergence en probabilité et topologie de Baxter–Chacon*, Sém. Probab. Strasbourg XII (Berlin), Lecture Notes in Math., no. 649, Springer Verlag, 1978, p. 424.

[dG69] Ennio de Giorgi, *Teoremi di semicontinuità nel calcolo delle variazioni*, Notes of a course held at the istituto nazionale di alta matematica, 1968–1969.

[DM75] Claude Dellacherie and Paul André Meyer, *Probabilités et potentiel. Chapitres I à IV*, Hermann, Paris, 1975.

[DM83] ———, *Probabilités et potentiel. Chapitres IX à XI, théorie discrète du potentiel*, Hermann, Paris, 1983.

[DM02] Jérôme Dedecker and Florence Merlevède, *Necessary and sufficient conditions for the conditional central limit theorem*, Ann. Probab. **30** (2002), no. 3, 1044–1081.

[Dou94] Paul Doukhan, *Mixing. Properties and examples*, Lecture Notes in Statistics, no. 85, Springer Verlag, New York, 1994.

[DRS93] J. Diestel, W. M. Ruess, and W. Schachermayer, *Weak compactness in* $L^1(\mu, X)$, Proc. Amer. Math. Soc. **118** (1993), 447–453.

[DU77] Joseph Diestel and J. J. Uhl, Jr, *Vector measures*, Mathematical Surveys, no. 15, American Mathematical Society, Providence, R. I., 1977.

[Dud66] R. M. Dudley, *Convergence of Baire measures*, Studia Math. **27** (1966), 251–268.

[Dud76] ______, *Probabilities and metrics*, Lecture Notes Series, no. 45, Matematisk Institut, Aarhus Universitet, Aarhus, Denmark, 1976.

[Dud89] ______, *Real analysis and probability*, The Wadsworth & Brooks/Cole Mathematics Series, Wadsworth & Brooks/Cole Advanced Books & Software, Cole, Pacific Grove, CA, 1989.

[Dud02] ______, *Real analysis and probability*, Cambridge University Press, Cambridge, 2002.

[Egg84] L. Egghe, *Stopping time techniques for analysts and probabilists*, London Mathematical Society Lecture Notes Series, no. 100, Cambridge University Press, Cambridge, 1984.

[EK72] Robert J. Elliott and Nigel J. Kalton, *The existence of value in differential games*, American Mathematical Society, Providence, R.I., 1972, Memoirs of the American Mathematical Society, No. 126.

[Eke72] I. Ekeland, *Sur le contrôle optimal de systèmes gouvernés par des équations elliptiques*, J. Functional Analysis **9** (1972), 1–62.

[Ell87] Robert J. Elliott, *Viscosity solutions and optimal control*, Pitman Research Notes in Mathematics Series, vol. 165, Longman Scientific & Technical, Harlow, 1987.

[Eng89] Ryszard Engelking, *General topology*, Heldermann Verlag, Berlin, 1989.

[ES84] L. C. Evans and P. E. Souganidis, *Differential games and representation formulas for solutions of Hamilton-Jacobi-Isaacs equations*, Indiana Univ. Math. J. **33** (1984), no. 5, 773–797.

[Fat99] H. O. Fattorini, *Infinite-dimensional optimization and control theory*, Encyclopedia of Mathematics and its Applications, vol. 62, Cambridge University Press, Cambridge, 1999.

[Fer67] Xavier Fernique, *Processus linéaires, processus généralisés*, Ann. Inst. Fourier **17** (1967), no. 1, 1–92.

[Fer94] ______, *Une caractérisation des espaces de Fréchet nucléaires*, Probability in Banach spaces, 9: Proceedings from the 9th International Conference on Banach Spaces, Held at Sandjberg, Denmark, August 16–21, 1993 (Boston) (J. Hoffmann-Jørgensen, ed.), Birkhäuser, 1994, pp. 173–181.

[FGH72] D. H. Fremlin, D. J. H. Garling, and R. G. Haydon, *Bounded measures on topological spaces*, Proc. London Math. Soc. III Ser. **25** (1972), 115–136.

[FGT00] Liviu C. Florescu and Christiane Godet-Thobie, *Quelques propriétés des mesures de Young*, Analele Ştiinţifice ale Universităţii "Al. I. Cuza" Iaşi, s. I a. Matematică **XLVI** (2000), no. 2, 393–412.

[Fie90] Raúl Fierro, *Domains of attraction for semi-martingales taking values in the tempered distributions space*, J. Theoret. Probab. **3** (1990), no. 1, 31–49.

[Fis67] Roger Fischler, *The strong law of large numbers for indicators of mixing sequences*, Acta Math. Acad. Sci. Hung. **18** (1967), no. 1–2, 71–81.

[Fis70] ______, *Suites de bi-probabilités stables*, Annales de la Faculté des Sciences de l'Université de Clermont **43** (1970), 159–167.

[Fis71] ______, *Stable convergence of random variables and the weak convergence of the associated empirical measures*, Sankhyā Ser. A **33** (1971), 67–72.

[Fis76] ______, *Convergence faible avec indices aléatoires*, Ann. Inst. Henri Poincaré – Sect. B **XII** (1976), no. 4, 391–399.

[FJW96] D. H. Fremlin, R. A. Johnson, and E. Wajch, *Countable network weight and multiplication of Borel sets*, Proc. Amer. Math. Soc. **124** (1996), no. 9, 2897–2903.

[Gal97] Florence Galdéano, *Convergence étroite de mesures définies sur un espace produit*, Ph.D. thesis, Université de Perpignan, 1997.

[Gam62] R. V. Gamkrelidze, *On sliding optimal states*, Dokl. Akad. Nauk SSSR **143** (1962), 1243–1245.

[Gam78] ______, *Principles of optimal control theory*, Mathematical Concepts and Methods in Science and Engineering, no. 17, Plenum Press, New York, London, 1978.

[Gap72] V. F. Gapoškin, *Convergences and limit theorems for sequences of random variables*, Theory Proba. Appl. **17** (1972), no. 3, 379–400, translation from Teor. Veroyatn. Primen. 17, 401–423 (1972).

[Gar79] D. J. H. Garling, *Subsequence principles for vector–valued random variables*, Math. Proc. Camb. Phil. Soc. **86** (1979), no. 2, 301–311.

[Gei81] R. Geitz, *Pettis integration*, Proc. Amer. Math. Soc. **82** (1981), 81–86.

[GH67] Alain Ghouila-Houri, *Sur la généralisation de la notion de commande d'un système guidable*, Rev. Française Informat. Recherche Opérationnelle **4** (1967), 7–32.

[GP84] R. J. Gardner and W. F. Pfeffer, *Borel measures*, Handbook of set-theoretic topology, North-Holland, Amsterdam, 1984, pp. 961–1043.

[Gro52] A. Grothendieck, *Critères de compacité dans les espaces fonctionnels généraux*, Amer. J. Math. **74** (1952), 168–186.

[Gro64] ———, *Espaces vectoriels topologiques*, third ed., Publ. Soc. Mat. Saõ Paulo, São Paulo, 1964.

[Gue97] Mohamed Guessous, *An elementary proof of Komlós–Revész theorem in Hilbert spaces*, J. Convex Anal. **4** (1997), no. 2, 321–332.

[Her81] Norbert Herrndorf, *Best Φ- and N_Φ-approximants in Orlicz spaces of vector valued functions*, Z. Wahrsch. Verw. Gebiete **58** (1981), no. 3, 309–329.

[HH80] P. Hall and C. C. Heyde, *Martingale limit theory and its application*, Probability and Mathematical Statistics, Academic Press, Inc., New York, London, 1980.

[HJ71] Jørgen Hoffmann-Jørgensen, *Existence of conditional probabilities*, Math. Scand. **28** (1971), 257–264.

[HJ72] ———, *Weak compactness and tightness of subsets of $M(X)$*, Math. Scand. **31** (1972), 127–150.

[HJ91] ———, *Stochastic processes on Polish spaces*, Various Publication Series, no. 39, Matematisk Institut, Aarhus Universitet, Aarhus, Denmark, 1991.

[HJ98] Jørgen Hoffmann-Jørgensen, *Convergence in law of random elements and random sets*, High dimensional probability (Oberwolfach, 1996), Progress in Probability, no. 43, Birkhäuser, Basel, 1998, pp. 151–189.

[HJ99] Karel Hrbacek and Thomas Jech, *Introduction to set theory*, third ed., Marcel Dekker Inc., New York, 1999.

[HK99] Petr Holický and Ondřej Kalenda, *Descriptive properties of spaces of measures*, Bull. Pol. Acad. Sci., Math. **47** (1999), no. 1, 37–51.

[Huf86] R. Huff, *Remarks on Pettis integration*, Proc. Amer. Math. Soc. **96** (1986), 402–404.

[Ibr62] I. A. Ibragimov, *Some limit theorems for stationary processes*, Theory Proba. Appl. **7** (1962), no. 4, 349–382.

[Iof77] Alexander D. Ioffe, *On lower semicontinuity of integral functionals.* I and II, SIAM J. Control and Optimization **15** (1977), 521–538 and 991–1000.

[IT69] C. Ionescu Tulcea, *Two theorems concerning the desintegration of measures*, J. Math. Anal. Appl. **26** (1969), 376–380.

[ITIT69] A. Ionescu Tulcea and C. Ionescu Tulcea, *Topics in the theory of liftings*, Springer Verlag, Berlin, 1969.

[Jac85] Jean Jacod, *Théorèmes limite pour les processus*, École d'Été de Probabilités de Saint–Flour XIII-1983 (Berlin) (P. L. Hennequin, ed.), Lecture Notes in Math., no. 1117, Springer Verlag, 1985, pp. 298–409.

[Jac97] ———, *On continuous conditional Gaussian martingales and stable convergence in law*, Séminaire de Probabilités, XXXI, Lecture Notes in Math., vol. 1655, Springer, Berlin, 1997, pp. 232–246.

[Jak86] Adam Jakubowski, *On the Skorokhod topology*, Ann. Inst. H. Poincaré Probab. Statist. **22** (1986), no. 3, 263–285.

[Jak88] ———, *Tightness criteria for random measures with application to the principle of conditioning in Hilbert spaces*, Probability and Mathematical Statistics **9** (1988), no. 1, 94–114.

[Jam64] Robert C. James, *Weakly compact sets*, Trans. Amer. Math. Soc. **113** (1964), 129–140.

[Jaw84] Abdelali Jawhar, *Compacité dans l'espace des mesures de transition et applications : étude de quelques problèmes de contrôle optimal*, Ph.D. thesis, Université de Montpellier II, 1984, Thèse de 3e cycle (Chapter 1 was published in Sém. Anal. Convexe Montpellier (1984), pages 13.1–13.62).

[Jec78] Thomas Jech, *Set theory*, Academic Press, New York, 1978.

[JM81a] Jean Jacod and Jean Mémin, *Existence of weak solutions for stochastic differential equations with driving semimartingales*, Stochastics **4** (1981), 317–337.

[JM81b] ———, *Sur un type de convergence intermédiaire entre la convergence en loi et la convergence en probabilité*, Sémin. de Probabilités XV, Univ. Strasbourg 1979/80 (Berlin) (Jacques Azéma and Marc Yor, eds.), Lecture Notes in Math., no. 850, Springer Verlag, 1981, pp. 529–546.

[JM83] ______, *Rectification à "Sur un type de convergence intermédiaire entre la convergence en loi et la convergence en probabilité"*, Sémin. de Probabilités XVII, proc. 1981/82 (Berlin), Lecture Notes in Math., no. 986, Springer Verlag, 1983, pp. 509–511.

[Kaw94] Jun Kawabe, *A criterion for weak compactness of measures on product spaces with applications*, Yokohama Math. J. **42** (1994), no. 2, 159–169.

[Kel55] John L. Kelley, *General topology*, Springer Verlag, Berlin, 1955.

[Kom67] J. Komlós, *A generalization of a problem of Steinhaus*, Acta Math. Acad. Sci. Hungar. **18** (1967), 217–229.

[Kön] Heinz König, *Measure and integration: an attempt at unified systematization*, To appear in Rend. Istit. Mat. Univ. Trieste, (Workshop di Teoria della Misura et Analisi Reale Grado, September 2001, Italia).

[Kön97] ______, *Measure and integration: An advanced course in basic procedures and applications*, Springer Verlag, Berlin, 1997.

[Kou81] G. Koumoullis, *Some topological properties of spaces of measures*, Pacific J. Math. **96** (1981), no. 2, 419–433.

[KP62] Mikhail I. Kadec and Aleksander Pełczyński, *Bases, lacunary sequences and complemented subspaces in the spaces L_p*, Studia Math. **21** (1962), 161–176.

[KP80] J. Kuelbs and Walter Philipp, *Almost sure invariance principles for partial sums of mixing B-valued random variables*, Ann. Probab. **8** (1980), no. 6, 1003–1036.

[KP94] David Kinderlehrer and Pablo Pedregal, *Gradient Young measures generated by sequences in Sobolev spaces*, J. Geom. Anal. **4** (1994), no. 1, 59–90.

[KR98] Piotr Kowalski and Zdisław Rychlik, *Limit theorems for maximal random sums*, Asymptotic Methods in Probability and Statistics. A volume in honour of Miklós Csörgő. ICAMPS 97, an International Conference at Carleton Univ., Ontario, Canada, July 1997. (Amsterdam) (Barbara Szyszkowicz, ed.), Elsevier, 1998, pp. 13–29.

[KS88] N. N. Krasovskiĭ and A. I. Subbotin, *Game-theoretical control problems*, Springer Series in Soviet Mathematics, Springer-Verlag, New York, 1988, Translated from the Russian by Samuel Kotz.

[KZ94] Ł. Kruk and W. Zięba, *On tightness of randomly indexed sequences of random elements*, Bull. Pol. Acad. Sci., Math. **42** (1994), 237–241.

[KZ95] ———, *A criterion of almost sure convergence of asymptotic martingales in a Banach space*, Yokohama Math. J. **43** (1995), 61–72.

[KZ96] Grzegorz Krupa and Wiesław Zięba, *Strong tightness as a condition of weak and almost sure convergence*, Comment. Math. Univ. Carolinae **37** (1996), no. 3, 641–650.

[Let98] Giorgio Letta, *Convergence stable et applications*, Atti Sem. Mat Fis. Modena (supplemento) **46** (1998), 191–211, Dedicated to Prof. C. Vinti (Perugia 1996).

[Lin73] Torgny Lindvall, *Weak convergence of probability measures and random functions in the function space* $D[0, \infty)$, J. Appl. Probability **10** (1973), 109–121.

[LP96] Giorgio Letta and Luca Pratelli, *Convergence stable vers un noyau gaussien*, Rend. Accad. Naz. Sci. XL Mem. Mat. Appl. **20** (1996), 205–213.

[LT91] Michel Ledoux and Michel Talagrand, *Probability in Banach spaces*, Springer Verlag, Berlin, New york, 1991.

[Maz33] Stanisław Mazur, *Über konvexe Mengen in linearen normierten Räumen*, Stud. Math. **4** (1933), 70–84.

[McS40] E. J. McShane, *Generalized curves*, Duke Math. J. **6** (1940), 513–536.

[Mey66] Paul André Meyer, *Le théorème de continuité de P. Lévy sur les espaces nucléaires (d'après X. Fernique)*, Séminaire Bourbaki 1965/1966, 1966, Exposé no 311, pp. 509–522.

[Mey78] ———, *Convergence faible et compacité des temps d'arrêt, d'après Baxter–Chacon*, Sém. Probab. Strasbourg XII (Berlin), Lecture Notes in Math., no. 649, Springer Verlag, 1978, pp. 411–423.

[Mic51] Ernest Michael, *Topologies on spaces of subsets*, Trans. Amer. Math. Soc. **71** (1951), 152–182.

[Mic66] E. Michael, $\aleph_0$*-spaces*, J. Math. Mech **15** (1966), 983–1002.

[Mit83] Itaru Mitoma, *Tightness of probabilities on* $C([0,1];\mathcal{S}')$ *and* $D([0,1];\mathcal{S}')$, Ann. Probab. **11** (1983), no. 4, 989–999.

[MM89] Manuel D. P. Monteiro Marques, *Minimization of integral functionals depending on Lipschitz domains*, Numer. Funct. Anal. Optim. **10** (1989), no. 9–10, 991–1002.

[MM93] ______, *Differential inclusions in nonsmooth mechanical problems, shocks and dry friction*, Progress in Nonlinear Differential Equations and their Applications, no. 9, Birkhäuser, Basel, 1993.

[Mog66] J. Mogyoródi, *A remark on stable sequences of random variables and a limit distribution theorem for a random sum of independent variables*, Acta Math. Acad. Sci. Hung. **17** (1966), no. 3–4, 401–409.

[Mor56] K. Morita, *On decomposition spaces of locally compact spaces*, Proc. Japan Acad. **32** (1956), 544–548.

[Mor77] Jean-Jacques Moreau, *Evolution problem associated with moving convex set in a Hilbert space*, J. Differential Equations **26** (1977), 347–374.

[MS48] E. Marczewski and P. Sikorski, *Measures in non-separable metric spaces*, Colloq. Math. **1** (1948), 133–139.

[Mus91] Kazimierz Musiał, *Topics in the theory of Pettis integration*, Rendiconti dell'istituto di matematica dell'Università di Trieste **23** (1991), 176–262, School on Measure Theory and Real Analysis Grado (Italy).

[MV02] Gérard Michaille and Michel Valadier, *Young measures generated by a class of integrands: a narrow epicontinuity and applications to homogenization*, J. Math. Pures Appl. (9) **81** (2002), no. 12, 1277–1312.

[Ole77] Czesław Olech, *A characterization of L_1-weak lower semicontinuity of integral functionals*, Bull. Pol. Acad. Sci., Math. **25** (1977), no. 2, 135–142.

[OW98] George L. O'Brien and Stephen Watson, *Relative compactness for capacities, measures, upper semicontinuous functions and closed sets*, J. Theor. Probab. **11** (1998), 577–588.

[Pad70] A. R. Padmanabhan, *Convergence in probability and allied results*, Math. Jap. **15** (1970), 111–117.

[Par67] K. R. Parthasarathy, *Probability measures on metric spaces*, Acad. Press, New York, London, 1967.

[Ped97] Pablo Pedregal, *Parametrized measures and variational principles*, Progress in Nonlinear Differential Equations and their Applications, vol. 30, Birkhäuser Verlag, Basel, 1997.

[Pel80] Jean Pellaumail, *Convergence en règle*, C. R. Acad. Sci. Paris Sér. I **290** (1980), 289–292.

[Pel81] ______, *Solutions faibles et semi-martingales*, Séminaire de Probabilités XV, 1979/80. Université de Strasbourg (Jacques Azéma and Marc Yor, eds.), Lecture Notes in Math., vol. 850, Springer Verlag, Berlin, 1981, pp. 561–586.

[Pro56] Yu. V. Prokhorov, *Convergence of random processes and limit theorems in probability theory*, Theory Proba. Appl. **1** (1956), 157–214.

[PRT00] René A. Poliquin, R. Tyrell Rockafellar, and Lionel Thibault, *Local differentiability of distance functions*, Trans. Amer. Math. Soc. **352** (2000), no. 11, 5231–5249.

[Pry66] J. D. Pryce, *Weak compactness in locally convex spaces*, Proc. Amer. Math. Soc. **17** (1966), 148–155.

[PS99] N. T. Parpieva and O. Sh. Sharipov, *The Central Limit Theorem for stationnary random fields with values in some Banach spaces*, Uzbek Mat. Zh. **1** (1999), 68–73, (in Russian).

[PU97] Magda Peligrad and Sergey Utev, *Central Limit Theorem for linear processes*, Ann. Probab. **25** (1997), no. 1, 443–456.

[PV95] Laurent Piccinini and Michel Valadier, *Uniform integrability and Young measures*, J. Math. Anal. Appl. **195** (1995), no. 2, 428–439.

[Rac91] Svetlozar T. Rachev, *Probability metrics and the stability of stochastic models*, Wiley, Chichester, New York, 1991.

[RdF03] Paul Raynaud de Fitte, *Compactness criteria in the stable topology*, Bull. Pol. Acad. Sci., Math. **51** (2003), no. 4, 343–363.

[RdFZ02] Paul Raynaud de Fitte and Wiesław Zięba, *On the construction of a stable sequence with given density*, Ann. Univ. Mariae Curie-Skłodowska Sect. A **56** (2002), no. 8, 77–84.

[Rén58] Alfred Rényi, *On mixing sequences of sets*, Acta Math. Acad. Sci. Hungar. **9** (1958), 215–228.

[Rén63] ———, *On stable sequences of events*, Sankhyā Ser. A **25** (1963), 293–302.

[Rén66] ———, *Calcul des probabilités*, Dunod, Paris, 1966.

[Rén70] ———, *Foundations of probability*, Holden Day, Inc., San Francisco, 1970.

[Res68] Yu. G. Reshetnyak, *Weak convergence of completely additive vector functions on a set*, Sibirsk Mat. Zh. **9** (1968), 1386–1394, (russian).

[Rio00] Emmanuel Rio, *Théorie asymptotique des processus aléatoires faiblement dépendants*, Mathématiques et Applications, no. 31, Springer Verlag, Paris, 2000.

[Roi84] Judy Roitman, *Basic S and L*, Handbook of set-theoretic topology, North-Holland, Amsterdam, 1984, pp. 295–326.

[Ros56] M. Rosenblatt, *A central limit theorem and a strong mixing condition*, Proc. Nat. Ac. Sc. U.S.A. **42** (1956), 43–47.

[Ros79] H. P. Rosenthal, *Topics courses*, Université de Paris VI (unpublished), 1979.

[Rou97] Tomáš Roubíček, *Relaxation in optimization theory and variational calculus*, de Gruyter Series in Nonlinear Analysis and Applications, no. 4, Walter de Gruyter, Berlin, 1997.

[RR58] A. Rényi and P. Révész, *On mixing sequences of random variables*, Acta Math. Acad. Sci. Hungar. **9** (1958), 389–393.

[RR98] S. T. Rachev and L. Rüschendorf, *Mass transportation problems. Volume I: Theory*, Probability and its Applications, Springer Verlag, New York, Berlin, 1998.

[Rze89] Tadeusz Rzeżuchowski, *Strong convergence of selections implied by weak*, Bull. Austral. Math. Soc. **39** (1989), no. 2, 201–214.

[Rze92] ______, *Impact of dentability on weak convergence in L^1*, Boll. Unione Mat. Ital. A Ser. VII **6** (1992), 71–80.

[SA76] Takuro Shintani and Tsuyoshi Ando, *Best approximants in L^1 space*, Z. Wahrsch. Verw. Gebiete **33** (1975/76), no. 1, 33–39.

[Saa98] Mohamed Saadoune, *A new extension of Komlós' theorem in infinite dimensions. Application: Weak compactness in L^1_X*, Portugaliae Mathematica **55** (1998), no. 1, 113–127.

[Sam84] Jorge D. Samur, *Convergence of sums of mixing triangular arrays of random vectors with stationnary rows*, Ann. Probab. **12** (1984), 390–426.

[SB74] M. F. Sainte Beuve, *On the extension of von Neumann–Aumann's theorem*, J. Funct. Anal. **17** (1974), 112–129.

[Sch66] Helmut H. Schaefer, *Topological vector spaces*, Macmillan, New York, 1966.

[Sch73] Laurent Schwartz, *Radon measures on arbitrary topological spaces and cylindrical measures*, Tata Institute of Fundamental Research Studies in Mathematics, Oxford University Press, London, 1973.

[Sch75] Manfred Schäl, *On dynamic programming: compactness of the space of policies*, Stochastic Processes Appl. **3** (1975), no. 4, 345–364.

[Sch82] Maria Elena Schonbek, *Convergence of solutions to nonlinear dispersive equations*, Comm. Partial Differential Equations **7** (1982), no. 8, 959–1000.

[Ser59] James Serrin, *On a fundamental theorem of the calculus of variations*, Acta Math. **102** (1959), 1–22.

[Sha88] Michael Sharpe, *General theory of Markov processes*, Pure and Applied Math., no. 133, Academic Press, Boston, MA, 1988.

[Sko56] A. V. Skorokhod, *Limit theorems for stochastic processes*, Theory Proba. Appl. **1** (1956), 261–290.

[Sła85] Marek Słaby, *Strong convergence of vector-valued pramarts and subpramarts*, Probability and Math. Stat. **5** (1985), no. 2, 187–196.

[SP75] Jean Saint Pierre, *Désintégration d'une mesure non bornée*, Ann. Inst. Henri Poincaré Probab. Stat. **11** (1975), 275–286.

[SP76] ———, *Une remarque sur les espaces sousliniens réguliers*, C. R. Acad. Sci. Paris Sér. A **282** (1976), 1425–1427.

[SS78] Lynn A. Steen and J. Arthur Seebach, Jr, *Counterexamples in topology*, Holt, Rinehart and Winston, Inc., New York, 1978, Second edition. Dover Publications paperback 1996.

[Sto76] A. H. Stone, *Topology and measure theory*, Measure theory (Proc. Conf., Oberwolfach, 1975) (Berlin), Springer, 1976, pp. 43–48. Lecture Notes in Math., Vol. 541.

[SV95a] Mohammed Saadoune and Michel Valadier, *Convergence in measure. Local formulation of the Fréchet criterion*, C. R. Acad. Sci. Paris Sér. I **320** (1995), no. 4, 423–428.

[SV95b] ———, *Convergence in measure. the Fréchet criterion, from local to global*, Bull. Pol. Acad. Sci., Math. **43** (1995), no. 1, 47–57.

[SV95c] ———, *Extraction of a "good" subsequence from a bounded sequence of integrable functions*, J. Convex Anal. **2** (1995), no. 1-2, 345–357.

[Syc98] M. Sychev, *Young measure approach to characterization of behaviour of integral functionals on weakly convergent sequences by means of their integrands*, Ann. Inst. H. Poincaré Anal. Non Linéaire **15** (1998), no. 6, 755–782.

[Syc99] M. A. Sychev, *A new approach to Young measure theory, relaxation and convergence in energy*, Ann. Inst. H. Poincaré Anal. Non Linéaire **16** (1999), no. 6, 773–812.

[Tal84] Michel Talagrand, *Weak Cauchy sequences in L^1_E*, Amer. J. Math. **106** (1984), 703–724.

[Tar78] L. Tartar, *Une nouvelle méthode de résolution d'équations aux dérivées partielles non linéaires*, Journées d'Analyse Non Linéaire (Proc. Conf., Besançon, 1977), Lecture Notes in Math., vol. 665, Springer, Berlin, 1978, pp. 228–241.

[Tar90] Luc Tartar, *H-measures, a new approach for studying homogenization, oscillations and concentrations effects in partial differential equations*, Proc. Roy. Soc. Edimburgh Sect. A **115** (1990), 193–230.

[Tat02] Hiroshi Tateishi, *On the existence of equilibria of equicontinuous games with incomplete information*, Adv. Math. Econ. **4** (2002), 41–59.

[Thi99] Lionel Thibault, *Sweeping process with regular and nonregular sets*, Tech. report, Université Montpellier II, 1999.

[Top70a] Flemming Topsøe, *Compactness in spaces of measures*, Studia Mathematica **36** (1970), 195–212.

[Top70b] ———, *Topology and measure*, Lecture Notes in Math., no. 133, Springer Verlag, Berlin, 1970.

[Top74] ———, *Compactness and tightness in a space of measures with the topology of weak convergence*, Math. Scand. **34** (1974), 187–210.

[Trè67] François Trèves, *Topological vector spaces, distributions and kernels*, Pure and Applied Mathematics, no. 25, Academic Press, New York, 1967.

[Ülg91] Ali Ülger, *Weak compactness in $L^1(\mu, X)$*, Proc. Amer. Math. Soc. **103** (1991), 143–149.

[Ute92] S. A. Utev, *A method for investigating the sums of weakly dependent random variables*, Siberian Math. J. **32 (1991)** (1992), no. 4, 675–690, translation from the Russian: Sibirsk Mat. Zh. **32** (1991) no 4, 165–183, 229.

[Val70] Michel Valadier, *Contribution à l'analyse convexe*, Université de Montpellier, Montpellier, 1970, (Thèse de Doctorat ès–Sciences – Mathématiques présentée à la Faculté des Sciences de Paris pour obtenir le grade de Docteur ès–Sciences. Secrétariat de Mathématiques de la Faculté de Montpellier, 1970, Publication no 92).

[Val71] ———, *Multiapplications mesurables à valeurs convexes compactes*, J. Math. Pures Appl., IX Sér. **50** (1971), 265–297.

[Val72] ______, *Comparaison de trois théorèmes de désintégration*, Travaux du Séminaire d'Analyse Convexe, Vol. II, Exp. No. 10, U.E.R. de Math., Univ. Sci. Tech. Languedoc, Montpellier, 1972, pp. pp. 10.1–10.21, Secrétariat des Math., Publ. No. 122.

[Val73] ______, *Désintégration d'une mesure sur un produit*, C. R. Acad. Sci. Paris Sér. I **276** (1973), A33–A35.

[Val84] M. Valadier, *La multi-application médianes conditionnelles*, Z. Wahrsch. Verw. Gebiete **67** (1984), no. 3, 279–282, (first published in Séminaire d'Analyse Convexe **12** no. 2 (1982, 21.1–21.15, Montpellier, France).

[Val89] Michel Valadier, *Différents cas où, grâce à une propriété d'extrémalité, une suite de fonctions intégrables faiblement convergente, converge fortement*, Sém. Anal. Convexe **19** (1989), 5.1–5.20.

[Val90a] ______, *Application des mesures de Young aux suites uniformément intégrables dans un Banach séparable*, Sém. Anal. Convexe **20** (1990), 3.1–3.14.

[Val90b] ______, *Young measures*, Methods of Nonconvex Analysis (Berlin) (A. Cellina, ed.), Lecture Notes in Math., no. 1446, Springer Verlag, 1990, pp. 152–158.

[Val94] ______, *A course on Young measures*, Rendiconti dell'istituto di matematica dell'Università di Trieste **26, suppl.** (1994), 349–394, Workshop di Teoria della Misura et Analisi Reale Grado, 1993 (Italia).

[vdVW96] Aad W. van der Vaart and Jon A. Wellner, *Weak convergence and empirical processes. With applications to statistics*, Springer Series in Statistics, Springer Verlag, Berlin, 1996.

[Vis84] Augusto Visintin, *Strong convergence results related to strict convexity*, Comment. Partial Differential Equation **9** (1984), no. 5, 439–466.

[VR59] V. A. Volkonskiĭ and Yu. A. Rozanov, *Some limit theorems for random functions I*, Theory Proba. Appl. **4** (1959), 178–197, translated from the Russian: Teor. Veroyatnost. i Primenen **4** (1959), 186–207.

[VT78] N. N. Vahanija and V. I. Tarieladze, *Covariance operators of probability measures in locally convex spaces*, Teor. Verojatnost. i Primenen. **23** (1978), no. 1, 3–26, English translation in Theory Probab. Appl. **23** (1978), no. 1, 1–21.

[VTC87] N. N. Vakhania, V. I. Tarieladze, and S. A. Chobanyan, *Probability distributions on Banach spaces*, Mathematics and Its Applications (Soviet Series), D. Reidel Publishing Company, Dordrecht, 1987.

[War67] Jack Warga, *Functions of relaxed controls*, SIAM J. Control **5** (1967), no. 4, 628–641.

[War72] ______, *Optimal control of differential and functional equations*, Academic Press, New York, London, 1972.

[Whe83] Robert F. Wheeler, *A survey of Baire measures and strict topologies*, Exposition. Math. **1** (1983), no. 2, 97–190.

[Wój87] Małgorzata Wójcicka, *The space of probability measures on a Prohorov space is Prohorov*, Bull. Pol. Acad. Sci., Math. **35** (1987), no. 11-12, 809–811.

[Xue91] Xue, *On the principle of conditioning and convergence to mixtures of distributions for sums of dependent random variables*, Stochastic Processes Appl. **37** (1991), 175–186.

[You37] L. C. Young, *Generalized curves and the existence of an attained absolute minimum in the Calculus of Variations*, C. R. Soc. Sc. Varsovie **30** (1937), 212–234.

[Zię85] Wiesław Zięba, *On some criterion of convergence in probability*, Prob. Math. Stat. **6** (1985), no. 2, 225–232.

Subject Index

Index of Notations